Power Distribution Planning Reference Book

POWER ENGINEERING

Series Editor

H. Lee Willis

*ABB Power T&D Company Inc.
Cary, North Carolina*

1. Power Distribution Planning Reference Book, *H. Lee Willis*

ADDITIONAL VOLUMES IN PREPARATION

Protective Relaying: Principles and Applications, Second Edition, *J. Lewis Blackburn*

Electrical Power Equipment Maintenance and Testing, *Paul Gill*

Electrical Insulation in Power Systems, *Nazar Hussain Malik, A. A. Al-Arainy, and Mohammad Iqbal Qureshi*

Transmission Network Protection: Theory and Practice, *Yeshwant Paithankar*

Power Distribution Planning Reference Book

H. Lee Willis

*ABB Power T&D Company Inc.
Cary, North Carolina*

MARCEL DEKKER, INC.   NEW YORK · BASEL · HONG KONG

Library of Congress Cataloging-in-Publication Data

Willis, H. Lee.
   Power distribution planning reference book / H. Lee Willis.
     p. cm.
   Includes bibliographical references and index.
   ISBN 0-8247-0098-8 (hc : alk. paper)
    1. Electric power distribution—Planning—Handbooks, manuals, etc.
I. Title.
TK3091.W55   1997
621.319—dc21
                                                                                       97-13116
                                                                                           CIP

The publisher offers discounts on this book when ordered in bulk quantities. For more information, write to Special Sales/Professional Marketing at the address below.

This book is printed on acid-free paper.

Copyright © 1997 by MARCEL DEKKER, INC. All Rights Reserved.

Neither this book nor any part may be reproduced or transmitted in any form or by any means, electronic or mechanical, including photocopying, microfilming, and recording, or by any information storage and retrieval system, without permission in writing from the publisher.

MARCEL DEKKER, INC.
270 Madison Avenue, New York, New York 10016
http://www.dekker.com

Current printing (last digit):
10 9 8 7 6 5 4 3 2 1

PRINTED IN THE UNITED STATES OF AMERICA

Series Introduction

Power engineering is the oldest and most traditional of the various subgenres within electrical engineering, yet no other area of modern technology is currently undergoing a more dramatic revolution in both technology and industry structure. This first book in Marcel Dekker's Power Engineering Series covers the portion of the power system closest to the electric consumer and most critical to achieving high quality. The *Power Distribution Planning Reference Book* provides a modern source of information for distribution planners and engineers who must meet demands for ever greater performance while working in an environment of intense cost containment and regulatory scrutiny.

Not since Westinghouse's famous "green book" (last revised in 1965) has any reference text addressed the layout and design of power distribution systems in such a comprehensive manner, from subtransmission through the service level. While that and other, even older, distribution texts contain a wealth of information still useful to modern distribution engineers, many of their concepts and approaches are outdated. All of the classical distribution systems books focus on electrical performance, while treating cost and reliability as important but clearly only consequential aspects of design. In contrast, the *Power Distribution Planning Reference Book* emphasizes economy as the primary goal of distribution design, and examines in great detail how distribution systems can be designed to achieve adequate performance and reliability at the lowest possible cost, and how cost interacts with electrical performance, reliability and customer service quality. It also covers the application of computerized analysis and optimization methods that were only a dream four decades ago, when most of the classic texts on distribution systems were written. In addition, it covers a host of very recent concepts, including value-based planning, budget-constrained planning, spatial T&D load forecasting, multiscenario planning, and deregulated utility planning.

As both the author of the *Power Distribution Planning Reference Book* and the editor of the Power Engineering Series, I am proud that this book was chosen to inaugurate this new and highly important group of books. Like all the books planned for Marcel Dekker's Power Engineering Series, this first is designed to provide modern technology in a context of proven, practical application and to be useful as a reference book as well as for self-study and advanced classroom use. The Power Engineering Series will eventually include books covering the entire field of power engineering, in all of its specialties and sub-subgenres, all aimed at providing practicing power engineers with the knowledge and techniques they need to meet the electric industry's challenges in the 21st century.

H. Lee Willis

Preface

This book is both a reference book and a tutorial guide for the planning of electric utility distribution systems. Distribution planning involves developing a schedule of future additions and changes that will assure the utility's goals for electric distribution are met. Usually, those goals are to provide satisfactory service at the lowest possible cost. Regardless, distribution planners must accomplish three tasks. First, they must identify the goals for their system. Exactly what constitutes "satisfactory performance?" How is that measured? What does "lowest cost" mean and when and how should they defer today's expenses now even if tomorrow's rise as a result? Unambiguous, quantitative targets for all goals must be established and used throughout the planning process.

Second, planners must understand how differences in distribution system design and equipment will affect the achievement of these goals. Distribution systems are complicated entities, whose performance and economy depend on the coordinated interaction of tens of thousands of individual equipment and circuit elements. Worldwide, there are several fundamentally different "design philosophies" for laying out a distribution system and engineering it to work well -- what might be called differing design paradigms. All get the job done. But while each paradigm has its plusses and minuses, for any given situation, there is usually one best design to achieve the desired end result.

Finally, the planners must find that best design, every time. Their planning methodology must be comprehensive and complete, assuring that nothing is overlooked and that every opportunity for savings or quality improvement is fully exploited.

This book addresses each of these three tasks in turn. The first part of the book consists of comprehensive single-chapter tutorial reviews on key aspects of distribution performance. Chapter 1 provides a brief introduction to distribution systems, their mission, the rules that govern their behavior, their equipment, and their performance and economics.

Chapter 2 discusses customer demand and electric load as seen by the distribution system, focusing particularly on coincidence of load (the manner in which local peak loads sum to area and regional peak loads).

Power quality, including availability (system reliability as seen from the customer's perspective), is covered in Chapter 3, which discusses power quality issues, giving for each a *précis* of its causes, customer concerns, and cures.

Planning criteria and engineering standards define the minimum requirements in equipment selection, design, loading, and application that planners must achieve in their distribution design. Chapter 4 looks at them and the differences that exist throughout the industry in standards and their application.

Chapter 5 reviews basic reliability definitions, analysis methods, rules of thumb, and reliability measurement and planning methods. A majority of utilities address reliability through application of contingency criteria to distribution design. These are examined in order to build a foundation for contingency planning guidelines and methods in later chapters.

Chapter 6 presents a review of engineering economics from the perspective of the distribution planner. It focuses mostly on time-value-of-money concepts and their pragmatic application as decision-making tools.

The second part of the book, Chapters 7 through 13, constitutes a detailed look at the design, performance, and economics of a power distribution system, based on two overriding principles. First, the *systems approach*: sub-transmission, substation, feeder, and service levels are integrated layers of the system. Good performance and economy come as much from sound coordination of these levels with one another, as they do from optimization of any one level. Second, *minimum cost:* this book's focus is on reducing cost as much as possible, while retaining satisfactory service quality. Chapters 7 through 13 take a bottom up approach to an examination of the distribution system, beginning with basic conductor and transformer economic selection, and working up through feeder system and substation design to the strategic layout of the entire system.

Chapter 7 looks at conductor and transformer economic sizing in some detail, beginning with basic conductor sizing economics. This will be familiar to most planners, but its extension to conductor set design, which involves finding the optimal *set* of line types to use in building a distribution system, will be new to many. Also new may be the concept of load reach, a power delivery distance criterion for conductor selection, used to assure the cost-minimized distribution system can deliver power over the distances required.

Primary feeder circuit layout is examined in Chapters 8 and 9. Every feeder is an assemblage of line segments. The various philosophies on how to combine line segments into feeders are examined and compared, including "American" or "European" layout; large trunk and multi-branch; loop, radial, or network systems; single or dual-voltage; and other variations. Also covered are design for contingencies, multi-feeder layout, and expansion and reinforcement strategy.

Chapter 10 looks at substations, the anchor points of the distribution system, along with the sub-transmission lines that provide them with power. Substations can be composed of a variety of types of equipment, laid out in many different ways. Cost, capacity, and reliability vary depending on equipment and design.

The systems approach highlights Chapter 11's look at the performance and economics of the combined sub-transmission/substation/feeder system. Optimal performance comes from a properly coordinated selection of voltages, equipment types, and layout at all three levels. This balance of design, and its interaction with load density, geographic constraints, and other design elements, is explored in a series of sample system designs exploring distribution system performance and economy.

No one aspect of layout is more important than assuring that the substations are in the right places with the needed capacity. Chapter 12 focuses on substation siting and sizing, the issues involved, the types of costs and constraints, and the ways in which decisions and different layout rules change the resulting system performance and economy.

Chapter 13 looks at the service (secondary) level. Although composed of relatively small "commodity" elements, cumulatively this closest-to-the-customer level is surprisingly complex, and represents a sizable investment, one that benefits from careful equipment specification and design standards.

The final portion of the book discusses distribution planning methods. Chapter 14 begins by looking at planning itself. Planning involves setting goals, identifying and evaluating options, and selecting the best one. Each of these steps is examined in detail, along with its common pitfalls. Short- and long-range planning are defined and their different purposes and procedures studied. Finally, the functional steps involved in T&D planning are presented, in detail.

Chapter 15 looks at how future capacity requirements are identified, which is usually done with a distribution load forecast. Characteristics of load growth that can be used in forecasting are presented, along with rules of thumb for growth analysis. Various forecasting procedures are delineated, along with a comparison of their applicability, accuracy, and data and resource requirements.

Feeder analysis methods are covered in Chapter 16. Mostly applied to short-range planning, feeder analysis involves load flow analysis and can include other functions, like reliability analysis. The chapter focuses on application, not algorithms. It covers the basic concepts, models, and approaches used to represent feeders and evaluate their performance against objectives.

Automated planning tools (computerized planning) are covered in Chapter 17. Used mostly in medium- and long-term planning, computerized, automated tools can streamline the process and lead to better distribution plans. Most are based on application of optimization methods, which while mathematically complicated, are simple in concept. The chapter presents a brief, practical, tutorial on optimization, on the various methods, and on how to pick a method that will work for a given problem. As in Chapter 16, the focus is on practical application, not on algorithms. Automated methods for both feeder-system planning and substation siting and sizing are discussed, along with typical pitfalls and recommended ways to avoid them.

As utilities change from fully regulated entities to participants in the competitive process, their sense of identify and purpose will change. Chapter 18 looks at several different paradigms, or identity and value systems, that apply to electric utilities. It compares the role of distribution planning in each of them, looking at how the results of similar planning evaluations would lead to different decisions depending on the paradigm in place.

This book, along with a companion volume spun off during its completion (*Spatial Electric Load Forecasting,* Marcel Dekker, 1996), took more than a decade to complete. I wish to thank many good friends and colleagues, including Hahn Tram and James Northcote-Green, for their support and advice during the past two decades; Mike Engel at Midwest Energy for his friendship, good humor and valuable suggestions; Walter Scott, for his jovial tolerance of strange questions and his constant helpful advice; and my present co-workers, Amir Khalessi, Doug Wall, Rafael Ochoa, John Day and Bill Rutz, for their encouragement and willing support. I also want to thank Rita Lazazzaro and Jennifer Kelly at Marcel Dekker, Inc., for their involvement and efforts to make this book a quality effort. Most of all, I wish to thank my wife, Lorrin Philipson, for the many hours of review and editorial work she unselfishly gave in support of this book and for her constant, loving support, without which this book would never have been completed.

H. Lee Willis

Contents

Series Introduction   iii

Preface   v

1 Power Delivery Systems   1

    1.1 Introduction   1
    1.2 T&D System's Mission   2
    1.3 The "Laws of T&D"   4
    1.4 Levels of the T&D Systems   7
    1.5 Utility Distribution Equipment   18
    1.6 T&D Costs   24
    1.7 Types of Distribution System Design   34
    1.8 Conclusion   46
    References   48

2 Electrical Load and Consumer Demand for Power   49

    2.1 Introduction   49
    2.2 Customer Demand   50
    2.3 Peak Load, Coincidence, and Load Curve Behavior   58
    2.4 Measuring and Modeling Load Curves   69
    References   73

3 Availability and Power Quality — 75

- 3.1 Introduction — 75
- 3.2 Interruption of Electric Service — 76
- 3.3 Voltage Variations — 80
- 3.4 Voltage Surges — 87
- 3.5 Harmonics — 89
- 3.6 Customer Value of Availability and Power Quality — 108
- 3.7 End-Use Modeling of Customer Power Quality Issues — 122
- 3.8 Summary — 125
- References — 127

4 Planning Criteria — 129

- 4.1 Introduction — 129
- 4.2 Criteria and Standards Must Be Met, Not Exceeded — 130
- 4.3 Voltage and Customer Service Standards — 130
- 4.4 Operating and Safety Standards — 145
- 4.5 Standard Equipment and Design Criteria — 150
- 4.6 Conclusion — 153
- References — 153

5 Reliability and Contingency Criteria — 155

- 5.1 Introduction — 155
- 5.2 Outages Cause Interruptions — 158
- 5.3 Reliability Indices — 164
- 5.4 Reliability and Contingency Criteria for Planning — 171
- 5.5 Comparison of Reliability Indices Among Utilities — 175
- 5.6 Setting Reliability Goals — 178
- 5.7 Conclusion and Summary — 183
- For Further Reading — 183

6 Economic Evaluation — 185

- 6.1 Introduction — 185
- 6.2 Costs — 186
- 6.3 Time Value of Money — 193
- 6.4 Cost-Effectiveness Evaluation — 213
- 6.5 Variability of Costs — 220
- 6.6 Conclusion — 227
- References and Further Reading — 228

Contents

7	**Line Segments and Transformer Economics and Set Design**	**229**
	7.1 Introduction	229
	7.2 Distribution Lines	230
	7.3 Basic Line Type Economics	241
	7.4 Voltage Drop and Load Reach	251
	7.5 Performance of the Conductor Set	259
	7.6 Conductor Set Design	266
	7.7 Transformer Selection Economics	275
	7.8 Substation Transformer Sizing and Loading	283
	7.9 Conclusion	284
	References	284
8	**Distribution Feeder Layout**	**285**
	8.1 Introduction	285
	8.2 The Feeder System	286
	8.3 Radial and Loop Feeder Layout	306
	8.4 Contingency Support Considerations	319
	8.5 Conclusion	340
	References	340
9	**Multi-Feeder Layout and Volt-VAR Correction**	**341**
	9.1 Introduction	341
	9.2 How Many Feeders in a Substation Service Area?	343
	9.3 Planning the Feeder System	347
	9.4 Planning for Load Growth	354
	9.5 Formula to Estimate Feeder System Cost and Its Application to Indicate Design Goals	362
	9.6 Volt-VAR Control and Correction	367
	9.7 Summary of Key Points	389
	References	390
10	**Distribution Substations**	**391**
	10.1 Introduction	391
	10.2 High-Side Substation and Equipment Layout	394
	10.3 Transformer Portion of a Substation	406
	10.4 Low-Side Portion of a Substation	415
	10.5 The Substation Site	420
	10.6 Substation Costs, Capacity, and Reliability	424

	10.7 Other Substation Planning Considerations	426
	10.8 Planning with "Transformer Units"	429
	10.9 Conclusion	429
	References and Bibliography	429

11 Distribution System Layout — 431

11.1 Introduction — 431
11.2 The Whole T&D System — 433
11.3 Design Interrelationships — 445
11.4 Example of a System Dominated by Voltage Drop, Not Capacity — 477
11.5 Conclusion and Summary — 486
References and Bibliography — 487

12 Substation Siting and System Expansion Planning — 489

12.1 Introduction — 489
12.2 Substation Location, Capacity, and Service Area — 490
12.3 Substation Siting and Sizing Economics — 495
12.4 Substation-Level Planning: The Art — 515
12.5 Guidelines for Substation Site and Size Application to Achieve Low Cost — 519
12.6 Substation-Level Planning: The Science — 523
12.7 The Most Important Point About Substation-Level Planning — 535
References — 535

13 Service Level Layout and Planning — 537

13.1 Introduction — 537
13.2 The Service Level — 538
13.3 Types of Service Level Layout — 539
13.4 Load Dynamics and Coincidence, and Their Interaction with the Service Level — 545
13.5 Service-Level Planning and Layout — 552
13.6 Conclusion — 563
References — 564

14 Planning and the T&D Planning Process — 565

14.1 Introduction — 565
14.2 Planning: Finding the Best Alternative — 566

Contents

14.3 The Functions of Short- and Long-Range Planning	581
14.4 Uncertainty and Multi-Scenario Planning	587
14.5 The T&D Planning Process	593
14.6 Summary and Recommended T&D Planning Process	610
References	613

15 Forecasting T&D Load — 615

15.1 Spatial Load Forecasting	615
15.2 Load Growth Behavior	618
15.3 Important Elements of a Spatial Forecast	625
15.4 Trending Methods	634
15.5 Simulation Methods for Spatial Load Forecasting	650
15.6 Selecting a Forecast Method	670
References and Bibliography on Spatial Load Forecasting	680

16 Distribution Feeder Analysis — 685

16.1 Introduction	685
16.2 Models, Algorithms and Computer Programs	688
16.3 Circuit Models	690
16.4 Models of Electric Load	700
16.5 Models of Electrical Behavior	707
16.6 Coincidence and Load Flow Interaction	718
16.7 Conclusion	726
References	727

17 Automated Planning Tools and Methods — 729

17.1 Introduction	729
17.2 Fast Ways to Find Good Alternatives	730
17.3 Automated Feeder Planning Methods	745
17.4 Substation-Level and Strategic Planning Tools	758
17.5 Application of Planning Tools	768
References	776

18 Traditional Versus Competitive Industry Paradigms — 777

18.1 Changing Environments Often Require New Value Systems	777
18.2 Paradigm Shifts	778
18.3 Least-Cost Present Worth Planning	783

18.4 Budget-Constrained Service Value Planning	784
18.5 Profit-Based Planning Paradigms	798
18.6 Conclusion	802
References	802
Index	*803*

Power Distribution Planning
Reference Book

1
Power Delivery Systems

1.1 INTRODUCTION

Retail sale of electric energy involves the delivery of power in ready to use form to the final consumers. Whether marketed by a local utility, load aggregator, or direct power retailer, this electric power must flow through a power delivery system on its way from power production to customer. This transmission and distribution (T&D) system consists of thousands of transmission and distribution lines, substations, transformers, and other equipment scattered over a wide geographical area and interconnected so that all function in concert to deliver power as needed to the utility's customers.

This chapter is an introductory tutorial on T&D systems and their design constraints. For those unfamiliar with power delivery, it provides an outline of the most important concepts. But in addition, it examines the natural phenomena that shape T&D systems and explains the key physical relationships and their impact on design and performance. For this reason experienced planners are advised to scan this chapter, or at least its conclusions, so that they understand the perspective upon which the rest of the book builds.

In a traditional electric system, power production is concentrated at only a few large, usually isolated, power stations. The T&D system moves the power from those often distant generating plants to the many customers who consume the power. In some cases, cost can be lowered and reliability enhanced through the use of distributed generation -- numerous smaller generators placed at strategically selected points throughout the power system in proximity to the customers. This and other distributed resources -- so named because they are distributed throughout the system in close proximity to customers -- including storage systems and demand-side management, often provide great benefit.

But regardless of the use of distributed generation or demand-side management, the T&D system is the ultimate distributed resource, consisting of thousands, perhaps millions, of units of equipment scattered throughout the service territory, interconnected and operating in concert to achieve uninterrupted delivery of power to the electric consumers. These systems represent an investment of billions of dollars, require care and precision in their operation, and provide one of the most basic building blocks of our society -- widely available, economical, and reliable energy.

This chapter begins with an examination of the role and mission of a T&D system -- why it exists and what it is expected to do. Section 1.3 looks at several fundamental physical laws that constrain T&D systems design. The typical hierarchical system structure that results, and the costs of its equipment are summarized in sections 1.4 and 1.5. In section 1.6, a number of different ways to lay out a distribution system are covered, along with their advantages and disadvantages. The chapter ends with a look at the "systems approach -- perhaps the single most important concept in the design of retail delivery systems which are both inexpensive and reliable.

1.2 T&D SYSTEM'S MISSION

A T&D system's primary mission is to *deliver* power to electrical consumers at their *place* of consumption and in *ready-to-use* form. The system must deliver power to the customers, which means it must be dispersed throughout the utility service territory in rough proportion to customer locations and demand (Figure 1.1). This is the primary requirement for a T&D system, and one so basic it is often overlooked -- *the system must cover ground* -- reaching every customer with an electrical path of sufficient strength to satisfy that customer's demand for electric power.

That electrical path must be *reliable*, too, so that it provides an uninterrupted flow of stable power to the utility's customers. Reliable power delivery means delivering all of the power demanded, not just some of the power needed, and doing so all of the time. Anything less than near perfection in meeting this goal is considered unacceptable -- 99.9% reliability of service may sound impressive, but it means nearly nine hours of electric service interruption each year, an amount that would be unacceptable in nearly any first-world country.

Power Delivery Systems

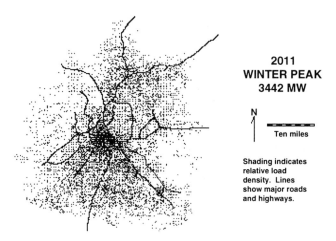

Figure 1.1 Map of electrical demand for a major US city shows where the total demand of more than 2,000 MW peak is located. Degree of shading indicates electric load distribution. The T&D system must cover the region with sufficient capacity at every location to meet the customer needs there.

Beyond the need to deliver power to the customer, the utility's T&D system must also deliver it in ready-to-use form -- at the utilization voltage required for electrical appliances and equipment, and free of large voltage fluctuations, high levels of harmonics or transient electrical disturbances (Engel et al., 1992).

Most electrical equipment in the United States is designed to operate properly when supplied with 60 cycle alternating current at between 114 and 126 volts, a plus or minus five percent range centered on the nominal *utilization voltage* of 120 volts (RMS average of the alternating voltage). In many other countries, utilization standards vary from 230 to slightly over 250 volts, at either 50 or 60 cycles AC.[1] But regardless of the utilization voltage, a utility must

---

[1] Power is provided to customers in the United States by reversed alternating current legs (+120 volts and -120 volts wire to ground). This scheme provides 240 volts of power to any appliance that needs it, but for purposes of distribution engineering and performance, acts like only 120 volt power.

maintain the voltage provided to each customer within a narrow range centered within the voltages that electric equipment is designed to tolerate.

A ten percent range of delivery voltage throughout a utility's service area may be acceptable, but a ten percent range of fluctuation in the voltage supplied to any one customer is not. An instantaneous shift of even three percent in voltage causes a perceptible, and to some people, disturbing, flicker in electric lighting. More important, voltage fluctuations can cause erratic and undesirable behavior of some electrical equipment.

Thus, whether high or low within the allowed range, the delivery voltage of any one customer must be maintained at about the same level all the time -- normally within a range of three to six percent -- and any fluctuation must occur slowly. Such *stable voltage* can be difficult to obtain, because the voltage at the customer end of a T&D system varies inversely with electric demand, falling as the demand increases, rising as it decreases. If this range of load fluctuation is too great, or if it happens too often, the customers may consider it poor service.

Thus, a T&D system's mission is to:

1. cover the service territory, reaching all customers
2. have sufficient capacity to meet the peak demands of its customers
3. provide highly reliable delivery to its customers
4. provide stable voltage quality to its customers

And of course, above everything else, to achieve these four goals at the very *lowest cost* possible.

1.3 THE "LAWS OF T&D"

The complex interaction of a T&D system is governed by a number of physical laws relating to the natural phenomena that have been harnessed to produce and move electric power. These interactions have created a number of "truths" that dominate the design of T&D systems:

1. It is more economical to move power at high voltage. The higher the voltage, the lower the cost per kilowatt, to move power any distance.
2. The higher the voltage, the greater the capacity and the greater the cost of otherwise similar equipment. Thus, high voltage lines, while potentially economical, cost a great deal more than low voltage lines, but have a much greater capacity. They are only economical in practice if they can be used to move a lot of power in one block -- they

Power Delivery Systems

are the giant economy size, but while always giant, they are only economical if one truly needs the giant size.

3. Utilization voltage is useless for the transmission of power. The 120/240 volt single-phase utilization voltage used in the United States, or even the 250 volt/416 volt three-phase used in "European systems" is not equal to the task of economically moving power more than a few hundred yards. The application of these lower voltages for anything more than very local distribution at the neighborhood level results in unacceptably high electrical losses, severe voltage drops, and astronomical equipment cost.

4. It is costly to change voltage level -- not prohibitively so, for it is done throughout a power system (that's what transformers do) -- but voltage transformation is a major expense, which does nothing to move the power any distance in and of itself.

5. Power is more economical to produce in very large amounts. Claims by the advocates of modern distributed generators notwithstanding, there *is* a significant economy of scale in generation -- large generators produce power more economically than small ones. Thus, it is most efficient to produce power at a few locations utilizing large generators.[2]

6. Power must be delivered in relatively small quantities at low (120 to 250 volt) voltage level. The average customer has a total demand equal to only 1/10,000th or 1/100,000th of the output of a large generator.

An economical T&D system builds upon these concepts. It must "pick up" power at a few, large sites (generating plants), and deliver it to many, many more small sites (customers). It must somehow achieve economy by using high voltage, but only when power flow can be arranged so that large quantities are moved simultaneously along a common path (line). Ultimately, power must be subdivided into "house-sized" amounts, reduced to utilization voltage, and

---

[2] The issue is more complicated than just a comparison of the cost of big versus small generation, as will be addressed later in this book. In some cases, distributed generation provides the lowest cost overall, regardless of the economy of scale, due to constraints imposed by the T&D system. Being close to the customers, distributed generation does not carry with it the costs of adding T&D facilities to move the power from generation site to customer. Often this is the margin of difference, as will be discussed later in this book.

routed to each business and home via equipment whose compatibility with individual customer needs means it will be relatively quite inefficient compared to the system as a whole.

Hierarchical Voltage Levels

The overall concept of a power delivery system layout that has evolved to best handle these needs and "truths" is one of hierarchical voltage levels as shown in Figure 1.2.

As power is moved from generation (large bulk sources) to customer (small demand amounts) it is first moved in bulk quantity at high voltage -- this makes particular sense since there is usually a large bulk amount of power to be moved out of a large generating plant. As power is dispersed throughout the service territory, it is gradually moved down to lower voltage levels, where it is moved in ever smaller amounts (along more separate paths) on lower capacity equipment until it reaches the customers. The key element is a "lower voltage and split" concept.

Thus, the 5 kW used by a particular customer -- Mrs. Rose at 412 Oak Street in Metropolis City -- might be produced at a 750 MW power plant more than three hundred miles to the north. Her power is moved as part of a 750 MW block from plant to city on a 345 kV transmission line, to a switching substation. Here, the voltage is *lowered* to 138 kV through a 345 to 138 kV transformer, and

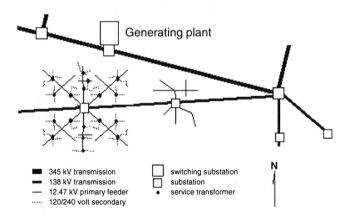

Figure 1.2 A power system is structured in a hierarchical manner with various voltage levels. A key concept is "lower voltage and split" which is done from three to five times during the course of power flow from generation to customer.

Power Delivery Systems

immediately after that, the 750 MW block is *split* into five separate flows in the switching substation buswork, each of these five parts being roughly 150 MW. Now part of a smaller block of power, Mrs. Rose's electricity is routed to her side of Metropolis on a 138 kV transmission line that snakes 20 miles through the northern part of the city, ultimately connecting to another switching substation. This 138 kV transmission line feeds power to several distribution substations along its route,[3] among which it feeds 40 MW into the substation that serves a number of neighborhoods, including Mrs. Rose's. Here, her power is run through a 138 kV/12.47kV distribution transformer.

As it emerges from the low side of the substation distribution transformer at 12.47 kV (the primary distribution voltage) the 40 MW is split into six parts, each about 7 MW, with each 7 MVA part routed onto a different distribution feeder. Mrs. Rose's power flows along one particular feeder for two miles, until it gets to within a few hundred feet of her home. Here, a much smaller amount of power, 50 kVA (sufficient for perhaps ten homes), is routed to a service transformer, one of several hundred scattered up and down the length of the feeder. As Mrs. Rose's power flows through the service transformer, it is reduced to 120/240 volts. As it emerges, it is routed onto the secondary system, operating at 120/240 volts (250/416 volts in Europe and many other countries). The secondary wiring splits the 50 kVA into small blocks of power, each about 5 kVA, and routes one of these to Mrs. Rose's home along a secondary conductor to her service drops -- the wires leading directly to her house.

Over the past one hundred years, this hierarchical system structure has proven a most effective way to move and distribute power from a few large generating plants to a widely dispersed customer base. The key element in this structure is the "reduce voltage and split" function -- a splitting of the power flow being done essentially simultaneously with a reduction in voltage. Usually, this happens between three and five times as power makes its way from generator to customers.

1.4 LEVELS OF THE T&D SYSTEM

As a consequence of this hierarchical structure, a power delivery system can be thought of very conveniently as composed of several distinct levels of equipment, as illustrated in Figure 1.3. Each level consists of many units of fundamentally similar equipment, doing roughly the same job, but located in different parts of the system. For example, all of the distribution substations

---

[3] Transmission lines whose sole or major function is to feed power to distribution substations are often referred to as "sub-transmission" lines.

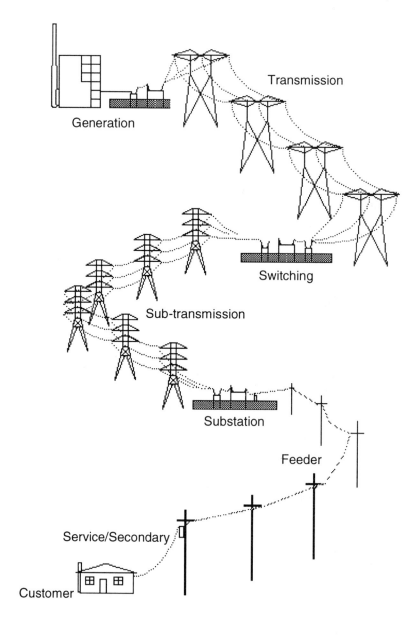

Figure 1.3 A T&D system consists of several levels of power delivery equipment, each feeding the one below it.

Power Delivery Systems

are planned and laid out in approximately the same manner and do roughly the same job. Likewise, all feeders are similar in equipment type, layout, and mission, and all service transformers have the same basic mission and are designed with similar planning goals and to similar engineering standards.

Power can be thought of as flowing "down" through these levels, on its way from power production to customer. As it moves from the generation plants (system level) to the customer, the power travels through the transmission level, to the sub-transmission level, to the substation level, through the primary feeder level, and onto the secondary service level, where it finally reaches the customer. Each level takes power from the next higher level in the system and delivers it to the next lower level in the system. While each level varies in the types of equipment it has, its characteristics, mission, and manner of design and planning, all share several common characteristics:

- Each level is fed power by the one above it, in the sense that the next higher level is electrically closer to the generation.

- Both the nominal voltage level and the average capacity of equipment drops from level to level, as one moves from generation to customer. Transmission lines operate at voltages of between 69 kV and 1,100 kV and have capacities between 50 and 2,000 MW. By contrast, distribution feeders operate between 2.2 kV and 34.5 kV and have capacities somewhere between 2 and 35 MW.

- Each level has many more pieces of equipment in it than the one above. A system with several hundred thousand customers might have fifty transmission lines, one hundred substations, six hundred feeders, and forty thousand service transformers.

- As a result, the net capacity of each level (number of units times average size) increases as one moves toward the customer. A power system might have 4,500 MVA of substation capacity but 6,200 MVA of feeder capacity and 9,000 MVA of service transformer capacity installed.[4]

- Reliability drops as one moves closer to the customer. A majority of service interruptions are a result of failure (either due to aging or to damage from severe weather) of transformers, connectors, or conductors very close to the customer, as shown in Figure 1.4.

---

[4] This greater-capacity-at-every-level is deliberate and required both for reliability reasons and to accommodate coincidence of load, which will be discussed in Chapter 3.

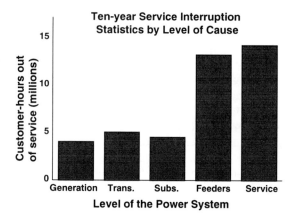

Figure 1.4 Ten years of customer interruptions for a large electric system, grouped by level of cause. Interruptions due to generation and transmission often receive the most attention because they usually involve a large number of customers simultaneously. However, such events are rare whereas failures and interruptions at the distribution level create a constant background level of interruptions.

Table 1.1 gives statistics for a typical system. The net effect of the changes in average size and number of units is that each level contains a greater total capacity than the level above it -- the service transformer level in any utility system has considerably more installed capacity (number of units times average capacity) than the feeder system or the substation system. Total capacity increases as one heads toward the customer because of non-coincidence of peak load (which will be discussed in Chapter 3) and for reliability purposes.

Table 1.1 Equipment Level Statistics for a Medium-Sized Electric System

Level of System	Voltage kV	Number of Units	Avg. Cap. MVA	Total Cap MVA
Transmission	345, 138	12	150	1,400
Sub-transmission	138, 69	25	65	1,525
Substations	139/23.9, 69/13.8	45	44	1,980
Feeders	23.9, 13.8	227	11	2,497
Service Trans.	.12, .24	60,000	.05	3,000
Secondary/Service	.12, .24	250,000	.014	3,500
Customer	.12	250,000	.005	1,250

Power Delivery Systems

The Transmission Level

The transmission system is a network of three-phase lines operating at voltages generally between 115 kV and 765 kV. Capacity of each line is between 50 MVA and 2,000 MVA. The term "network" means that there is more than one electrical path between any two points in the system (Figure 1.5). Networks are laid out in this manner for reasons of reliability and operating flow--if any one element (line) fails, there is an alternate route and power flow is (hopefully) not interrupted.

In addition to their function in moving power, portions of the transmission system -- the largest elements, namely its major power delivery lines, are designed, at least in part, for stability needs. The transmission grid provides a strong electrical tie between generators, so that each can stay synchronized with the system and with the other generators. This arrangement allows the system to operate and to function evenly as the load fluctuates and to pick up load smoothly if any generator fails -- what is called stability of operation. (A good deal of the equipment put into transmission system design, and much of its cost, is for these stability reasons, not solely or even mainly, for moving power.)

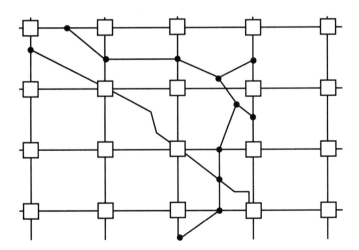

Figure 1.5 A network is an electrical system with more than one path between any two points, meaning that (if properly designed) it can provide electrical service even if any one element fails.

The Sub-transmission Level

The sub-transmission lines in a system take power from the transmission switching stations or generation plants and deliver it to substations along their routes. A typical sub-transmission line may feed power to three or more substations. Often, portions of the transmission system -- bulk power delivery lines, lines designed at least in part for stability as well as power delivery needs -- do this too, and the distinction between transmission and sub-transmission lines becomes rather blurred.

Normally, sub-transmission lines are in the range of capacity of 30 MVA up to perhaps 250 MVA, operating at voltages from 34.5 kV to as high as 230 kV. With occasional exceptions, sub-transmission lines are part of a network grid -- they are part of a system in which there is more than one route between any two points. Usually, at least two sub-transmission routes flow into any one substation, so that feed can be maintained if one fails.[5]

The Substation Level

Substations, the meeting point between the transmission grid and the distribution feeder system, are where a fundamental change takes place within most T&D systems. The transmission and sub-transmission systems above the substation level usually form a network, as discussed above, with more than one power flow path between any two parts. But from the substation on to the customer, arranging a network configuration would simply be prohibitively expensive. Thus, most distribution systems are radial -- there is only one path through the other levels of the system.

Typically, a substation occupies an acre or more of land, on which the various necessary substation equipment is located. Substation equipment consists of high and low voltage racks and busses for the power flow, circuit breakers for both the transmission and distribution level, metering equipment, and the "control house," where the relaying, measurement, and control equipment is located. But the most important equipment -- what gives this substation its capacity rating, are the substation *transformers*, which convert the incoming power from transmission voltage levels to the lower primary voltage for distribution.

Individual substation transformers vary in capacity, from less than 10 MVA to as much as 150 MVA. They are often equipped with tap-changing

---

[5] Radial feed -- only one line -- is used in isolated, expensive, or difficult transmission situations, but for reliability reasons is not recommended.

Power Delivery Systems

mechanisms and control equipment to vary their windings ratio so that they maintain the distribution voltage within a very narrow range, regardless of larger fluctuations on the transmission side. The transmission voltage can swing by as much as 5%, but the distribution voltage provided on the low side of the transformer stays within a narrow band, perhaps only ± .5%.

Very often, a substation will have more than one transformer. Two is a common number, four is not uncommon, and occasionally six or more are located at one site. Having more than one transformer increases reliability -- in an emergency, a transformer can handle a load much over its rated load for a brief period (e.g., perhaps up to 140% of rating for up to four hours). Thus, the T&D system can pick up the load of the outaged portions during brief repairs and in emergencies.

Equipped with from one to six transformers, substations range in "size" or capacity from as little as five MVA for a small, single-transformer substation, serving a sparsely populated rural area, to more than 400 MVA for a truly large six-transformer station, serving a very dense area within a large city.

Often T&D planners will speak of a *transformer unit,* which includes the transformer and all the equipment necessary to support its use -- "one fourth of the equipment in a four-transformer substation." This is a much better way of thinking about and estimating cost for equipment in T&D plans. For while a transformer, itself, is expensive (between $50,000 and $1,000,000); the buswork, control, breakers, and other equipment required to support its use can double or triple that cost. Since that equipment is needed in direct proportion to the transformer's capacity and voltage, and since it is needed *only* because a transformer is being added, it is normal to associate it with the transformer as a single planning unit -- add the transformer, add the other equipment along with it.

Substations consist of more equipment, and involve more costs, than just the electrical equipment. The land (the site) has to be purchased and prepared. Preparation is non-trivial. The site must be excavated, a grounding mat (wires running under the substation to protect against an inadvertent flow during emergencies) laid down, and foundations and control ducting for equipment must be installed. Transmission towers to terminate incoming transmission must be built. Feeder getaways -- ducts or lines to bring power out to the distribution system -- must be added.

The Feeder Level

Feeders, typically either overhead distribution lines mounted on wooden poles, or underground buried or ducted cable sets, route the power from the substation throughout its service area. Feeders operate at the primary distribution voltage.

The most common primary distribution voltage in use throughout North America is 12.47 kV, although anywhere from 4.2 kV to 34.5 kV is widely used. Worldwide, there are primary distribution voltages as low as 1.1 kV and as high as 66 kV. Some distribution systems use several primary voltages -- for example 23.9 kV and 13.8 kV and 4.16 kV.

A feeder is a small transmission system in its own right, distributing between 2 MVA to more than 30 MVA, depending on the conductor size and the distribution voltage level. Normally between two and 12 feeders emanate from any one substation, in what has been called a *dendrillic* configuration -- repeated branching into smaller branches as the feeder moves out from the substation toward the customers. In combination, all the feeders in a power system constitute the *feeder system* (Figure 1.6). An average substation has between two and eight feeders, and can vary between one and forty.

The main, three-phase trunk of a feeder is called the *primary trunk* and may branch into several main routes, as shown in the diagram on the next page. These main branches end at open points where the feeder meets the ends of other feeders -- points at which a *normally open switch* serves as an emergency tie between two feeders.

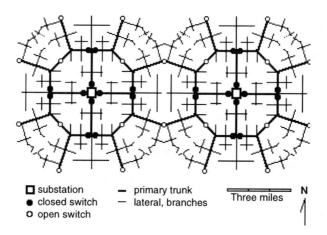

□ substation   — primary trunk
● closed switch — lateral, branches
○ open switch

Three miles   N

Figure 1.6 Distribution feeders route power away from the substation, as shown (in idealized form -- configuration is never so evenly symmetric in the real world) for two substations. Positions of switches make the system electrically radial, while parts of it are physically a network. Shown here are two substations, each with four feeders.

Power Delivery Systems

In addition, each feeder will be divided, by normally closed switches, into several switchable elements. During emergencies, segments can be re-switched to isolate damaged sections and route power around outaged equipment to customers who would otherwise have to remain out of service until repairs were made.

By definition, the feeder consists of all primary voltage level segments between the substations and an open point (switch). Any part of the distribution level voltage lines -- three-phase, two-phase, or single-phase, that is switchable, is considered part of the primary feeder. The primary trunks and switchable segments are usually built using three phases, with the largest size of distribution conductor (typically this is about 500-600 MCM conductor, but conductor over 1,000 MCM is not uncommon, and the author has feeders with 2,000 MCM conductor) being used for reasons other than just maximum capacity (e.g., contingency switching needs). Often a feeder has excess capacity because it needs to provide back-up for other feeders during emergencies.

The vast majority of distribution feeders worldwide and within the United States are overhead construction, wooden pole with wooden crossarm or post insulator. Only in dense urban areas, or in situations where esthetics are particularly important, can the higher cost of underground construction be justified. In this case, the primary feeder is built from insulated cable, which is pulled through concrete ducts that are first buried in the ground. Underground feeder costs from three to ten times what overhead does.

Many times, however, the first several hundred yards of an overhead primary feeder are built underground even if the system is overhead. This underground portion is used as the *feeder get-away*. Particularly at large substations, the underground get-away is dictated by practical necessity, as well as by reliability and esthetics. At a large substation, ten or 12 three-phase, overhead feeders leaving the substation mean from 40 to 48 wires hanging in mid-air around the substation site, with each feeder needing the proper spacings for electrical insulation, safety, and maintenance. At a large-capacity substation in a tight location, there is simply not enough overhead space for so many feeders. Even if there is, the resulting tangle of wires looks unsightly, and perhaps most important, is potentially unreliable -- one broken wire falling in the wrong place can disable a lot of power delivery capability.

The solution to this dilemma is the underground feeder getaway, usually consisting of several hundred yards of buried, ducted cable that takes the feeder out to a riser pole, where it is routed above ground and connected to overhead wires. Very often, this initial underground link sets the capacity limit for the entire feeder -- the underground cable ampacity is the limiting factor for the feeder's power transmission.

The Lateral Level

Laterals, short stubs or line segments that branch off the primary feeder, represent the final primary voltage part of the power's journey from the substation to the customer. A lateral is directly connected to the primary trunk and operates at the same nominal voltage. A series of laterals tap off the primary feeder as it passes through a community, each lateral routing power to a few dozen homes.

Normally, laterals do not have branches, and many laterals are only one or two-phase -- all three phases are used only if a relatively substantial amount of power is required, or if three-phase service must be provided to some of the customers. Normally, single and two-phase laterals are arranged to tap alternately different phases on the primary feeder, as shown below; an attempt by the distribution planning engineer to balance the loads as closely as possible.

Typically, laterals deliver from as little as 10 kVA for a small single-phase lateral, to as much as 2 MVA. In general, even the largest laterals use small conductors (relative to the primary size). When a lateral needs to deliver a great deal of power, the planner will normally use all three phases, with a relatively small conductor for each, rather than employ a single-phase and use a large conductor. This approach avoids creating a significant imbalance in loading at the point where the lateral taps into the primary feeder. Power flow, loadings and voltage are maintained in a more balanced state if the power demands of a "large lateral" are distributed over all three phases.

Laterals (wooden poles) are built overhead or underground. Unlike primary feeders and transmission lines, single-phase laterals are sometimes buried directly. In this case, the cable is placed inside a plastic sheath (that looks and feels much like a vacuum cleaner hose), a trench is dug, and the sheathed cable is unrolled into the trench and buried. Directly buried laterals are no more expensive than underground construction in many cases.

The Service Transformers

Service transformers lower voltage from the primary voltage to the utilization or customer voltage, normally 120/240 volt two-leg service in most power systems throughout North America. In overhead construction, service transformers are pole mounted and single-phase, between 5 kVA and 166 kVA capacity. There may be several hundred scattered along the trunk and laterals of any given feeder; since power can travel efficiently only up to about 200 feet at utilization voltage, there must be at least one service transformer located reasonably close to every customer.

Power Delivery Systems

Passing through these transformers, power is lowered in voltage once again, to the final utilization voltage (120/240 volts in the United States) and routed onto the secondary system or directly to the customers. In cases where the system is supplying power to large commercial or industrial customers, or the customer requires three-phase power, between two and three transformers may be located together in a transformer bank, and interconnected in such a way as to provide multi-phase power. Several different connection schemes are possible for varying situations.

Underground service, as opposed to overhead pole-mounted service, is provided by padmount or vault type service transformers. The concept is identical to overhead construction, with the transformer and its associated equipment changed to accommodate incoming and outgoing lines that are underground.

The Secondary and Service Level

Secondary circuits, fed by the service transformers, route power at utilization voltage within very close proximity to the customer, usually in an arrangement in which each transformer serves a small radial network of utilization voltage secondary and service lines, which lead directly to the meters of customers in the immediate vicinity.

At most utilities, the layout and design of the secondary level is handled through a set of standardized guidelines and tables, which are used by engineering technicians and clerks to produce work orders for the utilization voltage level equipment. In the United States, the vast majority of this system is single-phase. In European systems, much of the secondary is three-phase, particularly in urban and suburban areas.

What is Transmission and What is Distribution?

Definitions and nomenclature defining "transmission" and "distribution" vary greatly among different countries, companies, and power systems. Generally, three types of distinction between the two are made:

By voltage class: transmission is anything above 34.5 kV; distribution is anything below that.

By function: distribution includes all utilization voltage equipment, plus all lines that feed power to service transformers.

By configuration: transmission includes a network; distribution is all the radial equipment in the system.

Generally, all three definitions apply simultaneously, since in most utility systems, any transmission above 34.5 kV is configured as a network, and does not feed service transformers directly, while all distribution is radial, built of only 34.5 kV or below, and does feed service transformers. Substations -- the meeting places of transmission lines (incoming) and distribution lines (outgoing) -- are often included in one or the other category, and sometimes are considered as separate entities.

1.5 UTILITY DISTRIBUTION EQUIPMENT

The preceding section made it clear that a power delivery system is a very complex entity, composed of thousands, perhaps even millions, of components, which function together as a *T&D system*. Each unit of equipment has only a small part to play in the system, and is only a small part of the cost, yet each is critical for satisfactory service to at least one or more customers, or it would not be included in the system.

T&D system planning is complex because each unit of equipment influences the electrical behavior of its neighbors, and must be designed to function well in conjunction with the rest of the system, under a variety of different conditions, regardless of shifts in the normal pattern of loads or the status of equipment nearby. While the modeling and analysis of a T&D system can present a significant challenge, individually its components are relatively simple to understand, engineer, and plan. In essence, there are only two major types of equipment that perform the power delivery function:

- transmission and distribution lines, which move power from one location to another

- transformers, which change the voltage level of the power

Added to these three basic equipment types are two categories of equipment used for a very good reason:

- protective equipment which provides safety and "fail safe" operation

- voltage regulation equipment, which is used to maintain voltage within an acceptable range as the load changes. Monitoring and control equipment, used to measure equipment and system performance and feed this information to control systems so that the utility knows what the system is doing and can control it, for both safety and efficiency reasons.

Transmission and Distribution Lines

By far the most omnipresent part of the power distribution system is the portion devoted to actually moving the power flow from one point to another. Transmission lines, sub-transmission lines, feeders, laterals, secondary and service drops, all consist of electrical conductors, suitably protected by isolation (transmission towers, insulator strings, and insulated wrappings) from voltage leakage and ground contact. It is this conductor that carries the power from one location to another.

Electrical conductors are available in various capacity ranges, with capacity generally corresponding to the metal cross section (other things being equal, a thicker wire carries more power). Conductors can be all steel (rare, but used in some locations where winter ice and wind loadings are quite severe), all aluminum, or copper, or a mixture of aluminum and steel. Underground transmission can use a various types of high-voltage cable. Capacity of a line depends on the current-carrying capacity of the conductor or the cable, the voltage, the number of phases, and constraints imposed by the line's location in the system.

The most economical method of handling a conductor is to place it overhead, supported by insulators, on wooden poles or metal towers, suitably clear of interference or contact with persons or property. However, underground construction, while generally more costly, avoids esthetic intrusion of the line and provides some measure of protection from weather (it also tends to reduce the capacity of a line slightly due to the differences between underground cable and overhead conductor). Suitably wrapped with insulating material in the form of underground cable, the cable is placed inside concrete or metal ducts or surrounded in a plastic sheath.

Transmission/sub-transmission lines are always three-phase -- three separate conductors for the alternating current -- sometimes with a fourth neutral (unenergized) wire. Voltage is measured between phases -- a 12.47 kV distribution feeder has an alternating current voltage (RMS) of 12,470 volts as measured between any two phases. Voltage between any phase and ground is 7,200 volts (12.47 divided by the square root of three). Major portions of a distribution system -- trunk feeders -- are as a rule built as three-phase lines, but lower-capacity portions may be built as either two-phase, or single-phase.[6]

Regardless of type or capacity, every electrical conductor has an impedance (a resistance to electrical flow through it) that causes voltage drop and electrical losses whenever it is carrying electric power. Voltage drop is a reduction in the

---

[6] In most cases, a single-phase feeder or lateral has two conductors: the phase conductor and the neutral.

voltage between the sending and receiving ends of the power flow. Losses are a reduction in the net power, and are proportional to the square of the power. Double the load and the losses increase by four. Thus, 100 kilowatts at 120 volts might go in one end of a conductor, only to emerge at the other as 90 kilowatts at 114 volts at the other end. Both voltage drop and losses vary in direct relation to load -- within very fine limits if there is no load, there are no losses or voltage drop. Voltage drop is proportional to load -- double the load and voltage drop doubles. Losses are *quadratic*, however -- double the load and losses quadruple.

Transformers

At the heart of any alternating power system are transformers. They change the voltage and current levels of the power flow, maintaining (except for a very small portion of electrical losses), the same overall power flow. If voltage is reduced by a factor of ten from high to low side, then current is multiplied by ten, so that their overall product (voltage times current equals power) is constant in and out.

Transformers are available in a diverse range of types, sizes, and capacities. They are used within power systems in four major areas: at power plants, where power which is minimally generated at about 20,000 volts is raised to transmission voltage (100,000 volts or higher); at switching stations, where transmission voltage is changed (e.g., from 345,000 volts to 138,000 volts before splitting onto lower voltage transmission lines); at distribution substations, where incoming transmission-level voltage is reduced to distribution voltage for distribution (e.g., 138 kV to 12.47 kV); and at service transformers, where power is reduced in voltage from the primary feeder voltage to utilization level (12.47 kV to 120/240 volts) for routing into customers' homes and businesses.

Larger transformers are generally built as three-phase units, in which they simultaneously transform all three phases. Often these larger units are built to custom or special specifications, and can be quite expensive -- over $3,000,000 per unit, in some cases. Smaller transformers, particularly most service transformers, are single-phase -- it takes three installed side by side to handle a full three-phase line's power flow. They are generally built to standard specifications and bought in quantity.

Transformers experience two types of electrical losses -- no-load losses (often called core, or iron, losses) and load-related losses. No-load losses are electrical losses inherent in operating the transformer -- due to its creation of a magnetic field inside its core -- and occur simply because the transformer is connected to an electrical power source. They are constant, regardless of whether the power flowing through the transformer is small or large. Core losses are typically less than one percent of the nameplate rating. Only when the

Power Delivery Systems 21

transformer is seriously overloaded, to a point well past its design range, will the core losses change (due to magnetic saturation of the core).

Load-related losses are due to the current flow through the transformer's impedance and correspond very directly with the level of power flow -- like those of conductors and cables they are proportional to current squared, quadrupling whenever power flow doubles. The result of both types of losses is that a transformer's losses vary as the power transmitted through it varies, but always at or above a minimum level set by the no-load losses.

Switches

Occasionally, it is desirable to be able to vary the connection of line segments within a power delivery system, particularly in the distribution feeders. Switches are placed at strategic locations so that the connection between two segments can be opened or closed. Switches are planned to be normally closed (NC) or normally open (N.O.), as was shown in Figure 1.6.

Switches vary in their rating (how much current they can vary) and their load break capacity (how much current they can interrupt, or switch off), with larger switches being capable of opening a larger current. They can be manually, automatically, or remotely controlled in their operation.

Protection

When electrical equipment fails, for example if a line is knocked down during a storm, the normal function of the electrical equipment is interrupted. Protective equipment is designed to detect these conditions and isolate the damaged equipment, even if this means interrupting the flow of power to some customers. Circuit breakers, sectionalizers, and fused disconnects, along with control relays and sensing equipment, are used to detect unusual conditions and interrupt the power flow whenever a failure, fault, or other unwanted condition occurs on the system.

These devices and the protection engineering required to apply them properly to the power system are not the domain of the utility planners and will not be discussed here. Protection *is* vitally important, but the planner is sufficiently involved with protection if he or she produces a system design that *can* be protected within standards, and if the cost of that protection has been taken into account in the budgeting and planning process. Both of these considerations are non-trivial.

Protection puts certain constraints on equipment size and layout -- for example in some cases a very large conductor is too large (because it would permit too high a short circuit current) to be protected safely by available equipment and cannot be used. In other cases, long feeders are too long to be protected (because they have too low a short circuit current at the far end). A good deal of protective equipment is quite complex, containing sensitive electromechanical parts, (many of which move at high speeds and in a split-second manner), and depending on precise calibration and assembly for proper function. As a result, the cost of protective equipment and control, and the cost of its maintenance, is often significant -- differences in protection cost *can* make the deciding difference between two plans.

Voltage regulation

Voltage regulation equipment includes line regulators and line drop compensators, as well as tap changing transformers. These devices vary their turns-ratio (ratio of voltage in to voltage out) to react to variations in voltage drop -- if voltage drops, they raise the voltage, if voltage rises, they reduce it to compensate. Properly used, they can help maintain voltage fluctuation on the system within acceptable limits, but they can only reduce the range of fluctuation, not eliminate it altogether.

Capacitors

Capacitors are a type of voltage regulation equipment. By correcting power factor they can improve voltage under many heavy loads (hence large voltage drop) cases. Power factor is a measure of how well voltage and current in an alternating system are in step with one another. In a perfect system, voltage and current would alternately cycle in conjunction with one another -- reaching a peak, then reaching a minimum, at precisely the same times. But on distribution systems, particularly under heavy load conditions, current and voltage fall out of phase -- both continue to alternate 60 times a second, but during each cycle voltage may reach its peak slightly ahead of current -- there is a slight lag of current versus voltage, as shown in Figure 1.7.

It is the precise, simultaneous peaking of both voltage and current that delivers maximum power. If out of phase, even by a slight amount, effective power drops, as does power factor -- the ratio of real (effective) power to the maximum possible power (if voltage and current were locked in step).

Power engineers refer to a quantity called VARs (Volt-Amp Reactive) that is caused by this condition. Basically, as power factors worsen (as voltage and current fall farther apart in terms of phase angle) a larger percent of the electrical flow is VARs, and a smaller part is real power. The frustrating thing is that the voltage is still there, and the current is still there, but because of the shift in their timing, they produce only VARS, not power. The worse the power factor, the

Power Delivery Systems

higher the VAR content. Poor power factor creates considerable cost and performance consequences for the power system: large conductor is still required to carry the full level of current even though power delivery has dropped, and because current is high, the voltage drop is high, too, further degrading quality of service.

Unless one has worked for some time with the complex variable mathematics associated with AC power flow analysis, VARs are difficult to picture. A useful analogy is to think of VARs as "electrical foam." If one tried to pump a highly carbonated soft drink through a system of pipes, turbulence in the pipes, particularly in times of high demand (high flow) would create foam. The foam would take up room in the pipes, but contribute little of value to the net flow -- the equivalent of VARs in an electrical system.

Poor power factor has several causes. Certain types of loads create VARs -- in simple terms loads which cause a delay in the current with respect to voltage as it flows through them. Among the worst offenders are induction motors, particularly small ones as almost universally used for blowers, air conditioning compressors, and the powering of conveyor belts and similar machinery. Under heavy load conditions, voltage and current can get out of phase to the point that power factor can drop below 70%. In addition, transmission equipment itself can often create this lag and "generate" a low power factor.

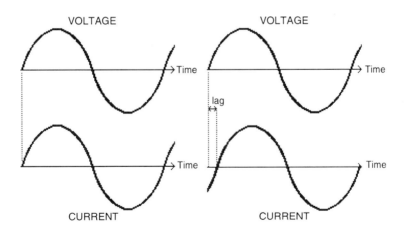

Figure 1.7 Current and voltage in phase deliver maximum power (left). If current and voltage fall out of phase (right), actual power delivered drops by very noticeable amounts -- the *power factor* falls.

Capacitors correct the poor power factor. They "inject" VARs into a T&D line to bring power factor close to 1.0, transforming VAR flow back into real power flow, regaining the portion of capacity lost to poor power factor. Capacitors can involve considerable cost depending on location and type. They tend to do the most good if put on the distribution system, near the customers, but they cost a great deal more in those locations than if installed at substations.

1.6 T&D COSTS

A T&D system can be expensive to design, build, and operate. Equipment at every level incurs two types of costs. Capital costs include the equipment and land, labor for site preparation, construction, assembly and installation, and any other costs associated with building and putting the equipment into operation. Operating costs include labor and equipment for operation, maintenance and service, taxes and fees, as well as the value of the power lost to electrical losses. Usually, capital cost is a one-time cost (once it's built, the money's been spent). Operating costs are continuous or periodic.

Electrical losses vary depending on load and conditions. While these losses are small by comparison to the overall power being distributed (seldom more than 8%), they constitute a very real cost to the utility, and the present worth of the lifetime losses through a major system component such as a feeder or transformer can be a significant factor impacting its design and specification, often more than the original capital cost of the unit. Frequently, a more costly type of transformer will be selected for a certain application because its design leads to an overall savings due to lower losses, or a larger capacity line (larger conductor) will be used than really needed due to capacity requirements, purely because the larger conductor will incur lower losses costs.

Cumulatively, the T&D system represents a considerable expense. While a few transmission lines and switching stations are composed of large, expensive, and purpose-designed equipment, the great portion of the sub-transmission-substation-distribution system is built from "small stuff" -- commodity equipment bought mostly "off the shelf" to standard designs. Individually inexpensive, they amount to a significant cost when added together.

Transmission Costs

Transmission line costs are based on a per mile cost and a termination cost at either end of the line associated with the substation at which it is terminated. Costs can run from as low as $50,000/mile for a 46 kV wooden pole sub-transmission line with perhaps 50 MVA capacity ($1 per kVA-mile) to over

$1,000,000 per mile for a 500 kV double circuit construction with 2,000 MVA capacity ($.5/kVA-mile).

Substation Costs

Substation costs include all the equipment and labor required to build a substation, including the cost of land and easements/ROW. For planning purposes, substations can be thought of as having four costs:

1. Site cost -- the cost of buying the site and preparing it for a substation.

2. Transmission cost -- the cost of terminating the incoming sub-transmission lines at the site.

3. Transformer cost -- the transformer and all metering, control, oil spill containment, fire prevention, cooling, noise abatement, and other transformer related equipment, along with typical buswork, switches, metering, relaying, and breakers associated with this type of transformer, and their installation.

4. Feeder buswork/getaway costs -- the cost of beginning distribution at the substation, includes getting feeders out of the substation.

Often, as an expedient in planning, estimated costs of feeder buswork and getaways are folded into the transformer costs. The feeders to route power out of the substation are needed in conjunction with each transformer, and in direct proportion to the transformer capacity installed, so that their cost is sometimes considered together with the transformer as a single unit. Regardless, the transmission, transformer, and feeder costs can be estimated fairly accurately for planning purposes.

Cost of land is another matter entirely. Site, and easements or ROW into a site, have a cost that is a function of local land prices, which vary greatly, depending on location and real-estate markets. Site preparation includes the cost of preparing the site (grading, grounding mat, foundations, buried ductwork, control building, lighting, fence, landscaping, and access road).

Substation costs vary greatly depending on type, capacity, local land prices and other variable circumstances. In rural settings where load density is quite low and minimal capacity is required, a substation may involve a site of only several thousand square feet of fenced area, a single incoming transmission line (69 kV), one 5 MVA transformer, fusing for all fault protection, and all "buswork" built with wood poles and conductor, for a total cost of perhaps no more than $90,000. The substation would be applied to serve a load of perhaps 4 MW, for a cost of $23/kW. This substation in conjunction with the system around it would probably provide service with about ten hours of service interruptions per year under average conditions.

However, a typical substation built in most suburban and urban settings would be fed by two incoming 138 kV lines feeding two 40 MVA, 138 kV to 12.47 kV transformers, each feeding a separate low side (12.47 kV) bus, each bus with four outgoing distribution feeders of 9 MVA peak capacity each, and a total cost of perhaps $2,000,000. Such a substation's cost could vary from between about $1.5 million and $6 million, depending on land costs, labor costs, the utility equipment and installation standards, and other special circumstances. In most traditional vertically-integrated, publicly-regulated electric utilities, this substation would have been used to serve a peak load of about 60 MVA (75% utilization of capacity), meaning that at its nominal $2,000,000 cost works out to $33/kW. In a competitive industry, with tighter design margins and proper engineering measures taken beforehand, thus could be pushed to a peak loading of 80 MVA (100% utilization, $25/kW). This substation and the system around it would probably provide service with about two to three hours of service interruptions per year under average conditions.

Feeder System Costs

The feeder system consists of all the primary distribution lines, including three-phase trunks and their lateral extensions. These lines operate at the primary distribution voltage -- 23.9 kV, 13.8 kV, 12.47 kV, 4.16kV or whatever -- and may be three, two, or single-phase construction as required. Typically, the feeder system is also considered to include voltage regulators, capacitors, voltage boosters, sectionalizers, switches, cutouts, fuses, any intertie transformers (required to connect feeders of different voltage at tie points, as, for example, 23.9 and 12.47 kV) that are installed on the feeders (i.e., not at the substations or at customer facilities).

As a rule of thumb, construction of three-phase overhead, wooden pole crossarm type feeders of normal, large conductor (about 600 MCM per phase) at a medium distribution primary voltage (e.g., 12.47 kV) costs about $150,000/mile. However, cost can vary greatly due to variations in labor, filing and permit costs among utilities, as well as differences in design standards, and terrain. Where a thick base of top soil is present, a pole can be installed by simply auguring a hole for the pole. In areas where there is rock close under the surface, holes have to be jackhammered or blasted, and cost goes up accordingly. It is generally less expensive to build feeders in rural areas than in

Power Delivery Systems

suburban or urban areas. Thus, while $150,000 is a good average cost, a mile of new feeder construction could cost as little as $55,000 in some situations and as much as $500,000 in others.

A typical distribution feeder (three-phase, 12.47 kV, 600 MCM/phase) would be rated at a thermal (maximum) capacity of about 15 MVA and a recommended economic (design) peak loading of about 8.5 MVA peak, depending on losses and other costs. At $150,000/mile, this capacity rating gives somewhere between $10 to $15 per kW-mile as the cost for basic distribution line. Underground construction of three-phase primary is more expensive, requiring buried ductwork and cable, and usually works out to a range of $30 to $50 per kW-mile.

Lateral lines -- short primary-voltage lines working off the main three-phase circuit, are often single or two-phase and consequently have lower costs but lower capacities. Generally, they are about $5 to $15 per kW-mile overhead, with underground costs of between $5 to $15 per kW-mile (direct buried) to $30 to $100 per kW-mile (ducted).

Cost of other distribution equipment, including regulators, capacitor banks and their switches, sectionalizers, line switches, etc., varies greatly depending on the specifics of each application. In general, the cost of the distribution system will vary from between $10 and $30 per kW-mile.

Service Level Costs

The service, or secondary, system consists of the service transformers that convert primary to utilization voltage, the secondary circuits that operate at utilization voltage and the service drops that feed power directly to each customer. Without exception these are very local facilities, meant to move power no more than a few hundred feet at the very most and deliver it to the customer "ready to use."

Many electric utilities develop cost estimates for this equipment on a per-customer basis. A typical service configuration might involve a 50 MVA pole-mounted service transformer feeding ten homes, as shown in Figure 1.8. Costs for this equipment might include:

Heavier pole & hardware for transformer application	$250
50 kW transformer, mounting equipment, and installation	$750
500 feet secondary (120/240 volt) single-phase @ $2/ft.	$1,000
10 service drops including installation at $100	$1,000
	$3,000

For a cost of about $300 per customer, or about $60/kW of coincident load.

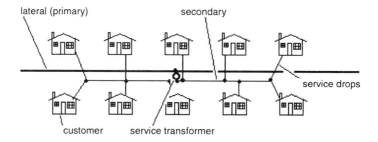

Figure 1.8 Here, a service transformer, fed from a distribution primary-voltage lateral, feeds in turn ten homes through secondary circuit operating at utilization voltage.

Maintenance and operating costs

Once put into service, T&D equipment must be maintained in sound, operating function, hopefully in the manner intended and recommended by the manufacturer. This will require periodic inspection and service, and may require repair due to damage from storms or other contingencies. In addition, many utilities must pay taxes or fees for equipment (T&D facilities are like any other business property). Operating, maintenance, and taxes are a continuing annual expense.

It is very difficult to give any generalization of O&M&T costs, partly because they vary so greatly from one utility to another, but mostly because utilities account for and report them in very different ways. Frankly, the author has never been able to gather a large number of comparable data sets from which to produce even a qualitative estimate of average O&M&T costs.[7] With that caveat, a general rule of thumb: O&M&T costs for a power delivery system probably run between 1/8 and 1/30 of the capital cost, annually.

---

[7] For example, some utilities include part of O&M expenses in overhead costs; others do not. A few report all repairs (including storm damage) as part of O&M, others accumulate major repair work separately. Still others report certain parts of routine service (periodic rebuilding of breakers) as a type of capital cost because it extends equipment life or augments capacity; others report all such work as O&M, even when the rebuilding upgrades capacity or voltage class.

The Cost to Upgrade Exceeds the Cost to Build

One of the fundamental factors affecting design of T&D systems is that it costs more to upgrade equipment to a higher capacity than to build to that capacity in the original construction. For example, a 12.47 kV overhead, three-phase feeder with a 9 MW capacity (336 MCM phase conductor) might cost $120,000/mile to build ($13.33 per kW-mile). Building it with 600 MCM conductor instead, for a capacity of 15 MVA, would cost in the neighborhood of $150,000 ($10/kW-mile).

However, upgrading an existing 336 MCM, 9 MW capacity line to 600 MCM, 15 MVA capacity could cost $200,000/mile -- over $30 per kW-mile for the 6 MW of additional capacity. This is more expensive because it entails removing the old conductor and installing new conductor along with brackets, crossarms, and other hardware required to support the heavier new conductor. Typically, this work is done hot (i.e., with the feeder energized), which means the work must be undertaken with extreme care and following a number of safety-related restrictions on equipment and labor.

Thus, T&D planners have an incentive to look at their long-term needs carefully, and to "overbuild" against initial requirements if growth trends show eventual demand will be higher. The cost of doing so must be weighed against long-term savings, but often T&D facilities are built with considerable margin (50%) above existing load to allow for future load growth.

The very high cost per kW for upgrading a T&D system in place creates one of the best perceived opportunities for DSM and DG reduction. Note that the capital cost/kW for the upgrade capacity in the example above ($33/kW) is nearly three times the cost of similar new capacity. Thus, planners often look at areas of the system where slow, continuing load growth has increased load to the point that local delivery facilities are considerably taxed, as areas where DSM and DG can deliver significant savings.

In some cases, distributed resources *can* reduce or defer significantly the need for T&D upgrades of the type described above. However, this does not assure a significant savings, for the situation is more complicated than an analysis of capital costs to upgrade may indicate. If the existing system (e.g., the 9 MW feeder) needs to be upgraded, then it is without a doubt highly loaded, which means its losses may be high, even off-peak. The upgrade to a 600 MCM conductor will cut losses 8,760 hours per year. Losses cost may drop by a significant amount, *enough in many cases to justify the cost of the upgrade alone*. The higher the annual load factor in an area, the more likely this is to occur, but it is often the case even when load factor is only 40%. However, DSM and in some cases DG also lower losses, making the comparison quite involved, as will be discussed later in this book.

Electrical Losses Costs

Movement of power through any electrical device, be it a conductor, transformer, regulator, or whatever, incurs a certain amount of electrical loss due to the impedance (resistance to the flow of electricity) of the device. These losses are a result of inviolable laws of nature. They can be measured, assessed, and minimized through proper engineering, but never eliminated completely.

Losses are an operating cost

Although losses do create a cost (sometimes a considerable one) it is not always desirable to reduce them as much as possible. Perhaps the best way to put them in proper perspective is to think of T&D equipment as *powered by electricity* -- *t*he system that moves power from one location to another runs on electric energy itself. Seen in this light, losses are revealed as what they are -- a necessary operating expense to be controlled and balanced against other costs.

Consider a municipal water department, which uses electric energy to power the pumps that drive the water through the pipes to its customers. Electricity is an acknowledged operating cost, one accounted for in planning, and weighed carefully in designing the system and estimating its costs. The water department could choose to buy highly efficient pump motors, ones that command a premium price over standard designs but provide a savings in reduced electric power costs, and to use piping that is coated with a friction-reducing lining to promote rapid flow of water (thus carrying more water with less pump power), all toward reducing its electric energy cost. Alternatively, after weighing the cost of this premium equipment against the energy cost savings it provides, the water department may decide to use inexpensive motors and piping and simply pay more over the long run. The point is that the electric power required to move the water is viewed merely as one more cost that had to be included in determining what is the lowest "overall" cost. *It takes power to move power.*

Since electric delivery equipment is powered by its own delivery product, this point often is lost. However, in order to do its job of delivering electricity, a T&D system must be provided with power, itself, just like the water distribution system. Energy must be expended to move the product. Thus, a transformer consumes a small portion of the power fed into it. In order to move power 50 miles, a 138 kV transmission line similarly consumes a small part of the power given to it.

Initial cost of equipment can always be traded against long-term losses costs. Highly efficient transformers can be purchased to use considerably less power to perform their function than standard designs. Larger conductors can be used in any transmission or distribution line, which will lower impedance, and thus

Power Delivery Systems

losses for any level of power delivery. But both examples here cost more money initially -- the efficient transformer may cost three times what a standard design does; the larger conductor might entail a need for not only large wire, but heavier hardware to hold it in place, and stronger towers and poles to keep it in the air. In addition, these changes may produce other costs -- for example, use of larger conductor not only lowers losses, but a higher fault duty (short circuit current), which increases the required rating and cost for circuit breakers. Regardless, initial equipment costs can be balanced against long-term losses costs through careful study of needs, performance, and costs, to establish a minimum overall (present worth) cost.

Load-related losses

Flow of electric power through any device is accompanied by what are called load-related losses, which increase as the power flow (load) increases. These are due to the impedance of the conductor or device. Losses increase as the *square* of the load -- doubling the power flowing through a device quadruples the losses. Tripling power flow increases the losses by nine.

With very few exceptions, larger electrical equipment always has a lower impedance, and thus lower load-related losses, for any given level of power delivery. Hence, if the losses inherent in delivering 5 MW using 600 MCM conductor are unacceptably large, the use of 900 MCM conductor will reduce them considerably. The cost of the larger conductor can be weighed against the savings in reduced losses to decide if it is a sound economic decision.

No-load losses

"Wound" T&D equipment -- transformers and regulators -- have load-related losses as do transmission lines and feeders, but, in addition, they have a type of electrical loss that is constant, regardless of loading. No-load losses constitute the electric power required to establish a magnetic field inside these units, without which they would not function. Regardless of whether a transformer has any load -- any power passing through it at all -- it will consume a small amount of power, generally less than 1% of its rated full power, simply because it is energized and "ready to work." No-load losses are constant, and occur 8,760 hours per year.

Given similar designs, a transformer will have no load losses proportional to its capacity -- a 10 MVA substation transformer will have twice the no-load losses of a 5 MVA transformer of similar voltage class and design type. Therefore, unlike the situation with a conductor, selection of a larger transformer does not always reduce net transformer losses, because while the larger transformer will always have lower load-related losses, it will have higher no-load losses, and this increase might outweigh the reduction in load-related losses.

Again, low-loss transformers are available, but cost more than standard types. Lower-cost-than-standard, but higher loss, transformers are also available (often a good investment for backup and non-continuous use applications).

The costs of losses

The electric power required to operate the T&D system -- the electrical losses-- is typically viewed as having two costs, demand and energy. Demand cost is the cost of providing the peak capacity to generate and deliver power to the T&D equipment. A T&D system that delivers 1,250 MW at peak might have losses during this peak of 100 MW. This means the utility must have generation, or buy power at peak, to satisfy this demand, whose cost is calculated using the utility's power production cost at time of peak load. This is usually considerably above its average power production cost.

Demand cost also ought to include a considerable T&D portion of expense. Every service transformer in the system (and there are many) is consuming a small amount of power in doing its job at peak. Cumulatively, this might equal 25 MW of power -- up to 1/4 of all losses in the system. That power must not only be generated by the utility but transmitted over its transmission system, through its substations, and along its feeders to *reach* the service transformers. Similarly, the power for electrical losses in the secondary and service drops (while small, are numerous and low voltage, so that their cumulative contribution to losses is noticeable) has to be moved even farther, through the service transformers and down to the secondary level.

Demand cost of losses is the total cost of the capacity to provide the losses and move them to their points of consumption.

Losses occur whenever the power system is in operation, which generally means 8,760 hours per year. While losses vary as the square of load, so they drop by a considerable margin off-peak. Their steady requirement every hour of the year imposes a considerable energy demand over the course of a year. This cost is the cost of the energy to power the losses.

Example: Consider a typical 12.47 kV, three-phase, OH feeder, with 15 MW capacity (600 MCM phase conductor), serving a load of 10 MW at peak with 4.5% primary-level losses at peak (450 kW losses at peak), and having a load factor of 64% annually. Given a levelized capacity cost of power delivered to the low side bus of a substation of $10/kW, the demand cost of these losses is $4,500/year. Annual energy cost, at 3.5¢ /kWhr, can be estimated as:

450 kW losses at peak x 8,760 hours x (64% load factor)^2 x 3.5¢ = $56,500

Power Delivery Systems

Thus, the losses' costs (demand plus energy costs) for this feeder are nearly $60,000 annually. At a present worth discount factor of around 11%, this means losses have an estimated present worth of about $500,000.[8] If the peak load on this feeder were run up to its maximum rating (about 15 MW instead of 10 MW) with a similar load factor of 64%, annual losses and their cost would increase to $(15/10)^2$ or $1,250,000 dollars.[9]

This feeder, in its entirety, might include four miles of primary trunk (at $150,000/mile) and thirty miles of laterals (at $50,000/mile), for a total capital cost of about $2,100,000. Thus, *total losses cost are on the order of magnitude of original cost of the feeder itself,* and in cases where loading is high, can approach that cost. Similar loss-capital relations exist for all other levels of the T&D system, with the ratio of losses' costs/capital cost increasing as one nears the customer level (lower voltage equipment has higher losses/kW).

Total T&D Costs

What is the total cost to deliver electric power? Of course this varies from system to system as well as from one part of a system to another -- some customers are easier to reach with electric service than others (Figure 1.9). Table 1.2 shows the cost of providing service to a "typical" residential customer.

Table 1.2 Cost of Providing Service to a Typical Residential Customers

Level	Cost Components	Cost
Transmission	4 kW x 100 miles x $.75/kW mile	$300
Substation	4 kW x $60/kW	$240
Feeder	4 kW x 1.5 miles x $10/kW-mile	$60
Service	1/10th of 50 kVA local service system	$300
	Total Initial cost (Capital)	$900
	Operations, Maintenance, and Taxes (PW next 30 years) =	$500
	Cost of electrical losses (PW next 30 years) =	$700
Estimated cost of power delivery, 30 years, PW		$2,100

---

[8] This computation used a simplification -- squaring the load factor to estimate load factor impact on losses -- which tends to underestimate losses' costs slightly. Actual losses costs probably would be more in the neighborhood of $565,000 PW.

[9] Most feeders are designed with a considerable margin between expected peak load (10 MW) and thermal (maximum continuous capacity) for reasons of reliability margin (so that there is capacity to pick up customers on a nearby feeder should it go out of service), and for losses' economics reasons -- losses are too expensive at close to the thermal limit.

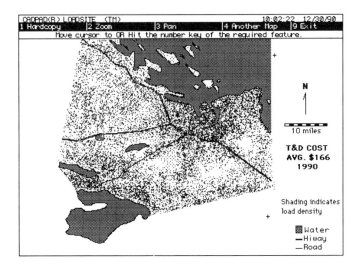

Figure 1.9 Cost of power delivery varies depending on location. Shown here are the annual capacity costs of delivery evaluated on a ten-acre basis throughout a coastal city of population 250,000. Cost varies from a low of $85/kW to a high of $270/kW.

1.7 TYPES OF DISTRIBUTION SYSTEM DESIGN

There are three fundamentally different ways to lay out a power distribution system used by electric utilities, each of which has variations in its own design. As shown in Figure 1.10, radial, loop, and network systems differ in how the distribution feeders are arranged and interconnected about a substation.

Most power distribution systems are designed as *radial distribution systems*. The radial system is characterized by having only one path between each customer and a substation. The electrical power flows exclusively away from the substation and out to the customer along a *single path,* which, if interrupted, results in complete loss of power to the customer. Radial design is by far the most widely used form of distribution design, accounting for over ninety-nine percent of all distribution construction in North America. Its predominance is due to two overwhelming advantages: it is much less costly than the other two alternatives and it is much simpler in planning, design, and operation.

In most radial plans, both the feeder and the secondary systems are designed and operated radially. Each radial feeder serves a definite service area (all customers in that area are provided power by only that feeder).

Power Delivery Systems

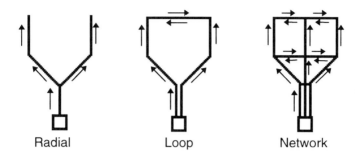

Figure 1.10 Simplified illustration of the concepts behind three types of power distribution configuration. Radial systems have only one electrical path from the substation to the customer, loop systems have two, and networks have several. Arrows show most likely direction of electric flows.

Most radial feeder systems are built as networks, but operated radially by opening switches at certain points throughout the physical network (shown earlier in Figure 1.6), so that the resulting configuration is electrically radial. The planner determines the layout of the network and the size of each feeder segment in that network and decides where the open points should be for proper operation as a set of radial feeders.

A further attribute of most radial feeder system designs, but not an absolutely critical element, is the use of single-phase laterals. Throughout North America, most utilities use single- and two-phase laterals to deliver power over short distances, tapping off only one or two phases of the primary feeder, minimizing the amount of wire that need be strung for the short segment required to get the power in the general vicinity of a dozen or so customers. These laterals are also radial, but seldom, if ever, end in a switch (they just end). There are many utilities, particularly urban systems in Europe, Africa, and Asia, that build every part of the radial distribution system, including laterals, with all three phases.

Each service transformer in these systems feeds power into a small radial system around it, basically a single electrical path from each service transformer to the customers nearby.

Regardless of whether it uses single-phase laterals or not, the biggest advantages of the radial system configuration, in addition to its lower cost, is the simplicity of analysis and predictability of performance. Because there is only one path between each customer and the substation, the direction of power flow is absolutely certain. Equally important, the load on any element of the system can be determined in the most straightforward manner -- by simply adding up all the customer loads "downstream" from that piece of equipment.

Before the advent of economical and widely available computer analysis, this alone was an overwhelming advantage, for it allowed simple, straightforward, "back of the envelope" design procedures to be applied to the distribution system with confidence that the resulting system would work well. The simplicity of analysis, and confidence that operating behavior is strictly predictable are still great advantages.

Because load and power flow direction are easy to establish, voltage profiles can be determined with a good degree of accuracy without resorting to exotic calculation methods; equipment capacity requirements can be ascertained exactly; fault levels can be predicted with a reasonable degree of accuracy; and protective devices -- breaker-relays and fuses -- can be coordinated in an absolutely assured manner, without resorting to network methods of analysis. Regulators and capacitors can be sized, located, and set using relatively simple procedures (simple compared to those required for similar applications to non-radial designs, in which the power flow direction is not a given).

On the debit side, radial feeder systems are less reliable than loop or network systems because there is only one path between the substation and the customer. Thus, if any element along this path fails, a loss of power delivery results. Generally, when such a failure occurs, a repair crew is dispatched to re-switch temporarily the radial pattern network, transferring the interrupted customers onto another feeder, until the damaged element can be repaired. This minimizes the period of outage, but an outage still occurred because of the failure.

Despite this apparent flaw, radial distribution systems, if well designed and constructed, generally provide very high levels of reliability. For all but the most densely populated areas, or absolutely critical loads (hospitals, important municipal facilities, the utility's own control center) the additional cost of an inherently more reliable configuration (loop or network) cannot possibly be justified for the slight improvement that is gained over a well-designed radial system.

An alternative to purely radial feeder design is a *loop system* consisting of a distribution design with two paths between the power sources (substations, service transformers) and every customer. Such systems are often called "European," because this configuration is the preferred design of many European utilities, as well as European electrical contractors, when called upon to lay out a power distribution system anywhere in the world. Equipment is sized and each loop is designed so that service can be maintained regardless of where an open point might be on the loop. Because of this requirement, whether operated radially (with one open point in each loop), or with closed loops, the basic equipment capacity requirements of the loop feeder design do not change.

Power Delivery Systems

Some urban areas in Europe and Asia are fed by multiple hierarchical loop systems: a 100+kV sub-transmission loop routes power to several substations, out of which several loop feeders distribute power to service transformers, which each route powers through a long loop secondary.

In terms of complexity, a loop feeder system is only slightly more complicated than a radial system -- power usually flows out from both sides toward the middle, and in all cases can take only one of two routes. Voltage drop, sizing, and protection engineering are only slightly more complicated than for radial systems.

But if designed thus, and if the protection (relay-breakers and sectionalizers) is also built to proper design standards, the loop system is more reliable than radial systems. Service will not be interrupted to the majority of customers whenever a segment is outaged, because there is no "downstream" portion of any loop.

The major disadvantage of loop systems is capacity and cost. A loop must be able to meet all power and voltage drop requirements when fed from only one end, not both. It needs extra capacity on each end, and the conductor must be large enough to handle the power and voltage drop needs of the entire feeder if fed from either end. This makes the loop system inherently more reliable than a radial system, but the larger conductor and extra capacity increase cost.

Distribution networks are the most complicated, most reliable, and in very rare cases, also the most economical method of distributing electric power. A network involves multiple paths between all points in the network. Power flow between any two points is usually split among several paths, and if a failure occurs it instantly and automatically re-routes itself.

Rarely does a distribution network involve primary voltage-level network design, in which all or most of the switches between feeders are closed so that the feeder system is connected between substations. This is seldom done because it proves very expensive and often will not work well.[10] Instead, a "distribution network" almost always involves "interlaced" radial feeders and a network secondary system -- a grid of electrically strong (i.e., larger than needed to feed customers in the immediate area) conductor connecting all the customers together at utilization voltage. Most distribution networks are underground (simply because they are employed only in high density areas, where overhead space is not available).

---

[10] The major reason is that this puts feeder network paths in parallel with transmission between substations, which results in unacceptable loop and circular flows and large dynamic shifts in load on the distribution system.

38   **Chapter 1**

   In this type of design, the secondary grid is fed from radial feeders through service transformers, basically the same way secondary is fed in radial or loop systems. The feeders are radial, but laid out in an interlaced manner -- none has a sole service area, but instead they overlap. The interlaced configuration means that alternate service transformers along any street or circuit path are fed from alternate feeders, as shown in Figure 1.11.

   While segments from two feeders always run parallel in any part of the system, the same two feeders never overlap for all of their routing. The essence of the interlaced system (and a design difficulty in any practical plan) is to mix up feeders so that each feeder partially parallels quite a few other feeders. Thus, if the feeder fails, it spreads its load out over quite a few other feeders, overloading none severely (Figure 1.12).

   At a minimum, distribution networks use an interlacing factor of two, meaning that two feeders overlap in any one region, each feeding every other service transformer. But such a system will fail, when any two feeders are out of service. Interlacing factors as high as five (four overlapping feeders, each feeding every fourth consecutive service transformer) have been built. Such systems can tolerate the loss of any three feeders (the other two in any area picking up the remaining load, although often very overloaded) without any interruption of customer service. If an element fails, the power flow in the elements around it merely re-distributes itself slightly.[11]

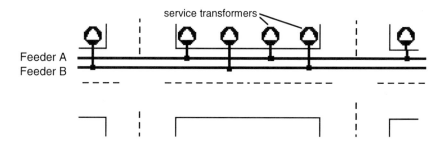

Figure 1.11 To obtain an interlacing factor of 2, two feeders are routed down each street, with alternating network transformers fed from each.

---

[11] So slightly, that a real problem can be determining when failures occur, for if the damage is not visible (most networks are underground systems) and no alarm or signal is given, the utility may not know a failure occurred until months later, when a second failure nearby puts a strain on the system or causes an outage.

Power Delivery Systems

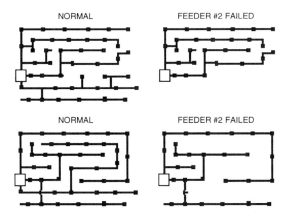

Figure 1.12 Top: a non-interlaced feeder system experiences the loss of one feeder, and all transformers in the lower right part of the system are lost -- service is certain to be interrupted. Bottom: the same system, but interlaced. Loss of the feeder is a serious contingency, but can be withstood because the feeder losses are distributed in such a way that each transformer out of service is surrounded by transformers still in service.

Networks are more expensive than radial distribution systems, but not greatly so in some instances. In dense urban applications, where the load density is very high, where the distribution must be placed underground, and where repairs and maintenance are difficult because of traffic and congestion, networks may cost little more than loop systems. Networks require only a little more conductor capacity than a loop system. The loop configuration required "double capacity" everywhere to provide increased reliability. A distribution network is generally no worse and often needs considerably less than that if built to a clever design.

Networks have one major disadvantage. They are *much* more complicated, than other forms of distribution, and thus much more difficult to analyze. There is no "downstream" side to each unit of equipment in a radial or loop system, so that the load seen by any unit of equipment cannot be obtained by merely adding up customers on one side of it, nor can the direction of power flow through it be assumed. Loadings and power flow, fault currents and protection, must be determined by network techniques such as those used by transmission planners. However, even more sophisticated calculation methods than those applied to transmission may be required, because a large distribution network can consist of 50,000 nodes or more -- the size of the very largest transmission-level power pool. Distribution network load flows are often more difficult to solve than transmission systems because the range of impedances in the modeled circuits is an order of magnitude wider.

In densely populated regions, such as the center of a large metropolitan area, networks are not inherently more expensive than radial systems designed to serve the same loads. Such concentrated load densities require a very large number of circuits anyway, so that their arrangement in a network does not inherently increase the number of feeder and secondary circuits, or their capacity requirements. It increases only the complexity of the design.

But in other areas, such as in most cities and towns, and in all rural areas, a network configuration will call for some increase (in kVA-feet of installed conductor) over that required for a radial or loop design. The excess capacity cost has to be justifiable on the basis of reliability.

Large-Trunk vs. Multi-Branch Feeder Layout

Figure 1.13 illustrates the basic concept behind two different ways to approach the lay-out of a radial distribution system. Each has advantages and disadvantages with respect to the other in certain situations, and neither is superior to the other in terms of reliability, cost, ease of protection, and service quality. Either can be engineered to work in nearly any situation. Most planning engineers have a preference for one or the other -- in fact, about 20% of utilities have standardized on the large-trunk design as their recommended guideline while another 20% prefer the multi-branch layout. Beyond showing that there are significantly different ways to lay out a distribution system, this brings to light an important point about distribution design: major differences in standards exist among electric utilities, as a result of which comparison of statistics or practice from one to the other is often not valid. Feeder layout types and practices are discussed in much greater detail in Chapters 8 and 9.

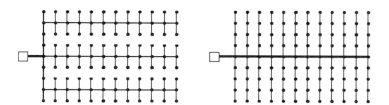

Figure 1.13 Two ways to route a radial feeder to 108 service transformers. Left, a "multi-branch" configuration. Right, a "large trunk" design.

Power Delivery Systems

Service Areas

As mentioned earlier, in most power systems, each substation is usually the sole provider of electrical service to the region around it -- its service area. Similarly, feeders and distribution networks also have distinct service areas. Usually, the service area for a substation, feeder, or other unit of equipment is the immediate area surrounding it, and usually these service areas are contiguous (i.e. not broken into several parts) and exclusive -- no other similar distribution unit serves any of the load in an area. An example, Figure 1.14 shows a map of substation service areas for a rectangular portion of a power system. Each distribution substation exclusively serves all customers in the area containing it.

Cumulatively, the customers in a substation or feeder's service territory determine its load, and their simultaneous peak demand defines the maximum power the substation must serve. Within a power system, each individual part, such as a substation or service transformer, will see its peak load at whatever time and in whatever season the customers in its service area generate their cumulative peak demand. One result of this is that the peak loads for different substations often occur at different seasons of the year or hours of the day. But whenever the peak occurs, it defines the maximum power the unit is required to deliver. Peak demand is one of the most important criteria in designing and planning distribution systems. Usually it defines the required equipment capacity.

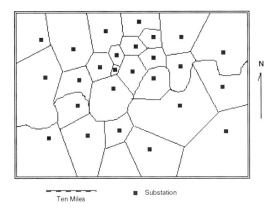

Figure 1.14 A power system is divided by substation service boundaries into a set of substation service areas, as shown.

Dynamic Service Area Planning

By making switching changes in the distribution system, it is possible to expand or shrink a substation or feeder's service area significantly, increasing or decreasing its net load. This is an important element of T&D planning, as shown in Figure 1.15, which illustrates a very typical T&D expansion situation. Two neighboring substations, A and B, each have a peak load near the upper limit of their reliable load-handling range. Load is growing slowly throughout the system, so that in each substation annual peak load is increasing at about 1 MW per year. Under present conditions, both will need to be upgraded soon. Approved transformer types, required to add capacity to each, are available only in 25 MVA or larger increments, costing $500,000 or more.

Both substations do not need to be reinforced. A new 25 MVA transformer and associated equipment are added to substation A, increasing its ability to handle a peak load by about 20 MVA. Ten MW of substation B's service area is then transferred to A. The result is that each substation has 10 MW of margin for continued load growth -- further additions are not needed for 10 years.

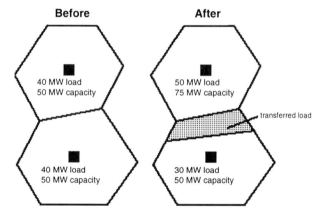

Figure 1.15 Load in both substations is growing at about 1 MW per year. Each substation has sufficient capacity to handle present load within contingency criteria (a 25% margin above peak) but nothing more. By transferring load as shown, only one substation has to be reinforced with an additional (25 MVA) capacity, yet both end up with sufficient margin for another ten years' growth. Service area shifts like this are how expansion costs are kept down in spite of the fact that equipment like transformers is available only in large, discrete sizes.

Power Delivery Systems

This type of planned variation in service areas is a major tool used to keep T&D expansion costs low, a key element in building a reliable, economical schedule of expansion. Optimization of this particular aspect of planning can be a challenge, not because of any inherent difficulty in analysis, but simply because there are so many parts of the system and constraints to track at one time. This is one reason for the high degree of computerization in distribution planning at many utilities. Balancing the myriad requirements of many substations and their design and operating constraints is an ideal problem for numerical optimization techniques.

The Systems Approach

One complication in determining the most economical equipment for a power system is that its various levels--transmission, substation, and distribution--are interconnected, with the distribution, in turn, connected to the customers (and hence a function of the local load density). The best size and equipment type at each level of the system is a function of the types of equipment selected for the other levels of the system and the load density. In general, the equipment is so interconnected that it is impossible to evaluate any one aspect of design without taking all others into account.

Consider the question of substation spacing -- determining how far apart substations should be, on average, for best utilization and economy. Within any service territory, if substations are located farther apart, there will be fewer of them, saving the utility the cost of buying and preparing some substation sites, as well as reducing the cost of building a large number of substations. Thus, overall substation cost is a function of spacing, but how it varies depends on land and equipment costs, which must be balanced carefully. With fewer substations, however, each substation must serve a larger area of the system. Hence, each substation will have a larger load, and thus require a larger capacity, meaning it must have more or larger transformers.

The larger substations will also require a larger amount of power to be brought to each one, which generally calls for a higher sub-transmission voltage. Yet, there will be fewer sub-transmission lines required (because there are fewer substations to which power must be delivered). All three aspects of layout are related -- greater substation spacing calls for larger substations with bigger transformers, and a higher transmission voltage, but fewer lines are needed -- and all three create better economies of scale as spacing is increased. Thus, transmission costs generally drop as substation spacing is increased.

Nevertheless, there is a limit to this economy. The distribution feeder system is required to distribute each substation's power through its service area, moving power out to the boundary between each substation's service area and that of its neighbors. Moving substations farther apart means that the distribution system must move power, on average, a greater distance. Distributing power over these longer distances requires longer and more heavily loaded feeders. This situation increases voltage drop and can produce higher losses, all of which can increase cost considerably. Employing a higher distribution voltage (such as 23.9 KV instead of 13.8 KV) improves performance and economy, but regardless, it costs more to distribute power from a few larger substations farther apart, than to distribute it from many smaller substations close together. Feeder costs go up rapidly as substation spacing is increased.

The major point of this section is that all four of the above aspects of system design are interconnected -- (1) substation spacing in the system, (2) size and number of substations, (3) transmission voltage and design, and (4) distribution feeder voltage and design. One of these factors cannot be optimized without close evaluation of its interrelationship with the other three. Therefore, determining the most cost-effective design guideline involves evaluating the transmission-substation-feeder system design as a whole against the load pattern, and selecting the best combination of transmission voltage, substation transformer sizes, substation spacing, and feeder system voltage and layout.

This economic equipment sizing and layout determination is based on achieving a balance between two conflicting cost relationships:

Higher voltage equipment is nearly always more economical on a per-MW basis.

Higher voltage equipment is available only in large sizes (lots of MW).

In cases where the local area demands are modest, higher voltage equipment may be more expensive simply because the minimum size is far above what is required -- the utility has to buy more than it needs. How these two cost relationships play against one another depends on the load, the distances over which power must be delivered, and other factors unique to each power system, such as the voltages at which power is delivered from the regional power pool, and whether the system is underground or overhead.

Figure 1.16 illustrates the difference that careful coordination of system design between levels of the power system can have in lowering overall cost. Shown are the overall costs from various combinations of T&D system layout for a large metropolitan utility in the eastern United States. Each line connects a set of cost computations for a system built with the same transmission and distribution voltages (e.g., 161 kV transmission and 13.8 kV distribution) but varying in substation sizes (and hence, implicitly, their spacing).

Power Delivery Systems

In all cases, the utility had determined it would build each substation with two equally sized transformers (for reliability), with none over 75 MVA (larger transformers are too difficult to move along normal roads and streets, even on special trailers). Either 161 kV or 69 kV could be used as sub-transmission, either 23.9 kV or 13.8 kV could be used as distribution voltage. Any size transformer, from 15 MVA to 75 MVA, could be used, meaning the substation could vary from 30 MVA to 150 MVA in size. (Peak load of such substations can normally be up to 75% of capacity, for a peak load of from 23 to 100 MW). Substation spacing itself is implicit and not shown. Given the requirement to cover the system, determining transmission voltage, distribution, and substation size defines the system design guidelines entirely.

Overall, the ultimate lowest cost T&D system guidelines are to build 120 MVA substations (two 60 MVA transformers) fed by 161 kV sub-transmission and distributing power at 23.9 kV. This has a levelized cost (as computed for this utility) of about $179/kW. In this particular case, a high distribution voltage is perhaps the most important key to good economy -- if 13.8 kV is used instead of 23.9 kV as the primary voltage, minimum achievable cost rises to $193/kW.

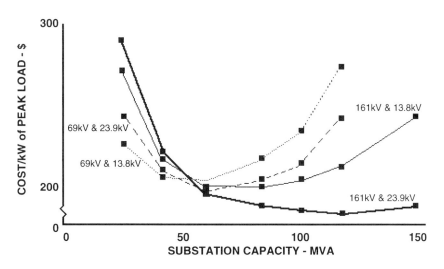

Figure 1.16 Overall cost of T&D system depends on the balanced design of the sub-transmission, substation, and feeder level, as described in the text. Cost can vary by significant margins depending on how well performance and economy at various levels of the system are coordinated.

The very worst design choices plotted in Figure 1.16, from an economic standpoint, would be to build 25 MVA substations fed by 161 kV sub-transmission and feeding power to 23.9 kV feeders ($292/kW). This would require many small substations, each below the effective size of both the transmission and distribution voltages, and lead to high costs. Overall, 161 kV and 23.9 kV are the correct choices for economy, but only if used in conjunction with a few, large substations. If substations are to be 25 MVA, then 69 kV and 13.8 kV do a much more economical job ($228/kW), but still don't achieve anything like the optimum value. The point: Achieving economy in power delivery involves coordinating the interactions, performance, and economies of the multiple system levels. Chapters 11 and 12 discuss the issues of and techniques for such coordinated multi-level planning..

1.8 CONCLUSION

A transmission and distribution system moves power from a utility's power production and purchase points to its customers. The T&D system's mission of delivering power to the customers means that it must be composed of equipment spread throughout the service territory, arranged locally so that capacity is always in proportion to local electrical demand, with the facilities in each neighborhood sized and configured both to fit well into the whole and to serve the local needs. In the author's opinion, engineering a system to meet this challenge is seldom easy, but engineering a system to do so at the minimum possible cost is always extremely challenging.

A T&D system is composed of several interconnected, hierarchical levels, each of which is required for completion of the power delivery task. These levels are:

- transmission
- sub-transmission
- substation
- feeder
- secondary
- customer

To a certain extent, the burden of power delivery, and costs, can be shifted from one level to another through changes in the specifications of equipment, lay-out standards, and design of the various levels. For example, costs can be pared at the substation level by using fewer, larger substations, but this means feeders in each substation area will have to carry power farther, and perhaps carry more power per feeder as well, increasing feeder costs. Low overall design cost is achieved by balancing these factors.

Power Delivery Systems 47

The performance and economy, and thus the design, of a power system are dominated by a number of constraints due to physical laws, and further shaped by a number of practical considerations with regard to equipment, layout, and operating requirements. The more important of these are:

The T&D System Must Cover Ground. This is *the* rule about T&D -- ultimately the electric system must "run a wire" to every customer. A significant portion of the cost of a distribution system is due to this requirement alone, independent of the amount of peak load or energy supplied.

Interconnectivity of Levels. The sub-transmission-substation-feeder triad comprises a highly interconnected system with the electrical and economic performance at one level *heavily* dependent on design, siting, and other decisions at another level. To a certain extent the T&D system must be planned and designed as a whole; its parts cannot be viewed in isolation.

Discrete Equipment Sizes. In many cases equipment is available only in certain discrete sizes. For example, 69 kV/12.47kV transformers may be available only in 5, 10, 20, and 22 MVA sizes. Usually, there is a large economy of scale -- the installed cost of a 20 MVA unit being considerably less than two times that for a 10 MVA unit.

Dynamic Service Area Assignment is a Powerful Planning Tool. Linearity of expansion costs by and utilization in the face of upgrades that are only available in discrete sizes is obtained by arranging service areas. Given that a substation can be built in only one of several discrete sizes, the planner obtains an economical match of capacity to load by varying service areas. When reinforced with new capacity additions, a substation might pick up load of surrounding substations, effectively spreading the capacity addition among several substations.

Losses Cost Can Be Significant. In some cases, the present worth of future losses on a line or heavily loaded transformer can exceed its total capital cost. In most cases, this does not happen. But in most cases, the present worth of losses is greater than the difference in cost between most of the choices available to the planner, meaning that losses are a major factor to be considered in achieving overall least-cost design.

Cost to Upgrade is Greater than the Cost to Build. For example, one mile of a 12.47 kV, three-phase feeder using 600 MCM conductor (15 MW thermal capacity) might cost $150,000 to build, and one mile of 9 MW feeder (336 MCM conductor) might cost only $110,000. But the cost to upgrade the 336 MCM feeder to 600 MCM wire size at a later date would be about $130,000, for a cumulative total of $240,000. Therefore, one aspect of minimizing cost is to determine size for equipment not on the basis of immediate need, but by assessing the eventual need and determining whether the present worth of the eventual savings warrants the investment in larger size now.

Coincidence of Peaks. Not all customer loads occur at the same time. This has a number of effects. First, peak load in different parts of the system may occur at different times. Second, the system peak load will always be less than the sum of the peak loads at any one level of the system -- for example in more power systems the sum of all substation peak loads usually exceeds system total peak by 3-8%. Diversity of peak loads means that considerable attention must be paid to the pattern and timing of electric load if equipment needs (and consequently costs) are to be minimized.

Reliability is Obtained through Contingency Margin. Traditionally, T&D reliability in a static system (one without automation, and hence incapable of recognizing and reacting to equipment outages and other contingencies by re-configuring and/or re-calibrating equipment) is assured by adding an emergency, or contingency margin throughout the system. At the T&D level this means that an average of 20% to 50% additional capacity is included in most equipment, so that it can pick up additional load if neighboring equipment fails or is overloaded.

A Complicated System

In conclusion, a power T&D system provides the delivery of an electric utility's product. It must deliver that product to every customer's site, with high reliability and in ready to use form. While composed of equipment that is individually straightforward, most T&D systems are quite complex due to the interactions of thousands, even millions, of interconnected elements. Achieving economy and reliability means carefully balancing a myriad of mutual interactions and conflicting cost trade-offs.

REFERENCES

M. V. Engel et al., editors, *Tutorial on Distribution Planning*, IEEE Course Text EHO 361-6-PWR, Institute of Electrical and Electronics Engineers, Hoes Lane, NJ, 1992.

A. C. Monteith and C. F. Wagner, *Electrical Transmission and Distribution Reference Book*, Westinghouse Electric Company, Pittsburgh, PA, 1964.

2
Electrical Load and Consumer Demand for Power

2.1 INTRODUCTION

A power delivery system exists because consumers want electric power. The amount and type of electric energy demanded, and the value that customers place on qualities such as reliability and power quality, largely define what the electric system is expected to do and set limits on the equipment and standards that can be applied in its design. Customer demand has two aspects. The first relates to quantity: electric "demand" itself -- the magnitude of the electricity consumed by the customer and its what, when, where, how much, why. This defines the load the power system must supply, the schedule of loading by season and hour of day, and other characteristics like power factor and voltage sensitivity.

In addition, customers have definite expectations about quality; the availability of power, the stability of the voltage provided, and the quantity (or lack thereof) of harmonics and noise.

This chapter looks at both quantity and quality from the customer demand standpoint. Section 2.2 examines electric demand -- customers and customer loads -- as well as how load "looks" to the power system when measured and recorded at the distribution level. Section 2.3 examines customer expectations and needs with respect to availability (reliability of supply), voltage regulation, harmonics, and other qualities. The chapter concludes with a series of guidelines and customer/load rules of thumb given in Section 2.4.

2.2 CUSTOMER DEMAND

Electricity is Purchased for End-Uses

An electric utility's customers purchase electricity as a means to some end-use for which electricity is only an intermediate means. No customer wants electric energy itself: They want the products it can produce -- a cool home in summer, a warm one in winter, hot water on demand, cold beer in the fridge, and 48 inches of dazzling color with stereo commentary during Monday night football. Electricity is purchased as a means towards some final, non-electrical product -- no one wakes up in the morning saying to himself or herself, "I just feel like using up 5 kWhr today."

These different goals and needs are called *end-uses,* and they span a wide range of applications. Some end-uses are unique to electric power (the author is aware of no manufacturer of natural-gas powered TVs or stereo radios), while in others it predominates to the point of a virtual monopoly market share (although there *are* gasoline-powered refrigerators, and natural gas for interior lighting and air conditioning). But for many other end-uses, electricity is only one of several possible energy sources (water heating, home heating, cooking, clothes drying).

Each end-use -- the need for lighting is a good example -- is satisfied through the application of appliances or devices that convert electricity into the desired end product. For lighting, a wide range of illumination devices can be applied, from incandescent bulbs to fluorescent tubes, to sodium vapor and high-pressure mono-chromatic gas-discharge tubes, and lasers. Each uses electric power to produce visible light. Each has differences from the others that give it an appeal to some customers or for certain types of applications. But regardless, each requires electric power to function, creating an electric load when it is activated.

Different types of customer purchase electricity: Among them are homeowners, who will vary in their needs, and brands and types of appliances they own, and their daily household activity patterns. Commercial businesses, both large and small, also buy electricity, sharing some end-use needs with residential customers (heating and cooling, lighting) and having some needs unique to them (cash register/inventory systems, office machinery, neon store display lighting). In addition, industrial facilities buy electricity to power process such as pulp heating, compressor motor power, and a variety of manufacturing applications.

As a result, the load on an electric system, or on any part of a T&D system, such as specific line, substation transformer, distribution feeder or lateral, can be

Electric Load and Consumer Demand for Power

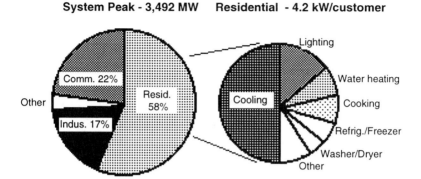

Figure 2.1 Electric peak demand of a utility in the southern United States, broken out by customer class, and within the residential class, by contribution to peak for the major uses for which electricity is purchased at time of peak by the residential class.

characterized by how many and what type of customers are buying power delivered by that unit, and what end-uses are they fulfilling with that power. Figure 2.1 shows this breakdown of the peak load for an entire power system.

End-use analysis

End-use analysis -- the study of the basic causes of electric demand by customer type, time, end-use category, and type of appliance -- is by far the best way to study load usage in order to develop a thorough understanding and model of customer usage. It is the predominant method that will be used throughout this book for discussion of customer demand, customer requirements, and load analysis and growth forecasting.

End-use analysis is done on a customer class and end-use category basis, by recognizing that customers of different classes -- residential, commercial, and industrial and sub-classes within those -- differ in their needs, activity patterns, and end-use priorities. Within each class, electric demand is categorized by end-use application, and within that, often further distinguished by appliance type (e.g., incandescent versus fluorescent lighting). Analytical techniques for quantitative end-use analysis, including computer models and algorithms for its implementation, are discussed in Chapter 15. However important such models may be, the concept of end-use analysis and its understanding and application on an intuitive basis is even more important.

Every electric system planner should understand who is buying their company's product, and why -- information of the type and detail shown in Figure 2.1. Whether power production, system, T&D, or integrated resource planners, a basic understanding of why electricity is bought and who is using it will help them better understand how to satisfy their company's customer needs.

Appliances convert electricity to end-uses

The term load refers to the electrical demand of a device that is connected to and draws power from the T&D system for the purpose of accomplishing some task (opening a garage door) or converting that power to some other form of energy (light, heat). These devices are called appliances, whether they are a commonly regarded household appliance (e.g., a refrigerator), or a lamp, garage door opener, paper shredder, electric fence to keep cattle confined, or anything else.

Electrical loads are usually rated by the level of power they require, measured in units of volt-amperes, called watts. Large loads are measured in kilowatts (thousands of watts) or megawatts (millions of watts). Power ratings of loads and T&D equipment refer to the device at a specific *nominal voltage.* For example, an incandescent light bulb might be rated at 75 watts and 1,100 lumens at 120 volts, at which voltage it consumes 75 watts and produces 1,100 lumens of light. If provided more or less voltage, its load (and probably its light output) would vary. Loads can be single-phase or multi-phase; they can have real (resistive only) or complex impedance (reactance), too.

Market share

The electric load within any one end-use category -- for example lighting -- will depend on the types of devices being used to convert electricity to the end-use and the market penetration (percent of customers who use it) of each type. Lighting load will be higher if most customers, say 95%, are using incandescent lighting to meet their needs, than if a large portion uses fluorescent lamps instead. Using information on the market share of each appliance type in each class, it is possible to assemble a much more detailed picture, or model, of electric demand in a power system or served by a particular unit of the system, as shown in Figure 2.2.

Electric Load and Consumer Demand for Power

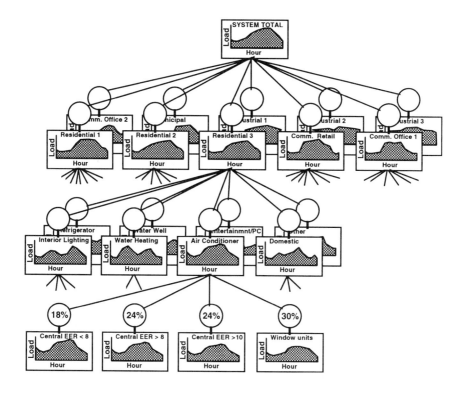

Figure 2.2 Overall structure of an end-use model of the type necessary to provide a good foundation for effective load study and forecasting. The model is hierarchical and computes each load curve shown by addition of all the curves below it in the model's tree structure (except those on the bottom level of the model, which are input). Only a small portion of the overall model -- one class and one end-use within that class -- is shown. Squares represent load curve data on a per appliance basis. Circles represent computations involving weighting factors (market penetrations, customer counts, as appropriate -- see text). Market penetration (portion of customers using each type of appliance) are shown inside the circles for the bottom appliance level, in this case representing the various types of air conditioners in the residential-2 class.

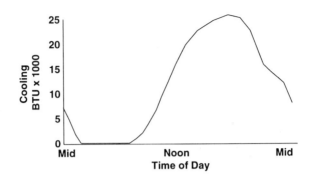

Figure 2.3 Average demand for BTU of cooling among houses in the author's neighborhood on a typical weekday in June.

The electric load created by a utility's customers within any one end-use category will depend on a number of factors. Chief among them is the timing of demand for the particular end-use itself. In most households, demand for lighting is highest in mid-evening, after dusk but before most of the household members have gone to bed. Demand may vary by time of day and season, too. Lighting needs are greater at times of the year when the sun sets earlier in the day, and on weekends, when activity often lasts later into the evening. Some end-uses are seasonal: heating demand generally occurs only in winter, being greatest during particularly cold periods and when family activity is at a peak -- early morning and early evening. The electric load stemming from a particular application or end-use will vary as a function of time, depending on the activity patterns and demand of the utility customer base, as shown in Figure 2.3.

Load Curves

The load on a power system, or on any unit of equipment within that power system, represents the accumulated electrical demand of all the customers being served by that system or unit of equipment. With few exceptions, the load will vary from hour to hour, from day to day, and from season to season as shown in Figure 2.4. Diagrams of load as a function of time are called *load curves*.

Typically, the value of most interest to the planner is the annual peak load, the maximum demand seen during the year. This peak is important because it is the maximum amount of power that must be delivered, and thus defines, either

Electric Load and Consumer Demand for Power

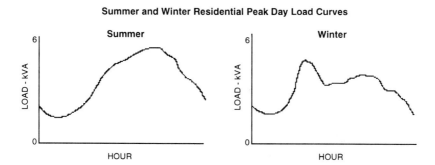

Figure 2.4 Electric load varies from hour to hour and season to season. These curves are for all-electric residential class (per customer, coincident load curves) from a utility system in northern Florida.

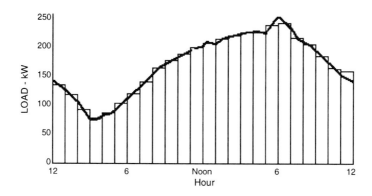

Figure 2.5 Demand on an hourly basis (blocks) over 24 hours, and the actual load curve for a feeder segment serving 53 homes. Demand measurement averages load over each demand interval (in this case each hour) missing some of the detail of the actual load behavior (solid black line). In this case the actual peak load (263 kW at 6 PM) was not seen by the demand measuring, which "split the peak," averaging load on an hourly basis and seeing a peak demand of only 246 kW, 7% lower. As will be discussed later in this chapter, an hourly demand period is too lengthy for this application.

directly or indirectly, the capacity requirements for equipment. In a growing power system, not only does the system-wide peak grow from year to year, but the peak loadings on many units of equipment, such as substations and feeders within that system, also increase annually as load increases.

Demand and Demand Periods

Demand is the average value of load over a period of time known as the demand interval. Very often demand is measured on an hourly basis as shown in Figure 2.5, but it can be measured on any interval basis -- seven seconds, one minute, 30 minutes, daily, and monthly. The average value of power during the demand interval is given by dividing the kilowatt-hours accumulated during the demand interval by the length of the interval. Demand intervals vary among power companies, but those commonly used in collecting data and billing customers for "peak demand" are 15, 30, and 60 minutes.

Load curves may be recorded, measured, or applied over some specific time, for example, a load curve might cover one day. If recorded on an hourly demand basis, then the curve consists of 24 values, each the average demand during one of the 24 hours in the day, and the peak demand is the maximum hourly demand seen in that day. Load data can and are gathered and used on a monthly basis and on an annual basis.

Load Factor

Load factor is the ratio of the average to the peak demand. The average load is the total energy used during the entire period (e.g., a day, a year) divided by the number of demand intervals in that period (e.g., 24 hours, 8760 hours). The average is then divided by the maximum demand to obtain the load factor, as

$$LF = \frac{KWH/Hrs}{KW\ Demand} = \frac{KW\ Average}{KW\ Demand} \quad (2.1)$$

$$= \frac{KWH}{(KW\ Demand) \times (Hrs)} \quad (2.2)$$

Load factor gives the extent to which the peak load is maintained during the period under study. A high load factor means the load is at or near peak a good portion of the time.

Electric Load and Consumer Demand for Power

Customer Class Load Curves

Most utility systems serve customers of several fundamentally different types -- residential, commercial, and industrial. These classes are very similar to -- often identical to -- the rate classes which determine the schedule of prices for different customers. While every customer is somewhat different in his electrical usage, customers within each class tend to be somewhat alike in usage patterns, and customers from different classes tend to be different from one another, both in the amount of their average power demands and in the time of day and year that their electrical demand peaks. Therefore, most electric utilities distinguish load behavior on a class by class basis, characterizing each class with a "typical daily load curve" showing the average or expected pattern of load usage for a customer in that class on the peak day, as shown in Figure 2.6. These curves describe the most important points concerning the customers' loads from the distribution planner's standpoint -- the amount of peak load per customer, the time and duration of peak, and the total energy (area under the curve). Such load modeling by class is almost universal.

Detail and quality of information in load curve data vary considerably from one electric utility to another. A few frequently update their databases and use rigorous systems for quality control. Others have data of much less quality. Regardless, very often the distribution planner will have little information on customer usage patterns beyond that given the customer class curve shapes and peak loads.

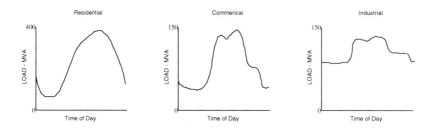

Figure 2.6 Different customer classes have different electrical demand characteristics, particularly with regard to how demand varies with time. Here are summer peak day load curves for the three major classes of customer from the same utility system.

2.3 PEAK LOAD, COINCIDENCE, AND LOAD CURVE BEHAVIOR

Most often, a utility will use smooth, 24 hour load curves like those shown in Figure 2.6 to represent the "average behavior" of each customer in each class. For example, the utility system whose data is used in Figure 2.6 has approximately 60,400 residential customers. Its analysts will take the total residential customer class load (leftmost curve, peaking at about 396 MW) and divide it by 60,400 to obtain a "typical residential load" curve for use in planning and engineering studies, a curve with a shape identical to that shown but with a peak of 6.59 kW (1/60,400the of 396 MW). This is common practice, and results in "customer class" load curves used at nearly every utility.

Actually, no residential customer in any utility's service territory has a load curve that looks anything like this average representation. Few concepts are as important as understanding why this is so, what actual load behavior looks like, and why the smooth representation is "correct" in many cases, but not in any that apply to the distribution level.

The load curve shown in Figure 2.7 is actually typical of what most residential customer load looks like over a 24 hour period. *Every residential customer's* daily load behavior looks something like that shown, with sharp "needle peaks" and erratic shifts in load as major appliances such as central heating, water heaters, washer-dryers, electric ranges, and other devices switch on and off.

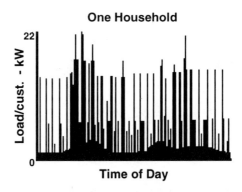

Figure 2.7 Actual winter peak day load behavior for an individual household looks like this, dominated by high "needle peaks" caused by the joint operation of major appliances.

Electric Load and Consumer Demand for Power

Appliance Duty Cycles

The reason for the erratic "needle peak" load behavior shown in Figure 2.7 is that a majority of electric devices connected to a power system are controlled with what is often called a "bang-bang" control system. A typical 50 gallon residential electric water heater is a good example. The water heater holds 50 gallons of water, which it keeps warm by turning on its heating elements anytime the temperature of the water, as measured by the thermostat, dips below a certain value. When the temperature has been raised sufficiently, the elements are turned off by the thermostat, as shown in Figure 2.8.

The water heater cycles on and off in response to the thermostat -- "bang," it is on until the water is hot enough, then "bang," it is off and remains so until the water is cold enough to cause the thermostat to cycle the elements back on.

Daily load curves for several residential water heaters are shown in Figure 2.9. Most of the time, a water heater is off, and for a majority of the hours in the day it may operate for only a few minutes an hour, making up for thermal losses (heat gradually leaking out of the tank). Only when household activity causes a use of hot water will the water heater elements stay on for longer than a few minutes each hour. Then, the hot water use draws heated water out of the tank,

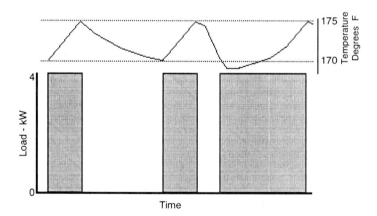

Figure 2.8 Top curve indicates the water temperature inside a 50 gallon water heater whose thermostat is set at 172 degrees over a period of an hour. The thermostat cycles the heater's 4 kW element on and off to maintain the water temperature near the 172 degree setting. Resulting load is shown on the bottom of the plot.

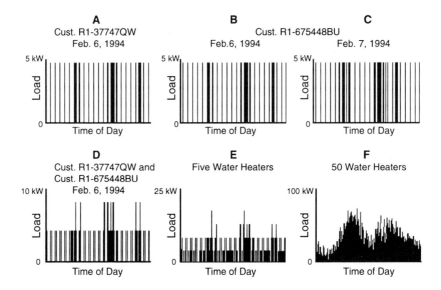

Figure 2.9 Curves A and B are water heater loads at neighboring houses on the same day (Feb. 6), displaying similar but not identical behavior. Curve C is the second household's water heater curve for Feb. 7, showing that random differences and slight shifts in usage cause different timing to that day's needle peaks. Curve D shows the sum of the two water heaters' curves for Feb. 6. For the most part, there is little likelihood that they are both on at the same time, but occasionally around peak time they do.

Assembling 50 water heaters results in a smoother load curve, one where load peaks during peak periods of water heater usage -- the average cycle time of the water heaters is longer (needle is *wider*) -- meaning there is more likelihood that many overlap, and hence add to a high peak value.

Electric Load and Consumer Demand for Power 61

which is replaced by cool water piped in to re-fill the tank, lowering the temperature in the tank significantly, and requiring the element to remain on for a long time thereafter to "manufacture" the replacement hot water. During that hour, its duty cycle -- the percent of the time it is on -- might be high, but the rest of the day it is quite low.

Figure 2.9 shows several water heater loads, as well as plots representing the sum of the load for a number of water heaters over a 24 hour period. It illustrates several important points. First, all water heaters exhibit the same overall on-off behavior, but differ slightly as to the timing of the cycles. Second, if enough load curves are added together, their sum begins to look more like a smooth curve (similar to Figure 2.3) than the typical blocky individual device curve.

Almost all the large loads in every residence operate in a manner similar to the water heaters discussed here. Air conditioners and heaters operate by control of thermostats. So do refrigerators and freezers and electric ranges. As a result, the household's total load curve (Figure 2.7) is the sum of a number of individual appliance on-off curves that each look something like those at the top of Figure 2.9.

Duty Cycles of Air Conditioners and Heat Pumps

Figure 2.10 shows the ambient (outside) temperature, and the electric load of an air conditioner set to 68°F inside temperature over a 24 hour period. It illustrates a slightly more complex behavior and will help to examine appliance duty cycle performance in slightly more detail than the preceding section.

Note that, like the water heater, the air conditioner unit shown here cycles on and off. As outside temperature increases, and the burden of removing heat from inside the house becomes greater, the AC unit stays on longer every time it turns on, and stays off for a shorter time. During the hottest period of the day, its duty cycle (percent of the time it is on during the hour) is above 90%. The particular unit shown in Figure 2.10 was selected for this house based on a 35 degree difference criterion -- at 100% duty cycle it can maintain a 35°F difference between outside and inside. At the 100°F ambient temperate reached, it is providing a 32°F difference for the 68°F thermostat setting, near its maximum.

Note also that here, unlike in the case of the water heater, the amount of load (the height of the load blocks when the unit is on) varies over the day. Connected load of the AC unit increases as the ambient temperature increases.

The mechanical burden of an air conditioner or heat pump's compressor will be higher when the temperature gradient the unit is opposing is higher (due to the nature of the system's closed cycle design causing higher internal pressures and

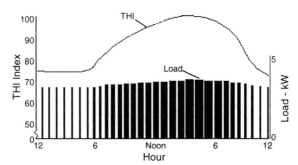

Figure 2.10 Daily cycle of THI (temperature-humidity-illumination index) and air conditioner load as it cycles between on and off under thermostatic control (see text). As THI rises throughout the morning, demand for cooling increases, and the air conditioner's duty cycle (% of time the unit is operating) increases, until at peak it is operating all but a few minutes in every hour.

hence more pumping load). When the pressure is higher, the electric motor must work harder, and hence draws more power (i.e., has a higher load). The net result is that an air conditioner's operating load increases as outside temperature gets warmer, and a heat pump's load increases as outside temperature gets colder. Thus, during the heat of mid-day, the AC unit shown in Figure 2.10 not only stays on most of the time, but has a load about 10% higher than it would when ambient temperature is cooler. Most AC and HP units, and many other mechanical pumping loads, display this characteristic.

Coincident Load Behavior

Suppose one were to consider one hundred homes, for example one hundred houses served by the same segment of a distribution feeder. Every one of these homes is full of equipment that individually cycles on and off like appliances just described -- water heaters, air conditioners, heaters, and refrigerators that all are controlled by thermostats.

Thus, each household will have an individual daily load curve similar to the erratic, choppy daily load curve shown in Figure 2.7, although each will be slightly different, because each home has slightly different appliances, is occupied by people with slightly different schedules and usage preferences, and because the times that individual thermostats activate appliances are usually

Electric Load and Consumer Demand for Power 63

random, happening at slightly different times throughout the day for each appliance.

Thus, one particular household might peak at 22 kVA between 7:38 AM and 7:41 AM, while another peaks at 21 kVA between 7:53 AM and 8:06 AM, while another peaks at 23 kVA between 9:54 AM and 10:02 AM. These individual peaks are not additive because they occur at different times. The individual peaks, all quite short, do not all occur simultaneously. They are *non–coincident*.

As a result, when one adds together two or more of these erratic household curves, the load curve needles usually interleave, much as the overlapping teeth of two combs might interleave. As yet more customer load curves are added to the group, more and more "needles" interleave, with only an occasional situation where two are absolutely coincident and add together. The load curve representing the sum of a group of households begins to take on the appearance of a smoother curve, as shown in Figure 2.11, in exactly the same way and for the same reasons that the water heater load curves in Figure 2.9 did.

By the time there are twenty-five homes in a group, a smooth pattern of load behavior begins to emerge from within the pattern of erratic load shifts, and by the time the loads for one hundred homes have been added together, the curve looks smooth and "well-behaved."

This is how coincidence changes load curve shapes and peak loads. Individual household demand consists of load curves with erratic swings as major appliances switch on and off at the bequest of their thermostats and other control devices. Distribution equipment that serves only one customer -- service drops or a dedicated service transformer for example -- sees a load like this. But distribution equipment that serves large numbers of customers -- a distribution feeder for example -- is looking at many households and hundreds of appliances at once. Individual appliance swings do not make a big impact. It is their sum, when many add together, that causes a noticeable peak for the feeder. This happens when many duty cycles overlap.

Peak load per customer drops as more customers are added to a group. Each household has a brief, but very high, peak load -- up to 22 kW in this example for a southern utility with heavy air-conditioning loads. These needle peaks seldom overlap, and under no conditions would all in a large group peak at exactly the same time.[1] As a result, the group peak occurs when the combination of the individual load curves is at a maximum, and this group peak load is usually

---

[1] Except under certain types of emergency operations during outages, due to something called "cold load pickup," when all appliances having been without power during a long outage, all want full power as soon as service is restored, creating a larger than normal demand.

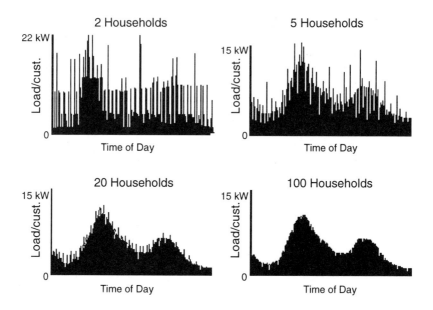

Figure 2.11 Daily load curves for groups of two, five, twenty, and one hundred homes in a large suburban area. Note vertical scale is in "load per customer" for each group. *Peak load per customer decreases as the number of customers in the group becomes larger.* This is coincidence of load seen on a load curve basis.

Electric Load and Consumer Demand for Power

substantially less than the sum of the individual peaks, as can be seen by studying Figure 2.11.

This tendency of observed peak load per customer to drop as the size of the customer group being observed increases is termed *coincidence* and is measured by the *coincidence factor,* the fraction of its individual peak that each customer contributes to the group's peak.

$$C = \text{coincidence factor} = \frac{\text{(observed peak for the group)}}{\Sigma \text{ (individual peaks)}} \quad (2.3)$$

Peak load per customer is generally a strictly decreasing value as a function of the number of customers in the group. Often, the value of C for large values is only .3 to .25.

Thus, the coincidence factor, C, can be thought of as a function of the number of customers in a group, n:

$$C(n) = \frac{\text{observed peak load of a group of N customers}}{\Sigma \text{ (their individual peaks)}} \quad (2.4)$$

where n is the number of customers in the group,
$1 < n < N$ = number of customers in the utility

C(n) has a value between 0 and 1, and varies with the number of customers in identical fashion to how the peak load varies, so that a curve showing C(n) and one showing peak load as a function of the number of customers are identical except for the vertical scale.

Distribution engineers sometimes use the inverse of C to represent this same phenomenon. The diversity factor, D, measures how much higher the customer's individual peak is than its contribution to group peak.

$$D = \text{diversity factor} = 1/\text{coincidence factor} \quad (2.5)$$

The coincidence curve shown in Figure 2.12 is typical of residential peak load behavior in the United States -- C(n) for large groups of customers is typically between .5 and .33, and may fall to as low as .2. Coincidence behavior varies greatly from one utility to another. Each utility should develop its own coincidence curve data based on studies of its own customer loads and system.

Estimates of equipment loads are often based on load data from coincidence curves. Very often, T&D equipment sizes are determined by using such coincidence curves to convert load research data (or whatever data are available) to estimates of the equipment peak loads. For example, the "coincident peak"

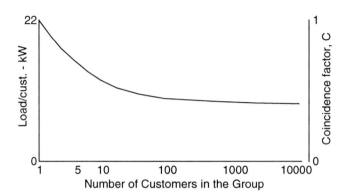

Figure 2.12 The peak load per customer drops as a function of the number of customers in a group.

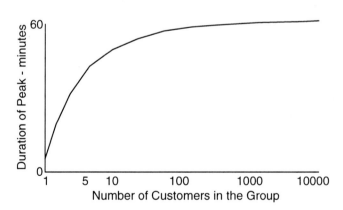

Figure 2.13 The peak period as a function of the number of customers in a group. Larger groups have longer peak periods.

Electric Load and Consumer Demand for Power

for a group of customers served by a feeder segment, transformer, or secondary line can be estimated from individual ("instantaneous") peak load data, as

Group peak for n customers =
$$C(n) \times n \times \text{(average individual peak load)} \quad (2.6)$$

Thus, the peak load for a transformer serving eight houses, each with an estimated individual peak load of 22 kW would be

$$\text{Transformer peak load} = C(8) \times 8 \times 22 \text{ kW} \quad (2.7)$$

Using a typical value of C(8), which is about .6, gives 106 kW as the estimated peak load for the transformer.

Often, reliable estimates of the individual customer peak loads (the 22 kW figure in these examples) may not be available. In this example, that would mean the planner did not know the 22 kW value of individual peak load but instead had only the "contribution to peak" value of 7.9 kW/customer. The transformer load can still be estimated from the coincidence curve as:

$$\text{Transformer peak load} = C(N)/C(8) \times 8 \times 7.9 \text{ kW} \quad (2.8)$$
$$\text{where } N>>>8$$

Here, using .36 as C(N) gives the same 106 kW.

Peak period lasts longer for larger groups of customers. The erratic, up and down load curve of a single customer may have a very high peak load, but that peak lasts only a few minutes at any one time, although it is reached (or nearly reached) several times a day. By contrast, the load curve for one hundred homes reaches its peak only once during a day but stays at or near peak much longer -- perhaps for over an hour. Figure 2.13 shows the peak load duration -- the longest continuous period during the day when the peak load is at or within 5% of maximum -- for various numbers of customers.

Coincident Load Curve Behavior Can Be Viewed as the Expectation of a Single Household's Load

What does the smooth, average residential customer curve, one computed from Figure 2.6 as described earlier, indicate? As stated, it is the average, coincident customer load curve, 1/60,400 of the load from 60,400 households. While no customer within the entire utility service territory has an individual load curve that looks anything like this one, the application of a smooth load curve computed in this way has two legitimate interpretations on a per customer basis:

1. *The curve is an individual customer's contribution to system load.* On the average, each customer (of this class) adds this much load to the system. Add ten thousand new customers of this type, and the system load will increase by ten thousand times this smooth curve.

2. *The curve is the <u>expectation</u> of an individual customer's load.* Every customer in this class has a load that looks something like the erratic behavior shown on Figure 2.7, but each is slightly different and each differs slightly from day to day. The smooth curve gives the expectation, the probability-weighed value of load that one would expect a customer in this class, selected at random, to have at any one moment, as a function of time. The fact that the expectation is smooth while actual behavior is erratic is a result of the unpredictable randomness of the timing of individual appliances.

The behavior discussed above is *intra-class coincidence.* Commercial and industrial customers have a behavior similar to that discussed above for residential loads, but the shape of their coincidence curves may be different. Qualitatively the phenomenon is the same, quantitatively it is usually quite different.

One can also speak of coincidence *between* classes. Figure 2.6 showed average, or typical coincident curves for residential, commercial, and industrial customer classes in a summer peaking utility in the southern United States. These various classes peak at different times of the day. Commercial load peaks in mid-afternoon, and residential load peaks in late afternoon or early evening. A substation or feeder serving a large number of customers of both types will not see a peak equal to the sum of the coincident residential and commercial peaks for the customers it serves. It will see a somewhat lower peak load, because these two classes of customer experience their peak load, *as classes,* at different times.

Coincidence and its effects on planning will surface again and again throughout this book, and we will make reference to coincidence, and discuss its impact on planning when we discuss how to determine feeder and transformer capacity; what loads to use in feeder voltage drop and losses studies; how to site and size substations; which load curves are appropriate for certain special applications; and how to forecast load growth. Therefore, it is vital that the reader understand the concept and phenomenon of load coincidence, both within each customer class, and among the classes.

Electric Load and Consumer Demand for Power

2.4 MEASURING AND MODELING LOAD CURVES

The way in which load curve data is obtained and processed can change the way it looks, to the point that some apparently valid data can be filled with error. Sampling rate and sampling method both have a big impact on what load curve data look like, how accurate they are, and how appropriate they are for various planning purposes. Sampling rate refers to the frequency of measurement -- the number of times per hour that load data are recorded. Sampling method refers to exactly what quantity is measured -- instantaneous load or total energy used during each period.

Most load metering equipment measures the energy used during each period of measurement (called demand sampling). If the load is measured every fifteen minutes, the equipment measures and records the total energy used during each 15 minute period, and the resulting 96 point load curve gives the energy used during each quarter hour of the day. Plotted, this forms a 24 hour demand curve. This method of recording is *sampling by integration* -- the equipment integrates the area under the load curve during each period and stores this area, even if the actual load is varying up and down on a second-by-second basis.

By contrast, *discrete instantaneous sampling* measures and records the actual load at the beginning of each period. Essentially, the equipment opens its eyes every so often, records whatever load it sees, and then goes to sleep until the beginning of the next period. Some load monitoring equipment uses this type of sampling.

Discrete instantaneous sampling often results in erratically sampled data that dramatically misrepresents load curve behavior. The right side of Figure 2.14 shows a data curve for a single home, produced by a discrete instantaneous sampling technique applied on a fifteen minute basis to the individual household data first shown in Figure 2.7. This "load curve" is hardly representative of individual, group, or average load behavior. It does not represent hourly load behavior in any reasonable manner. *This data collection error cannot be corrected or counteracted with any type of subsequent data processing.*

What happened is that the very instant of measurement, which occurred once every fifteen minutes, sometimes happened upon the very moment when a sharp load spike of short duration was at its peak. Other times, the sampling stumbled upon a moment when energy usage was low. This type of sampling would yield a good representation of the load behavior if the sampling were more rapid than the rapid shifts in the load curve. But sampling at a fifteen minute interval is much too slow -- the load curve can shoot back and forth from maximum to minimum several times within fifteen minutes. As a result, it is just random luck where each period's sampling instant happens to fall with respect to the actual load curve behavior -- and a portion of the sampled load curve data is essentially randomly selected data of no real consequence or value.

Automatic load metering equipment seldom uses discrete instantaneous sampling. Almost all load recorders use demand sampling (period integration). However, many sources of load data do start with instantaneous discrete sampled data, particularly because human beings usually do use this approach when gathering data.

For example, data written down by a technician who reads a meter once an hour, or by a clerk who takes hourly load values off a strip chart by recording the reading at each hour crossing tic mark, have been sampled by this method. So, too, have data that have been pulled from a SCADA monitoring system on an hourly basis by programming a system macro to sample and record the power flow at each bus in the system at the top of each hour. In all such cases, the planner should be aware of the potential problem and correct for it, if possible. (For example, the technician taking hourly values off the strip chart can be instructed to estimate the average of the plotted value over the entire hour).

As was discussed earlier, larger groups of customers have smoother and more well-behaved load curves. Thus, the problem discussed here is not as serious when the load being metered is for a feeder or substation as when the load being measured is for only a few customers, although it can still produce noticeable error at the feeder level. At the system level it seldom makes any difference.

A telltale sign that discrete sampling data errors are present in load data is erratic load shifts in the resulting load curves, like those shown on the right side of Figure 2.14. However, lack of this indicator does not mean the data are valid. Hourly, half hour, or even fifteen minute load research data gathered via discrete instantaneous sampling should be used only when absolutely necessary, and the planner should be alert for problems in the resulting load studies.

Sampling by integration, by contrast, always produces results that are valid within its context of measurement -- hourly load data gathered by period integration for an individual customer will accurately reflect that customer's average energy usage on an hourly basis. Whether that sampling rate is sufficient for the study purposes is another matter, but the hourly data will be a valid representation of hourly behavior.

Figure 2.15 shows the daily load curve for one household (that from Figure 2.7), demand sampled (period integration) at rates of one quarter hour, one half hour, one hour, and two hours. None give as accurate a representation of peak load and peak period as the original curve (which was sampled at one minute intervals), but each is an accurate recording usage averaged to its temporal sampling rate. Such is not the case with discrete sampling, where the recorded values may be *meaningless for any purpose* if sampled at too low a rate.

Electric Load and Consumer Demand for Power 71

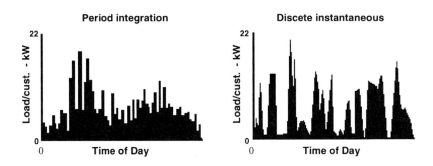

Figure 2.14 Load for a single household (the load curve in Figure 2.7) sampled on a fifteen minute interval by both period integration (left) and discrete instantaneous sampling (right). Neither curve is fully representative of actual behavior, but the period integration is more accurate for most purposes. The discrete sampling is nearly worthless for planning and load study applications.

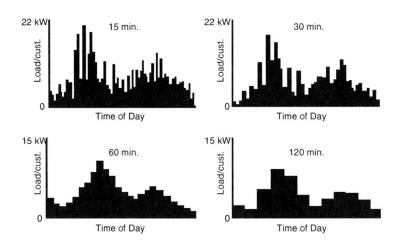

Figure 2.15 Daily load shape for the household load curve in Figure 2.7 demand sampled (i.e., using period integration) on a 15, 30, 60, and 120 minute basis.

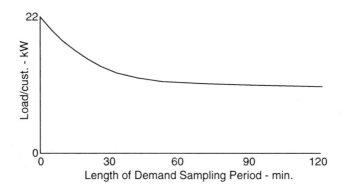

Figure 2.16 Observed peak load is a function of time period of sampling for the single household's daily load curve, as a function of the sampling period. Only high sampling rates can see the needle peaks and erratic load curve behavior common to individual customers and small-groups of customers. As a result, too low of a sampling rate can lead to underestimation of equipment capacities and losses costs, particularly at the service level, as will be discussed in Chapter 13.

As the sampling period of demand sampling is lengthened, the observed peak load drops, as shown in Figure 2.16. The sharp needle peaks are being averaged with all behavior over longer periods, reducing the measured "peak demand." Figure 2.16's curve of peak load versus sampling rate bears a striking resemblance to the peak coincidence curve plotted in Figure 2.12, an interesting potential source of confusion about coincidence and how much peak load levels diminish as a function of load group.

High Sampling Rates Are Needed for Small Groups of Customers

A smooth load curve, for example one representing 1,000 customers, or even that shown in Figure 2.11, for 100 homes, does not need to be sampled as rapidly as the load curve for one, two, or only a few customers. High sampling rates are needed only when studying the non-coincident load behavior of small groups of customers, but they are essential when trying to capture or characterize accurately the load curve behavior of individual customers. Analytical methods exist to determine what type of sampling rate is required for specific load modeling needs, and to estimate the approximation errors introduced by using lesser sampling rates. In addition, the analysis methods applied to small groups of customers and to analysis of "individual customer" load curves must also be selected and applied with regard to sampling rate compatibility -- typical load manipulation analysis methods are not appropriate.[2]

Distribution load behavior is dominated by coincidence, the fact that peak loads do not occur simultaneously. Individual customer load curves are erratic, formed of fast, sharp shifts in power usage. As customers are combined in groups, as when the planner analyzes load for groups served by equipment such as service transformers, laterals, and feeders, the erratic load curves add together with the sharp peaks intermingling and forming a smoother curve. As a result, when comparing the load curves for different size groups of customers, as the number of customers increases, the group load curve becomes smoother, the peak load per customer decreases, and the duration of the peak increases. Also important is the manner in which the load curves are measured. Where possible, particularly when working with small groups of customers, data should be collected and analyzed using the period integration method, and the sampling period should be quite short, generally less than 15 minutes.

REFERENCES

Electric Power Research Institute, "DSM: Transmission and Distribution Impacts," Volumes 1 and 2, Report CU-6924, Palo Alto CA, Electric Power Research Institute, August 1990.

H. L. Willis, T. D. Vismor, and R. W. Powell, "Some Aspects of Sampling Load Curves on Distribution Systems," *IEEE Transactions on Power Apparatus and Systems,* November 1985, p. 3221.

---

[2] For a more thorough discussion of load curve sampling and modeling, particularly of analytical interactions and requirements, see *Spatial Electric Load Forecasting,* by H. L. Willis, Marcel Dekker, 1996.

3
Availability and Power Quality

3.1 INTRODUCTION

Electric consumers also have definite needs with regard to the quality of electric service they are provided. Like demand, this need varies from one customer to another, and with time of day, day of week, and season of year. Consumers want electric power whenever they need it, and they want it to do the jobs they require. They want nothing more: interruptions that occur when they are not using power are unimportant, as for example an interruption of service during the weekend at a business that operates only on weekdays. Voltage spikes, dips, and harmonics are unimportant to them if they cause no problems.

Thus, from their perspective, they want nothing less than perfect service. Interruptions that do disturb their routine or business are considered unacceptable. So are voltage sags, transients, or harmonics that interfere with the proper operation of machinery or produce even nuisance levels of static, noise, or operational problems. The real issue, however, is not what customers want, but rather what type of service they are willing to *pay* for. An electric utility should always try to provide good value for their customers' money -- sloppy service at a high price is inexcusable -- but customers must understand that high quality of service and low price are mutually incompatible. For example, a SMES (Superconducting Magnetic Energy Storage) unit along with a fast start diesel generator and the proper associated electronic "waveforming" control equipment can provide availability on the order of less than one expected interruption per century, and flawless 60-cycle alternating current at an unvarying voltage. SMES service can also costs up to 50 times what the equipment for normal power delivery can cost. Thus, the really important aspect is determining how much customers value uninterrupted availability of power and high power quality -- how much they need to the extent they are willing to pay the cost. The fact that many SMES systems are in service is proof that there are customers for whom availability and power quality are important, valuable issues. That the vast majority of electrical consumers do not have such equipment, or want it, is proof that for a majority of customers, such levels of service quality are not cost justifiable.

This chapter begins by looking at the various electric problems where service can fall short of customer expectations. This is putting the cart before the horse is some cases, because many of the equipment and engineering issues related to these problems are not introduced until later in the book. However, the point of this discussion is to explain the types of problems that customers see, and some explanation of the engineering issues is required. Section 3.2 discussions interruption of power, along with the most typical causes, and the ways in which such problems are "fixed." In turn, Sections 3.3, and 3.4 then similarly treat voltage variation problems, and harmonics. Section 3.5 looks at the customer impact of these problems and at ways that are used to study customer quality needs and assess their value in T&D planning.

3.2 INTERRUPTION OF ELECTRIC SERVICE

An "interruption" is a cessation of electric service. By contrast an "outage" is the failure of one or more components of the electric system to do their job, either because of actual equipment failure, damage by weather or other causes, or because they have been switched out of service, either deliberately or by mistake or failure of control equipment.

This is certainly the worst type of "power quality problem" that can occur, but complete cessation of service is generally considered in a separate category from "power quality." (Power quality is most often used to refer to attributes of the power when there is some being delivered). Interruption of power -- a lack of availability -- is the inverse of reliability as seen by and considered from the utility perspective. Interruptions are usually caused by the outage of one or more parts of the T&D system -- a downed line, a cable that failed, a transformer damaged by lightning, etc. They can also be caused by failures in the customer's equipment -- open wiring inside a home, corroded switchgear at an industrial plant substation, and similar situations.

Regardless, to an electrical consumer an interruption is an interruption. The

Availability and Power Quality

cause is largely irrelevant -- power is not available. Chapter 5 discusses power system reliability, and provides a host of definitions and formulae for interruption and outage analysis.

Three, Two, or Single-Phase Interruptions

A complexity of interruptions has to do with phasing. Most power system equipment, feeders, and lines consist of three phases, with overhead and underground feeders consisting of three or four wires depending on whether they are delta or wye connected, respectively. Many outages do not involve all three or four wires, and thus one can get combinations of some phases out and some phases still in service. Many times these outages lead to both interruptions and voltage problems, as will be discussed in Section 3.3.

When only one phase of a three-phase feeder fails (perhaps a tree branch fell, taking it to the ground), the other two remain in service. Some residential and small commercial customers who receive single-phase service may still have acceptable service in this instance. Given that adjacent single-phase customers often have service from different phases, this can leave service in a neighborhood very spotty -- some customers with, some without -- which is confusing to customers.

Larger commercial and industrial customers may complain of "partial service." Some large facilities with three-phase service use a portion of the three-phase power in single-phase fashion -- a three story department store might use phase A for lighting on the ground floor, phase B for the floor above, and so forth, in addition to which large motors for air conditioning and similar service would be three-phase. Failure of one or two phases leave them with most equipment out of service, but some still operating.

On the other hand, some three-phase commercial buildings provided with three phase, 120 volt-to-ground power run lighting and other equipment in delta-connected fashion, as 208 volts between phases. Loss of one or two phases can result in unusual service voltages. In addition, loss of the neutral in wye-connected systems, whether in conjunction with or separate from loss of any of the phase conductors, can result in unusual voltages at the customer location. Further, the exact nature of the "post-disturbance" voltage depends on exactly what type of event caused the outage -- if a downed line is faulted at high impedance to ground, and the fault has not been cleared, the voltage will be far different than if the line is open.. Thus, some "interruptions" appear from the customer's perspective to be "voltage variation" power quality problems of the type to be described in Section 3.3.

"Fixes" for Interruptions

When speaking of "fixes" for power interruption problems it is important to distinguish exactly what interruption problem is being addressed. First, anytime an outage occurs which results in interruption of electric service to one or more customers, the immediate priority is to restore service. Often this requires repair or replacement of the outaged equipment. However, in urban and suburban systems, service can usually be restored in minutes and almost always within less than an hour by re-switching primary-voltage feeders to isolate the outaged equipment and minimize the number of customers without service.[1]

Recurring interruptions on a part of the system indicate some operational problem that needs to be addressed in operations and maintenance. By far the most prevalent is tree-trimming. If not kept a good distance from overhead lines, trees can brush up against bare conductor, causing intermittent interruptions, or broken branches can fall on conductor, bringing it down. This cause of interruptions can be greatly mitigated by a sound tree-trimming program which calls for periodic inspection and trimming of trees -- for example a program that cuts back "three years of growth" every three years.

Similarly, some underground circuits experience high rates of failure due to frequent problems with old cable. In such cases nothing but replacement of all the potentially troublesome cable will correct the problem.

Interruption frequency and duration can be influenced by the type of design and layout of the distribution system. The standard, radial configuration used for 99%+ of power distribution is prone to interruptions whenever equipment fails or is otherwise out of service, because it provides only one avenue of power flow to the customer. The "first level" of design for high reliability involves sound design of the standard radial configuration. Use of equipment with good in-service records (low observed failure rates, etc.) and careful attention to loading and manufacturer's guidelines, result in fewer outages. Good application of standard distribution protection engineering -- arrangement of fused zones on the feeder and proper coordination of protection -- can keep the extent of interruptions caused by potential outages to a minimum. This can usually provide customer interruptions at something on the order of one to three interruptions and a total of one to four hours without service per year, depending on weather and other conditions unique to the electric utility service territory.

Further improvement over standard radial performance can be achieved by providing automatic rollover from another source. Whether T&D or DG

---

[1] Rural systems generally offer far fewer opportunities for switching. Repair is often the only option.

Availability and Power Quality

(distributed generation) a rollover source will reduce duration but not frequency of outages.[2]

Providing two or more simultaneous sources, each capable of serving the load at a particular site, reduces both frequency and duration of interruptions. This can be accomplished with any of a number of loop or network distribution configurations, or with certain types of distributed generation. Without exception these "design fixes" increase the cost of service by a considerable margin, as they require both more conductor and routes, and often necessitate more expensive types of equipment (e.g., network protectors). However, this increased cost may be entirely justifiable from the standpoint of customer value and willingness to pay a higher price for more reliable power.

Momentary Interruptions

Interruptions have different durations, depending on their cause, the effort required to restore service, the location and nature of the outage, and other outages that have to be repaired simultaneously. Longer duration interruptions generally cause more inconvenience to customers, as will be discussed in section 5. Chapter 5 provides a discussion of interruptions categorized by length of time and other characteristics important to analysis of reliability and system design. But one type of interruption worth considering as a separate type of service-quality problem is momentary interruptions, for which duration is not really the issue as far as both impact and application are concerned.

Momentary interruptions are quite short interruptions of power flow -- often lasting only a few seconds, and never more than a minute. They are caused by trees brushing against overhead conductor during high winds for just an instant, causing a (usually high impedance) fault, or by small animals that contact the conductor (they are killed and often vaporized instantly, thus ending the cause of the fault), or by lightning. In each case the disturbance lasts but a few seconds. Often it is cleared by a recloser, which is applied in the hope that the fault can be cleared without breaker or fuse operation that leads to a longer interruption of service for some or all customers. Reclosers usually open the circuit instantly upon detection of a fault, then re-close to restore service after a very short period

---

[2] Technically, it is possible to switch from one radial source to a backup radial source nearly instantaneously, so that the net interruption is so brief that equipment is not affected by the outage of the first source. In practice, this requires purely electronic switching, not mechanical, if dependable switching is desired At the time of this writing (1996) all-electronic distribution-level switching equipment is just becoming available, and its high cost -- until volume production brings the price down -- will be a factor in many applications.

-- often less than a second. If the fault remains, the recloser opens for a longer time -- perhaps 5 to 15 seconds on the second try, and longer each succeeding try. Thus, the recloser "blinks" the lights of all customers downstream four times trying to clear a fault that occurs somewhere on the feeder. This is generally regarded as one "interruption" or outage, but often causes customers confusion and nuisance as power seems to keep failing just as they are trying to re-initialize appliances and plant equipment, etc.

Reclosing of momentary interruptions is a traditional approach to fault protection, which were clearly preferable when developed during the era prior to the heavy dependence modern electronic and digital equipment have put on continuity of service. Traditionally, given the "cost" of a very short outage to customers, this approach seemed like a reasonable trade of inconvenience: most customers on a feeder experience several momentary outages in an often successful attempt to prevent a sustained outage of service to some customers.

However, given modern requirements for service, some utilities are re-thinking the application of reclosers. Modern customer impact from an interruption can be considered to have a "fixed cost" that is independent of its length and a variable component that rises as the time of outage continues. (This cost is possibly non-linear, as will be discussed in section 3.6). Limited survey data collected by the author in 1992 indicates that momentary interruptions are considered about equal to roughly eight minutes of interruption.[3] Regardless, it is clear that the traditional balance of inconvenience is not optimal in many cases. Some utilities are changing policy with respect to instantaneous activation of reclosers upon the first detection of a fault -- they delay both because the fault might be very brief, and to give fuses time to clear the fault if located on laterals and small branches. Others are simply doing away entirely with instantaneous tripping of breakers and cutting back on the use of reclosers.

3.3 VOLTAGE VARIATIONS

Customers depend on stable alternating voltage much more than they often realize. Most of the appliances used by residential, commercial, and industrial customers operate only within a narrow band of operating voltages (usually about +/-10%). When provided with voltage outside of that narrow range many types of equipment and appliances will not operate, and may even be damaged.

In addition, a surprisingly large portion of electrical equipment provides an output that is noticeably voltage sensitive. Motors speed up or slow down slightly as voltage is increased or decreased about the nominal design voltage.

---

[3] An interruption "of less than a minute" was considered to be half as inconvenient as one lasting fifteen minutes.

Availability and Power Quality

Lighting of several types varies its lumen output noticeably as voltage is changed. As a result, variations in voltage, particularly if they occur rapidly, can alter equipment performance, at best creating a nuisance and at worst causing equipment damage.

Variations in voltage tend to be grouped into fuzzily defined categories by the duration that voltage strays from nominal.

Voltage Control Problems

Voltage at the customer site is most often kept within standards by control equipment installed at the primary distribution voltage level -- load tap changing transformers, line drop compensators, or distribution line regulators -- which adjust their voltage boost to raise or lower voltage at the substation low side bus. In company with sound distribution design and well-maintained equipment, this is generally sufficient to assure that voltage at all points on the distribution feeder stays within prescribed limits, and that variations, when they occur, are tolerable. Chapters 4 and 8 provide details on voltage control guidelines and typical criteria, and design practices used by operating utilities.

Voltage sags of up to 20%, that last for minutes or hours, can result from failure of these voltage control schemes -- either excessive voltage drops caused by overloading of the feeder beyond its design limit, or from the failure of voltage control equipment or their control systems during normal load conditions. Overloading and the accompanying low voltage it causes most often occur during cold load pickup, when after a sustained outage, service is restored to loads that have lost their natural diversity of operation. Low voltage can also occur during contingency operation, when a feeder may be used to serve much more than its normal maximum load. Low voltages can also occur at other times due to higher than expected demand -- as for example heating or cooling loads on a very cold day or a very hot day.

Higher than nominal voltages are possible, but rare, at times of very light load on certain circuits, when line charging can result in quadrature-boosted voltage that can reach as much as 20% above normal maximum. High voltages are also possible due to failure of voltage control equipment's controls (e.g., LTC and line regulators set themselves to maximum boost during light load conditions).

In most cases, voltage problems traceable to loading act upon all three phases in roughly the same proportion, as do failures affecting "gang-operated equipment." Therefore, the relative impact on all phases is an important diagnostic fact in determining cause.

Unbalanced Voltage problems

Unbalanced loads, faults, or equipment failures (e.g., single-phase interruptions) can result in low, high, or combinations of low and high voltages among the three phases on a circuit, depending on the nature of the layout (delta, wye), load, and fault or failure. In many cases, the cause is either an open ground or a single-phase, high-impedance fault that has not been cleared.

Very often, unbalanced voltage situations are due to multiple causes: a failure or an unbalanced load which would normally cause only a minor imbalance -- 5% or less -- leads to very excessive voltage differences the phases -- up to 25% -- because of a heretofore unnoticed open ground, etc. Or perhaps the situation is reversed and a sudden failure of good contact to ground reveals a load or equipment imbalance that had been previously tolerated and unnoticed.

"Fixes" for voltage control problems are largely within the domain of operations and maintenance. Operating the T&D system within designed loadings will avoid the over loading situations that lead to low voltages, and the low load situations that can lead to high voltages. Routine inspection can detect incipient failures and identify many problems before they cause poor service for customers. Correct maintenance of equipment is necessary to assure that equipment works as intended -- *"if you can't afford to maintain it, don't build it."*

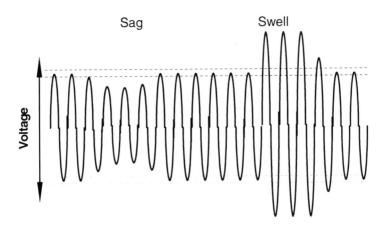

Figure 3.1 Voltage swells and sags are deviations of voltage outside the nominal range (dotted lines) which often last only a few cycles or seconds.

Voltage Sags and Swells

Sags and swells are brief periods in which voltage exceeds (swells) or drops below (sags) nominal voltage level (Figure 3.1), much briefer than the minutes or hours-long periods of off-nominal voltage discussed in section 3.2 . Sags and swells may last less than a second. Although the low or high voltage itself may cause a problem, often it is the change in voltage (e.g., a dimming of the lights) that is first noticed. Regardless, swells -- momentary excursions of source voltage above nominal voltage can damage equipment -- often quickly and occasionally rather spectacularly (see Chapter 4, Table 4.2). Sags -- momentary dips -- while seldom causing catastrophic equipment failure, often damage equipment in more subtle, slower ways. Both generally garner complaints not because of equipment failures (although they may eventually lead to such) but because they often cause disruptions of operation. Many appliances and machines, digital control equipment and heavily loaded synchronous motors among them, cannot tolerate low voltage for even a few cycles, and will shut down or trip off line.

Beyond this, even if appliances and equipment remain in operation, many have end-use performance characteristics that are quite sensitive to voltage. A majority of people can perceive the change in lumen output that a 3% shift in voltage will render in incandescent lighting. If variations of this amount or greater occur rapidly and frequently, it proves tiring and distracting. Industrial equipment such as compressor motors, pumps, conveyers, rippers and stamping machines, power looms and rollers, often are sensitive to voltage to the extent that rapid variations affect their performance. As a result, sags and swells can cause air pressure and water volume to "hiccup," conveyer speed to "stutter," power looms to vary their weave rate, and so forth, resulting in unacceptable performance or even damaged plant product.

Voltage Deviation vs. Duration Envelopes

The Computer and Business Equipment Manufacturers Association (CBEMA) has established the recommended voltage variation versus duration envelope shown in Figure 3.2. This establishes a standard for voltage variations with respect to time -- greater excursion from nominal voltage is permitted if it lasts a shorter time. Most but not all computer and business equipment is designed to function as long as voltage stays within this envelope. Therefore, this serves as a target for T&D delivery system performance.

Regardless, the concept of deviation versus time envelopes can be used to evaluate customer needs and delivery performance, as shown in Figure 3.3. In general, about 40% of all voltage sags lie below the lower curve of the envelope.

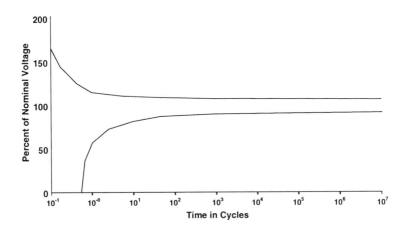

Figure 3.2 CBEMA curve of voltage deviation versus period of deviation.

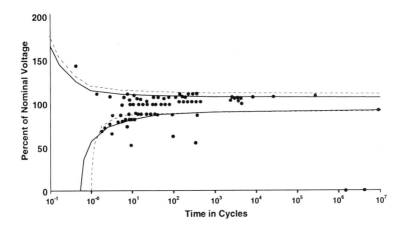

Figure 3.3 Data from 67 disturbances which occurred over a two-year period at a clothing factory, plotted against the CBEMA curve (solid line) and the actual requirement envelope of a digitally-controlled hosiery loom (dotted line). In many systems, about 40% of all voltage sags and 10% of all voltage swells seem to lie outside of the CBEMA envelope.

Causes of Voltage Sags and Swells

Voltage sags are generally caused by faults in nearby parts of the system, the low voltage occurring for only the brief time between the fault and the breaker or fuse operation. As an example, a fault on a distribution feeder will cause a severe voltage sag, perhaps on the order of 60%, in feeders connected to the same bus. In general, roughly 40% of voltage sags caused by faults fall outside of the CBEMA envelope, and roughly 85% of voltage sags outside the CBEMA envelope are due to faults of some type.

Starting large motors can also cause reduced voltage, often lasting for far longer times than fault-induced sags, but generally the voltage dip is not nearly as severe. Motor-induced voltage sags are generally on the order of 10%, and rarely worse than 20%, partly because of utility standards that try to prevent customers from installing motors that cause flicker problems, and also because the motors, themselves, will not start under severe low voltages. (Motor starting is often called "flicker" -- see motor flicker criteria in Chapter 4).

In somewhat the opposite manner of motor flicker causing voltage sags, distributed generation (DG) located on the distribution system can cause voltage swells. Improper load-following or operation can contribute to this, but the most common situation of co-generation-induced overvoltage is when connectivity to the rest of the power system is lost. DG units feeding more power onto a feeder than the feeder load (i.e., with a net injection of power into the substation bus may take a moment to adjust if connectivity to the system is lost. During that period they can produce over-voltages as high as 50% above nominal. (Short et al., 1993).

In addition, ferroresonance can cause high over-voltages -- as much as 4.0 per unit but typically on the order of 1.5 to 2.0 per unit -- during some types of switching operations on distribution feeders. In addition, certain types of transformers can cause ferroresonance over voltages also.[4]

But the most frequent cause of voltage swells are most often attributable to single line to ground faults on a three-phase circuit -- theoretically voltage on the unfaulted phases can reach 1.73 per unit (on delta systems) or 1.25 per unit (on wye systems) during such faults, although it is unusual for voltage to attain these theoretical upper limits. Very rarely, voltage swells can also be caused by unusual equipment failures -- faults/open circuits in only one winding of a large motor, or from overspeed conditions when a very large motor suddenly loses its mechanical load.

---

[4] In particular, "five leg" three-phase transformers in grounded wye-systems. For an excellent discussion of ferroresonance and its effects and cures using station over-voltage arrester (see Short et al.).

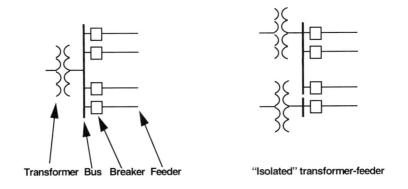

Figure 3.4 Isolating an individual feeder through transformers (right) results in less propagation of low voltage sags to it in the case of faults on the other feeders, compared to the normal practice of using one transformer for several feeders (left).

"Fixes" for Sags and Swells

The occurrence of sags and swells on the distribution and subtransmission system is best reduced by assuring that the events that lead to them -- generally single-phase faults of one type or another, and ferroresonance, do not occur. In this respect, the "solutions" are the same as those given above for voltage control problems. The "extent" (number of customers affected) of fault-induced sags and swells can be reduced by various designs (Ward et al.). Using individual transformers for a feeder (Figure 3.4), so that the only common electrical path between them is through two transformers will greatly reduce the voltage variation seen on one if the other is faulted, but is an expensive "fix."

Another approach is to limit the severity of faults that do occur. Various types of fault-limiting protection and equipment schemes approach the problem in this manner, including fault-limiting fuses installed in selected locations, and series inductors, but these often limit capacity or induce other operating problems of their own.

Finally, sensitivity of customer equipment can be reduced by changes to power supply equipment (transformers, smoothing filters). In the case of digitally-controlled motor-drive systems, control software is often set by the manufacturer to trip the system off line at even slight voltage sags: re-programming can increase tolerance (reducing the number of sags that cause the system to shut down) without adversely affecting performance.

Availability and Power Quality

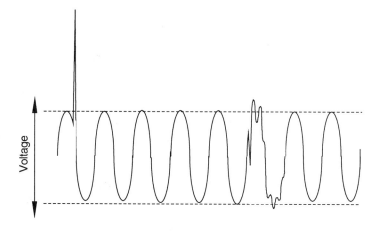

Figure 3.5 A lightning-induced transient (left) and a surge caused by capacitor switching (right).

3.4 VOLTAGE SURGES

The term "voltage surge" refers to a very transient swing in voltage -- a spike or other quick shift -- which does not last over a number of cycles like a voltage swell or a sag. Figure 3.5 shows two of the more common types of surges, a lightning induced spike, and "ringing" due to capacitor switching. Voltage surges, particularly those from lightning, cause equipment damage, often quite severe.

Causes of Voltage Surges

Although conclusive data is unavailable, the largest source of voltage surges seen by an electric utility's customers is probably switching of customer equipment itself.[5] In general, many switched motors can cause voltage spikes that propagate locally within residential, commercial, and industrial sites,

---

[5] In the late 1980s, the author had a refrigerator which routinely caused voltage spikes of between 1.4 and 1.6 whenever its relay cycled it off. This led to no discernible problems at the time, although it has always been blamed for the early failure (after only one year) of a digital clock radio connected to the same outlet.

particularly if grounding of the customer's wiring or equipment is faulty. One study indicated that almost 2.5% of customers see voltage trasients above ten per unit due to appliance or equipment switching within their location (Ward, 1993).

Switching problems at a customer facility cause problems within, but rarely outside of, that site. Although surges on the T&D system frequently travel through service transformers and secondary circuits into a customer site, rarely are customer-side surges strong enough to travel the other direction to make any discernible impact on the system or on other customers -- any switching surge powerful enough to travel through a customer facility's wiring and up onto the T&D system where it would trigger an arrester or damage nearby customer facilities would leave the customer's electrical facilities in an absolute shambles.

Lightning is the cause of most very severe voltage spikes on the T&D system, which can reach 20 or more times nominal voltage. It is probably the leading cause of overvoltage damage worldwide. Lightning surges can be caused either by *direct* strokes, where the lightning stroke actually hits the line, or indirect strokes where the strike is nearby. Interestingly, on distribution and subtransmission systems, indirect strokes account for a majority of line failures caused by overvoltage surges, causing up to four times as many as direct strikes (Cinieri and Muzi).

Switching, particularly of capacitors, is by far the most predominant non-weather-related cause of voltage surges created on the T&D system, which may be transmitted to the customer level. Capacitors are switched frequently, and are often installed at locations close to customers.

"Fixes" for Voltage Surges

Installation of lightning arresters (LA) or transient voltage surge suppressors (TVSS) can reduce the number of overvoltage surges seen by the T&D system and passed on to customers. Generally they are installed at a customer entrance if the intention is to protect the customer from surges on the utility system, or at individual appliance loads within a household or business if the desire is to protect them from transients in the customer's wiring.

LA and TVSS both limit the amount of overvoltage seen by clipping the voltage above some predetermined level. This limits the damage overvoltage causes but does not eliminate it altogether. Generally, the coordinated use of arresters will make a noticeable reduction in surges and momentary interruptions (Ward et al., 1990). Yet, arresters do little to reduce the overvoltages caused by capacitor switching (Short et al., 1993), and they will do nothing to reduce lower voltage surges caused by switching of customers' equipment. However, customer equipment does not have to cause significant switching surges. Various types of

Availability and Power Quality 89

current-limiting or voltage-clipping circuits are available and often used in standard appliances.

Installation of shield wires on overhead lines and UG terminal poles can reduce the incidence of lightning-caused overvoltage damage by very significant margins. Shielding does little to reduce the number of direct strokes causing damage, but cuts damage from indirect strokes by up to 80% (Cinieri and Muzi).

3.5 HARMONICS

An electric power system, and the appliances and equipment connected to it on both utility and customer sides, is designed to operate at 50 or 60 hertz, with sinusoidal voltage and current. For a variety of reasons, additional electrical flows at frequencies other than 50 or 60 hertz can occur on parts of the power system or within the wiring at a customer site. This non-system frequency power flowing through system and equipment can cause operating problems on both

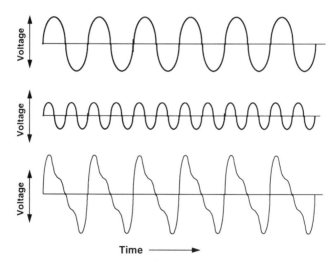

Figure 3.6 Top, 60-cycle power. Middle, the second harmonic, 120-hertz, with magnitude .5 per unit. Bottom, their sum, with peak voltages more than 30% above nominal. Every wavelength is affected in the same manner.

the utility and customer sides and usage problems on the customer side, for a combination of four reasons:

1) The non-system frequency power flow voltages and currents superimpose themselves on the standard 50 or 60 hertz flow (Figure 3.6) creating higher-voltages and insulation stress, and increased losses and thermal stress. Even tolerable levels of harmonics (in the sense that they cause no malfunctions or dramatic equipment failures) are thought to increase costs due to losses and loss of life by 10% PW.

2) Many appliances are designed to accept and run well with power provided at 50 or 60 hertz but do not respond well to significant amounts of non-standard frequency power. These may cause noise or static in electrical equipment, stuttering or buzzing of mechanical action, and in extreme cases almost immediate equipment failure.

3) Harmonics generated on the electric system can create electrical noise levels sufficient to render nearby telephone service inoperative.

4) Non-standard frequency voltage and current are often (usually) not detectable by standard T&D monitoring, measurement, and control equipment, so their presence goes unnoticed. For example, standard single-phase residential meters do not "see" frequencies much above 60 cycles.[6] Very often, the first indication that significant harmonics are present at a site is when they cause an operating problem of lead to an equipment failure.

Harmonic Frequencies

Any frequency of power flow other than the system frequency is generally unwanted on the power system, the exception being carrier or ripple control signals. Harmonic frequencies, which are the source of most problems generated by non-standard frequency power flow, have integer multiples of the basic system frequency -- in a 60 hertz system they are 120, 180, 240, 300, 360 hertz

---

[6] A notorious but effective way to "steal" power from an electric utility is to half-wave rectify a standard residential electric storage water heater. This halves its heat recovery rate, but the unit still provides plenty of hot water -- it just takes twice as long to recover after heavy usage. The half-wave rectification transforms most of the water heater's load into third harmonics and higher, which a standard residential watt-hour meter *does not* see. Thus, the energy for water heating is "free." (This also causes severe harmonic flow that disrupts televisions, VCRs, and digital equipment, all of which the homeowner deserves from his dishonesty).

Availability and Power Quality

and so forth. These harmonics are dubbed respectively the second harmonic, third harmonic, and so forth. (The system frequency is the "first harmonic.") Harmonics as high as the 100th harmony (6000 Hz) have been measured on power systems

By contrast, non-harmonic frequencies, for example one of 217 hertz, generally have to be generated and injected into the system by some special purpose device. Whenever present, they have generally been produced deliberately even if they have strayed onto the power system inadvertently.[7] Often, non-harmonic frequency flow is more difficult to diagnose, because it does not alter every wavelength in the same way (meaning it will not show up as stable waveshape changes on an oscilloscope). However, once detected their source is usually much easier to identify because it is almost certainly an unusual and deliberate generator of non-harmonic frequencies.

"Zeroeth" harmonic

It is worth noting that direct current flow (frequency of zero) is the "zeroeth" harmonic, and that DC contamination of the power system is a part of a theoretically complete look at the series of all harmonics, whether done in the time domain (DC is the zeroeth harmonic) or frequency domain (DC has a frequency of zero). Generally, detectable DC current or voltage is a sign of poor grounding, severe load imbalance, or damaged equipment. Even a small component is indication that whatever other problems may exist, ground, neutral flow, or internal balance is not quite right in the system.

Harmonic frequencies result in a waveform distortion

Since they are integer multiples of the system frequency, harmonic frequencies always stay in step with the system frequency (i.e., their phase remains constant with respect to the system frequency) and they impact each succeeding waveform in basically the same was, as is shown in both Figures 3.6 and 3.7.

---

[7] The greatest non-standard frequency power flow of which the author knows is that used in ripple control systems widely used in Europe for decades to control distributed automation and load control equipment. Ripple control provides both signal and power for the control equipment with power injected into the T&D system at a non-multiple of 50 or 60 cycles (e.g., 217 cycles). Injections of 10 to 20 MW are not uncommon.

On the other hand, the author knows of one situation where the output of equipment being tested at a manufacturing plant for military power systems (400 hertz) managed to flow onto the local power system, causing more than just a nuisance to neighboring customers -- deliberately generated but inadvertently placed on the system.

This means that the distortion harmonics added to every system wavelength is the same, i.e., they result in a distortion of the normal 50 or 60 hertz waveshape.

> *When distorted by harmonics, every power cycle has the same non-sinusoidal shape. A sure sign of harmonics is a non-sinusoidal 60-hertz wave shape.*

Harmonic frequencies are the most common type of non-standard frequency flow on a power system, because they are generated passively by any electric equipment which has a non-linear load (i.e., impedance is not constant -- this will be discussed at length later in this chapter). Non-linear loads include transformers and motors and other "wound" devices which are overloaded, as well as electronic power supplies (AC-DC), and many other types of loads which work on the basis of clipping or triggering based on voltage level.

Causes of Harmonics

Harmonics are generated passively in the presence of standard power flow, by any device which exhibits, either permanently or temporarily, a non-linear load. A non-linear load means the impedance is not constant. While there are many ways to represent or think about non-linearity of load, for harmonics analysis it is often most useful to view "non-linear load" as meaning that the impedance -- the ratio between the current and the voltage passing through a device -- is itself, a function of voltage.

Regardless, strictly speaking, there are two categories of harmonic generators. The first are simply *non-linear loads* in which the current flowing through the device is not proportional to voltage. As a result, when fed from a purely sinusoidal voltage of one frequency, the resulting current flow will not be purely mono-frequency. Transformers, regulators, and other wound T&D equipment on the utility system can exhibit somewhat non-linear load behavior when overloaded. So will certain types of multi-phase wye-wye transformer banks with unbalanced loads and/or imperfect grounding. Diodes, semiconductor devices of nearly every type, and deliberately saturated transformers are examples of this type of harmonics generator used in many modern appliances. Invariably, this category of harmonic generators will produce harmonics whenever they are energized with alternating voltage. They are the original sources of harmonics generated on the power system.

The second type of device that can generate harmonics has *a frequency dependent impedance*. At any one frequency it may have a very constant impedance -- for example 3 ohms regardless of voltage level at 60 Hz, but its impedance varies as a function of frequency -- 3 ohms at 60 Hz, 5 ohms at 120

Availability and Power Quality

Hz, etc. Electrical and electronic filters and certain types of servo- and variable-speed motor drives have this characteristic. So, for that matter, do distribution feeders with shunt or series capacitors. These types of devices will not generate any harmonics if energized with a single frequency of voltage. However, they will distort the input waveshape if there is anything more than one frequency present; if harmonics are already present, they will alter the harmonic content. They may mitigate the harmonic content, or worsen it. In combination with harmonics generators of the first category, these frequency dependent impedances can result in complex interactions in which harmonic energy is transformed or even multiplied from one frequency to others.

Table 3.1 lists some of the more common generators of harmonics on a power system. While in some circumstances overloaded or damaged T&D equipment can be responsible for the creation of harmonics, the bulk of harmonics on most power systems are generated by customer equipment, in particular AC to DC power supplies and other electronics, which invariably contain diodes or their equivalent.

Residential, business, and industrial customer sites are all replete with such appliances ranging from microwave ovens to computers to robotic control systems to televisions, VCRs, stereos, stepper motor systems, and more. All have power supplies whose quality and harmonic contributions vary greatly, but invariably produce some harmonic output. Frankly, because there is no requirement or even uniform, worldwide standards to limit harmonic output of appliances, many such power supplies are designed without any features to mitigate harmonics effects, in an effort to reduce cost of the appliance as much

Table 3.1 Sources of Non-Standard Frequency Power

Sources of Harmonics

AC-DC power conversion (any type)	Adjustable frequency motor drives
Arc furnaces - AC or DC	DC motor drives
Fluorescent lighting ballast	EMI-fluorescent lamps
Induction motors, if highly loaded	Low-frequency oscillators
Multi-phase conversion (15 phase)	Poor or damaged ground/neutral
Saturated magnetic devices	Series capacitors
Shunt capacitors	Transformer inrush current
Voltage clipping devices	Wye-wye transformers

Sources of Non-Harmonic Frequencies

Adjustable Speed Drives	AC frequency converters (60-25 Hz)
Doubly-Fed Induction Motors	Poorly grounded motor-generators

as possible. In addition, numerous other customer loads create harmonics -- light dimmers use voltage clipping, variable-speed motor drives cause numerous harmonics, microwave ovens and electric ranges can generate strong harmonics, too. Even electric fans, and other simple induction motors run at near overload, will contribute to the problem. And the outputs of these multiple sources can be additive -- i.e., harmonics produced by an overloaded transformer might add together with those produced by switched power supplies. While individually not troublesome as sources, in combination they may pose a problem.

Harmonics are attenuated through the normal manner that electric power is absorbed. In rare cases they contribute real power to motors and so forth, but rarely and not to any useful degree -- they are transformed to heat as they pass through wiring and equipment and other appliances. Thus, the harmonics generated by a stereo system, which might measure 20% total distortion at the outlet feeding the stereo, may never make it out of that particular household circuit, being attenuated in light bulbs and other loads on the same circuit. On the other hand, and particularly when combined with harmonics generated by other sources throughout a customer's site, they might travel quite far.

Different Voltage and Current Waveshapes

One result of a non-linear load is that the voltage and current waveforms for power flow have a very different shape, as shown in Figure 3.7. In fact, since the presence of harmonics indicates that load *is* non-linear, the voltage and current waveshapes are almost certain to be different, perhaps significantly different. Thus, it is important to note whether harmonic distortion is measured as current

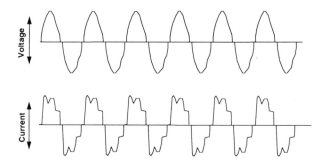

Figure 3.7 Harmonics are caused by non-linear loads, making voltage and current waveshapes dissimilar and often far different, as shown here.

Availability and Power Quality

or as voltage. Generally, harmonic distortion for a power system is measured and reported on the basis of voltage, since the power system is designed and expected to operate as a constant voltage source. However, because power systems are an almost constant voltage source, non-linear loads usually result in far more current distortion than voltage distortion, as shown in Figure 3.7, where the voltage seen at a digital AC-DC power supply has 20% distortion while the current has a 220% distortion. Measuring distortion in current results in a more sensitive test and generally provides a more dramatic statistic for the level of contamination. and evaluation.

Total Harmonic Distortion Index

The most widely used measure of harmonic contamination in a power system is total harmonic distortion (THD)

$$\text{THD} = \left(\sqrt{\sum C_n^2}\right) / C_1 \quad (3.1)$$

where C_n is the value of the nth harmonic
and C_1 being the value of standard frequency

C_n can be calculated as either current or voltage, depending on whether distortion is to be measured as voltage or current distortion.

There are at least two other indices used in harmonic analysis, generally applicable to special circumstances. These include the telephone influence factor which compares harmonic content in relation to the telephone system "C-message" curve[8] and the K-factor index useful for estimating harmonics impact on losses. However, in the majority of cases where harmonics are being studied on the power system to identify their source or to design ways to get rid of them, THD is the most appropriate distortion index. Usually THD is measured with respect to both voltage and current, with the two reported as separate values.

Frequency Domain Perspective

Analysis of harmonics in the frequency domain is sometimes useful, both to analyze content (Figure 3.8) and the response of various equipment (Figure 3.9) to passage of harmonics, and to evaluate the feasibility of different equipment applications of filter designs to mitigate harmonic propagation.

---

[8] A weight index of the frequencies over which human hearing is sensitive and for which a telephone system needs to operate.

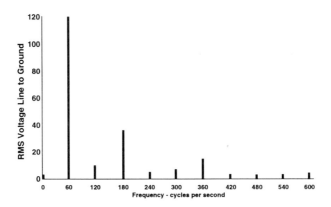

Figure 3.8 Frequency domain examination of voltage shows the fundamental frequency and the presence of several harmonics which vary in magnitude.

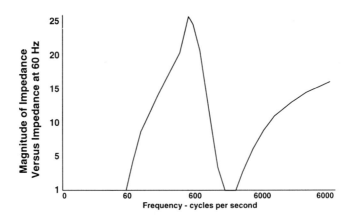

Figure 3.9 Frequency response of a distribution feeder with one capacitor bank.

Availability and Power Quality

Harmonics Curves and Harmonics Diversity

So many electrical devices generate harmonics that some amount of harmonic distortion of voltage is certain to exist within the wiring of nearly every household, business, and industrial location. With no equipment in place to block or trap their propagation onto the T&D system, this means the customer level is generally contaminated with measurable amounts of harmonics. Thus, even if a particularly customer produces no harmonics, one of more of his neighbors is sure to have an abundance to share with him.

Like load itself, harmonics generation within a household, business, or industrial plant varies as a function of time, and can be plotted as "harmonics curves" much like the load curves discussed in Chapter 2, as shown in Figure 3.10. This variation in harmonics output is caused by the same end-use schedules that create the load profiles themselves: usage of appliances -- the source of harmonics -- varies with time of day, day of week, and season as the demand for the product of the appliance varies. However, harmonics is not proportional to load or necessarily a function of appliance usage, but often depends on other factors including changes in system impedance, as shown below.

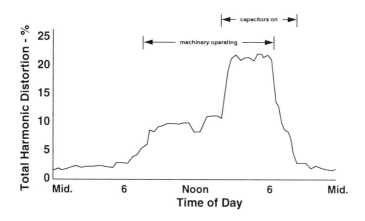

Figure 3.10 Daily schedule of total harmonic distortion (voltage) measured at the meter of a small business that operates its equipment from 8 AM to 6 PM, as shown. Harmonics are a problem whenever equipment is operating, but are worse in the afternoon after 3 PM, during a period when a time-switched distribution capacitor bank nearby has been activated, creating resonances that increase the harmonics.

The harmonics added to a power system in a neighborhood or on a feeder are subject to a behavior very similar to the coincidence (diversity) discussed for peak demand levels in Chapter 2. As with peak demand, the peak harmonics created by a group of customers is not equal to the sum of the individual peak harmonics levels they can cause, but less, due to three phenomena:

Temporal diversity. This is identical in concept and application to coincidence or diversity of peak load: different customers produce their peak harmonics output at different times. Like load, when examined at high temporal resolution, peak harmonics periods may last only a few minutes at a time (while the offending appliance is on), cycling up and down in level with the cycling of the appliance.

Peak harmonics level for a group of customers occurs when the sum of the customer peaks is at a maximum, and because all may not peak at the same time, this may be much less than the sum of the individual customer harmonics peaks.

Phase diversity. Suppose that two otherwise identical switched power supplies connected to the same power outlet produce equal amounts of 3rd-order harmonics, but due to triggering differences or differences in their circuitry, in reverse polarity to one another (i.e., with phases 180 degrees apart). Their harmonic outputs would cancel completely.

Given the manifold sources of harmonics on a power system, there are a wide range of phase relationships in the harmonics created, and a certain amount of cancellation occurs. However, all switched power supplies see the same waveform, which tends to synchronize their switching (they all trigger on voltage rise) somewhat. Similarly, many other sources are synchronized somewhat, with diversity in phasing occurring due to differences in equipment specifications, etc.

Attenuation can be thought of as the harmonic devices' impact on their own ability to generate harmonics. If a sufficient number of similar harmonics-causing devices are installed, to the point they begin to be significant with respect to the system impedance as seen at their location, they distort the voltage they see (the voltage which drives them to produce harmonics) and in many cases this reduces their individual harmonic contributions. For example, a single 120 W switched power supply (as in a typical desktop PC) might typically create about .9 amp of 3rd harmonics current injection (along with currents at other harmonics, too). However, if several hundred are installed at one site (perhaps a heavily computerized office building) -- enough so that total

Availability and Power Quality

current draw of these devices exceeds 1% of the short circuit current levels at that location -- the 3rd harmonic current injected will average only .80 amps per device, an 11% reduction. The devices distort the voltage waveform in a manner counter to producing more harmonics, thus limiting their harmonics capability *en masse*. Higher order harmonics are reduced even more by attenuation -- under similar circumstances the fifth harmonic is reduced by a factor of roughly 40% and the ninth by nearly 66%.

As a result of these three phenomena, harmonic contribution per device decreases as more devices are included, as shown in Figure 3.11. In general, temporal and phase diversity tend to be the major reasons why lower order harmonics decrease with increasing "group size," whereas attenuation is the major phenomenon that reduces the magnitude at higher harmonic levels.

A harmonics coincidence factor (HC) can be formed to measure the coincidence of harmonic injection as

$$HC = \frac{\text{observed THD of a group of customers}}{\Sigma \text{ (their individual THD)}} \quad (3.2)$$

In general, in the presence of very many similar sources, the level of harmonics in a system added by each device attains a saturation level "fully coincident" value. which reduces the contribution of further sources to a "fully diversified" value, usually about 70% of individual contribution.

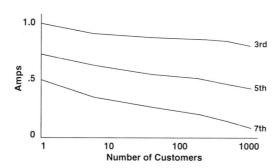

Figure 3.11 Net additional harmonic current produced per PC as a function of number of PCs added to an office building with a load of 1,600 kVA.

Standards on Harmonic Content

Harmonics become a problem only when they interfere with proper operation of equipment, increase current levels to the point of saturating or overheating equipment or causing other similar problems. Common *perceivable* problems caused by harmonics are "buzzing" in radios and stereo systems, and poor picture quality on televisions and VCRs due to the harmonic distortion passing through inadequate filtering into the appliance circuitry. Additionally, with proper equipment *detectable* problems are increased losses and higher thermal and electrical stresses on equipment.

Considering the ubiquitous sources of harmonics, it is surprising that more customers do not have problems with them. The real worry in cases where there are not immediately identified problems, however, is that there may be either or both of two types of inchoate problems. First, harmonics mostly do damage to equipment by causing overheating of windings and circuitry, an action that can destroy equipment through accelerated loss of life -- damage may be taking place while not yet recognized. Second, the level of harmonics present might be just below some threshold, above which it would cause problems, and above which it might rise at any moment. The operative concern here is that harmonics sensitivity is not linear, but instead there is some threshold below which it is tolerable and above which it is not. This is often the case, although there is no hard and fast rule.

IEEE Standard 519-1992 provides recommended guidelines for the amount of harmonics that can be produced as well as what amount should be permitted to flow on the power system. It specifies recommended limits on how much harmonic injection customers should be allowed to make into the system, as shown in Table 3.2.

Table 3.2 Harmonic Current Limits (IEEE Standard 519-1992) for Customer Current Injection into the System as a Function of the Ratio of Short Circuit Current to Normal Load Current - %

I_{sc}/I_{load}	2-11	11-16	17-22	23-34	> 34	Total
<20	4.0	2.0	1.5	.6	.3	5.0
20 - 50	7.0	3.5	2.5	1.0	.5	8.0
50 - 100	10.0	4.5	4.0	1.5	.7	12.0
100 - 1000	12.0	5.5	5.0	2.0	1.0	15.0
> 1000	15.0	7.0	6.0	2.5	1.4	20.0

Availability and Power Quality

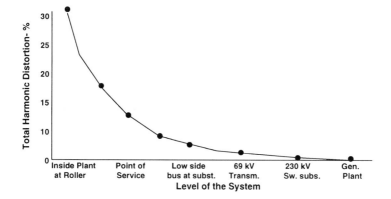

Figure 3.12 A certain amount of harmonics tend to be everywhere within a power system, but they are usually concentrated nearest their source. Shown is the measured total harmonic distortion at points along a backpath from a precision rolling mill inside a sheet metal factory through the power system to the nearest generating station.

Standard 519-1992 also sets limits on the total harmonic distortion that should exist anywhere on the power system, and on the amount of any one harmonic that can be present anywhere on the system. It recommends no more than 5.0% total harmonic distortion (THD) on the distribution level (all voltages between 2.3 and 69 kV), with no more than 3% distortion attributable to any one harmonic. Recommended THD level is lower at higher voltages -- 2.5% on 69-138 kV portions of the system with no one harmonic contributing more than 1.5%, and 1.5% total with no more than 1% for any one harmonic at higher voltages. In general, THD levels on a power system are lower at high voltages, because these are farther (electrically) from the harmonic sources and except in rare cases harmonic energy is dissipated as it makes its way through the system (Figure 3.12).

Harmonic Power Flow

Harmonic power flows from its generating source(s) through the power system to nearby loads, obeying exactly the same laws of natural physical behavior as 50 or 60 cycle power. It is often a "counter flow" moving from the customer up onto the higher voltage T&D system. Although harmonics are "higher" in frequency than 60 cycle power, they are quite low with respect to the frequency bandpass limits of utility equipment, so that they pass through transformers, motors of all types, and most other equipment with little attenuation. The exception is equipment built specifically to block or absorb harmonic distortion, and certain types of transformer combinations (delta-wye) which will force certain harmonics to "cancel themselves out" due to phase differences.[9]

In addition, feeders with shunt or series capacitors, high cable line charging, or severe unbalance, and long transmission lines with significant series capacitance can amplify harmonics. The capacitance causes a resonance at certain frequencies, and as a result these lines can carry several times the harmonic currents injected into them (Mahmoud and Shultz, McGranaghan et al.). The feeder whose impedance was plotted in Figure 3.9 clearly had an impedance that varies with frequency (making it a harmonic generator in the second category), and in addition has a resonance near the 13th harmonic to the extent that it propagates that in amplified form throughout part of its length.

Harmonic Analysis Methods

There are a wide variety of analytical methods used to study harmonics and evaluate potential "fixes." All harmonics analysis methods employ approximations, linearizations, or other computational "short-cuts" of one type or another, and each has advantages and disadvantages in comparison to the other methods: none is best in all situations. Occasionally, two or more methods will give slightly different analytical results when applied to the same problem, and in rare situations may even give contra-indicating recommendations on how to reduce the harmonics. In general, these methods can be grouped into four broad categories:

> *Frequency scanning methods* study the harmonics behavior of a circuit, appliance, or system in a series of discrete frequency steps, in each step using an appropriately "tuned" frequency-dependent model of the system or bus being analyzed. In this approach, a system might be analyzed for harmonics potential at 60, 120, 180, 240, 300 Hz and so forth. This method is the most appropriate for analysis of possible resonance conditions that augment harmonics, and for analysis of filter design (Arrigula et al., Jiang and Gole).

---

[9] Delta-wye transformer combinations attenuate third harmonic power flow, but not that of all other harmonics. The third harmonic is the primary product of voltage clipping and other switched AC-DC power supplies, and thus quite prevalent, making this "curve" often quite effective.

Frequency scanning can be done in conjunction with a load flow analysis applied on an individual harmonic-basis, for example individually analyzing the flow of the 60, the 120, 180 and 240 hertz levels by applying the correct harmonic impedance's of equipment and circuit elements at each discrete frequency, while representing harmonic generation at its source. While approximate, this approach has the advantage that normal load flow programs can be used to analyze harmonic flow, and sometimes it will help identify the extent of flow of harmonics through the system.

Unfortunately, frequency scanning often fails to properly diagnose harmonic problems for a number of reasons -- among others, while harmonic flows and normal power flows are additive, superposition of harmonics to a system already loaded to near maximum by "normal" power flow can cause saturated and other non-linear load problems. The frequency scanning method seems to be best for evaluation of circuits and systems to identify if harmonics might be a problem, where resonances or problems might exist.

Linearized analysis, usually applied with current injection techniques, often represents harmonic sources as composed of various square waves and the equipment and circuitry that must be represented as series and shunt connected sets of (sometimes switched) linear devices (or at least linear within each frequency range). Once linearized in pieces, the system state is formed as the superposition of a series or set of such "harmonic sources."

This method's advantages are is relative simplicity, and a reasonably good representation of why and how harmonics are created and how they propagate. Models of this type are often built after the general nature of harmonics at a particular site are known, as a model to study harmonic behavior and propagation in a more detailed manner.

Non-linear time domain analysis directly applies non-linear load and equipment temporal (time domain) simulation models, in programs such as EMTP (Electro-Magnetic Transients Program) as well as what are called time simulation harmonic power flow models, which compute power flow spectrum using non-linear load models and rather standard electric equipment line models with their impedance represented throughout the spectrum as needed. (EPRI, 1983). Regardless of whether implemented in EMTP or another approach, this type of analysis method "steps" through a simulation of the system being studied in time, as discrete intervals, modeling the transition of voltages, currents, fields, and energy from one to the next.

While the EMTP is overkill in many cases (transient analysis is not required), EMTP is often the best approach for analysis of severe problems, because that approach has a very wide bandwidth of model validity and its ability to represent complex interactions of equipment and energy. It is a preferred method when evaluating "transient" harmonics such as those caused by transformer inrush currents, etc.

Wavelet methods apply analytical techniques based on wavelet theory. Like frequency domain analysis, wavelet analysis applies a series of mathematically orthogonal wave shapes to the representation of signal shape and sensitivity, but these differ in scale, not frequency. Wavelets have both time and frequency domain representations, if one wants to look at them in such a way, and thus the method can provide analysis in both domains (Pillay and Bhattacharjee).

Often, it is best to apply two or more of the analytical methods discussed above to the evaluation of a harmonics problem -- results do differ, and while it can be frustrating to have to deal with two or more different models that give different "answers," this often identifies both the limits of knowledge about the problem and a wider range of possible solutions to be explored.

"Fixes" for Harmonics

Usually, the solution to a harmonics problem is to eliminate the symptoms, not the source of the harmonics, for while the devices creating harmonics often constitute only a small portion of the load as measured in kW, they arguably constitute a majority of the *value* contributed by electric usage: eliminating their usage is unfeasible. Modifying them so that they do not cause harmonics is usually unaffordable. What remains is to reduce the symptoms, either by increasing the tolerance of equipment and system to harmonics, or by modifying circuitry and systems to reduce their impact, or trapping or blocking them with filters. Of course, there are exceptions. In cases of overloaded or damaged equipment, or inappropriate design, the cause of harmonics can be fixed. Similarly, if investigation leads to a single device or appliance that is producing unusually high levels of harmonics, it can be modified or replaced.

Modified appliance design

Often it would cost very little -- only pennies -- to modify the design of an electronic device or its power supply before it is built so that the harmonic contamination it causes is reduced by an order of magnitude or more. In most consumer and small business electronic products, full wave rectification of

Availability and Power Quality 105

power with larger than minimum-size capacitors, and simple analog filters and traps, could reduce harmonic output, particularly of the 3rd harmonic and its multiples, by more than an order of magnitude.

However, there are no standards or enforceable requirements on harmonics output in most consumer and business product categories (to the author's knowledge, certain medical and military equipment are the only exceptions, largely by virtue of shielding and accuracy standards that imply low harmonic content). The drive for low cost means that few appliance manufacturers will accommodate such changes.[10] And while such modifications cost only a few pennies if done in the original design, retrofitting them can be much more costly, and is generally not a viable option.

Overall, the industry recognizes recommendations that appliances and electrical devices be able to function in the presence of a 5% voltage THD. However, some manufacturers do not follow this guideline. Given that the recommended standards for utility distribution-level harmonics are 5% or less, this tolerance level for appliances leaves no margin for any additional harmonics created inside a customer's site. In addition, the 5% targets on both sides are only recommendations. While many appliances tolerate much more than 5% THD, others will malfunction on even less.

Increasing the harmonic tolerance of an appliance by modifying its design to withstand a higher ambient harmonics level is usually slightly more expensive than installing additional components or modifying its design to reduce its own harmonic output. The reason is that "hardening" the device means protecting it against significant levels of harmonics over a wide range of frequencies, whereas mitigating its own output means merely reducing its specific template of harmonics output, which is generally a lesser target.

Perhaps the most impressive fact about harmonics is that so few customers experience significant harmonics problems, even though THD levels in the average home or small commercial office may reach 10% (of voltage) or over 200% (of current). However, when problems occur they are usually severe -- overheating, malfunctioning equipment, and even failure due to overcurrent, and modifying the appliance(s) is not a viable solution to the problem.

There may be more than one cause

One point often overlooked is that harmonics may have recently become a problem due to the additive or multiplicative effect of two causes, either of

---

[10] The author is not holding the appliance manufacturer's to blame -- would consumers pay even a small amount more for "harmonics friendly" appliances if they were offered on the market? Most likely, no.

which would not cause the problem individually, and one of which only recently occurred. For example, the harmonic distortion caused by a loaded induction motor using air circulation within an agricultural packaging plant may have been tolerated for many years, yet may suddenly cause flicker problems because the ground of the in-plant circuit feeding it has corroded open. In the author's experience, two or more contributing factors are often linked in any case of severe harmonics problems, particularly when no new equipment has recently been added. In cases where recent equipment changes have been made, those changes are the suspect. In addition, when there are multiple-simultaneous causes, usually only one is the major generator of harmonics, and the other factors merely contribute by creating resonances or aiding in their propagation.

Maintenance and inspection

The recommended first step in any search for harmonics (or power quality problems in general, for that matter) is thorough inspection of existing equipment and circuitry. Many harmonics problems are either caused or worsened by unbalanced loads, poor grounds and neutrals, inappropriately applied equipment (i.e., an induction motor's reduced voltage starting system that has been modified and is used as a "speed control") and other equipment-related problems. These can be identified by careful inspection of appropriate equipment. The author's experience is that inspection and maintenance identifies damaged or weak equipment as a contributor in about 25% of all harmonics cases.

Improved grounding and neutral capacity

From the standpoint of both voltage surges and harmonics, increasing the grounding (reducing impedance to ground and increasing neutral ampacity) often cures power quality problems, including harmonics, even when detailed analytical evaluation is uncertain as to cause or curve. Poor or inadequate grounding is a contributor in 33-40% of all power quality problems. The proliferation of DC electronics equipment and switched power suppliers in modern loads means that traditional grounding requirements and expected ground currents are inadequate (Andres).

Filters and shunts

Filters to block or trap harmonic energy so that it does not flow into equipment or onto the system are among the most popular harmonic "solutions," largely because filters are problem specific and they can be retro-fitted where necessary.

Availability and Power Quality

It is worth noting that filters, by their very nature, are devices whose impedance varies with frequency. Thus, they have the potential to create and amplify harmonics problems unless carefully located and tuned -- in some cases poor diagnosis and design has led to the solution being worse than the problem.

Passive filters are the simplest, least expensive, but least flexible and effective type of harmonics filter. They are unpowered (i.e., purely passive power) devices, often applied by the utility as shunt circuits across the customer service entrance with a goal of preventing harmonics generating inside the customer site from flowing onto the T&D system. Alternatively, filters may be installed directly on a particular piece of equipment if the harmonics generated by it are severe, and the desire is to prohibit their flow to the rest of the circuitry inside the customer location.

The performance of passive filters is quite sensitive to the system impedance for which they must be exactly tuned. System impedance can change over time as volt/VAR equipment alters its status, and it is difficult to estimate accurately even with the benefit of substantial measurements. Thus, passive filters often do not provide satisfactory performance. Under some cases they can also cause unwanted resonances on the system.

Active filters are powered devices, which work by applying a shunt power converter to produce harmonic currents equal to those in the load current, effectively routing them off the path into the system (Figure 3.13). Their harmonics reduction depends only on their correctly measuring the harmonic

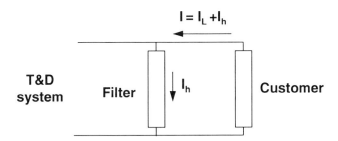

Figure 3.13 Filters are usually applied as a shunt across the customer site or the equipment creating harmonics, as shown here. Both passive and active filters remove the harmonics current (I_h) by routing it off the path into the system, leaving only the load current (I_L) to flow into the system: passive filters by creating a very low impedance shunt path to attract it, active filters by "copying" the harmonic flow current with a current they generate themselves, essentially forcing it along their shunt path.

current emanating from the load, and is not a function of the system impedance. Therefore, they prove much more robust in application. Unfortunately, they can be quite expensive, they consume noticeable amounts of power, and can create quite high electromagnetic interference (EMI) levels.

Hybrid filters, in which active and passive filters are placed in series as a shunt across the customer load, often combine robust performance with much lower cost and power consumption.

System or Customer-Side Re-Design

Generally, installation of larger capacity T&D equipment, particularly larger transformers, reduces harmonics problems by cutting source impedance and proportionally reducing the harmonics voltage distortion. In wye-connected systems neutral capacity should be sized to $\sqrt{3}$ times (173%) of the phase rating, due to the possibility of significant harmonics flows along the shared neutral path. Use of a delta-connected transformer will provide a circulating path for the 3rd harmonic and all its multiples (6th, ninth, etc.), shunting any harmonics of order 3×N generated on the customer side into the transformers windings. Harmonics losses (heat) generating in the transformer windings as a result can be significant and the transformer must be sized appropriately. As 3×N harmonics are a significant by-product of switched power supplies, this approach often works well in cases where computer or electronic equipment is the cause of harmonics generation.

In cases where investigation and analysis shows that the T&D system propagates harmonic flow, modifications may have to be made. Generally, line capacitors and charging capacitance of cable are the major contributors.

3.6 CUSTOMER VALUE OF AVAILABILITY AND POWER QUALITY

As mentioned in this chapter's introduction, the central issue in customer value of service analysis is matching availability and power quality against cost. T&D systems with near perfect availability and power quality can be built, but their high cost will mean electric prices that the utility customers may not want to pay, given the savings an even slightly less reliable system would bring. All types of utilities have an interest in achieving the correct balance of quality and price. The traditional franchised monopoly utility, in its role as the "electric resource manager" for the customers it serves, has a responsibility to build a system whose quality and cost balances its customers' needs. A competitive retail distributor of power wants to find the best quality-price combination: only in that way will it gain a large market share.

Availability and Power Quality

While it is possible to characterize various power quality problems in an engineering sense, characterizing them as interruptions, voltage sags, dips, surges, or harmonics, the customer perspective is somewhat different. Customers are concerned with only two aspects of service quality:

- They want power when they need it.

- They want the power to do the job.

If power is not available, it fails to provide both of these. If it is available but the quality is low, it fails to do the second.

Assessing Value of Quality by Studying the Cost of a Lack of It

In general, customer value of reliability and service quality are studied by assessing the "cost" that something less than perfect reliability and service quality creates for customers. Electricity provides a value, and interruptions or poor power quality decrease that value. This value reduction -- cost -- occurs for a variety of reasons, some of whose costs are difficult if not impossible to estimate: re-scheduling of household activities or lack of desired entertainment when power fails;[11] or flickering lights that make reading more difficult.

But often, very exact dollar figures can be put on interruptions and poor power quality: food spoiled due to lack of refrigeration; wages and other operating costs at an industrial plant during time without power; damage to product caused by the sudden cessation of power; lost data and "boot up" time for computers; equipment destroyed by harmonics; and so forth. Figure 3.13 shows two examples of such cost data.

Value-Based Planning

To be of any real value in utility planning, information of the value customers put on quality must be usable in some analytical method that can determine the best way to balance quality against cost. Value-based planning (VBP) is such a method: it combines customer-value data of the type shown in Figure 3.14 with data on the cost to design the T&D system to various levels of reliability and power quality, in order to identify the optimum balance. Figure 3.15 illustrates the central tenet of value-based planning. The cost incurred by the customer due

---

[11] No doubt, the cost of an hour-long interruption that began fifteen minutes from the end of a crucial televised sporting event, or the end of a "cliffhanger" movie, would be claimed to be great.

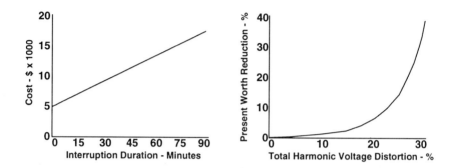

Figure 3.14 Left, cost of a week-day interruption of service to a pipe rolling plant in the southeastern United States, as a function of interruption duration. An interruption of any length costs about $5,000 -- lost wages and operating costs to unload material in process, bring machinery back to "starting" position and restart -- and a nearly linear cost thereafter. At the right, present worth of the loss of life caused by harmonics in a 500 horsepower three-phase electric motor installed at that same industrial site, as a function of harmonic voltage distortion.

Figure 3.15 Concept of value-based planning. The customer's cost due to poorer quality (left) and the cost of various power delivery designs with varying levels of quality (center) are computed over a wide range. When added together (right) they form the total cost of quality curve, which identifies the minimum cost reliability level (point A).

Availability and Power Quality 111

to various levels of reliability or quality, and the cost to build the system to various levels of reliability, are added, to get the total cost of power delivered to the customer, as a function of quality.[12] The minimum value is the optimum balance between customer desire for reliability and aversion to cost. This approach can be applied for only reliability aspects, i.e., value-based reliability planning, or harmonics, or power quality overall. Generally, what makes sense is to apply it on the basis of whatever qualities (or lack of them) impact the customer -- interruptions, voltage surges, harmonics, etc. -- in which case it is comprehensive value-based quality of service planning.

Cost of Interruptions

The power quality issue that affects the most customers, and which receives the most attention, is cessation of service, often termed "service reliability." Over a period of several years, almost all customers served by any utility will experience at least one interruption of service. By contrast, a majority will never experience serious harmonics, voltage surge, or electrical noise problems. Therefore, among all types of power quality issues, interruption of service receives the most attention from both the customers and the utility, and a great deal more information is available about cost of interruptions than about cost of harmonics or voltage surges.

Voltage Sags Cause Momentary Interruptions

The continuity of power flow does not have to be completely interrupted to disrupt service: if voltage drops below the minimum necessary for satisfactory operation of an appliance, power has effectively been "interrupted" as illustrated in Figure 3.16. For this reason many customers regard voltage dips and sags as momentary interruptions -- from their perspective these are interruptions of the end-use *service they desire,* if not of voltage.

Much of the electronic equipment manufactured in the United States, as well as in many other countries, have been designed to meet or exceed the CBEMA (Computer and Business Equipment Manufacturer's Association) recommended curves for power continuity, shown earlier in Figure 3.2. If a disturbance's voltage deviation and duration characteristics are within the CBEMA envelope, then normal appliances should operate normally and satisfactorily despite the it.

---

[12] Figure 3.15 illustrates the concept of VBP. In practice, the supply-side reliability curves often have discontinuities and significant non-linearities that make application difficult. These and other details will be discussed in Chapter 5.

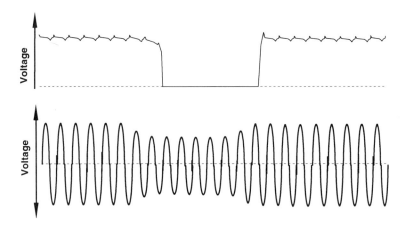

Figure 3.16 Output of a 5.2 volt DC power supply used in a desktop computer (top) and the incoming AC line voltage (nominal 113 volts). A voltage sag to 66% of nominal causes power supply output to cease within three cycles.

However, many appliances and devices in use will not meet this criterion at all. Others will fail to meet it under the prevailing ambient electrical conditions (i.e., line voltage, phase unbalance power factor and harmonics may be less than perfect).

The manner of usage of an appliance also affects its voltage sag sensitivity. The voltage sag illustrated in Figure 3.16 falls just within the CBEMA curve. The manufacturer probably intended for the power supply to be able to withstand nearly twice as long a drop to 66% of nominal voltage before ceasing output. However, the computer in question had been upgraded with three times the standard factory memory, a second and larger hard drive, and optional graphics and sound cards, doubling its power usage and the load on the power supply. Such situations are common and mean that power systems that deliver voltage control within recommended CBEMA standards may still provide the occasional momentary interruption.

For all these reasons, there are often many more "momentary interruptions" at a customer site than purely technical evaluation based on equipment specifications and T&D engineering data would suggest. Momentary interruptions usually cause the majority of industrial and commercial interruption

Availability and Power Quality 113

problems. In addition, they can lead to one of the most serious customer dissatisfaction issues. Often, utility monitoring and disturbance recording equipment does not "see" voltage disturbances unless they are complete cessations of voltage, or close to it. Many events that lie well outside the CBEMA curves and definitely lead to unsatisfactory equipment operation, are not recorded or acknowledged. As a result, a customer can complain that his power has been interrupted five or six times in the last month, and the utility will insist that its records show power flow was flawless. *The utility's refusal to acknowledge the problem irks some customers more than the power quality problem itself.*

Frequency and Duration of Interruptions Both Impact Cost

Traditional power system reliability analysis recognizes that service interruptions have both frequency and duration (See Chapter 5). Frequency is the number of times during some period (usually a year) that power is interrupted. Duration is the time power is out of service. Typical values for urban/suburban power system performance in North America are 2.2 interruptions per year with 100 minutes total duration.

Both frequency and duration of interruption impact the value of electric service to the customer and must be appraised in any worthwhile study of customer value of service availability. A number of reliability studies and value-based planning methods have tried to combine frequency and duration in one manner or another into "one dimension." A popular approach is to assume all interruptions are of some average length (e.g., 2.2 interruptions and 100 minutes is assumed to be 2.2 interruptions per year of 46 minutes each). Others have assumed a certain portion of interruptions are momentary and the rest of the duration is lumped into one "long" interruption (i.e., 1.4 interruptions of less than a minute, and one 99-minute interruption per year). Many other approaches have been tried (see References and Bibliography). But all such methods are at best an approximation, because frequency and duration impact different customers in different ways -- no single combination of the two aspects of reliability can fit the value structure of all customers.

Figure 3.17 shows four examples of the author's preferred method of assessing interruption cost, which is to view it as composed of two components, a fixed cost (Y intercept) caused when the interruption occurred, and a variable cost that increases as the interruption continues. As can be seen in Figure 3.17, customer sensitivity to these two factors varies greatly. The four examples are:

A pipe-rolling factory (upper left). After an interruption of any length, material in the process of manufacturing must be cleared from the rolling

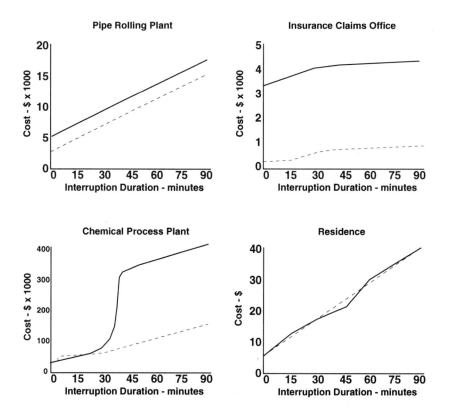

Figure 3.17 The author's recommended manner of assessing cost of interruptions includes evaluation of service interruptions on an event basis. Each interruption has a fixed cost (Y-intercept) and a variable cost which increases as the interruption continues. Examples given here show the wide range of customer cost characteristics that exist. The text gives details on the meaning of solid versus dotted lines and the reasons behind the curve shape for each customer.

Availability and Power Quality 115

and welding and polishing machinery, all of which must be reset and raw material feed set up to begin the process again. This takes about 1/2 hour and sets a minimum cost for an interruption. Duration longer than that is simply a linear function of the plant operating time (wages and rent, etc., allocated to that time). Prior to changes made by a reliability study, the "re-setting" of the machinery could not be done until power was restored (i.e., time during the interruption could not be put to use preparing to re-start once it was over). The dotted line shows the new cost function after modifications to machinery and procedure were made so that preparations could begin *during* the interruption.

An insurance claims office (upper right) suffers loss of data when power fails, equivalent to roughly one hour's processing. An unexpected power interruption causes loss of about one hour's work, according to the site supervisor, who estimates that "another half hour" is lost due to interruption of steady work flow and the impact of any interruption on the staff -- thus the fixed cost of each interruption is equivalent to about ninety minutes of work. After one-half hour of interruption, the supervisor's policy is to put the staff to work "on other stuff for a while," making cost impact lower (some productivity); thus, variable interruption cost goes down. The dotted line shows the cost impact of interruptions after installation of a UPS on the computer system, which permits orderly shut-down in the event of an interruption.

An acetate manufacturing and processing plant (lower left) has a very non-linear cost curve. Any interruption of service causes $38,000 in lost productivity and after-restoration set-up time, and cost rises slowly for about half an hour. At that point, molten feedstock and interim ingredients inside pipes and pumps begins to cool, requiring a day-long process of cleaning sludge and hardened stock out of the system. The dotted line shows the plant's interruption cost function after installation of a diesel generator, started whenever interruption time exceeds five minutes.

Residential interruption cost function (lower right), estimated by the author from a number of sources including a survey of customers made for a utility in the northeastern United States in 1992, shows roughly linear cost as a function of interruption duration, save for two interesting features: a) a fixed cost equal to about eight minutes of interruption at the initial variable cost slope which reflects "the cost to go around and re-set our digital clocks," as one respondent put it, along with similar inconvenience costs, and b) a jump in cost between 45 and 60 minutes, which reflects inconsistencies in human reaction to outage time on questionnaires (Apparently, homeowners tend to answer questions while thinking of short interruptions as not having a high cost, but an hour seems to exceed the minimum period that qualifies as "long" and thus cost rises substantially in their mind). The dotted line shows the relation the author uses in his analysis, which makes adjustments thought reasonable to account for these inconsistencies.

This recommended analytical approach, in which cost is represented as a function of duration on a per event basis, requires more information and more analytical effort than simpler "one-dimensional" methods, but the results are much more valid.

Interruption Cost is Lower if Prior Notification is Given

Given sufficient time to prepare for an interruption of service, most of the momentary interruption cost (fixed) cost and a great deal of the variable cost, can be eliminated by many customers. Figure 3.18 shows the interruption cost figures from Figure 3.17 adjusted for "24 hour notification given."

Cost of Interruption Varies By Customer Class

Cost of power interruption varies among all customers, but there are marked distinctions among classes, even when cost is adjusted for "size" of load by computing all cost functions on a per KW basis. Generally, the residential class has the lowest interruption cost per kW, and commercial the highest. Table 3.3 gives the cost/kW of a one-hour interruption of service by customer class, obtained using similar survey techniques for three utilities in the United States: 1) a small municipal system in the central plains, 2) an urban/suburban/rural system on the Pacific Coast, and 3) an urban system on the Atlantic coast.

Table 3.3 Typical Interruption Costs by Class for Three Utilities -- Daytime, Weekday (dollars per kilowatt hour)

Class	1	2	3
Agricultural	3.80	4.30	7.50
Residential	4.50	5.30	9.10
Retail Commercial	27.20	32.90	44.80
Other Commercial	34.00	27.40	52.10
Industrial	7.50	11.20	13.90
Municipal	16.60	22.00	44.00

Availability and Power Quality

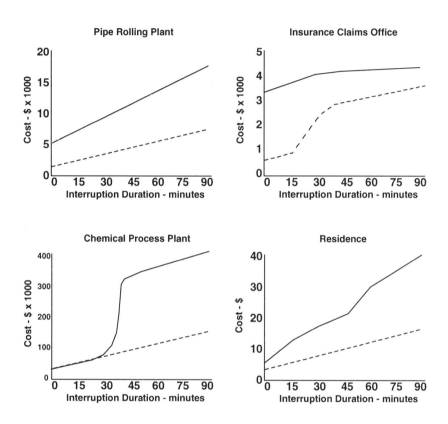

Figure 3.18 If an interruption of service is expected, customers can take measures to reduce its impact and cost. Solid lines are the interruption costs (the solid lines from Figure 3.17). Dotted lines show how 24-hour notice reduces the cost impact in each case.

Cost Varies From One Region to Another

Interruption costs for apparently similar customer classes can vary greatly depending on the particular region of the country or state in which they are located. There are many reasons for such differences. The substantial difference (47%) between industrial costs in utilities 1 and 3 shown in Table 3.3 is due to differences in the type of industries that predominate in each region. The differences between residential costs of the regions shown reflect different demographics and varying degrees of economic health of their respective regions.

Cost Varies Among Customers Within a Class

The figures given for each customer class in Table 3.3 represent an average of values within those classes as surveyed and studied in each utility service territory. Value of availability can vary a great deal among customers within any class, both within a utility service territory and even among neighboring sites. Large variations are most common in the industrial class, where different needs can lead to wide variation in the cost of interruption, as shown in Table 3.4. Although documentation is sketchy, indications are that *the major differing factor is the cost of a momentary interruption* -- some customers are very sensitive to any cessation of power flow, others are impacted mainly by something longer than a few cycles or seconds.

Table 3.4 Interruption Costs by Industrial Sub-Class for One hour, Daytime, Weekday (dollars per kilowatt)

Class	$/kW
Bulk plastics refining	38
Cyanide plant	87
Weaving (robotic loom)	72
Weaving (mechanical loom)	17
Automobile recycling	3
Packaging	44
Catalog distribution center	12
Cement factory	8

Availability and Power Quality

Cost of Interruption Varies as a Function of Time of Use

Cost of interruption will have a different impact depending on the time of use, usually being much higher during times of peak usage, as shown in Figure 3.19. However, when adjusted to a per-kilowatt basis, the cost of interruption can sometimes be higher during off-peak than during peak demand periods, as shown. There are two reasons. First, the data may not reflect actual value. A survey of 300 residential customers for a utility in New England revealed that customers put the highest value on an interruption during early evening (Figure 3.19). There could be inconsistencies in the values people put on interruptions (data plotted were obtained by survey).

However, there is a second, and valid, reason for the higher cost per kilowatt off-peak: only essential equipment, such as burglar alarms, security lighting and refrigeration is operating -- end-uses that have a high cost of interruption. Regardless, while it is generally safe to assume that the cost of interruption is highest during peak, the same cannot be assumed about the cost per kilowatt.

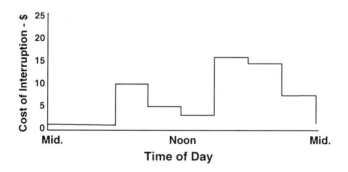

Figure 3.19 Cost of a one-hour interruption as a function of when it occurs, as determined by surveys and interviews with 300 customers of a utility in the northeastern United States, was determined on a three-hour period basis. A high proportion of households in this survey have school age children at home and thus perhaps weighed interruption costs outside of school hours more than during school hours. However, in general, most households rate interruption cost higher in early morning and early evening.

Recommended Method of Application of Customer Interruption Cost Data

As mentioned earlier, the recommended analytical approach to value-based planning of power delivery includes assessment of customer costs using functions that acknowledge both a fixed cost for any interruption, not matter how brief, and a variable cost as a function of duration. It is also important to acknowledge the differences in value of service among the different customer classes, and at times differences within those customer classes. In the case of large industrial customers, interruption cost can be significant enough that specific site studies of the customer's location, T&D system, and delivery history can be justified.

Ideally, reliability and service quality issues should be dealt with using a value-based reliability or power-quality planning method, with customer interruption cost data obtained through statistically valid and unbiased sampling of a utility's own customers (see Sullivan et al.), and not with data taken from a reference book, report, or technical paper describing data on another utility system. However, in many cases for initial planning purposes the cost and time of data collection are not affordable.

Figure 3.20 provides a set of costs of typical interruption curves that the author has found often match overall customer values in a system, but it is worth stressing that major differences can exist in seemingly similar utility systems, due to cultural and economic differences in the local customer base. These are not represented as average, or "best" for use in value-based planning studies, but they are illustrative of the type of cost functions usually found and they provide a guide for the preparation of approximate data from screening studies and initial survey data.

Cost in Figure 3.20 is given in terms of "one hundred times nominal price." *It is worth noting that in many surveys and studies of interruption cost, the cost per kW of interruption is on the order of magnitude one hundred times the normal price (rate) for a kWhr.* Generally, if a utility has low rates, its customers report a lower cost of interruption than if it has relatively higher rates. No reliable data about why this correlation exists has been forthcoming.[13]

---

[13] It could be that value of continuity is worth more in those areas where rates are high (generally, more crowded urban areas). However, the author believes that a good part of this correlation is simply because in putting a value on interruptions, respondents to surveys and in focus groups base their thinking on the price they pay for electricity. Given that a typical residential customer uses roughly 1,000 kWhr/month, they may simply be valuing an interruption as about "one tenth of my monthly bill."

Availability and Power Quality

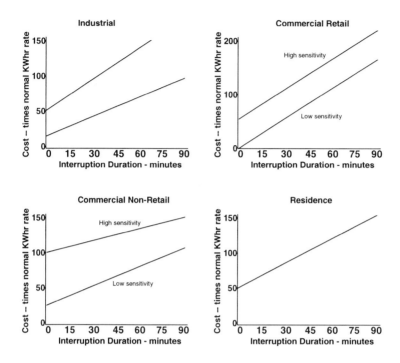

Figure 3.20 Typical interruption cost characteristics for customer classes.

Cost of Surges and Harmonics

Far less information is available on customer costs of harmonics and voltage surges as compared to that available on the cost of interruptions. What data made available in publications, and most of the reported results and interpretations in the technical literature, were obtained in very customer-specific case studies, most done on a single customer basis. As a result there is very little information available on average of typical costs of voltage surge sensitivity and harmonics sensitivity.

Few customers suffer from voltage dip (voltage problems lasting minutes or hours), surge, harmonics, electrical noise, and similar problems. For this reason most studies of cause, effect, and curve are done on a single-customer or specific area basis. Here, results reported in the technical literature are the best guide.

3.7 END-USE MODELING OF CUSTOMER POWER QUALITY NEEDS

The customer-class, end-use basis for analysis of electric usage, discussed in Chapters 2 and 15, provides a reasonably good mechanism for study of the service reliability and power quality requirements of customers, just as it provides a firm foundation for analysis of requirements for the amount of power. Reliability and power quality requirements vary among customers for a number of reasons, but two predominate:

> End-usage patterns differ: the timing and dependence of customers' need for lighting, cooling, compressor usage, hot water usage, machinery operation, etc., varies from one to another.

> Appliance usage differs: the appliances used to provide end-uses will vary in their sensitivity to power quality. For example, many fabric and hosiery manufacturing plants have very high interruption costs purely because the machinery used (robotic looms) is quite sensitive to interruption of power. Others (with older mechanical looms) put a much lower cost on interruptions

End-use analysis can provide a very good basis for detailed study of power quality needs. For example, consider two of the more ubiquitous appliances in use in most customer classes: the electric water heater and the personal computer. They represent opposite ends of the spectrum from the standpoint of both amount of power required and cost of interruption. A typical 50-gallon storage electric water heater has a connected load of between 3,000 and 6,000 watts, a standard PC a demand of between 50 and 150 watts. Although it is among the largest loads in most households, an electric water heater's ability to provide hot water is not impacted in the least by a one-minute interruption of power, and in most cases a one-hour interruption does not reduce its ability to satisfy the end-use demands put on it.[14] On the other hand, interruption of power to a computer, for even half a second, results in serious damage to its "product." Often there is little difference between the cost of a one-minute outage and a one-hour outage.

It is possible to characterize the sensitivity of most end-uses in most customer classes by using an end-use basis -- this is in fact how detailed studies of

---

[14] Utility load control programs offer customers a rebate in order to allow the utility to interrupt power flow to water heaters at its discretion. This rebate is clearly an acceptable value for the interruption, as the customers voluntarily take it in exchange for the interruptions. In this and many other cases, economic data obtained from market research for DSM programs can be used as a starting point for value analysis of customer reliability needs on a value-based planning basis.

Availability and Power Quality

industrial plants are done in order to establish the cost-of-interruption statistics which they use in VBP of plant facilities and in negotiations with the utility to provide upgrades in reliability to the plant. Following the recommended approach, this requires distinguishing between the fixed cost (cost of momentary interruption) and variable cost (usually linearized as discussed above) on an end-use basis.

A standard end-use model used to study and forecast electric demand can be modified to provide interruption cost sensitivity analysis, which can result in "two-dimensional" appliance end-use models as illustrated in Figure 3.21. Generally, this approach works best if interruption costs are assigned to appliances rather than end-use categories -- in commercial and industrial classes different types of appliances within one end-use can have wildly varying power reliability and service needs. This requires an "appliance sub-category" type of end-use model. Modifications to an end-use simulation program to accommodate this approach are straightforward (see Willis, 1996, Chapter 11), and not only provide accurate representation of interruption cost sensitivity, but produce analysis of costs by time and location, as shown in Figures 3.22 and 3.23.

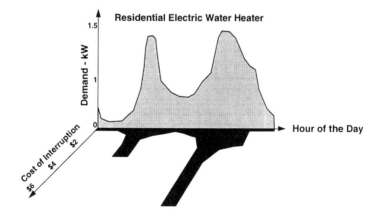

Figure 3.21 The simulation's end-use model is modified to handle "two-dimensional" appliance curves, as shown here for a residential electric water heater. The electric demand curve is the same data used in a standard end-use model of electric demand. Interruption cost varies during the day, generally low prior to and during periods of low usage and highest prior to high periods of use (a sustained outage prior to the evening peak usage period would result in an inability to satisfy end-use demand).

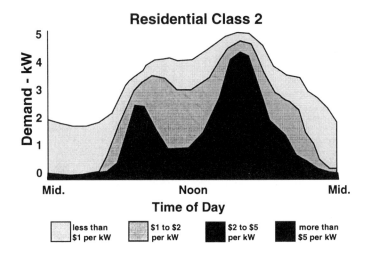

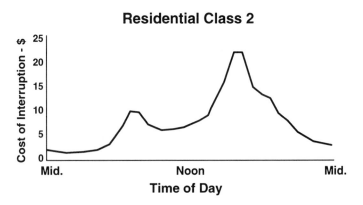

Figure 3.22 Result of a power quality evaluation using an end-use model. Top: the daily load curve for single family homes segmented into four interruption-cost categories. High-cost end-uses in the home are predominantly digital appliances (alarm clocks, computers) and home entertainment and cooking. Bottom: total interruption cost by hour of the day for a one hour outage -- compare to Figure 3.19.

Availability and Power Quality 125

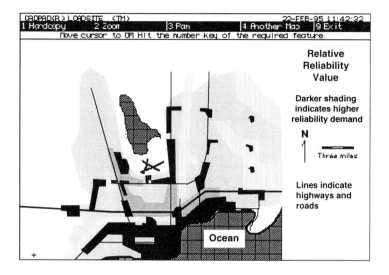

Figure 3.23 Map of average reliability needs computed on a 10-acre small area grid basis for a port city of population 130,000, using a combination of an end-use model and a spatial customer simulation forecast method, of the type discussed in Chapter 15. Shading indicates general level of reliability need (based on a willingness-to-pay model of customer value).

3.8 SUMMARY

In addition to having a demand for electric energy, electric consumers value availability and quality of power, too, seeking satisfactory levels of both. Each customer determines what is satisfactory based on their specific needs -- an interruption that occurs when a customer does not need power does not concern the; harmonics or surges that make no impact on appliances are not considered of consequence. Among the most important points in customer value analysis is:

Interruption of service is the most widely affecting power quality problem. A majority of electric customers in most utility system experience unexpected interruptions at least once a year.

Voltage dips and sags create momentary interruptions for many customers.

Harmonics create serious problems for only a small fraction of customers, but the problems are usually severe, the analysis can be involved, and solutions are often expensive.

Customer loads are the largest source of harmonics, which are created by many types of appliances and devices, particularly those with electronic power supplies. Harmonics are detectable within the internal wiring of most customer sites.

Analysis of customer needs for power quality is generally done by looking at the cost which less than perfect quality creates, not the value that near-perfect quality provides. Thus, availability of power is usually studied by analyzing the cost incurred due to interruptions, and harmonics is assessed by determining the costs created by them.

Interruption costs have two dimensions -- a fixed cost because an interruption occurred and a variable cost that increases the longer it lasts.

Momentary interruptions create a significant cost. There is a significant cost, irrespective of whether and how long the interruption of service may last.

Power quality costs and values vary greatly, depending on customer class, time of day and week, region of the country, and individual characteristics.

Data on customer costs may not be reliable. Most information about customer value of reliability and quality comes from survey and interview data. There have been cases, both in the electric industry and elsewhere, of customers replying one way in surveys and acting in quite another.

Price concerns often outweigh quality concerns. Throughout the late 1980s and early 1990s, several independent studies indicated that roughly "30-40% of commercial and industrial customers are willing to pay more for higher power quality" or something similar, a statistic that seems quite believable and realistic. This led to considerable focus on improving power system reliability and quality. However, it is worth bearing in mind that these same studies indicate that 20-30% are quite content with the status quo, while the remaining 30-40% would be interested in trading some existing quality for lower price.

Power quality value studies are best done with data specifically gathered on a utility's own customer base.

End-use analysis provides a useful tool for quantitative study of power quality issues and engineering.

REFERENCES

C. Andres, "Understanding Grounding Contamination," *Power Quality Assurance,* July 1995, p. 10.

J. Arrillaga et al., *Power System Harmonics,* John Wiley and Sons, New York, 1985.

E. Cinieri and F. Muzi, "Lightning Induced Overvoltages: Improvement in Quality of Service in MV Distribution Lines by Addition of Shield Wires," *IEEE Transactions on Power Delivery,* January 1996, p. 361.

Electric Power Research Institute, "Harmonic Power Flow Studies," EPRI EL-3300, November 1983.

IEEE Standard 519-1992, *Recommended Practices and Requirements for Harmonic Control in Electric Power Systems,* IEEE, New York, 1993

A. A. Mahmoud and R. D. Schultz, "A Method for Analyzing Harmonic Distribution in AC Power Systems," *IEEE Transactions on Power Apparatus and Systems,* June 1982, p. 1815.

M. F. McGranaghan, et al., "Distribution Feeder Harmonic Study Methodology," paper presented at IEEE/PES Transmission and Distribution Exposition, Kansas City, 1984.

P. Pillay and A Bhattacharjee, "Application of Wavelets to Model Short-Time Power System Disturbances," paper presented at 1996 IEEE Winter Power Meeting, Baltimore, and to appear in *Transactions on Power Systems.*

T. A. Short et al., Application of MOVs in the Distribution Environment, *IEEE Transactions on Power Delivery,* August, 1993.

M. Sullivan et al., "Interruption Costs, Customer Satisfaction and Expectations for Service Reliability," *IEEE Transactions on Power Systems,* May 1995, p. 989.

D. J. Ward et al., "Power Quality -- Two Different Perspectives," *IEEE Transactions of Power Delivery,* June, 1990.

H. L. Willis, *Spatial Electric Load Forecasting,* Marcel Dekker, New York, 1996.

4
Planning Criteria

4.1 INTRODUCTION

A viable distribution plan must not only provide good economy, but also satisfy a long list of standards and criteria for equipment, design, layout, loadings, or performance -- all of which the utility has determined are needed to achieve its goals. These standards and criteria, and their application to the planning process, depend somewhat on the utility's situation and its planning value system, or paradigm (See Chapter 18). Chief among traditional criteria are voltage and other service quality standards that assure the electrical needs of its customers will be met in a satisfactory manner. Safety is also a major consideration, and a number of standards or criteria on equipment loading, placement, clearance, application, and protection exist primarily to guarantee safe operation of the equipment and safe shut down and isolation in the event of failure. Protection and protection coordination have a large role in assuring reliability of service (reliability and reliability criteria will be discussed in Chapters 5 and 9), but a major function is to ensure as much as possible that equipment failures do not lead to any harm to the public or company personnel, damage to personal or company property, or damage to the equipment itself.

Additional criteria and standards exist for reasons of efficiency or the utility's convenience. These include specifications to ensure that the system is built with compatible equipment that will fit and function together, and that, when installed, will be maintainable in an economical and standardized manner.

Beyond these, utilities often implement a variety of other criteria and standards based on esthetics and other qualifying factors for fitting their system into the local community's needs.

Many criteria and standards directly address their reason for being, while others address their goals indirectly. For example, voltage standards directly define limits within which service is required. Operating voltage maintained within that limit *is* the goal of a voltage standard. By contrast, the loading criterion for a certain overhead conductor might specify that its nominal peak load be no more than 66% of its maximum thermal rating, because that level achieves the greatest economy of operation during normal situations (best balance of initial versus lower losses costs), and leaves a sufficient margin for unexpected load growth (planning flexibility), while providing a measure of extra capacity during emergency conditions (reliability). Many criteria are developed in this manner: their specification achieves one or more goals implicitly.

Together, the standards and criteria form a set of requirements against which the planning process can compare alternatives in the evaluation and selection phase. This chapter begins with a review of standards and criteria that address voltage, voltage spread, and fluctuation. It then examines how operating criteria and other design criteria influence distribution planning.

4.2 CRITERIA AND STANDARDS MUST BE MET, NOT EXCEEDED

Criteria and standards only have to be met, not exceeded, by any plan. What separates criteria and standards from other desirable *planning goals* is that they are fully satisfied if the distribution plan falls within their limits and there is no additional benefit perceived for doing anything more than satisfying their minimum or maximum requirements, as the case may be. By contrast, planning goals, such as *lower cost*, or *use as much of the system's existing margin as possible before building anything new,* are open-ended, and the planner is always challenged to do better no matter how well a plan rates versus these goals.

4.3 VOLTAGE AND CUSTOMER SERVICE STANDARDS

Reasons for Voltage Criteria: Customer Service Requirements

The fundamental objective of the distribution system is to deliver power to the utility's customers in usable form, literally ready to use without further transformation or re-fashioning. "Usable form" to the vast majority of electric consumers is defined by the electrical requirements of the motors, lighting, and

Planning Criteria

Table 4.1 Voltage Definitions and Terms

Base voltage - the voltage used as a common denominator for the ratings and analysis of all equipment in a study or plan. Distribution might be analyzed on a 120 volt base, even though much of the equipment is at 12.47 kV level. Values at the higher voltage are converted to the standard base for comparison and analysis.

Maximum voltage - the highest voltage observed at a point during a five minute period.

Minimum voltage - the lowest voltage observed at a point during a five minute period.

Nominal voltage - the voltage at which equipment is rated and on which operating characteristics and computations are based (e.g., 120 volts at the service level).

Rated voltage - the voltage at which equipment ratings are based. Usually this is the same as the nominal voltage, but not always. An electric water heater may be designed for a nominal voltage of 120 volts, but its "4,000 watt" rating determined at a rated voltage of 125 volts (at 120 volts it would provide only 3690 watts heating).

Service voltage - the voltage at the point where power moves from utility to customer facilities (usually at or near the meter box).

Utilization voltage - the voltage at the line terminals into the equipment that actually consumes the power (e.g. the air conditioner compressor motor).

Voltage - the root square mean of the phase to phase (or phase to neutral or ground as applicable) alternating voltage.

Voltage drop - the difference between voltage at two points along an electrical path, such as at the head and end of a distribution feeder. Voltage drop is *not* equal to the impedance drop (current times impedance) between the two points, but instead, is the difference between the *absolute magnitudes* of the voltages as measured.

Voltage range - the difference between the maximum and minimum voltages permitted by standards a particular level of the system (e.g., service entrance).

Voltage regulation - the percentage voltage drop along a conductor path or feeder as a function of the minimum voltage. Given V_s as maximum voltage on a feeder and V_m as the minimum voltage, voltage regulation is equal to $100 \times (|V_s| - |V_m|)/|V_m|$.

Voltage spread - the difference between the actual maximum and minimum voltages at a particular point in the system as conditions change (e.g., as seen by a particular customer as load shifts from peak to minimum time).

Voltage standards - minimum and maximum voltages within which voltage must be maintained by design and operation.

electronic equipment they employ to improve their lives and make themselves more productive. Table 4.1 provides definitions of terms that relate to voltage and voltage criteria.

Although utilization voltage standards vary, within any country or region all electric equipment and all electric consumers have equipment designed to operate within a narrow range of voltage around a standard utilization voltage. There are several utilization standards, including an "American" system centered on a nominal 120 volts, several European standards between 230 and 250 volts depending on country, and what is called a "100 volt" standard throughout Japan (it is actually a nominal 105 volts). Within any nation, electrical equipment on sale and in use will correspond to the local standard.

Most electrical equipment can operate satisfactorily within a range of supply voltages, usually wider than plus or minus five percent about its nominal utilization voltage. But the fact that equipment *will* function over a particular range of voltages does not mean that it will function well at the extreme ends of that range. Induction motors function best when provided with more than the minimum voltage at which they can function -- losses and heat build up decline and power output improves as voltage is raised. Incandescent lights provide more illumination, but last a shorter time, at higher voltage. Many other electronic devices have similar voltage sensitivities. Table 4.2 gives the range of voltage needs for some common types of equipment as recommended by the manufacturer and as measured by the author.

Table 4.2 Extreme Operating Voltage Ranges of Selected Household Equipment

Device	Recommended	Author's Tests
75 watt incandescent light bulb	110 - 125	75 - 160
23 watt compact fluorescent "bulb"	110 - 125	112 - 146
Clock Radio	- 115 -	104 - 137
13" B&W television (1967)	115 - 125	107 - 132
1/2 HP 1-ph motor air compressor	110 - 125	104 - 130
Handheld electric drill**	110 - 125	90 - 190
13" Color television	115 - 125	107 - 126
Personal Computer	114 - 126	108 - 126
VCR	110 - 125	106 - 128

*Appliances tested were the author's personal property. The top six devices were tested to actual failure at the high end. The bottom three might have functioned well at voltages beyond the highest one shown, but the author decided not to push his luck.

** Not handheld during these tests.

Planning Criteria

Voltage fluctuation impact on customer appliances

Although most equipment will operate over a range of voltage much wider than 10%, as shown in Table 4.2, this does not mean that it will provide customer satisfaction if the voltage fluctuates between the tolerable minimum and maximum limits. Equipment that is particularly well known for voltage sensitivity includes incandescent lighting and televisions. A majority of people can perceive the change in illumination that occurs from only a 3% change in supply voltage to an incandescent light, if that change occurs within a second or two. The image on some television sets shrinks dramatically as supply voltage is reduced.[1]

Other electrical equipment occasionally has voltage/performance sensitivity that is detectable and affects its performance. For example, the author's tests for Table 4.2 revealed that the clock radio required retuning to hold the same station as voltage changed (its analog tuner had to be moved slightly more than twice the width of its tuning needle). The air compressor (the 1/2 HP motor) required resetting of a regulator diaphragm to maintain acceptably smooth pressure for paint spraying as voltage varied back and forth within the ranges shown. In addition, the VCR had what would prove to be a frustrating characteristic for a customer who experienced a large spread in power supply voltage. It provided noticeably poorer picture quality upon playback below only 97% of nominal voltage, and playback tracking had to be reset whenever voltage dipped below 97%, even if replaying tapes it had recorded while voltage was higher.[2]

For all of the reasons given above, a utility desires to provide its customers with a voltage that is close to the nominal utilization voltage, that seldom varies from that, and then only slowly.

Reasons for Voltage Criteria: Utility Equipment Voltage Standards

The T&D equipment used by the utility also requires certain operating ranges both to maintain equipment within specified loading ranges, and to assure operation as expected. While some equipment (conductor) can conceivably work at any voltage, the majority of insulators, cables, PTs, CTs, breakers,

---

[1] The author tested a 13 inch-diagonal black and white television (circa 1967 manufacture) whose picture shrank to 11.5 inches at 93% of nominal voltage. However, the 13-inch color television (1987) indicated in Table 4.2 showed no variation in its picture size as voltage was varied from a high of 105% to a low of 92% of nominal.

[2] This poor performance was not due to harmonics added by the voltage variation method used (in the tests) a variable winding transformer. This admittedly low-end price VCR, just had a wretchedly noticeable sensitivity to operating voltage.

transformers, regulators, capacitors and other equipment have nominal voltages for which they were designed and specific ranges of voltage within which they are intended to function. Equipment, particularly transformers and regulators, have both limits within which it is permissible to operate them, and formulae for re-rating their capacity if operating voltage is varied even slightly from their nominal ratings. Distribution planning must assure that these voltage requirements are respected, too.

Voltage standards are established by electric utilities in order to define the level of service they provide and as criteria for planners, engineers, and operating personnel so that the system can be maintained within these limits. Voltage standards vary widely among utilities, both worldwide and within the United States. Utility standards in the United States are usually based on providing customer service entrance voltages within ± 5% of nominal voltage (126 to 114 volts on a 120 volt scale). This corresponds to Range A in ANSI standard C84.1-1989, which lists 126 to 110 volts for utilization voltage and 126 to 114 volts for service voltage.[3] However, practice varies, and there are some utilities that permit service entrance voltages as high as 127 volts and as low as 112. Few utilities permit standard minimum to dip below 112 volts under normal conditions, for although 110 volts provides satisfactory operation of most equipment with a nominal 120 volt rating, allowance must be made for voltage drop in the building wiring itself, which might require one or two volts.

What is important to the distribution planners is not their company's standard for service entrance voltages, but how that standard translates to design criteria for distribution planning. Table 4.3 gives primary feeder voltage design limits for ten utilities. In specifying voltage at the primary level their standards must take into account the further voltage drop as power passes through service transformers and secondary service before reaching the customer. Generally, this "transformer and service drop" reduction amounts to between one and four volts, depending on the utility's standards for secondary equipment and its layout (secondary layout will be covered in a later chapter). Beyond this, many utilities include a small allowance against unknown factors, inexact data, and errors due to approximations in their voltage drop computations.

As a result, the range of voltages permitted at the primary feeder level is considerably narrower than the range of standard service-entrance voltages. While most utilities allow a full ten percent range (± 5%) at the service entrance,

---

[3] Range A (105% to 95% of nominal) specifies the limits under which most operating conditions occur. Range B, (105.8% to 91.7%, or 127 to 110 volts at the service entrance, to 86.7% or 106 volts minimum for utilization) applies to certain temporary or infrequent operating conditions.

Planning Criteria

primary-level standards can be as narrow as 3.3%. The maximum primary voltage limit must take into account the minimum secondary voltage drop that could occur, usually only a volt, so that in most cases the maximum primary voltage limit is only a volt above the maximum service entrance voltage standard. The utility must also allow for the worst possible voltage drop in secondary and building wiring, so that minimum primary limits must be significantly above minimum service entrance standards. Table 4.3 lists primary level voltage standards as applied by ten utilities.

One interesting aspect of Table 4.3 is the lack of consistent pattern in voltage standards among utilities serving similar areas, due to very different layout and equipment standards. For instance, in rural areas utility ten has a minimum standard dipping of only 113 volts while utility five allows no less than 119 volts. This difference is due to how the two utilities lay out their secondary systems. Utility fives's secondary layout guidelines "take the primary right to the back door" of most rural customers, as one of their engineers put it, requiring a dedicated service transformer and no more than 100 feet of service drop to the customer's service entrance. There is no need to allow for more than one or two volts of secondary voltage drop in this case.

Table 4.3 Maximum and Minimum Voltage Design Standards at the Primary Distribution Level for Ten Electric Utilities (120 volt scale)

Utility	Service Area Type	Maximum	Minimum	Range %
1	Dense urban area	127	120	5.4
2	Dense urban area	126	117	7.5
3	Urban/suburban	126	114	10.0
4	Urban/suburban	125	115	8.3
5	Multi-state area, urban standard	127	123	3.3
	rural standard	127	119	6.6
6	Suburban & rural	125	113	10.0
7	Suburban & rural urban standard	125	116	7.5
	rural standard	127	112	10.8
8	Urban & rural	127	115	10.0
9	Rural, mountainous	126	116	8.3
10	Rural, mountainous	127	113	10.0

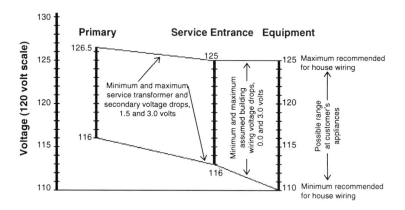

Figure 4.1 Voltage criteria for primary-level feeder planning are related to but not the same as electrical requirements for the utility's service voltage standards, as shown here for one utility's standards. Primary-level voltage standards reflect adjustment for assumptions about additional voltage drop in the secondary system as well as a small margin to allow for inexact data and unknown factors. Exactly what these adjustments are depends on the utility's standards and guidelines for laying out its system.

On the other hand, utility number five sometimes uses considerable lengths of secondary and service drop to serve two or more suburban or rural customers from one service transformer. The relatively long low-voltage conductor runs that result from this practice mean utility five must allow up to eight volts for service-level voltage drop. Working back from a service entrance minimum voltage standard of 111 volts, their rural primary-level minimum voltage must be 119 volts. A utility's standards at the primary level are related to its standards at the service entrance, and to assumptions about secondary level voltage drop, as shown in Figure 4.1.

Voltage drop is an unavoidable consequence of moving power through electrical equipment, and a topic that will be addressed at length in later chapters. What is important with regard to voltage standards and criteria is that they define a *range* of voltages occurring on the distribution system, and that the system must be designed so that maximum and minimum voltages fall within the company's voltage standards, as illustrated in Figure 4.2. Any distribution system will experience some drop in voltage from source to load, and voltage drop will increase as load increases.

Planning Criteria

Voltage drop can be mitigated by design and equipment changes but never eliminated altogether. The use of larger conductor, capacitor correction of power factor, and voltage regulation through application of tap changing transformers, line drop compensators, and line regulators can reduce both voltage drop and voltage range on any distribution feeder, but these measures have a cost all their own. The distribution planner's challenge is to determine the minimum cost plan that does not exceed maximum permitted voltage during light load conditions and does not drop below minimum voltage limits during peak conditions.

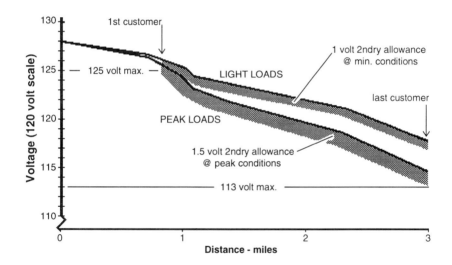

Figure 4.2 Voltage drop along a three mile feeder and its relation to the utility's voltage standards. Shading shows the voltage drop range assumed for the service level. Whatever distribution layout is selected must have maximum and minimum service voltages that fall within the utility's voltage standards, regardless of loading. The voltage profile shown above just barely satisfies this requirement at both ends of the voltage scale. Service entrance voltage at the first customer on the feeder, during light load conditions, is at the company's maximum 125 volt standard (120 volt nominal scale), and service entrance voltage at the last point on the feeder, during peak, reaches its minimum standard 113 volts.

Voltage Criteria Must Be Met, Not Exceeded

This discussion brings up an important point -- voltage drop is *not* a negative factor, as often viewed by engineers and technicians new to distribution planning. Instead, a distribution planner should view voltage drop as perhaps the single most useful tool available to achieve economy of design. Given otherwise identical situations, it is much easier and more economical to distribute power if allowed eight volts drop in the system, than if restricted to only six volts. Therefore, a distribution planner should *never* attempt to minimize voltage drop just for the sake of reducing voltage spread or regulation, because this nearly always incurs a capital or operating cost, or forces a trade-off in achieving other criteria and goals. If eight volts drop is permitted within company standards, all eight should be used. Nothing should be spent to reduce maximum voltage drop to six volts, seven, or even seven point nine volts.[4] Only if there is some other reason for the reinforcement should the planner spend resources for conductor and equipment that results in lower voltage drop.

> *Voltage criteria and standards only need to be met, not exceeded.*

Voltage Criteria Are Normally Applied at the Primary Distribution Level

In most distribution planning studies voltage is computed, analyzed, and applied as a criterion at the primary feeder level, with secondary voltage drops implicitly assumed based on prior "generic" analysis of secondary system layout standards. In the example shown in Figure 4.2, the planner's voltage analysis, performed on the distribution primary voltage level for the feeder, would show a maximum voltage of 126 volts (on a 120 volt scale) during light load conditions and a minimum of only 114.5 volts during peak conditions. Implicitly, due to assumptions about voltage drop through the secondary and service drops (shaded

---

[4] This comment assumes that the planner is taking all other factors, including future load growth, into account. If the feeder serves a new, growing area, the planner may have to allow for future growth, so that initially the feeder is lightly loaded with little voltage drop. But judged against the final design load, voltage drop should be aimed at the maximum allowed -- a full eight volts.

Planning Criteria 139

bands in Figure 4.2) these voltages correspond to computed customer service voltages at that location of 125 and 113 volts respectively after adjustment for secondary voltage drop (shaded bands).

Voltage spread

The difference between the highest and lowest voltages seen on a feeder at a particular customer point -- the change in voltage seen by a customer's equipment over time as the load shifts from lightest to heaviest, is called the voltage spread. In Figure 4.2, maximum spread from light to peak load conditions is 4 volts (3.33%) for the customer at the end of the feeder, quite acceptable if it does not occur too rapidly. Among utilities that have standards on voltage spread, anything up to 6% is generally considered acceptable if it occurs over several hours or more. Given the other voltage criteria (e.g., voltage must never stray outside of a ± 5% range about of the nominal voltage) this is generally quite easy to achieve.

Voltage flicker

A 6% voltage spread between light and peak conditions may be acceptable because it is expected to occur over several hours as the daily load cycle shifts from minimum to peak conditions. In fact, it generally takes a change of season to bring load from the annual minimum to peak conditions, so that the move from maximum to minimum load and voltage occurs over several months. A slow change in voltage over such a long period usually goes unnoticed.

A large shift in voltage within a very short time *is* easily noticeable. Most people can discern the change in illumination wrought by an instantaneous 3% fluctuation in voltage provided to an incandescent light. A 5% fluctuation causes a very noticeable change if it occurs within a second. Rapid voltage fluctuations -- often called flicker -- can cause changes in the output, sound and picture, or other performances of electrical equipment (as was discussed above with reference to the testing for Table 4.1), and can constitute a nuisance (flickering lights, shifts in operating speed of motors) or a major inconvenience (constant speed equipment isn't, motors overload, wear and tear increases and machinery accelerates and decelerates). For all these customer-service related reasons, most utilities maintain voltage flicker standards that require voltage change due to any sudden load shift to be less than some limit. Very often the major cause of flicker on a distribution system will be the starting of large, multi-phase motors. As a result these criteria are often called "motor start voltage standards" or "motor starting criteria" although they apply to all voltage drops and loads. Table 4.4 shows flicker criteria as applied by six utilities.

Table 4.4 Flicker Limits of Six Electric Utilities

Utility Number	Service Area Characteristics	Criteria Voltage or %
1	Dense urban area	3%
2	Urban/suburban	3%
3	Suburban & rural	3%
4	Urban & rural	3%
5	Rural, mountainous	4 volts
6	Rural, mountainous	none

While a majority of utilities surveyed define unacceptable flicker as anything greater than 3%, many differ in exactly their limit is defined, interpreted, and applied in the distribution planning process. These three standards are all considered a "3% flicker standards" the utilities applying them:

1) Flicker is excessive if it exceeds 3% *of nominal voltage* anywhere on the feeder. Here, the flicker limit is 3% of 120 volts, or 3.6 volts, and the criteria is applied as a voltage -- fluctuations due to motor start cannot exceed 3.6 volts anywhere on the feeder.

2) Flicker is excessive if it exceeds 3% *of the minimum permissible service entrance voltage* anywhere on the feeder. For example, if the utility's minimum service entrance voltage is 112 volts, this gives a flicker limit of 3.36 volts, which is applied as a voltage -- fluctuations cannot exceed 3.36 volts anywhere on the feeder.

3) Flicker is excessive if the voltage for any customer drops by more than 3% *of the actual service entrance voltage* at that point. Here, the criterion is applied as a percent, rather than as a voltage. In the example in Figure 4.3, this means the criterion varies from a high of 3.75 volts for the 1st customer on the feeder to 3.45 volts at the last customer.

Flicker is always interpreted as a rapid, nearly instant drop or rise in voltage. As mentioned above, the most common cause of sudden, noticeable shifts in load is the starting of large electric motors, although other industrial equipment (welders, etc.) can cause flicker, too, and switched utility equipment such as capacitors and phase shifters will often cause flicker. Almost any type of motor momentarily draws a starting current several times its normal (full speed) current. This high load occurs instantly upon the motor's being activated, and

Planning Criteria 141

(assuming the motor starts as expected) lasts but a few seconds, dropping rapidly to full-run current levels as the motor comes up to speed. The resulting short but high starting load can cause a severe but short voltage dip -- a drop and then a rise as the motor comes up to speed, that is perceivable and perhaps inconvenient to customers nearby. In extreme cases, the voltage dip may be severe enough that the motor cannot start. An important aspect of some motor-start studies is to determine if the motor's load can be supported at all -- if it will even start under the severe voltage dip that its starting current causes.

However, usually a "motor-start" or other voltage dip analysis is carried out to assure that the motor's starting will not disturb other customers on the feeder by causing flicker outside of acceptable limits. A motor start case usually involves analysis of two or three voltage profiles for the feeder, as shown in Figure 4.3. Usually the worst voltage fluctuation occurs upon starting, causing differences between the "no motor load" voltage profile and an immediate drop to the "starting" case profile.

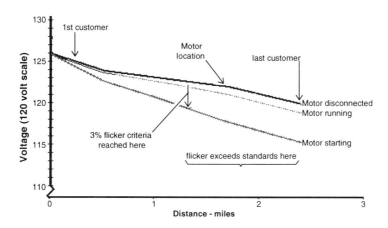

Figure 4.3 Profile of primary-level voltage drop due to a 550 HP 3-Ph. induction motor starting on a 12.47 kV distribution feeder. The motor is located about 1.7 miles from the substation. Its starting current (3.3 times the current drawn at full speed) causes up to a 4.4% dip in voltage while starting, and causes a momentary dip of 3% or more over nearly half the feeder's length. Minimum voltage anywhere on the feeder, even under these starting conditions, still satisfies the utility's overall standard for minimum voltage (115 volts). However this case does not meet the utility standard for flicker, which requires no sudden fluctuations over 3% of voltage.

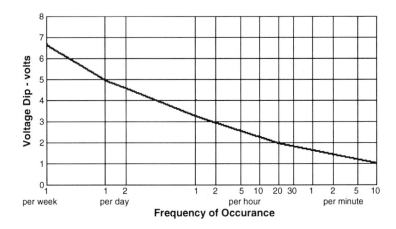

Figure 4.4 Flicker magnitude versus frequency of occurrence standard as applied by one utility allows larger voltage dips if they occur less often. Equipment that starts only once per hour (e.g., a large pump motor) and causes only a 3% voltage dip would be permissible (point is below the curve) whereas a 3% voltage dip occurring 5 times per hour would not be within standards.

Flicker can become a problem not only at the motor location, but both up and downstream of the motor location on the feeder, as indicated in Figure 4.3. When flicker is a problem, the planner's choice of cures may be limited, for the load and voltage shifts almost always occur too fast and too briefly for voltage regulation equipment to provide a solution.[5] The "cure" for such problems is larger conductor or transformers, various motor starting equipment (capacitors, etc.), or limitations on the size or the operation of the motor.[6]

Occasional flicker is not considered as objectionable as frequent flicker of the same magnitude. Therefore, many utilities apply a flicker magnitude versus frequency of occurrence guideline. An example graphic guideline is shown in Figure 4.4.

---

[5] Most voltage regulators and line drop compensators are designed to respond only after a certain short delay, so they do not overreact to such fluctuations but adjust only to changes in stable voltage that last more than several seconds.

[6] In some cases the utility will permit service only under special arrangements. For example, the owner of a 1,500 HP compressor motor might be required to start it only during daily minimum load period (2-4 AM) and to call the utility prior to starting.

Conservation Voltage Reduction (CVR)

A number of state utility regulatory agencies require conservation voltage reduction as a means of reducing energy consumption. Most electrical devices have a load that varies as a function of voltage. For pure impedance loads, such as incandescent lighting or resistive water heaters, power consumed is proportional to the square of the voltage (Power=$V^2 \times Z$) so that a 5% reduction in voltage results in a 10% reduction in load. However, other loads, such as induction motors, draw more current as voltage is reduced, so that their consumption remains constant or even increases slightly (if losses are considered) when voltage is reduced. Given the load characteristics common to most utility systems, CVR seems to reduce load linearly, with somewhere between a 1:1 and 1/2:1 ratio with respect to voltage (e.g., a 5% reduction in voltage gives between 5% and 2.5% reduction in load). Load reduction is not nearly what would be expected if all load were pure impedance, but CVR *does* reduce load by measurable amounts.[7]

Thus, to conserve energy, a particular CVR regulatory requirement may specify that no customer be provided with over 120 volts at the service entrance, although up to 125 volts is acceptable from an equipment and standards standpoint. Alternately, the regulatory requirement may specify that the feeder be operated at the lowest voltage that is permissible within voltage drop allowances, as shown in Figure 4.5. This requirement reverses the traditional voltage profile design methodology. Traditionally, the distribution engineer started with the highest permissible voltage at the substation and let voltage drop occur along the feeder as the design dictated, often ending up with end-of line voltage considerably above minimum standard. CVR rules allow the same voltage drop, but require that the feeder be operated at the lowest source voltage that keeps this end-of-line voltage within standards.

---

[7] Numerous tests have established this beyond a doubt, verifying that an instantaneous 3% drop in voltage at the substation will reduce load on a feeder by something approaching 2 to 3%. This does not mean that a corresponding reduction in energy consumption occurs, something not realized in most evaluations. In the case of an electric water heater rated at 4,000 watts at a nominal 230 volts, a reduction of 3% to 223.1 volts lowers its instantaneous load to 3764 watts, a drop of 6%. But, it still operates under thermostatic control to keep water at the same temperature, which means it compensates for its lower output by heating 6% longer. After a short period (one hour) of transient adjustment to the new voltage level, daily kWhr use of all electric water heaters affected, and the coincident contribution of any large group of water heaters, is exactly the same as before. (See Willis and Rackliffe, 1994, Section 3.4.)

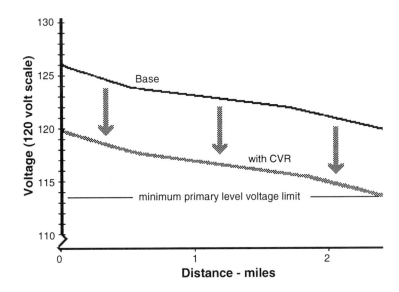

Figure 4.5 Conservation voltage reduction either restricts the highest allowable customer voltage level to less than traditional upper limits, or requires that feeders be operated at the lowest regulated voltage that meets utilization requirements, as shown here in a plot of primary level voltage profile for a feeder before and after CVR. An additional 1.5 volt drop is assumed in secondary circuits for a total of only 112 volts expected at the service entrance of customers near the end of the feeder.

CVR, if required by regulation or law, can reduce the amount of voltage drop that the planner has available for use. This can increase cost, because as mentioned earlier, distributing (roughly) the same amount of power with significantly less voltage drop requires larger conductor or tighter and more controlled voltage regulation equipment. While it is questionable if the reduction in energy consumption is worth the cost in some cases, if regulations have set such limits, they must be respected and treated as criteria.[8]

---

[8] In some cases, CVR regulations do not force the narrowed voltage spread on the planner, if it can be shown that the energy consumption savings (calculated according to a specified formula) are not justified when compared against the cost of additional T&D equipment required to distribute power with the lower voltage spread. In these cases, CVR is not a hard and fast criteria, but more or an additional attribute requiring additional planning effort to analyze cost against voltage drop.

Planning Criteria

Applying Voltage Standards and Criteria

In general, voltage standards are applied to distribution planning studies by computing voltages at the primary feeder level and comparing these computations to criteria for minimum and maximum primary voltage limits, levels developing from the service entrance standards by assuming certain voltage drop ranges for secondary and service transformer equipment. Usually, voltages for candidate designs are determined using a computerized voltage drop calculation method. Both the computational methods in use throughout the industry, and the data to which they are applied, vary in quality and accuracy.

Distribution planners should be aware of the characteristics of the particular analysis tools and data they are using, and the conditions under which they underestimate or overestimate voltage drop. Regardless of computational accuracy, some allowance should be made for additional error due to inexactness of the load data and possible phase imbalance. In almost all situations, the planner will use the feeder voltage drop analysis method to develop a peak load case to check for low voltages during peak conditions. Often a light load case will be run to determine if higher than standard voltages develop during light load conditions. Motor start studies are performed only when a large motor or other load is planned or suspected as the cause of operating problems.

4.4 OPERATING AND SAFETY STANDARDS

In addition to voltage standards, utilities have a number of other standards and criteria for their system equipment and its operation. These include a variety of requirements and limitations to assure safe and efficient service.

Voltage Imbalance

It is impossible to maintain perfect balance of load, voltage, and current on a distribution system. Many loads are single-phase, and as these cycle on and off, loading on a phase by phase basis will change slightly. Thus, a slight imbalance among phases is bound to occur and cannot be corrected. This load imbalance causes differences current, and hence voltage drop, among phases. Voltage imbalance is defined as:

$$\% \text{ imbalance} = \frac{\text{maximum deviation from average voltage} \times 100}{(V_{A\phi} + V_{B\phi} + V_{C\phi})/3} \quad (10.1)$$

where $V_{A\phi}$, $V_{B\phi}$ and $V_{C\phi}$ are the phase to neutral or ground voltages from phase A, B, and C, respectively.

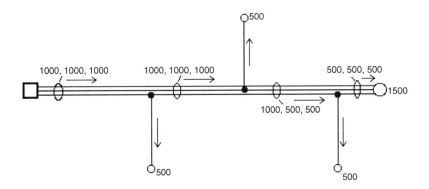

Figure 4.6 Intermediate segment voltage imbalance in a feeder with "perfect balance" at the substation and at a balanced load at the end of the feeder. Use of single phase laterals causes loading on two intermediate segments to be quite imbalanced. This may result in operating voltages that are imbalanced to the point they exceed voltage imbalance criteria, even though all voltages are within corporate standards for maximum and minimum voltage. To correct the problem, the distribution planner will have to re-design the feeder, using two or three phase laterals to balance load of the branches among phases.

Imbalanced phase loadings occur on many distribution systems not only because of minor fluctuations in load, but because of their layout and design. This is quite common in systems built to what are often called "American" design standards, which call for considerable use of single-phase laterals (in contrast to European systems which use three-phase in laterals and secondary). Figure 4.6 illustrates how a feeder with perfect load balance at the substation and end segment can nonetheless have portions with considerable imbalance in loading. Such situations can lead to significant voltage imbalance.

Voltage imbalance standards vary, with a 4% voltage imbalance being the most common limit. On systems with tight voltage standards (such as utility five in Table 4.3) voltage imbalance criteria is often much tighter, usually limited to one half of the maximum voltage standard range (e.g., 3.3% when voltage standards range is 6.6%). Imbalance criteria are always applied as voltages -- load imbalance *per se* is not objectionable, if the resulting voltages is within

Planning Criteria 147

limits. Only when load imbalance causes unacceptable voltage imbalance must the planner adjust system design.[9] Often, changing a single-phase lateral from one phase to another (e.g., from phase A to phase C) can reduce imbalance. However, the problem may be simply that too much load is being put on a single phase, and the lateral will have to be converted to a two or three-phase branch.

Fault Current Limits and Protection Criteria

One of the most important aspects of electrical system design is to assure that the system fails in a safe manner, minimizing the prospect of injury to people or property, and protecting as much equipment from additional harm as possible. Usually, protection is not part of the planning process, but rather the domain of engineering and detailed design. However, protection standards and criteria place some restrictions on the distribution planner, most notably on equipment sizing and feeder reach.

Increasing equipment size is often a preferred solution to overloading and voltage drop problems -- if 336 MCM conductor won't do, maybe 600 MCM conductor will. If not, maybe 795 MCM will. The larger conductor carries with it a larger fault current, and under some circumstances this can exceed the limits of protective devices in place or certified for use on the system. So, while the larger conductor solves an overloading problem, it may bring with it other design or equipment problems.

Similarly, fault currents associated with a particular alternative design can be too low or outside of a permissible range for protection. For example, other criteria may be satisfied if service to a rural customer is provided with a lateral run of several miles of single phase #6 AA conductor. However, fault current at the end of this lateral might be so small that fusing within company protection standards cannot cover the low fault range. A larger conductor might be required to meet this criterion, or some alteration in other parts of the feeder leading to the lateral, or perhaps a change in the type of protective devices.

Protection is much more important to the overall design and operation of a distribution system than indicated by the attention given it in this chapter. However, as mentioned above, these concerns are normally addressed as part of the design process. Distribution planning addresses protection only to the extent that the plan must meet criteria that ensure that fault currents fall within limits

---

[9] Balance is a desirable goal whenever it can be accomplished with little or no cost or effort, but significant cost should be incurred to correct imbalance only when voltage imbalance exceeds criteria or increases losses costs enough to justify the correction measures.

covered by the corporate protection standards. A distribution planner must do nothing more than provide a distribution plan that *can* be protected. Protection engineers take it from there.

Load Ratings and Loading Limits

One measure of any unit of electrical equipment is its capacity rating -- how many kW or kVA it can handle. Manufacturers establish "nameplate ratings," which are their rating of the equipment's capacity under a tightly specified set of conditions including voltage, power factor, ambient temperature, maintenance, and other operating factors. The equipment may be capable of carrying more or less load if conditions are altered from these nominal rating factors. For example, transformers and regulators must be de-rated if operated at higher or lower voltages than specified in their nominal rating.

Electrical equipment "wears out" in service depending on how heavily it is used. All electrical equipment, particularly wound devices such as transformers and regulators, but including buswork, switches, breakers, cable, and even conductor, has a longer lifetime if lightly loaded, if operating under relatively cool ambient conditions, and if never operated outside of its recommended voltage ranges. Generally, equipment is given a capacity rating such that routine loading to that rating, under expected circumstances, will result in a calculated expected lifetime of at least thirty and perhaps as long as fifty years.

Contingency ratings considerably above this normal rating allow additional margin for emergencies, but with a recognized impact on equipment lifetime. Such ratings usually are developed based on 1% loss of life -- operation for four hours at a transformer's emergency rating might subtract 1% off its expected lifetime of fifty years (equal to half a year -- 4380 hours -- or *1100 times* its normal rate of aging during those four hours).

Generally, regardless of the manufacturers' nameplate ratings for equipment, a utility will establish its own rating standards for the equipment is uses, ratings which reflect characteristics and needs unique to its territory, climate, and service standards. Differences between a manufacturer's rating and the utility's rating can be due to any number of reasons. The manufacturer may have rated equipment at a lower ambient temperature than is common in the utility's service area, requiring a de-rating to allow for the higher temperature. "Peak load" as used in rating the equipment might assume a constant maximum load lasting four hours, preceded by a twenty-hour period at 75% of that value. The peak day load might have a peak period of only two hours, preceded by a period where load averages only 66% of peak, permitting a higher rating than given by the manufacturer. Beyond this, a manufacturers' rating often assumes perfect balance of phases, whereas the utility may de-rate the equipment to allow for the

Planning Criteria

inevitable load and voltage imbalance that occurs in practice. Table 4.5 lists capacity ratings for the same 16/20/24 MVA substation transformer as applied by six utilities.

A triple rating (i.e., 16/20/24 MVA) refers to maximum MVA capacity under various operating conditions, *but there is no uniformity in what three capacity values written this way mean,* one more reason why a planner must be careful when comparing equipment standards and operating records from one utility to another. This triad of numbers often refers to the maximum capacity of the transformer under the respective conditions of:

1) operating in still air with no fans or oil pumps running and a maximum rise in temperature at the top of the transformer core of 55° C (sometimes referred to as NOA)

2) operating with internal pumps circulating oil through the transformer and to radiators, with a maximum temperature rise of 55° C (FO)

3) operating with internal pumps circulating oil and fans blowing air over the radiators, with a maximum temperature rise of 55° C (FOFA)

However, some utilities use a three number rating set indicating NOA/FOFA/EMERG, where the second of the three values indicates the forced oil and forced air rating (FOFA) as described above), and the third number indicates an even higher operating limit allowed for certain periods during contingency situations, with forced oil and forced air but at a higher core temperature rise (usually 10° C more). Several other utilities substitute an economic rating -- the peak load that gives the lowest capital versus losses cost economics for their annual load factor -- as the first value.

Table 4.5 Loading Standards for a "16/20/24" 138 kV/ 12.47 kV Substation Transformer as Used by Six Utilities, in MVA

Utility	Service conditions	Summer		Winter	
		Normal	Emerg.	Normal	Emerg.
1	High ambient, long peak conditions	21.0	26.0	24.0	32.0
2	High ambient, short peak conditions	22.8	29.6	25.2	33.3
3	Mild but long summer, high load factor	24.0	32.0	24.0	32.0
4	Mild climate, short peak, low load factor	25.2	32.3	26.1	33.8
5	Mild except cold winters, long win. peak	24.0	32.0	24.0	32.0
6	Extremely cold at peak	25.0	31.0	26.0	32.0

Regardless of how they are determined, loading standards generally define both normal and emergency peak loads for all equipment, often varying between winter and summer conditions. These values cannot be exceeded, at least in the planning stage.[10] Again, these are criteria. The planner's job is to use equipment fully, not to keep peak load far below normal rating.

Contingency Margin

Most utilities specify contingency margin and switching requirements in the design of their feeder systems. This allows sufficient capacity margin to cover equipment failure, and permits enough flexibility in switching of feeders to re-route them around most equipment outages. These requirements will be discussed in at length in many other locations throughout this book (see Index).

4.5 STANDARD EQUIPMENT AND DESIGN CRITERIA

For a variety of reasons, utilities have additional criteria and standards for the layout and design of their systems. Among the most important are those that specify what equipment can be used in the system -- the distribution planner must expand and add to the distribution system using only equipment drawn from a standard inventory of approved types and sizes. This assures that equipment used in the system meets other standards, such as maintainability -- if company personnel only have equipment, tools, and facilities to maintain certain types of breakers, it makes no sense to install others types on the system. Similarly, a utility will often test new types of equipment to assure itself that the equipment does meet claimed ratings and losses performance before certifying its use on the system.

The approved list of equipment and sizes generally provides a range of capabilities to meet most needs, but uses only a small portion of equipment that is available in the industry. For example, while there are literally dozens of different sizes and types of overhead conductors available to an electric utility, most standardize on the use of only between three and six, which will span a

---

[10] During severe emergencies or unusual circumstances, a utility may choose to operate equipment above even emergency ratings. This is an operating decision made at the time, weighing the possible consequences (excessive loss of life or even failure) against the benefit (restoration of service to many customers who would otherwise be without it under extreme weather conditions for an extended period of time), against a set of conditions outside those covered by normal planning criteria.

Planning Criteria

Table 4.6 Examples of Standard Equipment Types

Equipment	Voltage	Permitted sizes & sizes
Primary OH conductor	12.47 or 25 kV	ACSR: 336, 600, 795 MCM
	12.47 or 25 kV	AA: #2, 2/0,
Primary UG cable	12.47 or 25 kV	200, 400, 1150 MCM AA Hiplex
1Ø service trans.	12.47/120/240V	OH: 15, 25, 50, 75, 100 kVA
1Ø service trans.	12.47kV /120/240V	UG: 25, 50, 75, 100, 150 kVA
Substation transformers	69kV/ 12.47kV	5, 7.5, 10, 15 MVA
	138kV /12.47kV	12, 16, 24, 32 MVA
	138kV /25kV	20, 42, 32 MVA

range from small to very large. Using more than this small number of conductor types means keeping more types of conductor in inventory, purchasing a larger inventory of tools, clamps, supports, and splices; to fit many sizes, and training personnel in the installation and repair of a wider range of conductor types. For these reasons, using too many types of conductor increases operating costs.

On the other hand, the utility doesn't want to force its planners to use large conductor where smaller (and less expensive) conductor will fulfill requirements. Depending on factors specific to each utility, it takes between three and six well-selected conductor sizes to minimize the overall cost, balancing the availability of a wide range of conductor sizes against the need for standardization of equipment inventory (this will be discussed in detail in Chapter 7). Similarly, a standard set of service transformers, breaker types and sizes, and substation transformer types and sizes will be permitted with standards, several examples of which are shown in Table 4.6. The distribution planner must put together a future system using only these equipment types, and nothing more.

Design Criteria

Beyond being limited to a set of specific equipment types and size, the planner also must work within criteria that specify how that equipment can be used and how the "pieces are put together." Considerable engineering design criteria exist to specify precisely how equipment such as primary feeder lines are to be assembled, with suitable conductor spacings, with hardware heavy enough to handle the weight, wind, and ice loadings to be expected, and in compliance with all other requirements. Usually, such engineering criteria have little impact on the planning process, except that they may specify a wider or larger right-of-way, site, or easement that is available.

Beyond this, the distribution planner may have a number of other types of design limitations or criteria that the eventual plan must respect. Table 4.7 gives a sampling of some other design criteria in use by various utilities, and the reasons for their application. Many of them are unique to only one or a handful of utilities, others are widely used.

Esthetics are often one component of design criteria, as illustrated by several items listed in Table 4.7. Most distribution planners are sensitive to the fact that electrical equipment is not aesthetically pleasing, and that its impact should be minimized within limits permitted, while still achieving good economy and safe, efficient operation. Those limits often include criteria that specify design to minimize esthetic impact in certain situations. However, use of such criteria is often limited to specifically defined situations, in order not to show favoritism or increase cost.

Table 4.7 Examples of Design Criteria and Standard Rules

Criteria	Reason
Only post-insulator primary construction on streets with 4+ lanes and along highways.	Esthetics. In areas where equipment is very visible, use a "cleaner" design.
Only steel poles within 30 feet of street intersection	Safety and reliability. Less likely to fall when hit by car or truck.
No double-circuit feeders (two sets of feeder conductor on one pole.	Reliability. A car hitting a pole is less likely to outage two feeders at once.
12.47 kV primary distribution built to 34.5 kV standards.	Future flexibility. Feeders can be upgraded to higher capacity if needed.
No overhead construction in front of church, synagogue, mosque, or other house of worship.	Esthetics in a possibly sensitive, quality area with high public visibility.
OH capacitors, regulators, and switches must be located within fifteen feet of a street.	Maintenance. Difficult to get to them with a bucket truck if not on a street.
Cable vaults must be no closer than fifty feet to any street intersection.	Public convenience. Otherwise, it would tie up traffic on two streets during repair or enhancement work.
Only steel poles in wildlife protection areas of some areas in Africa.	Rhinoceros scratch their hide against wooden poles (thinking they're trees) knocking them over. They ignore steel.

Planning Criteria

4.6 CONCLUSION

A satisfactory distribution plan must meet a number of criteria, predominantly related to standards for acceptable electrical performance both under normal (voltage criteria) and abnormal (fault current, protection) operation. It must also meet addition criteria and standards with respect to loading, equipment type, maintenance, reliability, and esthetics. These constitute a list of requirements and planning criteria which the planner must check against alternatives.

Criteria need only be meet. It is sufficient if voltage is barely within standard limits, if fault current is very near but not over the maximum allowed, if loading at peak just equals capacity. The whole point of a criterion is that the system will function as required if the criterion is met. As a general rule, resources (money, flexibility, or attainment of other goals) should never be spent to do anything more than minimally satisfy any criterion.

REFERENCES

ANSI C.84.1-1989, American National Standard Ratings for Electric Power Systems and Equipment (60 Hz), American National Standards Institute, New York, 1989.

IEEE Recommended Practice for Electric Power Distribution for Industrial Plants (Red Book), Institute of Electrical and Electronic Engineers, New York, 1994.

H. L. Willis and G. B. Rackliffe, *Introduction to Integrated Resource T&D Planning,* ABB Guidebook Series, ABB Systems Control, Santa Clara, 1994.

5
Reliability and Contingency Criteria

5.1 INTRODUCTION

"Reliability" as normally applied to power distribution means continuity of service to the utility's customers. A reliable power system provides power without interruption. Almost anything less than perfection in this regard will garner complaints that the utility's system is "unreliable." An availability of 99.9% might sound impressive, but it means eight and three quarters hours without electric service each year, a level of service that would be considered unsatisfactory by nearly all electric consumers in North America.

This chapter reviews distribution reliability concepts in general and introduces the key components used in reliability analysis. It begins with a discussion of equipment outages and service interruptions, and the two aspects of reliability that measure both -- frequency and duration of failure. The relationship between equipment outages and interruptions, the extent of an outage's service interruptions is presented and identified as one of the key concepts for distribution planning. The most popular reliability indices are discussed, along with their usefulness to the distribution planner and the differences in their application among electric utilities. Finally, various ways of setting reliability standards and criteria for distribution are covered. Planning methods to design distribution systems to meet these or other reliability-related criteria are covered in several other chapters in this book (see Index). Table 5.1 provides a glossary of standard reliability-related terms.

Table 5.1 Reliability Terms and Definitions

Availability: the fraction of the time that service is available. The (steady-state) probability that power will be available.

Failure rate: the number of times a unit or set of equipment can be expected to fail during a period (usually a year), which may be much less than once per year (e.g. .0125 times/year.

Clear: to remove a fault or other disturbance-causing condition from the power supply source or from the electrical route between it and the customer.

Cut set: in reliability analysis, a set of components which when removed from service causes a lack of continuity in power flow, resulting in an interruption.

Duration: the total elapsed time of an interruption or outage, as the case may be.

Failure rate: the average number of failures of a component or unit of the system in a given time (usually a year).

Frequency: how often interruptions or outages, as the case may be, occur.

Interruption: a cessation of service to one or more customers, whether power was being used at that moment or not.

Instantaneous interruption: an interruption lasting only as long as it takes for completely automatic equipment to clear the disturbance or outage causing the interruption. Often less than one second.

Momentary interruption: an interruption lasting only as long as it takes automatic but manually supervised equipment to be activated to restore service, usually only a few minutes, sometimes less than fifteen seconds.

Planned interruption: an interruption of power due to a planned outage, of which customers were informed reasonably far in advance.

Temporary interruption: an interruption manually restored, lasting as long as it takes.

Interruption duration: the time from cessation of service until service is restored.

Interruption frequency: the average number of times interruption of service can be expected during some period (usually a year).

Reliability and Contingency Criteria 157

Table 5.1 cont.

Mean time between failures (MTBF): the average or expected time a unit of equipment is in service between failures.

Mean time to repair (MTTR): the average or expected time to repair an outaged component or unit of the system, once failed.

Minimal cut-set: a cut set (of components) that does not contain any other cut set as a subset.

Outage: the complete failure of part of the power supply system -- a line down, transformer out of service, or whatever else is intended to be in operation but is not -- whether due to unexpected or planned circumstances.

Repair time: the time required to return an outaged component or unit of the system to normal service. Related in a cause and effect way, but not necessarily the same as, interruption duration.

Reporting period: the period of time over which reliability, interruption, and outage statistics have been gathered or computed.

Restoration: return of electric service after an interruption, because of repair of the outage causing the interruption, or because of re-switching of supply source.

Restoration time: the time required to restore service to an interrupted customer or portion of the system. Also, in T&D operations, the time required to return equipment to service in a temporary configuration or status (i.e., not completely repaired) as for example an overhead feeder put back into service with stubbed poles and slack spans temporarily in place. It has been restored to service, but not repaired.

Service: supply of electric power in *sufficient amount and quality* to satisfy customer demand.

Switching time: the time required to operate devices to change the source of power flow to a circuit, usually to restore service where interrupted due to an equipment outage.

Trouble: anything that causes an unexpected outage or interruption including equipment failures, storms, earthquakes, automobile wrecks, vandalism, operator error, or "unknown causes."

5.2 OUTAGES CAUSE INTERRUPTIONS

The single most important distinction to make in distribution reliability analysis is between two words that are often used interchangeably: outage and interruption. They have far different meanings. "Outage" means a failure of part of the power supply system -- for example a line down, transformer out of service, or a breaker that opens when it shouldn't. "Interruption" means a cessation of service to one or more customers. Interruptions are almost always caused by outages, and this leads many people to confuse the two words and their meanings. However, to understand distribution reliability properly, and more importantly, set criteria and design a system to achieve high reliability, the distinction between outage (the cause of service problems) and interruption (the result) is critical. Electrical service to a customer is interrupted whenever equipment in the line of supply to that customer fails or is otherwise out of service. *Outages cause interruptions.*

Frequency and Duration

Two different aspects of reliability receive attention in any type of power distribution reliability analysis, whether of outages, or interruptions. These are the *frequency* (how often something occurs) and *duration* (how long it lasts). With respect to interruption, frequency refers to the number of times service is interrupted during a period of analysis -- once a decade, two times a year, five times a month, or every afternoon. Duration refers to the length of these interruptions; some last only a few cycles, others for hours, even days. Frequency and duration are used with regard to outages, too.

Extent

Frequency and duration are important factors in reliability analysis and design, but there is a third factor equally important to the distribution planner because it is the one over which the planner has considerable control, particularly in the design phase. The *extent* of an outage -- how many customers are interrupted by the outage of a particular line or unit of equipment -- is the key to the outage-interruption relationship. The layout of a distribution system greatly influences just how many customers are interrupted when outages occur.

Consider for a moment two hypothetical distribution systems, A and B. During a particular year, both experience the same number of distribution failures. In both, it takes just as long for repairs to be made and service to be restored. But suppose that in system A, the average equipment outage cuts off service to 150 customers, while in system B, it cuts off service to only 75. In this

Reliability and Contingency Criteria 159

case, system B's customer interruption statistics will be only half of A's, even though their equipment-level failures rates are the same.

The extent to which outages spread customer interruptions through a distribution system is very much under the control of the distribution planner -- among other things it is a function of how the distribution system is laid out. Various types of distribution designs and layouts behave differently in terms of extent of outages. In general, a system laid out to minimize extent of outages from any failure will be more expensive than one that is not, but for the planner this means that consideration of configuration and its influence on reliability is just one more alternative to be considered in planning. In many cases a higher-reliability layout may be less expensive than any other way of achieving the particular reliability goal.

Most distribution systems use some form of radial feeder and secondary layout, which means that a single failure anywhere in the electrical path between substation and customer will cause service interruptions to all customers downstream.[1] One way to improve reliability in such systems is to lay out the feeder trunk and branches, and arrange the protection (fusing) so that outages have a low extent, i.e., so that the average feeder segment failure does not interrupt service to very large numbers of customers.

Control of extent through layout design

Figure 5.1 and Table 5.2 show two different configurations for feeder layout to reach the same customer points. Although simplified, this example illustrates well both the concept and the typical magnitude of differences in costs and reliability between layouts. Configuration B gains a 6.76% improvement in customer interruptions over configuration A, even though it has a 4.8% *higher* expected rate of equipment outages.[2] Also quite important, the worst number of interruptions any customer on the feeder can expect (i.e., its worst case customer) is 12.6% less.

The reason for the difference is due to variations in the extent of outages that can be expected on the feeders. In configuration A, if the second trunk feeder segment fails, service to 300 customers is interrupted. That trunk is a mile long and expected to fail once a year. By contrast, in configuration B, the maximum extent of any outage is 200 customers, for either of the two major branches,

---

[1] There are distribution layouts so robust that one, two, or even three equipment failures will not lead to a single customer interruption -- either network service or the interlaced radial feeder/secondary network system as discussed in Chapters 1, 8, and 12.

[2] Expected outages are 4.8% higher because configuration B has 4.8% more miles of feeder length.

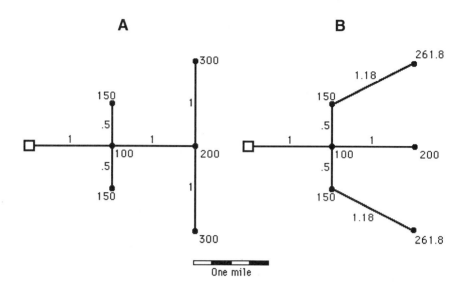

Figure 5.1 Simplified illustration of two ways a radial feeder can be laid out to serve the same 600 customers and how configuration alters reliability. Each node represents 100 customers (identical in both feeder layouts). Both feeders have a failure rate of once per year per mile. A fuse protects the rest of the feeder from outages behind every node. Numbers shown at the middle of each segment are the expected failures for that segment. Numbers shown at nodes are the total customer interruptions/year expected at that node assuming the protection works.

Table 5.2 Reliability-Related Differences Between the Feeder Layouts A and B shown in Figure 5.1

Statistic	A	B	Difference
Miles of conductor	5.00	5.24	+4.8%
Feeder failures/yr.	5.00	5.24	+4.8%
Customer interr./year	1200	1124	-6.76%
Maximum interr./cust.	3.00	2.62	-12.6%
Estimated relative cost	1.00	1.03	+3.0%

Reliability and Contingency Criteria

which have expected failures of .5 per year (for one mile total). This single difference in extent reduces expected customer outages by 100 customer interruptions per year. Configuration B has slightly longer branches to its most outlying nodes (1.18 miles final run instead of 1.0, which increases outage rate slightly, but overall its layout results in a net decrease of 76 customer interruptions per year (6.76% less), due only to its different layout. Configuration B has better reliability from the customer standpoint than configuration A, because, on average, fewer customer interruptions occur whenever an outage happens.

The difference in outage extent between configurations A and B really impacts only the frequency of outages. The average customer on feeder configuration B is 4.8% less likely to be interrupted. The expected total duration of customer interruptions -- the number of customer hours of expected customer interruptions annually -- is also lower by 4.8%, but only as a consequence of there being fewer customer interruptions in the first place. Changes in layout and fusing that reduce the expected extent of outages work on only the frequency of customer interruptions. They do little if anything to reduce duration of any particular interruption, or to shorten the time to repair or restore service (assumed to be an hour in all cases in Figure 5.1).

Changing the layout and protection scheme for a feeder to lower extent of outages is often the distribution planner's chief means to reduce frequency of interruptions. Other means of improving customer reliability, such as providing additional contingency margin, arranging multiple re-switching zones, and automation reduce duration; they do not reduce *frequency* of interruptions.

Types of Interruptions

Interruptions are classified by their duration. Table 5.3 lists typical interruption definitions, averages from the author's survey of usage throughout the power industry. Practice and interpretation vary from utility to utility. In the author's opinion these definitions are all somewhat unsatisfactory. The names imply that they relate to interruption duration, yet they are defined by equipment and operating factors, such as whether an interruption is restored automatically or manually. This, along with differences in exactly how the definitions are interpreted, means that the application of an index -- for example "momentary interruption," varies greatly from utility to utility, not only in definition, but in *how it is used.* In addition, the typical definitions shown in Table 5.3 strike the author as too lengthy to match their names. "Instantaneous" as used by many utilities encompasses interruptions of up to 20 seconds. In the author's opinion, "instantaneous" implies no more than a second. Similarly, "momentary" is used for durations of up to three minutes, whereas a minute seems more appropriate.

Table 5.3 Traditional Type Classification for Interruptions

Type	Definition
Instantaneous	An interruption restored immediately by completely automatic equipment, or a transient fault that causes no reaction by protective equipment. Typically less than 15 sec.
Momentary	An interruption restored by automatic, supervisory, or manual switching at a site where an operator is immediately available. Usually less than three minutes.
Temporary	An interruption restored by manual switching by an operator who is not immediately available. Typically, thirty minutes.
Sustained	Any interruption that is not instantaneous, momentary, or temporary. Normally more than an hour.

Perhaps more important, these distinctions mean nothing to electric consumers, so they do little to help a utility improve service from the customer's perspective. A four-minute interruption is a four-minute interruption, whether restored automatically (making it momentary) or manually (making it a "temporary" interruption).

Beyond the distinctions shown in Figure 5.2, interruptions are classified into scheduled interruptions (the utility scheduled the interruption for maintenance purposes) or forced interruptions (the interruption was not expected and not scheduled). Implicit in these definitions, and the most important factor in the author's mind, was whether customers were informed reasonably far in advance of the details of any "scheduled interruption." The combination of the distinctions with the duration classifications in Tables 5.2 and 5.3 creates such subcategories as "scheduled temporary outage," etc.

Every planner should pay close attention to the definitions used by his company and by other utilities to which his system's reliability is compared. What appear to be identical indices can vary because of differences in how interruptions are defined. Many utilities do not include scheduled outages, storms, and other "non-failure-related" interruption in their reporting procedure.[3] Others leave momentary interruptions out, or report them as a special category.

---

[3] Most utilities remove "storms" from their statistics, arguing that outages and interruptions caused by hurricanes, tornadoes, and earthquakes should be removed from statistics that are meant to reflect long-term trends in satisfying normal reliability needs. There is merit in this concept, but the removal of "storm-caused" interruptions has been carried to extremes, such as a utility which claimed that all line-down failures during the three months following a hurricane were due to stress put on overhead lines by the hurricane's high winds. As a result, statistics for that year were the best ever recorded.

Reliability and Contingency Criteria

Such differences in definitions, interpretations, and reporting practices can frustrate distribution planners who try to find some real meaning in comparisons of their company's reliability experience to that of other utilities. Worse, these differences often cause confusion and frustration among customers, regulators, and industry standards organizations. A very common situation that leads to frustrated customers is the following: a disgruntled factory manager calls his electric supplier to complain about six interruptions that have occurred during the last five weeks. The electric utility's representative stops by the next day with operating statistics that show that no interruption in the plant's service occurred during the last six months.

Both factory manager and utility representative are "correct," but the utility's definition counted only temporary and sustained interruptions, while the factory manager was concerned because interruptions lasting less than a minute could lock up robotic equipment and bring other computerized operations to a standstill. At the present, this leads to disgruntled customers. As the power industry switches to open access and competition between suppliers, it will lead to customers who switch to another supplier. This is one reason reliability will become more important in distribution planning and operations.

Another point every distribution planner must bear in mind when working with reliability statistics is that utilities which report a certain type of interruption statistic, such as number of momentary operations, vary widely in how they interpret that and other definitions, as shown by Table 5.4, which lists interruption definitions from eight utilities in North America.

Table 5.4 Duration Limits Used to Define Instantaneous, Momentary, and Temporary Outages at Eight Utilities

Utility	Instantaneous	Momentary	Temporary
1	not used	< 5 minute	5 min. to 45 min.
2	< 15 secs.	15 sec. to 1 min.	1 min. to 45 min.
3	not used	up to 30 seconds	30 sec. to 30 min.
4	< 15 seconds	15 sec. to 2 min.	2 min. to 1 hour
5	< 60 seconds	60 sec. to 5 min.	5 min. to 2 hours
6	not used	< 1 minute	1 min. to 52 min.
7	< 1 minute	not used	1 min. to 2 hours
8	"cleared by inst. relay"	up to 2 minutes	1 min. to 1 hour

5.3 RELIABILITY INDICES

In order to deal meaningfully with reliability as a design criterion for distribution, it is necessary to be able to measure it and set goals. A bewildering range of reliability indices, or measures are in use within the power industry. Some measure only frequency of interruption, others only duration. A few try to combine both frequency and duration into a single value, which proves to be a nearly impossible task. Some measures are system-oriented, looking at reliability averaged over the entire customer base. Others are customer or equipment oriented, meaning that they measure reliability only with regard to specific sets of customers or equipment. To complicate matters further, most utilities use more than one reliability index to evaluate their reliability. The large number of indices in use is compelling evidence that no one index is really superior to any other. In fact, the author is convinced that no one index alone is particularly useful.

The basic problem in trying to measure reliability is in how to relate the two quantities, frequency and duration. Are two one-hour interruptions equivalent to one two-hour interruption, or are two one-hour interruptions twice as bad as the one two hour interruption? Most people conclude that the correct answer lies somewhere in between, but no two distribution planners, and no two electric consumers for that matter, will agree on exactly how frequency and duration "add up" in importance. For one thing, the importance of frequency and duration vary tremendously from one electric consumer to another. There are some industries where a five minute outage is nearly as damaging to productivity as a one-hour outage -- their computer and robotic equipment requires hours to re-set and re-start, or they have a long production process so sensitive to power fluctuations that they have to start over at the beginning if there is even a slightly interruption. For these customers five five-minute outages are much more serious than a single five-hour outage.

But there are other customers for whom short outages cause no significant problems, but who experience inconvenience during a sustained outage. This category includes factories or processes where production is only damaged if equipment has time to "cool down." Interruption of power to a heater or tank pressurizer for a minute or two is not serious, but sustained interruption allows a cool-down or pressure drop that results in reduced production.

One thing is certain: frequency of interruption is becoming increasingly important to more customers. Before the widespread use of digital clocks and computerized equipment, few residential or commercial customers even noticed or cared if power was interrupted briefly (for a few minutes at most) in the early morning (1 AM to 4 AM) for switching operations. It was common practice for utilities to perform minor maintenance and switching operations during this

Reliability and Contingency Criteria 165

period, so as to minimize any inconvenience to their customers. Today, utilities find this practice leaves them with customers who are not amused, for these people wake up to a house full of blinking digital displays. Worse, many wake up later than they expected, because their alarm clocks did not to go off.

The amount of power interrupted is also an important factor. Some reliability calculations and indices are weighted proportionally to customer load at the time of interruption, or to the estimated energy (kWhr) *not* delivered during the interruption.

The four most popular indices used in reliability analysis make no distinction of customer size. They treat all customers alike regardless of peak demand, energy sales, or class. These indices are SAIFI and CAIFI, which measure only frequency, and SAIDI and CTAIDI which measure only duration. Usually these four are used in conjunction with one another -- four numbers that give a rough idea of what is really happening to reliability system-wide.

These four indices, and most others, are based on analysis of customer interruptions during some *reporting period,* usually the previous month or year. All count a customer interruption as the interruption of service to a single customer. If the same customer is interrupted three times in a year, that constitutes three customer interruptions. If one equipment outage caused simultaneous interruption of service of three customers, that, too, is three customer interruptions.

The "S" in SAIFI and SAIDI means these two indices average their interruption statistics over the entire customer base (the system), the "C" in CAIFI and CTAIDI means those indices refer to only customers who experienced interruptions. Customers who had uninterrupted service during the period are precluded from CTAIDI and CAIFI analysis.

System Average Interruption Frequency Index (SAIFI) is the average number of interruptions per utility customer during the period of analysis. This is just the total number of customer interruptions that occurred in the year, divided by the total number of customers in the system,

$$\text{SAIFI} = \frac{\text{number of customer interruptions}}{\text{total customers in system}} \qquad (5.1)$$

Customer Average Interruption Frequency Index (CAIFI) is the average number of interruptions experienced by customers who had at least one interruption during the period. While SAIFI might state that there where 1.3 interruptions per customer, perhaps half of the customers experienced no outage. This means the rest had an average of 2.6 outages/customer. CAIFI tries to account for this by considering as its denominator only those customers that had at least one interruption.

$$\text{CAIFI} = \frac{\text{number of customer interruptions}}{\text{number of customers who had at least one interruption}} \quad (5.2)$$

If there is a large difference between CAIFI and SAIFI, it means outages are concentrated in only certain parts of the system or on certain customers. This could be due to poor design, poor maintenance, differences in the weather among areas of the system, or just plain bad luck, and further investigation will be required to determine cause, and cure.

System Average Interruption Duration Index (SAIDI) is the average duration of all interruptions per utility customer during the period of analysis. Here, the total customer-minutes of interruption is added together and divided by the total number of customers in the system

$$\text{SAIDI} = \frac{\text{sum of the durations of all customer interruptions}}{\text{total customers in system}} \quad (5.3)$$

Customer Total Average Interruption Duration Index (CTAIDI)[4] is the average total duration of interruptions among customers who had at least one interruption, during the period of analysis

$$\text{CTAIDI} = \frac{\text{sum of the durations of all customer interruptions}}{\text{number of customers who had at least one interruption}} \quad (5.4)$$

Again, if there is a large difference between CTAIDI and SAIDI, it means outages are concentrated on only certain parts of the system of certain customers. This could be due to poor design, poor maintenance, or just plain bad luck. Further investigation will be required to determine cause and cure.

Momentary Average Interruption Frequency Index (MAIFI) is the average number of momentary (and sometimes instantaneous) interruptions per utility customer during the period of analysis. Typically, if a utility distinguishes

---

[4] This particular computation, total number of customer hours out of service divided by the number of customers who had at least one outage during the period, was traditionally called CAIDI - customer average interruption duration index, without the "T." However, recently many planners in the power industry have tended to compute CAIDI as the total number of customer hours of interruption divided by the total number of customer interruptions, which yields just the average duration of a customer interruption. The difference between CTAIDI and CAIDI is that a customer who has three outages is counted only once in the denominator of CTAIDI, but three times in the denominator of CAIDI. The value now associated with "CAIDI" is actually a statistic, not an index, but regardless, the more commonly used definition of "CAIDI" now corresponds to the average customer interruption duration. Note that CTAIDI/CAIFI = CAIDI.

Reliability and Contingency Criteria 167

momentary interruptions as a separate class, then they are not included in the SAIFI and CAIFI statistics and MAIFI may be computed separately as,

$$\text{MAIFI} = \frac{\text{number of customer momentary interruptions}}{\text{total customers in system}} \quad (5.5)$$

In such a situation, total interruptions (from the customer standpoint) are probably best estimated by the sum: SAIFI + MAIFI.

Load Curtailment Indices

In addition, a number of indices try to relate reliability to the size of customer loads, in order to weight the interruption of a large load more than a small load. These usually work with kVA-minutes of outage using,

$$\text{Customer load curtailment} = \text{duration of outage} \times \text{kVA unserved} \quad (5.6)$$

Customer Average Load Curtailment Index (CALCI) is the average interruption kVA duration interrupted per affected customer, per year

$$\text{CALCI} = \frac{\text{sum of all customer load curtailments}}{\text{number of customers who had at least one interruption}} \quad (5.7)$$

Often, a very important statistic is the interruption rate and duration for the worst-case customer or customers. While the system as a whole could have good reliability, a small group of customers could still receive substantially lower reliability. CTAIDI and CAIFI do not always highlight such a situation, because the worst-case statistics could be mixed in with hundreds, even tens of thousands, of other interruptions. For this reason, some utilities record and use the worst case situations for their system evaluations.

Maximum Individual Customer Interruption Frequency (MICIF) is the maximum number of interruptions experienced by any customer during the period. In some cases, to average the impact of extreme or unusual cases, MICIF will be computed as the average of the worst n customers with regard to number of outages during the period (n is typically 12 to 50).

Maximum Individual Customer Interruption Duration (MICID) is the maximum total interruption time experienced by any customer during a period. Similar to MICIF, this index is often averaged over the worst n customers in order to remove the impact of unusual circumstances.

Usually, the same customer(s) is not involved with both MICIF and MICID, because frequency and duration are tied to very different processes. The customer with the maximum duration time in a year may have had only one or two outages, but be in a very isolated location that takes quite some time to reach in order to effect repairs.

Availability Indices

Roughly 75% of utilities in North American also calculate and report monthly and annually the availability of power: the percent of time, percent of customers, or fraction of demand for which power was available. These statistics tell almost nothing about the system, because they are always very close to 1.0 and the difference between acceptable (e.g. 99.97%) and non-acceptable performance (99.94%) is negligible as measured. Availability indices expressed in this manner are popular chiefly because they look very impressive to the uninitiated -- as stated in the introduction of this chapter, a 99.9% availability statistic looks very good, and this number exceeds the "very best" that can be done in many other endeavors, but with respect to power it is unacceptable to most customers and industries. Indices such as SAIDI, SAIFI, CTAIDI, and CAIFI are much more useful at tracking reliability and indicating what and where improvement is needed.

Using Reliability Indices

Duration as given by SAIDI and CTAIDI and frequency as given by SAIFI, MAIFI and CAIFI are very different aspects of reliability, related separate aspects of the utility system and its operation. Frequency, whether defined by SAIFI, CAIFI, or some other index, measures how often failures occur. It is mostly a function of the *causes* of the outages and their *extent* -- the type and condition of T&D equipment, and the layout of the system. Duration, on the other hand, is almost exclusively a function of the utility's organization, management, and resources for T&D repair.[5] Generally, the planner has a great deal more influence over layout and specification of equipment, factors that relate to frequency, not duration. Therefore, distribution planners tend to focus mostly of SAIFI, CAIFI, MAIFI and MICIF. SAIFI is the simplest reliability

---

[5]For example, the duration indices can be cut dramatically by doubling the number of repair crews standing by, and spacing them throughout the system so that the probability is high one or another will be close to every outage. Time to reach outages and restore service will be cut, leading to a reduction in SAIDI and CTAIDI. However, frequency of outages and the SAIFI and CAIFI indices will be unaffected.

Reliability and Contingency Criteria 169

measure, the most intuitively understandable, and the one which holds the least counter-intuitive surprises. It is perhaps the best single indicator of a distribution system's relative "health" as regards equipment and design -- number of customers interrupted by is dependent on the layout of the system and the condition of its equipment more than it is on the utility's management of the restoration process.

By contrast, SAIDI and CTAIDI and duration indices in general are not so much a measure of the condition of the distribution system and equipment, as they are a measure of how quickly the utility can restore service once a failure occurs. Doubling the repair crews available to restore service and spacing them out over the system so that on average they are closer to potential trouble spots will reduce repair times, and thus SAIDI and CTAIDI, but it will do nothing to reduce SAIFI and CAIFI.

CAIFI and CTAIDI are only superficially similar to SAIFI and SAIDI. The point of these two statistics is to distinguish how evenly distributed interruptions are over the system, or how concentrated in only a few areas they may be. Note that for any one system and period:

$$\text{SAIFI} \leq \text{CAIFI} \quad \text{SAIDI} \leq \text{CTAIDI} \quad \text{CAIFI} \geq 1.0$$

Note that the ratio of SAIFI to CAIFI and SAIDI to CTAIDI is the same; each pair has the same numerator, and they differ only in (the same two) denominators.

$$\text{CTAIDI/SAIDI} = \text{CAIFI/SAIFI} = \text{fraction of customers who experienced at least one outage in the period}$$

Therefore, CTAIDI versus SAIDI tells the planner nothing more than CAIFI versus SAIFI. Although many utilities use both, it actually makes sense to use only one of either CTAIDI or CAIFI, not both.

Analysis Using Reliability Indices

Reliability indices are used to evaluate historical and recent data to reveal trends and patterns, expose problems, and indicate how and where reliability can be improved. They are used in predictive (analytical study) to evaluate how well proposed solutions are likely to solve the identified problems.

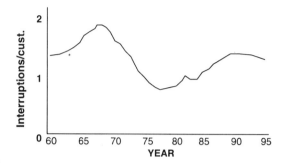

Figure 5.2 Three-year moving average of the annual number of interruptions for a region of a utility in the central United States. This area grew from open farmland into a developed metropolitan suburb during the period 1962 to 1970, during which time interruption rate rose due to the problems common to areas where new construction is present (digging into lines, initial high failure rates of new equipment and connections). Twenty years later, interruption rate began to rise again, a combination of the effects of aging cable (direct buried laterals in some residential areas) and trees, planted when the area was new, finally reaching a height where they brush against overhead conductor. The anomaly centered around 1982 was due to a single, bad storm.

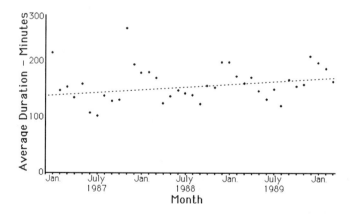

Figure 5.3 Average hours out of service per interruption on a monthly basis for a rural utility in the northern United States, varies due to random fluctuations, differences in the number of days in each month, and changes in weather. It also appears that average duration of interruption (dotted line), is slowly increasing over time, but further study is necessary before this conclusion can be confirmed.

Reliability and Contingency Criteria 171

Any of the reliability indices discussed above can be tracked over time to identify trends that indicate developing problems. Figure 5.2 shows frequency of customer interruption (SAIFI) for one suburban operating district of a metropolitan utility, on an annual basis over a twenty-eight year period. The operating district was open farmland (i.e., with no trees) up to 1962, when suburban growth of a nearby metropolitan region first spilled into the area, with much construction and the typically high interruption rates characteristic of construction areas (trenchers digging into underground lines, early equipment failures of new distribution facilities, etc.).

Over the following ten years, as the area plotted in Figure 5.2 gradually filled in with housing and shopping centers, and construction and system additions stabilized, frequency of interruption dropped. Twenty years after the area first began developing, interruption rate again began to rise, because of two factors: gradually increasing cable failure rates due to age, and the effects of trees, planted when the area first developed, now reaching a height where they brush against bare conductor on overhead lines.

Another obvious comparison to look at is the average duration of an outage, which is simply the total duration of all interruptions divided by the total number of interruptions that occurred during a period. The average duration is also given by the ratio SAIDI/SAIFI or CTAIDI/CAIFI. Average duration of interruption provides an indication of how the types of outages vary or how the utility is managing the entire process from trouble call to service restoration, as shown in Figure 5.3. Here, the increase in outage duration during winter for a utility in the northern United States is clearly visible (winter storms stress repair resources and make travel to repair sites lengthier). A possible upward trend in overall duration is also discernible. Without additional study, it is not possible to determine whether this is due to an increase in the number of outages, a change in type of outages occurring (perhaps "more serious" outages are occurring more often as equipment ages), or to a less effective (slower) repair process.

5.4 RELIABILITY AND CONTINGENCY CRITERIA FOR PLANNING

The best way to address reliability in distribution planning is to use it directly as a criterion in the evaluation of various alternatives. In this approach, frequency and duration of service interruption are computed for each potential alternative design (using a reliability analysis method of the type to be discussed in Chapter 16) and compared to corporate reliability criteria to assure compliance of all new designs, additions, and operating changes. The corporate reliability criteria set goals for the expected average frequency and duration of outages on any part of the system, as well as an upper limit on the worst case reliability that will be allowed for any one customer. Use of reliability values directly as design criteria is not a widely used practice at the time of this writing (1997). Instead, most utilities use indirect criterion based on contingency margin, switching zone counts, and other equipment or design-related criteria. However, use of reliability directly as a criterion can be expected to become widely used for three reasons:

1) It addresses reliability needs directly from the customer standpoint, rather than indirectly from the utility (equipment criteria standpoint), leading to improved quality assurance.

2) It imposes no restrictions on design or equipment selection other than that the final result meet criteria of customer performance. This can lead to lower costs.

3) Dependable, proven methods to accommodate reliability-based design are becoming widely available.

Reliability criteria, themselves, as well as estimated values made during the planning process are *expectations*, values that predict average behavior over a long period of time based on the probabilities of failure. Like any probabilistic phenomena, actual results can vary from what is expected, particularly over short periods of time.

Typically, utilities applying this approach will establish four criteria of interest to the planner:

SAIDI = System must satisfy this overall limit on expected duration of interruptions

SAIFI = System must satisfy this overall limit on expected number of interruptions

MICIF = Expected number of outages in a year should not exceed this for any customer

MICID = Expected duration of outages in a year should not exceed this for any customer.

In the design of any feeder, substation layout, or other addition to the distribution system, the planner uses these criteria as design constraints: any acceptable plan must have an expected average frequency and duration of interruption equal to or better than the corporate SAIDI and SAIFI goals (otherwise the plan would dilute the system's overall reliability). In addition, the expected maximum frequency and duration for any customer covered in the planning case must be less than MICIF and MICID. Since the SAIDI and SAIFI goals are averages for the system, feeder plans that fall slightly short of the

Reliability and Contingency Criteria 173

system's SAIDI and SAIFI targets can be deemed good enough, but they must go through a special review process to be judged acceptable.

CTAIDI and CAIFI are seldom used as criteria. In general, distribution planning studies focus on small portions of the system, at most a few substations at a time. If every such area of the system is planned so that its frequency and duration of interruption falls close to the corporate SAIDI and SAIFI targets, reliability will be as evenly distributed as practical throughout the system. The utility might still compute CTAIDI and CAIFI in its assessment of actual operating experience, to make certain results match its goals, but planners do not need to apply more than four criteria.

Tiered Reliability Areas

Often a utility will designate different frequency and duration targets for urban and rural areas. For example, its reliability criteria may specify part of the service territory as "urban" with SAIFI and SAIDI targets of 1.3 interruptions and 70 minutes/year for the system within that region. The rest of the system would be characterized as "rural" and designated with SAIFI and SAIDI rates of 2.2 and 150 minutes respectively. Such a tiered reliability design structure merely recognizes a common and long-standing characteristic of most utility systems -- service reliability is less in sparsely populated areas than in densely populated areas.

This difference in urban and rural standards does not imply that rural customers are less important, nor is it due to inattention on the part of the utility or to poorer design practices applied to rural areas. Theoretically, there is no reason why equivalent levels of reliability cannot be delivered to urban and rural areas alike. But pragmatically, in rural areas the cost of most reliability enhancement measures is high because of the great distances involved, and that must be spread over fewer customers per mile of feeder. This very high cost per customer is simply not acceptable to most of those affected. Most rate-design and regulatory processes do not consider it equitable to spread the cost of such rural upgrades over the entire rate base (i.e., to those who live in the cities, too). Thus, a higher frequency of interruption, and longer interruptions are one of the expected pleasures of rural life.

In some cases, a utility will further partition its service territory into rate-reliability tier areas, with as many as six different types of areas to reflect greater distinction in target reliability levels and expenses by location.[6]

---

[6] For example, from highest to lowest reliability these might be: downtown network, urban, high-use industrial, suburban, rural, and mountainous areas.

As mentioned above, reliability-based planning, whether with tiered criteria or not, is performed by only a handful of utilities at the present time (1996). Its rarity is due to two reasons. First, electric utilities have only recently developed the information systems needed to track customer data with the detail necessary for such analysis. Computing in detail, and accurately, the extent of an outage (the number of customers behind any point in the system) is the key factor in estimating the frequency and duration of interruptions for a feeder. Information systems capable of routinely providing this data to feeder planning and analysis programs have been available only in the last few years.[7]

Second, applying reliability criteria in distribution planning requires the computation of the expected frequency and duration of interruptions for each particular design alternative, using something that could be termed a "reliability load flow" -- a technique to compute expected outage rate and interruption statistics of the type shown earlier in Figure 5.1, for any arbitrary system design. Reliability assessment methods must analyze not only the distribution layout and its equipment, but such factors as the number, type, and proximity of repair crews. Methods to perform this type of analysis do exist, but while they are growing in popularity, they are not in widespread use at the time of this writing.

Contingency and Margin Criteria Are Indirect Reliability Standards

The vast majority of electric utilities apply reliability design concepts to distribution planning indirectly rather than directly, by interpreting their reliability goals into contingency margin standards and design criteria. These address frequency of interruption by specifying equipment standards, design and layout guidelines, and switching and fusing rules which will limit outages and the extent of outages to acceptable limits. The utility still maintains reliability goals, and still measures and reports SAIDI, SAIFI and similar indices, but these goals are not reflected directly in design criteria. Rather, they are achieved indirectly, by applying contingency margin criteria which have been interpreted from those SAIDI and SAIFI goals and which are hopefully appropriate.

Contingency margin standards are designed to allow sufficient leeway for any unit of equipment to pick up load interrupted due to the outage of a nearby unit,

---

[7] Another perceived barrier is a lack of equipment data needed for reliability analysis -- factors such as the expected failure rate of overhead feeders, underground conductor, transformers, and other equipment in the distribution system. Many utilities are under the impression that this information is unavailable or difficult to obtain. In fact, this is never a problem. Failure rate data sufficient for any conceivable practical planning application can be found without undue difficulty.

Reliability and Contingency Criteria 175

and to permit reasonably quick (e.g., 15 or 30 minutes to an hour) switching. For example, a utility may establish a policy that substations will always have two transformers of equal size, each loaded to no more than 80% of their rating. During the outage of either transformer, the other can pick up both loads with an overload of 60%, an amount it should be able to sustain for two to four hours, time enough it is hoped for the utility to take steps to handle the problem over a longer period if need be.

Such a policy establishes a certain type of design (two transformers per substation) and loading standard (transformer loading cannot exceed 80% of rating) in order to leave sufficient margin for equipment failure. Similar and compatible criteria are established for conductor loading, switch placement, and other design practices, all rules and standards that *imply* certain reliability results. Extent can be implicitly handled with limitations on the number of customers behind any fuse. For example, a criterion of "no more than 200 customers can be behind any fuse" would distinguish configuration A from B in Figure 5.1).

5.5 COMPARISON OF RELIABILITY INDICES AMONG UTILITIES

Comparison of SAIDI, SAIFI, and other reliability indices among different utilities can be an exercise in frustration for the distribution planner, because supposedly similar standard definitions such as SAIFI vary tremendously in practice from one utility to another, as alluded to in section 5.3. The definitions and formulae for the various indices are nearly always the same from utility to utility, but the characteristics of the component measurements and data sources that input into the formulae are different. As a result, comparison of SAIFI values between two utilities may show a large discrepancy when there is no significant difference in actual customer interruption levels. Despite this difficulty, and the questionable utility gained by such comparisons, most distribution planners are asked to perform such studies from time to time.

The usual reason that utilities compare reliability statistics are to determine how well they are doing versus the rest of the industry and to prove that their system is better than average. They may also be interested in determining what might be a reasonable target for their reliability criteria. Table 5.5 gives SAIDI and SAIFI statistics for six utilities, based on what the author believes to be fairly equivalent definitions and reporting methods. If nothing else, this table shows that reliability results differ tremendously because of real differences in service territory, climate, and demographics.

Table 5.5 Average Annual SAIFI and SAIDI for Six Utilities 1988-1992

Utility	Service Area	Climate	SAIFI	SAIDI
1	Dense urban area	Hot summers, bitter winters	.13	16
2	Urban/suburban	Hot nearly year round	2.7	34
3	Suburban & rural	High isochronic (lightning)	2.0	97
4	Urban & rural	Mild year round	2.1	122
5	Rural, mountainous	Mild seasons	1.1	168
6	Rural, mountainous	Bitter winters	2.3	220

Differences in How Reliability Indices Are Computed

The values shown in Table 5.5 were adjusted, as carefully as possible, to reflect a common basis of definitions, measurement, and reporting. The author believes they are roughly equivalent. However, as mentioned earlier, terms like SAIDI and SAIFI are in wide use throughout the power industry, but often with differences in exact interpretation and reporting of their component parts. For one thing, as was discussed earlier (Table 5.4), there are differences in how utilities define the types of interruptions they include in their reliability indices. This can materially affect the resulting values.

In addition, there is no commonality in how the customer count used in the denominator of most reliability indices are defined and ascertained. Many utilities consider a master-metered apartment building or shopping mall to be a single "customer," while a few count the individual units or stores as customers to get a more accurate assessment of interruption impact. Regardless of this difference, there are other variations in customer count. Most utilities define the number of customers system-wide as the average number during the year or use best estimates on a monthly basis. Others use the number as of January 1st, or some other date. Policy on counting seasonal customers (summer beach homes only served part of the year) varies too. The important point is that slight variations in system reliability indices, due to growth, seasonal change, and simple differences in how "customers" are counted, are common, and can account for differences of up to 5% in SAIDI and SAIFI statistics.

In addition, utility practice differs widely when estimating how many customers any particular outage actually interrupts, something that makes statistics like CTAIDI and CAIFI vary greatly from utility to utility. Some utilities use very comprehensive databases and distribution management systems (DMS) to determine precisely how many customers each equipment outage takes out of service. But many utilities use approximate methods to estimate the

Reliability and Contingency Criteria 177

customer count of each particular outage because they lack the data or resources to perform such analysis accurately.

For example, one utility in the midwest United States employees a reliability assessment procedure that *assumes* every feeder outage knocks out service to an average number of customers, in this case 1764 customers (the average customer per feeder over their entire system). Thus, every instance of a locked-out (completely out of service) feeder is assumed to have interrupted service to exactly 1764 customers, even though the number of customers per feeder varies from less than 600 to more than 2400. Furthermore, their procedure assumes that every partial feeder outage knocks out power to 294 customers (one sixth of an average feeder). Over a large system, and a full year, such approximations may make little difference in the computed SAIDI, SAIFI, and other indices. On the other hand, they may; and they do introduce a significant approximation error when computing indices over short periods, or within small regions.

Another difference in reporting practices is whether momentary interruptions caused by "cold switching" during restoration and maintenance are counted. Sometimes, to avoid closing a loop between substations or feeders, it is necessary to isolate a portion of a feeder first before switching it to a new feeder or substation source. Proper organization of distribution operations and personnel can reduce this interruption to as little as five seconds, but it does cause customer interruptions. Some utilities count these interruptions as momentary or temporary interruptions, but many don't.

Load estimation procedures for the various curtailment indices such as CALCI are another factor that varies tremendously in practice. Some utilities calculate the curtailed energy based on average load (annual billing divided by 8760 hours) for the customers involved, not the actual load at time of outage. Others try to estimate the actual demand using hourly load models for various customer types.

Estimated Durations of Interruptions

The single widest variation in reliability reporting practices relates to the definition of the duration of interruptions. Nearly every utility has a good record of when service is restored to each interrupted customer, but all but a few (those with real-time monitoring equipment in the field) estimate the time the interruption began. Breaker operations that lock out a feeder or a whole substation usually are recorded by SCADA or on-site equipment, so the timing is known precisely. But these outages constitute only a small part of all distribution failures, and there is no similar recording mechanism to identify the precise time when a line fuse operated, or when a lateral line fell, or if a service transformer failed. To compute the duration of the outage and the interruptions it caused, *this time of occurrence has to be estimated* by some means.

Many utilities define the "start" of each interruption as the time when it is first reported, even though it had to have occurred several minutes before the customer called in to report the trouble. Some utilities try to account for that delay by adding a minute or three minutes to the time when they recorded the call. However, no one wants to apply a procedure that lengthens the duration of the outage data, making the duration indices worse, as this does.

A very few utilities still apply a practice with the opposite effect, one common in the 1930s, defining the "start time" for interruptions as the moment when each is relayed to the distribution dispatcher, which can be from two to fifteen minutes after the trouble call was first received from the customer.

The point of the discussion above is that anyone comparing statistics between two or more utilities should delve deeply into how apparently similar statistics are calculated, and exactly how data was defined and collected for those calculations. They should expect major differences to exist in the statistics due only to definitions and recording differences. If they want to adjust these data to account for these differences so that they can be compared on a valid basis, they will have to be prepared to do a great deal of additional work, and they should not be surprised if this cannot be done well.

5.6 SETTING RELIABILITY GOALS

What Constitutes Good Reliability?

Basing corporate reliability goals on industry averages and statistics of other utilities is almost fruitless, both because of the differences in reporting practices discussed above (a utility could be comparing its apples to the industry's oranges) and because the difficulty of providing reliability will vary greatly from one area to another. For example, a utility in rural, northern New England averages roughly 200 to 300 minutes of customer interruption per year, while one in southern California has a reliability target of only 60 minutes per customer per year. Assuming for the moment that both utilities define interruptions on a common basis and compute reliability indices in a similar manner, does this mean the California utility is doing a better job of providing reliable service than the one in New England? Not necessarily, because the northern utility must contend with frequent and bitter ice-storms in the winter (a rarity in southern California). Those same storms inhibit rapid travel of repair crews during much of the year, and the utility has a large part of its system and customer base scattered through sparsely populated areas that are reachable only over secondary roads that remain snow-covered through much of the winter.

Reliability and Contingency Criteria 179

Beyond those reasons, there is absolutely nothing to indicate that the customer bases in Maine and southern California expect similar levels of service reliability. Each utility would be better off setting reliability goals based on local conditions, and in particular, local customer requirements and feedback. A problem with this approach is determining exactly what the customers want. In the author's experience, consumer surveys reveal little more than that electric customers would always prefer lower rates and perfect reliability. It does not appear easy to design a questionnaire that is simple but comprehensive in explaining the relationship between cost and reliability, so that customers understand the reliability/cost trade-offs they are being asked to consider. Focus groups and discussion at community meetings provide a better quality of feedback, but still give what is at best only qualitative, not quantitative, indications of customer reliability preferences.

Assessing Reliability Based on Past Performance

One practical way to set reliability goals is to base them on past results. For example, suppose a study of operations from 1987 to 1992 determined that customer service reliability was unacceptable in 1985, 1988, and 1991, but acceptable in the other years. Then SAIDI and SAIFI limits for the system fall somewhere between the worst and best annual statistics in this period, as shown in Figure 5.4. This figure illustrates the *concept* of setting reliability criteria and goals based on past results, but the actual analysis might include both more detail and more complicated analysis than a comparison based only on two statistics, as shown.

For example, with respect to the example cited in Figure 5.4, it was suggested by several of the utility's distribution engineers that since frequency and duration are related as regards customer impact in some way, rather than the column and row approach (shown shaded), a single diagonal line could be used to indicate a relationship between frequency and duration that separates good from poor reliability. (The reader will notice that a straight line drawn through the two points SAIFI=3.0, SAIDI=60 minutes, and SAIFI=1.7, SAIDI=120 minutes divides the results into acceptable and non-acceptable categories.) There is considerable merit in performing such analysis of possible relationships between frequency and duration, although in this particular case the result was considered too tentative and the implementation of a linear formula as a planning criterion was deemed unjustifiable. Ultimately, the utility established SAIFI and SAIDI limits of 2.2 interruptions per year and 105 minutes/year respectively.

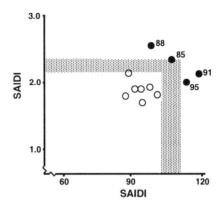

Figure 5.4 Annual interruption statistics for an electric membership cooperative located in the southern United States for 1985 through 1995 give an indication where its reliability targets should be set. Customer interruptions in 1985, 1988, 1991, and 1995 (solid dots) were considered to be "higher than desired" while other years (lighter dots) were considered satisfactory. Thus, the reliability design targets lie somewhere in the shaded column (duration) and row (frequency).

Similarly, interruption statistics studied on a feeder by feeder basis can be useful in setting standards, particularly if several problem feeders where reliability was not acceptable, and several good feeders where it definitely was, can be identified. It is best to use a long observation period (2 years at the least) for this type of feeder-level evaluation, since random occurrences can make almost any feeder have a poor reliability record for one year.

This particular method of comparing "good" and "bad" times and good and bad areas is certainly not perfect, but it has proven to be the only practical way the author has found of fixing reliability targets based on generally available historical data.

Value-Based Planning

Perhaps the most intellectually appealing method of setting reliability goals is to optimize the entire utility-customer value system, as illustrated in Figure 5.5. For the utility, the cost of building and maintaining the distribution system increases as greater reliability is sought, something that can be quantified and calculated if need be. For the customer, poor reliability has a cost -- inconvenience and disruption of schedule to a household, lost revenues for a busines, and lost wages,

Reliability and Contingency Criteria

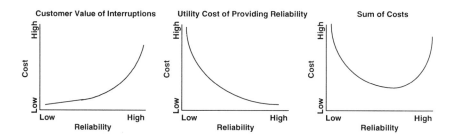

Figure 5.5 Value-based planning attempts to include both the utility (supply-side) and customer's (demand-side) costs of reliability (or lack of it) into a single cost-reliability assessment which is then optimized. This identifies the reliability target for the system.

production, and rent to a factory or similar industry. This cost penalty drops as reliability is improved. The total cost, utility cost to provide a certain level of reliability plus the customer costs due to imperfect reliability, has a minimum, which is the desired target point in VBP. Providing any more reliability than that would mean charging the utility's customers more for reliability than it is worth to them.

As good as the concept of value based planning is, it presents the distribution planner with a number of practical difficulties. The most insurmountable is that different customers put very different values on reliability. Two adjacent factories served by the same feeder may have far different reliability needs as regards either frequency or duration of interruptions. One may need very high reliability, the other may be content with low reliability if that is reflected in the price.[8] To satisfy the needs of the first customer, the utility must make improvements to the feeder and the level of service to both. Worse, only one of the two customers is willing to pay more for higher reliability (although perhaps not enough to pay for the improvements alone). For this reason, when applied to the planning of any entire utility or a customer class, value-based planning can do

---

[8] This is common. The first customer might be a factory with robotic control equipment, which not only has to re-initialized and set up after an interruption, but that often experiences mechanical transients as the power fails, destroying work in progress. The second customer may be a paper pulping or steaming/pressure process, which tolerates momentary and short duration power outages relatively well.

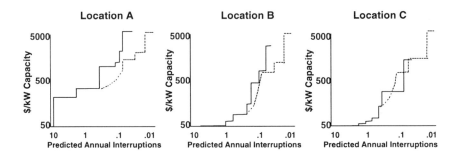

Figure 5.6 In practice, value-based planning often has to deal with discontinuous cost-versus reliability curves, which complicates application of the method. Shown here are the supply-side T&D costs to deliver power at certain reliability levels to three locations in a utility system in the south central United States (solid lines). An additional complication is that beyond a certain point, it is usually less expensive to improve reliability with customer-side equipment improvements (dotted lines) which includes equipment like standby generation, UPS, and measures to reduce interruption sensitivity.

little more than establish a general guideline of reliability criteria for the whole distribution system -- a common denominator that some customers will find too high and others too low.

A second difficulty with applying value-based planning is that quite often the reliability-cost curve for the supply-side is distinctly discontinuous, as shown in Figure 5.6. Large reliability changes can only be affected by major improvements that cause a leap in reliability (i.e., such measures as changing a radial feeder to a closed loop design, or provision for automatic rollover switching during an outage) with no in-between levels of reliability available. The concept of value-based planning still works, but not as smoothly or as clearly as might be hoped. If nothing else, the situation becomes more difficult to explain to customers.

Despite these and other difficulties, value-based planning is a method of balancing supply and demand-side costs against one another. This alone makes it a recommended exercise for all planners looking to establish or confirm the reliability goals for their company. In addition, it is of great use in the service planning for special or large customers, particularly in cases where the utility and customer are negotiating or working together to solve the customer's unique needs at a special price. Here, it can be used to balance the utility's combination of delivered reliability and price against the customer's needs.

5.7 CONCLUSION AND SUMMARY

Reliability of service is one of the major factors customers weigh in perceiving how well their electric utility is doing its job. A utility must set appropriate reliability goals and take measures to plan, engineer, and operate its system to achieve targeted levels of both frequency and duration of customer interruption. Various reliability indices can be used to track frequency and duration of customer interruptions over time and by location within the distribution system, in order to identify problems and ensure that goals are being met.

Equipment outages cause interruptions of service. Customers do not care about equipment outages *per se,* only the interruptions caused by these outages. Thus, one way to reduce reliability problems is to set standards that limit the extent of outages -- how widely outages spread interruption of service. The cost of this means of improving reliability must be weighed against the cost and effectiveness of other means. Most utilities make sure that reliability targets are met, not through detailed reliability analysis in the planning stage, but implicitly through contingency and design criteria that are based on assessment of their reliability impacts. Tracking of outages and interruptions on a monthly and annual basis is still necessary, however, to assure goals are being met and to detect early any anomalies or unexpected problems.

Comparison of reliability indices among different utilities is difficult due to differences in data gathering, definitions, and computation practices. Reliability will vary considerably among utilities, due to reasons of climate and terrain. The recommended way to set reliability goals for a system is to study recent local results and try to establish when and where these were met well versus when they were not, setting criteria based on these differences. Value-based planning, although difficult, is also useful, particularly in cases where reliability is an issue to large or special industrial and commercial customers.

FOR FURTHER READING

Albrecht, P.F., and H.E. Campbell, "Reliability Analysis of Distribution Equipment Failure Data," EEI T&D Committee Meeting, January 20, 1972.

Allan, et al., "A Reliability Test System for Educational Purposes -- Basic Distribution System Data and Results," *IEEE Transactions on Power Systems,* Vol. 6, No. 2, May 1991, pp. 813-821.

Billinton, R., and J. E. Billinton, "Distribution System Reliability Indices," *IEEE Transactions on Power Delivery,* Vol. 4, No. 1, January 1989, pp. 561-68.

Billinton R., and R. Goel, "An Analytical Approach to Evaluate Probability Distributions Associated with the Reliability Indices of Electric Distribution Systems," *IEEE Transactions on Power Delivery,* PWRD-1, No. 3, March 1986, pp. 245-51.

Bunch, J.B., H.I. Stalder, and J.T. Tengdin, "Reliability Considerations for Distribution Automation Equipment," *IEEE Transactions on Power Apparatus and Systems,* PAS-102, November 1983, pp. 2656 - 2664.

"Guide for Reliability Measurement and Data Collection," EEI Transmission and Distribution Committee, October 1971, Edison Electric Institute, New York.

Horton, W. F., et al., "A Cost-Benefit Analysis in Feeder Reliability Studies," *IEEE Transactions on Power Delivery,* Vol. 4, No. 1, January 1989, pp. 446 - 451.

Institute of Electrical and Electronics Engineers, *Recommended Practice for Design of Reliable Industrial and Commercial Power Systems,* The Institute of Electrical and Electronics Engineers, Inc., New York, 1990.

Patton, A.D., "Determination and Analysis of Data for Reliability Studies," *IEEE Transactions on Power Apparatus and Systems,* PAS-87, January 1968.

N. S. Rau, "Probabilistic Methods Applied to Value-Based Planning," *IEEE Transactions on Power Systems,* November 1994, p.4082.

Walker, A.J., "The Degradation of the Reliability of Transmission and Distribution Systems During Construction Outages," Int. Conf. on Power Supply Systems. IEE Conf. Publ. 225, January 1983, pp. 112 - 118.

White, H.B., "A Practical Approach to Reliability Design," *IEEE Transactions on Power Apparatus and Systems,* PAS-104, November 1985, pp. 2739 - 2747.

6
Economic Evaluation

6.1 INTRODUCTION

A major attribute of planning in almost all endeavors is reduction of cost. Criteria on service quality and standards must be met and guidelines must be followed, but within those limits the planner's goal is to minimize the cost. Every alternative plan contains or implies certain costs: equipment, installation labor, operating, maintenance, losses, and many others as well. Alternatives vary not only in the total cost, but often equally important, when the costs are incurred -- how much must be spent *now*, and how much *later?*

Traditionally, electric utilities have been given a monopoly franchise for electric service in a region, which carried with it both an obligation to serve and a requirement to work within a regulated price structure. Regulated prices are based on cost and regulated utility planning on cost minimization -- the utility can expect to make a reasonable return on its investment and recover all its costs, but it must work to reduce its costs as much as possible.

Thus, in traditional utility planning, expansion needs were defined by the obligation to serve, and planning goals were defined by a cost-based pricing environment. Utility planning focused on finding the lowest cost alternative to serve all the customers. In addition, as the sole provider of electric service the utility assumed a "resource portfolio management" function for its customers, meaning that it had an obligation to determine and implement the least-cost use of energy efficiency and other resources, which led to integrated resource planning.

In a de-regulated power industry, this situation will change very little for the distribution planner. The local distribution company (LDC) will still be regulated, and prices for delivery (if not power) will still be cost-based. Instead of an obligation to serve, the LDC will have an obligation to connect, or more specifically an *obligation to provide sufficient capacity*. Under retail wheeling, some other company's power may be flowing through the distribution system to the customers, but the distribution company is still required to provide sufficient capacity and to do so at the lowest possible cost.

This chapter looks at costs, costing analysis, and cost minimization concepts for planning purposes. Entire books have been written on the subject of engineering economics (i.e., cost analysis and comparison), and the purpose of this chapter is not to duplicate or summarize them. Instead, its focus is on what costs represent and how they are to be used in planning. The chapter begins with a look at cost and the various types of cost elements that are included in an analysis, in Section 6.1. Section 6.2 examines the time-value of money and the application of present worth analysis methods to planning. Cost-effectiveness tests are covered in Section 6.3. Section 6.4 concludes with a look at an important aspect of planning, particularly in a competitive marketplace: price-related variation of costs with respect to location, timing, and level of the system

6.2 COSTS

"Cost" is the total sacrifice that must be expended or traded in order to gain some desired product or end result. It can include money, labor, materials, resources, real estate, effort, lost opportunity, and anything else that is given up to gain the desired end. Usually, such a combination of many different resources and commodities is measured on a common basis -- such as money -- by converting materials, equipment, land, labor, taxes and permits, maintenance, insurance, pollution abatement, and lost opportunity costs to dollars, pounds, marks, yen, or whatever currency is most appropriate. In cases where all the elements of cost can be put onto a common basis, the subsequent planning can be done in a *single-attribute* manner, the goal being to minimize the monetary cost and the cost reduction basically trading the cost of one item against another to find the best overall mix. However, in rare cases, some costs cannot be converted to money -- for example rarely can esthetic impact or other "intangibles" be converted to monetary cost. In such cases, cost reduction must be done as *multi-attribute* planning and cost minimization, involving more complicated planning methods which are, frankly, often not entirely satisfactory.

Economic Evaluation

Initial and Continuing Costs

A distribution substation, feeder, or any other item has both an initial cost to create it and put it in place, and a continuing cost to keep it in operation, as illustrated in Figure 6.1. The initial cost includes everything needed to put the substation in place for its initial operation, and may include a very comprehensive list including engineering, permits, surveying, land, legal fees, site preparation, equipment, construction, testing, inspection, certification, incidental supplies and labor, and insurance. Continuing costs entail keeping it in operation -- inspection and maintenance, routine supplies and replacement parts, taxes, insurance, electrical losses, fuel, and perhaps other expenditures.

The initial cost is incurred only once, often over a period of several months or years during which the item (e.g., substation or transmission line) is built, but it is usually considered as a single budget expense item allocated to a specific year in the utility plan. The continuing costs persist as long as the substation exists or is in operation. Usually, continuing costs are recorded and studied on a periodic basis -- daily, monthly, or annually -- with yearly analysis generally being sufficient for most planning applications.

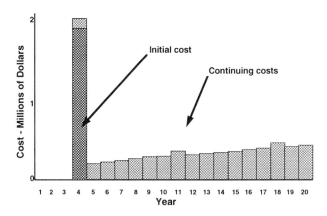

Figure 6.1 A new substation's costs are broken into two categories here, the *initial cost* -- a one-time cost of creating the new substation -- and its *continuing annual costs* which are required to maintain it in operation.

Fixed versus Variable Costs

The cost of a substation, feeder, or other element of a power system can also be viewed as composed of fixed and variable costs. Fixed costs are those that do not vary as a function of any variable element of the plan or engineering analysis. For example, the annual cost of taxes, insurance, inspection, scheduled maintenance, testing, re-certification, and so forth required to keep a 100 MVA substation in service do not vary depending on its loading -- whether it serves a peak load of 5, 32, 47, 60, or 92 MVA these costs would be the same. Similarly, the no-load (core) losses of its transformers create a cost that does not vary as a function of load, either. By contrast, the transformer's load-related (copper) losses do vary with peak load -- the higher the substation's loading the higher the losses costs. These are a variable cost, varying as a function of the amount of load served. Some types of maintenance and service costs may be variable, too -- highly loaded transformers need to be inspected and stress damage repaired more often than similar but lightly loaded ones. Figure 6.2 illustrates fixed and variable costs.

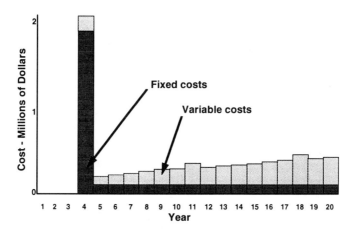

Figure 6.2 The new substation's costs can also be viewed as composed of both fixed costs -- those that are constant regardless of loading and conditions of its use -- and variable costs, that change depending on load, conditions, or other factors in its application as part of the power system. In this example, the variable costs increase nearly every year -- reflecting higher losses costs as the load served gradually increases due to load growth.

Economic Evaluation

In many engineering economic evaluations, fixed costs include all costs except variable costs of operation such as fuel and losses. However, there are a sufficient number of exceptions to this rule so that planners should both consider what they will designate as "fixed" and "variable" in a study, and also check any data or study upon which they build to make certain the meaning of those terms in that work is consistent with their own.

As an example of how "fixed" and "variable" costs can change depending on the planning context, consider a 100 MVA substation -- one composed of four 25 MVA transformers and associated equipment. After it is built and in place, its initial construction cost is always considered "fixed" in any study about how it might be utilized, expanded, or otherwise changed with regard to the plan. Variable costs include only those that change depending on its load or utilization: if more load is transferred to it so that it serve a higher load, its losses go up; if low-side voltage is regulated more tightly, its LTC maintenance costs will probably increase, etc.

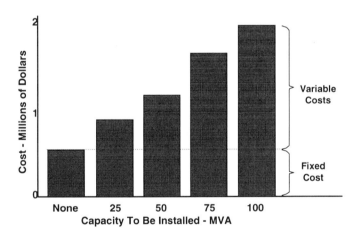

Figure 6.3 Definition of "fixed" and "variable" costs depends on the context. Although often the capital cost of a substation is labeled a "fixed" feature in engineering cost studies, here it consists of both fixed and variable costs: planners have four capacity options to consider for a substation -- 25, 50, 75, and 100 MVA -- representing installation of from one to four transformers. In this planning context the cost of the substation can be viewed as composed of a "fixed" component consisting of those items common to all four options, and a "variable cost" that depends on the installation of capacity from 25 to 100 MVA, as shown.

But in the initial planning stage, perhaps the planners considered the alternatives of building it with one, two, three or four transformers in place. In that situation, its "fixed" cost within the planners' minds might have included *only* those costs common to all four alternatives -- land, site prep, and so forth. In addition, the planners would consider a variable capacity cost, which changes as a function of adding one, two, three, or four transformers (Figure 6.3). This is in fact how some substation planning optimization programs accept cost data as input, as will be discussed in Chapter 12.

Sunk Cost

Once a cost has been incurred, even if not entirely paid, it is a sunk cost. For example, once the substation has been built (all 100 MVA and four transformers of it) it is a sunk cost, even if ten years later the utility still has the substation on its books as a cost, paying for it through depreciation or financing of its initial cost.

Embedded, Marginal, and Incremental Cost

Embedded cost is that portion of the cost that exists in the current planned system, or configuration, or level of use. Depending on the application, this can include all or portions of the initial fixed cost, and all or parts of the variable costs. Often, the "embedded" cost is treated as a fixed cost in subsequent analysis about how cost varies from the current operating point. Marginal cost is the slope (cost per unit) of the cost function at the current operating point (Figure 6.4). This point is usually (but not always) the point at which current embedded cost is defined.

Incremental cost is the cost per unit of a specific jump or increment -- for example the incremental cost of serving an additional 17 MVA from a certain substation, or the incremental cost of losses when load on a feeder decreases from 5.3 to 5.0 MVA . Marginal cost and incremental costs both express the rate of change of cost with respect to the base variable, but they can differ substantially because of the discontinuities and non-linearities in the cost relationships. Therefore, it is important to distinguish correctly and use the two. In the example shown in Figure 6.4, the marginal cost has a slope (cost per unit change) and an operating point (e.g., 45 MW in Figure 6.4). Incremental cost has a slope and both "from" and "to" operating points (e.g., 45 MVA to 55 MVA in Figure 6.4) or an operating point and an increment (e.g., 45 MW plus 10 MVA increase).

Economic Evaluation

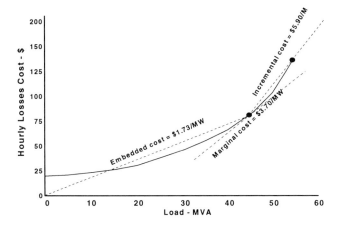

Figure 6.4 Hourly cost of losses for power shipped through a substation transformer as a function of the load. Currently loaded to 45 MVA, it has an embedded losses cost of 1.73/MVA, and a marginal cost of losses at that same point of $3.70/MVA. The incremental cost of losses for an increase in load to 55 MVA load is $5.90/MVA.

Revenue Requirements

In order to operate as a business, an electric utility must take in sufficient revenue to cover its continuing operating costs, pay for its equipment and system, cover its debt payments (loans and bonds), and provide earnings for its owners (shareholders), as shown in Figure 6.5. These revenues are the charges its customers must pay. Minimizing the total amount of revenue the utility needs is one way to keep customer charges at a minimum, a method used by a majority of utilities.

Minimum Revenue Requirement (MRR) planning is aimed at keeping customer bills as low as possible -- it seeks to minimize the amount of money the utility must collectively charge its customers in order to cover its costs. While cost reduction nominally contributes to revenue requirements reduction, often particular planning decisions not only incur or avoid costs, but also shift them from one column of the balance sheet to another. It is here where attention to revenue requirements can often reduce revenue even if not costs.

While similar in some cases, particularly capital expansion planning, revenue minimization and cost minimization will lead to slightly different decisions with regard to selection of alternatives and timing of expenses. Capital expansion

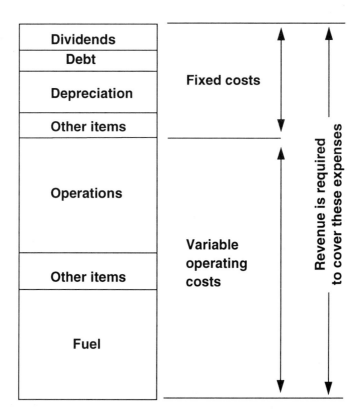

Figure 6.5 The total revenue includes what is needed to cover operations, fuel, debt, etc. Minimum Revenue Requirements planning seeks to minimize the total requirement for revenue (rightmost arrow).

Economic Evaluation 193

decisions -- such as when and how to build a new substation -- require capital which is often financed, while most continuing operating costs are not. For example, returning to the question poised earlier in this chapter, about whether to build a new feeder with larger conductor than the minimum necessary in order to reduce losses, the capital cost of building with the larger conductor may often be a financed cost to the utility -- an expense which it carries on its books and for which it pays interest.[1]

Thus, the present worth of a "present expense" (e.g., adding the larger conductor today) has a future element to it (finance costs over some number of future years). In order to reach the correct decision about whether the feeder should be built in this manner, the utility's planning evaluation must include in its assessment the correct evaluation of such costs. This can be done by adjusting the present worth factors used to compare "present" to "future" costs so that they are compatible with the utility's structure, as illustrated in Figure 6.5. If a utility has very high interest payments for its capital, it means it wants to raise the present worth factor for its capital investments, because the benefits (losses savings) of an investment made today (a larger feeder) must "pay for" the capital financing of the investment.

6.3 TIME VALUE OF MONEY

Any utility planner must deal with two types of time versus money decisions. The first involves deciding whether a present expense is justified because it cancels the need for a future expense of a different amount. For example, suppose it has been determined that a new substation must be built in an area of the service territory which is the site of much new customer growth. Present needs can be met by completing this new substation with only one 25 MVA transformer, at a total initial cost of $1,000,000. Alternatively, it could be built with two 25 MVA transformers installed, at a cost of $1,400,000. Although not needed immediately, this second transformer will be required within four years because of continuing growth. If added at that time, it will cost $642,000 -- a reflection of the additional start up cost for a new project, and of working at an already-energized and in-service substation rather than at a "cold" site. Which plan is best? Should planners recommend that the utility spend $400,000 now to save $642,000 four years from now?

A second and related cost decision involves determining if a present expense is justified because it will reduce future operating expenses by some amount. Suppose a new feeder is to be built along with the new substation. If built with

---

[1] Perhaps because as the utility views its business, it considers that all capital is financed.

336 MCM conductor at a cost of $437,000, the new feeder will be able to satisfy all loading, voltage drop, contingency and other criteria. However, if built with 600 MCM conductor, at a total cost of $597,000, in will lower annual losses costs every year in the future by an estimated $27,000. Are the planners justified in recommending that $160,000 be spent on the larger, based on the long-term continuing savings?

These decisions, and many others in distribution planning, involve comparing present costs to future costs, or comparing plans in which a major difference is *when* money is scheduled to be spent. To make the correct decision, the planner must compare these different costs in time on a sound, balanced basis, consistent with the electric utility's goals and financial structure.

Common Sense: Future Money Is Not Worth As Much As Present Money

Few people would happily trade $100 today for a reliable promise of only $100 returned to them a year from now. The fact is that $100 a year from now is not worth as much to them as $100 today. This "deal" would require them to give up the use of their money for a year. Even if they do not believe they will need to spend their $100 in the next year, conditions might change.

A utility has similar values, which means that when money has to be spent it can have a value of its own -- in general the later, the better. For example, suppose the planning process has determined that a new feeder must be built at a certain site within the next five years, in order to serve future load. It will cost $100,000 to build, regardless of whether built this year, next year, or five years from now. If there are no other reasons to build it any earlier, the best decision is to wait as long as practically possible -- in this case for five years. One hundred thousand dollars spent five years from now is future money, not present money, and while it is the same number of dollars it is not worth as much to the company at the moment:

> *The cost of money spent in the future is less than the cost of that same amount spent today.*

Of course, there could be sound reasons why it will benefit the utility to build the feeder earlier. Perhaps if added to the system now it will help lower losses, reducing operating cost over the next five years by a significant amount. Perhaps it will improve reliability and service quality, reducing the risk of customer outages and penalty-repayments under performance-based contracts. Thus, the sooner it can be built, the more the utility will save.

Or perhaps if built a year from now, during a predicted slump in the local construction industry, the utility believes it can negotiate particularly

Economic Evaluation

advantageous rates from the contractor it will hire to do the construction. Or perhaps some of the cable and equipment to be used is expected to rise in price over the next few years, so that if the utility delays even a year, the feeder will eventually cost a bit more than $100,000.

There are often many possible reason why it might be prudent to build earlier than absolutely necessary. A planners' job is to balance factors like these to determine how they contribute to the decision of when to build an item or actually commit to an expense, and to identify what alternative with regard to timing has the lowest overall cost (or alternately, the highest benefit) to the company.

Present Worth Analysis

Present worth analysis is a method of measuring and comparing costs and savings that occur at different times on a consistent and equitable basis for decision-making. It is based on the present worth factor, P, which represents the value of money a year from now in today's terms. The value of money at any time in the future can be converted to its equivalent present worth as:

$$\text{Value today of X dollars t years ahead} = X \times P^t \qquad (6.1)$$

where P is the *present worth factor*

For example, suppose that $P = .90$, then $100 a year from now is considered equal in value to today's money of

$$100 \cdot (.90) = \$90$$

and $100 five years from now is worth only

$$\$100 \cdot (.90)^5 = \$59.05$$

Present worth dollars are often indicated with the letters PW. Today's $100 has a value of $100 PW, $100 a year from now is $90 PW, $100 five years from now is $59.05 PW, and so forth.

Alternately, the present worth factor can be used to determine how much future money equals any amount of current funds, for example, to equal $100 present worth, one year from now the utility will need

$100/.90 = 111.11
and two years from now, one would need $100/(.90)^2 = 123.46 to equal a present worth of $100.

A continuing annual future cost (or savings) can be converted to a present worth by adding together the PW values for all future years. For example, the present worth of the $27,000 in annual losses savings discussed earlier in this chapter can be found by adding together the present worths of $27,000 next year, $27,000 the year after, and so forth

$$\text{PW of \$27,000/year} = \sum_{t=1}^{\infty} (\$27,000 * P^t) \quad (6.2)$$

$$= \$27,000 \times (\sum_{t=1}^{\infty} P^t)$$

$$\cong \$27,000 \times (\sum_{t=1}^{30} P^t)$$

$$= \$258,554$$

Discount rate

Present worth analysis *discounts* the value of future costs and savings versus today's costs and savings, as shown in Figure 6.6. The discount rate used in an analysis, d, is the perceived rate of reduction in value from year to year. The present worth factor is related to this *discount rate*

$$P(t) = 1/(1+d)^t \quad (6.3)$$

where d = discount rate
and t = future year

If d is 11.11%, it means that a year ahead dollar is discounted 11.11% with respect to today's dollar, equivalent to a present worth factor of $P = (1/1.111) = .90$. Therefore, $111.11 a year from now is worth $111.11/1.1111 = $100. A decision-making process based on the values of d = 11.11% and PWF = .90 would conclude that spending $90 to save $100 a year from now was a break-even proposition (i.e., there is no compelling reason to do it), while if the same $100 savings can be had for only $88, it has a positive value.

Economic Evaluation

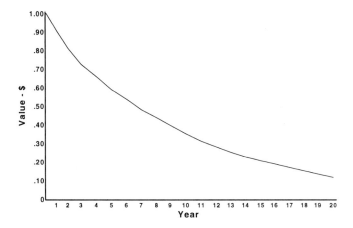

Figure 6.6 Present worth analysis discounts the value, or worth, of a dollar the further it lies in the future. Shown here is the value of one dollar as a function of future year, evaluated at a .90 present worth factor (11.1% discount rate).

A higher discount rate -- a lower present worth factor -- renders the decision-making process less willing to trade today's costs for future costs. If the discount rate is doubled, to 22.2%, P drops to .82. Now, $88 spent today to obtain a $100 cost reduction a year from now is no longer viewed as a good investment: that $88 must save at least $88/1.222 = $108 a year from now to be viewed as just break even.

Present worth analysis does not say "no" to truly essential elements of a plan

Power system planners, with a commitment to see that service is always maintained, often fall into a trap of wanting to build everything as early as practicable so as to have plenty of capacity margin. As a result, they come to view present worth analysis, and any other aspect of planning that may say "delay," as negative.

But this is not the case. If applied correctly, present worth analysis never says "no" to essential expenses -- for additions or changes which *must* be done *now*. For example, suppose a hurricane has knocked down ten miles of critically needed line. Then there is no question about the *timing* of this expense. It must be done as soon as other conditions permit, and present worth analysis is not even appropriate to apply to this decision.

Present worth analysis is used to evaluate and rank alternatives only when there is a difference in the timing of expenses -- the substation could be built this year or next year -- or when costs and savings that occur at different times must be balanced -- more spent today will lower tomorrow's costs. It is an essential element of keeping cost as low as possible and it should be applied in all situations where alternatives differ with respect to when expenses will be incurred. The present worth factor is the tool which makes these comparisons valid within the utility's value system.

How are Present Worth Factors and Discount Rates Determined?

There is no one rule or set of guidelines that rigorously defines what contributes to the present worth factor (or its companion, the discount rate). Likewise, there is no inviolate formula like Ohm's law or Schroedinger's equation which lays out completely and with no exceptions, exactly how to compute PWF from those elements that are determined to be relevant. Quite simply, PWF is merely a number -- often arrived at through careful analysis and computation, but sometimes it is an empirical estimate "that just works well." PWF is a decision-making tool that allows planners to compare future expenses to present ones.

> *The present worth factor is simply a value that sums up all the reasons why a company would prefer to spend money tomorrow rather than today.*

There can be many reasons why a utility may wish to limit spending today in favor of spending tomorrow, or in rare instances, wish to spend now instead of in the future. These reasons all influence the selection of a value for the present worth factor. Priorities vary greatly from one company to another and also change over time. Among them is one reason planners often find difficult to accept: "We just don't have the money, period." In most utilities, present worth factors and discount rates are determined by a utility's financial planners and based on the company's requirements for financing and use of money. The methods applied to fix these requirements quantitatively, and their relation to PWF are beyond the scope of this discussion, and available in other texts (*IEEE Tutorial on Engineering Economics*). However, it is worth looking at some of the primary influences on present worth factor and how and why they impact it.

Interest rate

The fact that it takes more future dollars to equal a dollar today is often attributed to interest rate: a person who has $100 today can invest it at the prevailing interest rate, i, so that a year from now it will be (1+i) times as much. Thus, $100 invested today at an annual rate of 5% interest, will be worth $105 a year from now and $128 five years from now. If the prevailing interest rate is 5%, then it is cost-effective to spend $100 only if it will save or return a value a year from now that exceeds $105. Otherwise it would be better to simply leave the money invested and drawing interest. A PWF of .952 can be applied in present worth analysis as described earlier, and will lead to decisions that reflect this concern.

But, in practice present worth factor clearly represents something more than *just* interest rate, because present worth factor as applied by most electric utilities is nearly always greater than what corresponds to the prevailing interest rate. For example, at the time of this writing, the inflation-adjusted interest rate on safe, long-term investments is about 5.5%, yet most utilities are using a present worth factor, of about .11, equivalent to a 12.4% interest rate.

The difference, 12.4% versus 5.5% in this case, is attributable to other factors. Although it may prove difficult or impossible for a distribution planner to determine and quantify all of them.

Inflation

Inflation is *not* one of the factors normally taken into account by the present worth factor, although this is often misunderstood and neophytes assume it is a part of the analysis. Inflation means that what costs a dollar today will cost more tomorrow -- 3% annual inflation would mean that same item or service will be priced at $1.03 a year from now. While inflation needs to be considered by a utility's financial planners, distribution planners can usually plan with constant dollars -- by assuming that there is no inflation.

The reason is that inflation raises the cost of everything involved in distribution planning cost analysis. If inflation is 3%, then on average a year from now equipment will cost 3% more, labor will cost more 3% more, as will paperwork and filing fees and legal fees and taxes and replacement parts and transportation and everything else. More than that, over time inflation will call for similar adjustments in the utility's rates, and in the value of its stock, its dividends, and everything associated with its expenses and financing (including hopefully the planners' salaries). It affects nearly every cost equally, at least to the extent that costs can be predicted in advance.[2]   From the standpoint of planning, inflation makes no impact on the *relative costs* of various components, so it can be ignored, making the planning process just a little bit easier.

Some utilities do include inflation in their planning and their present worth analysis. In such cases, an increment to account for inflation is added to the discount rate. Given a 5% interest rate and a 3% inflation rate discussed above, this would mean that the planners' present worth factor might be

$$P = 1/(1+5\%+3\%) = .926$$

Present worth analysis would now compare today's investment measured in today's dollars against tomorrow's investment measured in tomorrow's inflated dollars. This type of accounting of future costs must sometimes be done for budgeting and finance estimation purposes.[3]

But while inflation must be included in the analysis, all planning should be done with "constant dollars"

However, while "inflated dollars" present worth analysis sometimes has to be done to estimate budgets, it is rarely an effective planning tool. Planning and decision-making are facilitated when expenses and benefits are computed on a common basis -- by measuring costs from all years in *constant dollars*. Is plan A, which calls for $1.35 million in expenses today, worth more than plan B, which costs S1.43 million three years from now? If both are their respective present worth measured in constant dollars, then it is clear that plan A is the less costly of the two. But if inflation were a factor in the present worth analysis, then one has to do a further adjustment of those numbers to determine which is best (at 3% inflation, $1.35 million in three years will equal $1.48 million in then inflated currency, so plan B at $1.43 three-year ahead inflated dollars is the better plan). Beyond this, errors or unreasonable cost estimates for future projects are much more difficult to catch when expressed in non-constant dollars. For these reasons, present worth analysis in constant dollars is strongly

---

[2] In actuality, planners and executives know that inflation will not impact every cost equally, but small variations in individual economic sectors have proven impossible to predict. In cases where there is a clear indication that certain costs will rise faster or slower than the general inflation rate, that difference should be taken into account. This was mentioned at the start of this section, when a possible rise in cost of materials for the candidate feeder that could be built early was discussed as one reason that might lead the utility to decide to build it early.

[3] For example, when estimating a future budget, the costs often do have to be put in dollars for the years being estimated, and those are inflated dollars.

Economic Evaluation

recommended for all planning, even if non-constant analysis may need to be done for some other purposes.

Earnings targets

The present worth factor for a dollar under the utility's control should be higher than 1/(1 + interest rate), because *the utility must be able to do better with its earnings than the prevailing interest rate.* If a utility cannot better the prevailing interest rate with its own investment, then it should liquidate its assets and invest the results in earning that interest rate. Frankly, it won't get the chance to do so: its shareholders will sell their stock, take their money elsewhere and invest it in companies that *can* beat the prevailing interest rate through their investment of their stockholder's money.[4]

Therefore, the goal of any investor-owned utility must be to use its money to earn more than it could by other, equivalently safe means. As a result, while the prevailing interest rate may be 5%, the utility's financial planners may have determined that a 12% earnings potential on all new expenditures is desirable. Rather than a PWF of .952 (5% interest) the utility would use a PWF of .892 (12%) to interject into the decision-making process its unwillingness to part with a dollar today unless it returns at least $1.12 a year from now.[5] This is one tenet of MRR planning, that all investments must return a suitable "revenue" earning.

Risk

One hundred dollars invested with the promise of $108 payback a year from now may look good, but only if there is a very small chance that the investment will go wrong, with the loss of the interest *and* the $100 itself. Such cases are very rare, particularly if investments are made wisely, but there are other types of risks that a utility faces. For example, suppose that shortly after spending $100 to

---

[4] Planners from municipal utilities may believe that this does not apply to their company, but that is not necessarily true. If a municipal electric department cannot "earn" more from its electric system investment than other businesses could, it is costing the city and its voters money, designing a system and a "business" that needs to be subsidized. If such subsidized operation is the policy of the city government, the planners should make the most of it and try to get as sound a return as possible on what money is spent.

[5] This concept strikes many planners as bizarre, yet it is completely valid, and is in fact only common sense from a business perspective. If a company expects to earn 9% return on its money, then as a rule is should never invest in anything that is projected to earn or repay less than that rate. Since some of the unavoidable elements of its expenses may not have the potential to earn this rate, the expected return on the rest may have to be even higher.

save $108 a year from now, the utility gets hit by a severe storm which damages much of its system equipment, causing widespread outages. It may desperately wish it had that $100 to pay for repairs and rebuilding of its system, work that it simply has to have done. It would have no choice but to borrow the money it needed at short-term interest rates, which might be 12%. In retrospect, its expenditure of $100 to save $108 a year later would look like a poor choice.

In practice, a present worth factor often includes a bias or margin to account for this type of "bird in the hand" value of money not spent. By raising the PWF from 8% to 10%, the utility would be stating that yes, perhaps $108 is the year-ahead earnings goal, but simply breaking even with that goal is not enough to justify committing the company's resources: today's money will be committed only when there are very sound reasons to expect a better than minimum return.

Planning errors

In addition, the present worth factor often implicitly reflects a sobering reality of planning -- mistakes cannot be avoided entirely. The author has devoted considerable effort to the study of the distribution planning process itself, particularly with regard to how accurate it is, and how and why planning mistakes or non-optimal plans come about (Willis and Northcote-Green; Willis and Powell). Under the very best realistic circumstances, even the finest planning methods average about 1% "planning error" for every year the plan is extended into the future. Distribution expansion projects that must be committed a full five years ahead will turn out to spend about 5% more than could be arranged in retrospect, if the utility could somehow go back and do things over again, knowing with hindsight certainty exactly what is the minimum expenditure it needs to just get by.

Adding 1% -- or whatever is appropriate based on analysis of the uncertainty and the planning method being used -- to the PWF biases all planning decisions so that they reflect this imperfection of planning. This makes the resulting decision-making process a little more reluctant to spend money today on what appears to be a good investment for tomorrow, unless the predicted savings includes enough margin over the element of risk to account for the fact that the planning method is simply wrong.

Present Worth Factor is a Decision-Making Tool, Not a Financial Factor

Present worth analysis is a decision-making tool to determine not only which alternative is best but *when* money should be spent. It can embrace some or all of the factors discussed here, as well as others. Two equally prudent utilities might

Economic Evaluation

Table 6.1 Discount Rate "Computation" for Two Utilities

Factor	IOU	Muni	Comment
Prevailing interest rate	5.7%	5.1%	Municipal bond rate is lower than prime.
Inflation factor	-	-	Both do constant dollar planning.
Earnings target	5.5%	-	Municipal has no need for "earnings."
Risk	1.5%	3.0%	Political cost more serious than financial.
Planning error	1.0%	1.5%	This IOU has the better planning method.
"We just don't have funds"	-	5.0%	This municipality is nearly "broke."
Total discount	13.7%	14.6%	
Equivalent PWF	87.9%	87.3%	

might pick very different present worth factors, depending on their circumstances, as summarized in Table 6.1 with the inputs that determined the PWFs used in the mid 1990s by a large investor owned utility in the northwestern United States, and a municipality in the southwestern United States.

But while the present worth factor can be attributed to interest rate, risk, earnings goals, and other factors, the simple fact is that such determinations are irrelevant to the distribution planner.

> *Present worth factor is used to evaluate and rank alternatives based on* when *they call for expenditures.*

A relatively low PW factor means that the planning process will be more likely to select projects and plans that spend today in order to reduce costs tomorrow. As the PWF is increased, the planning process becomes increasingly unwilling to spend any earlier than absolutely necessary unless the potential savings are very great. A very high PW factor will select plans that wait "until the last moment" regardless of future costs.

Example Present Worth Analysis

Returning to the two time-value of money decisions that started this section, the recommendation on whether initially to build a new substation with one or two transformers can be evaluated by comparing the present worth of the cost versus the savings of adding the second transformer early. With a 90% PWF (discount rate of 11%) the two values are:

Present worth of spending $400,000 this year = $400,000
Present worth of saving $642,000 in four years = $421,200

The answer is yes, the additional initial cost is justified by the savings.

Often, it is useful to determine if the decision is particularly sensitive to the value of the PWF and discount rate use. Would a very slight change in the perceived value of future money have changed the recommendation? This can be determined as:

$$\text{Present worth decision limit} = (\text{cost in year } t_1)/(\text{cost in year } t_2)^{(1/(t_2-t_1))} \quad (6.4)$$

$$= (400,000/642,000)^{1/4}$$

$$= .888, \text{ equivalent to a discount rate of } 12.6\%$$

In this case, while a present worth decision limit (PWDL) of .888 seems quite close to the .90 present worth factor used in the analysis, a comparison of the corresponding discount rates -- 11.1% versus 12.5% -- indicates a substantial margin. The gulf between these values indicates that at least one factor of the order discussed earlier in this section (risk, planning error, etc.) would have to change substantially in order for a large enough shift to occur. Based on this analysis, the recommendation seems sound: the substation should be built with only one transformer in place.

Looking at the second planning decision that headed this section, is a $160,000 expense justified on the basis of a continuing $27,000 savings? Using the same 90% PWF (discount rate of 11%) the two costs are:

Present worth of spending $160,000 this year = $160,000
Present worth of saving $27,000 for next 30 years = $258,600

The answer is yes, the feeder's greater initial cost is more than justified by the continuing savings in losses. It should be built with the larger conductor.

Determining the present worth decision limit in this case is slightly more complicated, but can be accomplished iteratively in a few steps using a calculator or a spreadsheet. In this case it is .832, equivalent to a discount rate of 20.2%. Thus, this is a very strong recommendation.

Comprehensive Present Worth Example

In general, actual planning problems are more complicated with regard to the interrelationship of long-term and initial costs than either of the two examples above. Table 6.2 and Figure 6.7 show more cost details on the two-versus-one

Economic Evaluation

substation transformer decision, the type of details that often exist in actual planning situations. In the case shown here, the decision is being made now (in year zero), even though the substation will actually be built three years from now (the decision must be made now due to lead time considerations). Here as before, alternative A calls for building the new substation with two 25 MVA transformers at a total cost of $1,400,000 in year four. Alternative B defers the second transformer four years (to year 7), reducing the initial cost to $1,000,000. And again, the addition of the second transformer four years later is estimated to cost $642,000 because of project start up costs, separate filing and regulatory fees, and the fact that the work will have to be done at what will then be an energized site.

However, this decision involves other costs, which while of secondary importance, should be considered in any complete analysis. For example, no-load losses, taxes, and O&M will be greater during the first fours in Alternative A than in Alternative B. Table 6.2 and Figure 6.7 compare the expenditure streams for the eight-year period for both alternatives, including capital; operations, maintenance, and taxes (O&M&T); and losses.[6]

Table 6.2 Comparison of Yearly Expenses by Category for an Eight Year Period

		Alternative A Build initially with two			Alternative B Build initially with one		
Year	Load-MW	Capital	O&M&T	Losses	Capital	O&M&T	Losses
0							
1		20			20		
2		370			290		
3	12	1010	52	54	690	44	39
4	15.5		110	119		92	99
5	18.5		110	130		92	122
6	21.3		110	142	80	92	147
7	23.8		110	155	562	101*	164*
		1400	492	600	1642	317	666
			Total	= 2492		Total	= 2633

*Taxes and losses for the transformer added in year eight (alternative B) are pro-rated for that year, since it is assumed here that it is installed just prior to summer peak (six months into the year).

---

[6] The table and figure cover the only period during which there are differences that must be analyzed in the decision-making. At the end of the eight-year period, the two alternatives are the same -- either way the substation exists at that point with two 25 MVA transformers installed. Present worth thereafter is essentially the same.

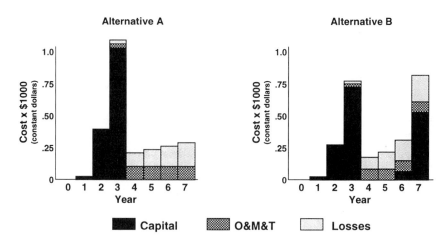

Figure 6.7 Yearly expenses for the two alternative plans for the new substation (constant dollars).

Note, also, that the capital for substation construction in either alternative is not spent entirely in the year the substation is completed. A small amount is spent two years earlier (for filing and preparing for the site), and the site itself and some equipment purchased a year earlier so that work can begin on schedule.

Comparing differences in the two alternatives, note that the second transformer in Alternative A increases O&M&T costs during the first four years of operation, because it has value (taxes must be paid on its additional value) and it must be maintained and serviced. The difference in losses costs is more complicated. Initially, losses are higher if two transformers are installed, because this means twice the no-load losses, and the twin transformers are initially very lightly loaded, so that there is no significant savings in load-related losses over having just one transformer. But by year six the two-transformer configuration's lower impedance (the load is split between both transformers) produces a noticeable savings in reduced losses.

Table 6.3 and Figure 6.8 show the various costs for Alternatives A and B converted to PW dollars using a present worth factor of .90 (equivalent to p = 11.11%). Note that based on the present worth of capital costs, Alternative A is rated the lower present worth cost ($1,054,000 versus $1,067,000), in keeping with the preliminary evaluation done earlier. The margin of present worth savings shown in Table 6.4 ($13,000) appears to be smaller than the $21,200

Economic Evaluation

difference computed several paragraphs above, largely because in this evaluation all the costs and savings are pushed three years into the future (the substation is built in year three, not year zero), and thus both costs and savings have been lowered by another four years of present worth reduction (i.e., $.9^4 = .73$).

However, more important is the fact that this more complete analysis shows that Alternative B has a lower present worth by $51,000 PW -- a margin in the other direction, and one nearly four times larger than the difference in present worth based on capital alone. What changed the balance in favor of alternative B was the continuing fixed operating costs -- maintenance, taxes, and no-load losses -- which are higher when two transformers are installed.

The present worth difference between these alternatives, $51,000, is slightly more than 3% of the total PW value of either alternative. Considering the straightforward nature of this analysis, and the amount of cost common to both alternatives, it is unlikely that the analysis of this difference is in error by anything like this amount. Therefore, this is very likely a dependable estimate: it is likely that alternative B's PW costs really will be lower over the long run.

Looking at the sensitivity of this decision to present worth factor, the two alternatives would evaluate as equal at a present worth factor of .937, equal to a discount rate of 6.77%. Present worth factor would have to be higher (i.e., the discount rate lower) for the future savings to outweigh the increased initial costs. Valid PWF values less than .92 are extremely uncommon, and therefore the recommendation to build the substation with one transformer seems quite dependable, regardless of the exact financial considerations.

Table 6.3 Comparison of Yearly Present Worth by Category for an Eight-Year Period

Year	Load-MW	Alternative A Build initially with two			Alternative B Build initially with one		
		Capital	O&M&T	Losses	Capital	O&M&T	Losses
0	1.0						
1	.90	18			18		
2	.81	300			235		
3	.73	736	38	39	503	32	28
4	.66		72	78		60	65
5	.59		65	77		54	72
6	.53		58	75	43	49	78
7	.48		53	74	269	48*	78*
		1054	286	343	1067	244	322
			Total = 1684			Total = 1633	

*Taxes and losses for the transformer added in year eight (alternative B) are pro-rated for that year, since it is assumed here that it is installed just prior to summer peak (six months into the year).

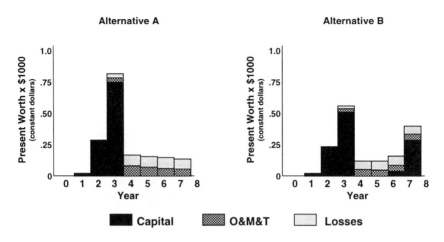

Figure 6.8 Present worth of expenses associated with the two alternative substation plans given in Table 6.3 and Figure 6.7.

Implied Present Worth Factor

A frustrating reality for many distribution planners is that occasionally projects that justify themselves on the basis of present worth analysis are rejected by management nonetheless, purely for budgetary reasons. For example, perhaps using the utility's PWF and approved method of analysis, a decision is made that a new substation should be built with two new transformers, while one would satisfy immediate needs, because the higher initial cost is more than made up for by the long-term present worth of the savings. Yet the recommendation may be rejected by management because "there simply isn't enough money in the budget." Implicitly, management's approval process is using a lower present worth factor (e.g., higher discount rate) than the official or planning present worth factor.

Often situations like this are a reality which cannot be avoided -- budgetary restrictions beyond those accounted for in the that worth factor can be forced on the utility by unexpected or unavoidable conditions. In such cases, planners may wish to analyze the present worth factor decision limit of projects that are approved and rejected, to determine what the implied present worth factor being used in the approval process actually is.

Economic Evaluation

Table 6.4 Use of Present Worth Decision Limit Analysis to Identify the Implicit Present Worth Factor

Recommended Project	PWDL	Equiv. Disc. Rate	Disposition
Upgrade Eastwood substation	.40%	150%	Approved
Build new Knoll Wood #6 feeder	.48%	108%	Approved
Upgrade Echo-Wouton 69kV line	.51	96%	Approved
Split bus & upgrade, Keen River subs.	.58	72%	Approved
Breaker upgrades, River Bend station	.59	69%	Approved
Add third 25 MVA transf. at Pine St.	.60	66%	Delayed
Larger conductor, new Knoll Wood # 5 feeder	.75	33%	Rejected
Reconductor Downtown #2 feeder	.80	25%	Rejected
Wharton subs. enhancement plan	.87	15%	Rejected

For example, the projects listed in Table 6.4 were decided upon as shown in 1993 by the municipal utility whose "discount rate computation" is given in column three of Table 6.1. While officially the discount rate in effect was 14.6% (PWF of .873), management was implicitly using a PWF of less than .60, equivalent to a discount rate greater than 66%. Knowledge of the payback that is expected in any project can help planners in not wasting their or their management's time with recommendations that have little chance of approval.

Levelized Value

Often, it is useful to compare projects or plans on the basis of their *average* annual cost, even if their actual costs change greatly from one year to the next. This involves finding a constant annual cost whose present worth equals the total present worth of the plan, as illustrated by Figure 6.9. Levelized cost analysis is a particularly useful way of comparing plans when actual costs within each vary greatly from year to year over a lengthy period of time, or when the lifetimes of the options being considered are far different. It also has applications in regulated rate determination and financial analysis. In general, the present worth, Q, levelized over the next n years is

$$\text{Levelized value of Q over n years} = Q (d \times (1+d)^n)/((1+d)^n - 1) \quad (6.5)$$

Thus, the steps to find the levelized cost of a project are: a) find the present worth of the project, b) apply equation 7.4 to the present worth. As an example, using Alternative A's calculated total present worth of $1,684,000 from Table 6.3, that value can be levelized over the seven-year period in the table as

$$= \$1,684,000 \, (.1111(1.1111)^7/(1.1111^7 - 1))$$
$$= \$358,641$$

By contrast, Alternative B has a levelized cost over the same period of $347,780. If the utility planned to pay for alternative A or B over the course of that seven-year period, the payments it would have to make would differ by a little more than $10,800 per year.

Lifetime levelized analysis

Rather than analyze planing alternatives over a short-term construction period -- for example the seven-year period during which the schedule of transformer installations may vary -- planners or budget administrators often wish to look at a new addition to the system over the period in which it will be in service or be financed.

For example, in the two-or-one transformer substation planning problem presented earlier, regardless of whether Alternative A or B was chosen, the substation would be planned on the basis of providing a minimum of 30 years of

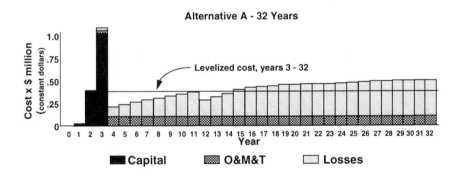

Figure 6.9 Cost of alternative A from Figure 6.7 over a 32-year period that includes its first 30 years of service. Present worth of all costs sums to slightly more than $3,000,000. Levelized over years three through 30 -- those in which the substation is planned to be in service -- this represents an annual cost of nearly $350,000 per year, which can be viewed as the annual cost of providing the substation service.

Economic Evaluation

service, and it would be carried as a depreciating asset on the company's books for 30 years or more (i.e., the company is basically financing it over that period). Therefore, planners might wish to both analyze cost over 30 years, and compare alternative plans on that same basis.

Figure 6.9 shows projected costs for Alternative A over a 32 year period, which includes the first through thirtieth years of service for the substation. The total present worth of the substation's costs over the period is just over $3 million (construction plus 30 years of O&M&T plus losses), which represents a levelized cost of $348,500/year. By contrast, Alternative B, (the less expensive of the two construction schedules, as determined earlier) has a levelized cost $342,200/year. It saves an average of $5,900/year or about $500/month in terms of a 30-year budget perspective.

Comparing cost over different equipment lifetimes

Very often the lifetimes or periods of cost being considered for two or more options are not equivalent, and yet it is desired to compare their costs on a valid basis. Levelized lifetime cost is a good way to do so. Table 6.5 compares the annual costs of providing 1000 lumens of illumination (about the equivalent of a standard 75 watt incandescent light bulb) for 2500 hours per year, using incandescent, halogen, or compact fluorescent lamps, which have expected lifetimes of three, five, and six years respectively. The initial costs, the annual operating costs, and the lifetimes of these three options all differ substantially. In each case, initial and continuing annual energy costs are evaluated in Table 6.5 on a present worth basis over the lifetime of the particular lamp; then the levelized cost over that period is determined. This permits a comparison that identifies the compact fluorescent as having less than half the annual cost of the incandescent.

An important point is that each was evaluated over a different period of time -- its lifetime, yet the results are directly comparable. The results, however, are identical to those that would have been obtained from analysis of all over the same, longer, period.

Suppose that the analysis in Table 6.5 had compared all three options over a 30-year period. That would have required a series of ten incandescent bulbs -- one every three years -- six halogen, and five compact fluorescent lamps. Each of the sections of Table 6.5 would have extended for 30 years, with appropriate PWF, computations, and sums, but each would have arrived at exactly the same levelized cost per year for each of the three lamps.

Table 6.5 Lifetime Analysis of Illumination

Year	PWF	Capital	Energy	Total	PW
Incandescent					
1	.90	$2.00	$23.00	$25.00	$22.50
2	.81		$23.00	$23.00	$18.63
3	.73		$23.00	$23.00	$16.77
		Total Present Worth			$57.90
			Levelized over 3 years =		$23.74
Halogen					
1	.90	$8.00	$13.00	$21.00	$18.90
2	.81		$13.00	$13.00	$10.53
3	.73		$13.00	$13.00	$9.48
4	.66		$13.00	$13.00	$8.53
5	.59		$13.00	$13.00	$7.69
		Total Present Worth			$55.11
			Levelized over 6 years =		$14.95
Fluorescent					
1	.90	$17.00	$8.00	$25.00	$22.50
2	.81		$8.00	$8.00	$6.48
3	.73		$8.00	$8.00	$5.83
4	.66		$8.00	$8.00	$5.25
5	.59		$8.00	$8.00	$4.72
6	.53		$8.00	$8.00	$4.25
			Total Present Worth		$49.04
			Levelized over 6 years =		$11.62

Economic Evaluation

6.4 COST-EFFECTIVENESS EVALUATION

How does a planner know that a particular project is cost effective, that it contributes to lower cost and greater efficiency? In many cases, once present worth analysis has adjusted all costs and savings with respect to equipment lifetime, schedule, and other time-value-of-money considerations, such decisions are straightforward -- in many cases trivial. But in other situations, determining what to do is not merely a matter of identifying the option with the lowest cost, or the option that produces the largest savings. Selection may be based on the relative merits of costs versus savings. In integrated resource situations, the decision on what to include in the utility's plans may also depend on the planner's (or the regulatory authority's) philosophy about what constitutes savings and who should benefit most from the utility's investments in the system.

Lowest Cost Among Alternatives

In many cases, planning consists of selecting from a set of alternatives, all of which satisfy the pertinent engineering and service criteria. The alternative with the lowest cost is the one selected for implementation. Here, if all pertinent costs (capital, losses, O&M&T, etc.) are considered, and the appropriate type of present worth adjustment is made for timing, lifetimes, and schedules, then the alternative with the lowest PW cost *is* the preferred option.

This almost trivially obvious approach is at the core of the majority of traditional utility planning methods. In such cases, the planning process is being driven by expansion needs -- there is no doubt that something must be done to satisfy the load growth, and the planner's challenge is to find an acceptable way to do this at the lowest possible cost. In such cases, planning consists of evaluating alternatives to assure that they satisfy requirements, and then selecting the alternative with the lowest *net present value* of expenses -- total present worth of all future costs.

But in many situations, one or all of the options being considered have both costs and benefits, and there are constraints or interrelated considerations with respect to one or both, that impact the selection of the plan. In such cases, planning consists of more than simply ranking the alternatives by cost and selecting the one at the top of the list.

Benefit-Cost Analysis

Often planning decisions are based on an evaluation of the expected gain versus the cost. For example, earlier it was shown that building a new feeder with larger conductor, at a cost of $160,000 in order to reduce annual losses by $27,000 per year, would provide a PW savings equal to $258,600. A planner can evaluate the benefit to cost ratio of this proposal

$$\text{Benefit/Cost} = \$258{,}600/\$160{,}000 = 1.62$$

Benefit-cost ratio is a simple concept: a ratio less than one indicates the project will lose more than it gains; the higher the benefit/cost ratio above 1.0, the greater the potential gain relative to its cost.

Benefit-cost ratio is generally applied as a ranking method, to prioritize projects that are viewed as discretionary or optional, or in cases where there simply isn't enough funding for all the projects that present worth analysis has indicated are worthwhile. Frankly, this is usually the case in any company, particularly a utility -- there are always worthy projects in which to invest, but the goal is to do so in a way that assures sound, and high, profits.

Generally, if there are more projects than can be fit into the budget, those with the highest benefit-cost ratios should be selected in order to maximize the gain. Table 6.6 shows the benefit/cost analysis, evaluated on a levelized lifetime basis, for ten demand-side management alternatives analyzed in terms energy savings for customers of a utility in New England. They vary from a B/C of more than two to one to just over one to one.

If the utility invested in all of these DSM options, it would spend a total of $5.1 million to achieve a projected gain of $8.2 million, for an overall benefit cost ratio of 1.59:1. However, if budget restrictions limited it to only $3.35 million for the DSM program, it would make sense for it to select the top five programs ($6,014 savings, $3,350 cost), obtaining a B/C ratio of 1.8:1.

In fact, the utility did something slightly different. Working backward from the values shown in the table, its management selected only the top four options. Added together ($4,258 savings, $2,120 cost) this four-option DSM program had a projected benefit-cost ratio of two to one. That figure not only has a nice political appeal ("DSM is saving two dollars for every dollar spent") but this value always provided a good margin against error in what was, at the time, a new and untried resource.[7]

Benefit/cost analysis is a useful alternatives-ranking method, but one appropriate to only a limited number of planning situations. Particularly in the presence of budget limits, it can lead to non-optimal spending (See Chapter 18).

---

[7] This proved a very prudent decision, because unforeseen problems, perhaps due to optimism in the original study, led to higher costs and lower savings than projected. The program ultimately had only a 1:14 benefit cost ratio as observed over the period 1989 through 1995.

Economic Evaluation

Table 6.6 Benefit Cost Analysis of Energy Efficiency and Load Control Alternatives -- Utility in New England (1987)

DSM Alternative	Cust. Class	($ × 1000) Savings	Cost	B/C Ratio
Insulation blankets on water heaters	resid.	3,200	1500	2.13
Direct water heater control	resid.	650	375	1.73
Direct water heater control	comm.	343	200	1.72
Direct water heater control	ind.	65	45	1.44
Building shell upgrades	comm.	1756	1230	1.43
Efficient interior lighting upgrade program	resid.	78	57	1.37
High-efficiency motor upgrade program	indus.	340	270	1.26
Efficient interior lighting upgrade program	comm.	1189	956	1.24
Insulation blankets on water heaters	comm.	400	350	1.14
Efficient exterior lighting upgrade program	comm.	<u>161</u>	<u>148</u>	<u>1.08</u>
		8182	5141	1.59

Payback Period

The period of time required for savings to recoup the initial investment is often used as a measure of the cost-effectiveness when evaluating investments or proposed expansion plans for T&D systems. Returning to the "small versus large conductor" feeder problem discussed earlier in this chapter, a losses savings of $27,000 per year will require six years to repay the $160,000 additional cost of the larger conductor. If the breakeven is computed based on PW dollars at a .90 PW factor, the payback of the original $160,000 takes ten years.

"Payback period" analysis is most often used informally, to provide an additional perspective on plans and financing. Rarely is it part of the formal planning criteria or a specific cost-effectiveness test called for by regulatory authority. However, it does address one question that can be quite important to any investor --"When will I get my money back?"

Payback period does matter, and can be independent of B/C ratio and present worth determination. Plans with net present worth savings (i.e., PW savings > PW cost) and good benefit-cost ratios, can have rather lengthy payback periods, if for one reason or another their costs are in the present and their savings, while substantial, are quite far into the future. This is often the case with projects that show a positive PW value from long-term losses cost reduction.

Integrated Resource Cost Effectiveness Tests

In cases where energy conservation, load control, high-efficiency appliances, fuel switching, or renewable energy sources are combined with T&D resources, the determination of "benefit" and "cost" can be quite involved. Often these must be done within very rigid protocols set by government regulation.

In most regulatory frameworks, the benefits of a project, program, change in policy, or other step that could be taken by the utility are the *avoided costs*. These are the power system costs which are *not* incurred by the utility as a result of the decision being considered, and are usually framed against some other scenario. For example, when evaluating energy efficiency applications, the avoided costs are the incremental costs over the future planning period (in this case 30 years) that would be required to provide equivalent service were the energy efficiency not implemented. These might include fuel expected to not be burned, new facilities not needed to be built, etc. These represent a savings.

On the other side of the energy efficiency B/C ratio, the costs are the expenses associated with the proposed energy efficiency program or plan and may include the costs borne directly by the participants as well as those paid for by the utility and/or government subsidization. These costs include such items as the technology (materials and installation), maintenance, and administrative expenses. In the case of the analysis of a new facility or option for T&D expansion, the savings might be the change in present worth of future expenses in equipment, O&M&T and losses, and the cost would be the present worth of all future costs associated with instituting the plan.

Very often, exactly what is to be included in both benefits and costs is prescribed by law or regulation. Two major types of cost-effectiveness tests are utilized to assess DSM options, the Total Resource Cost (TRC) test, the Rate Impact Measure (RIM) test.

Total resource cost test (TRC)

The TRC method is an analytical test which evaluates all of the direct costs and benefits to society associated with a specific resource. Under TRC, a resource option is cost-effective when the present worth of benefits over the planning period exceeds the present worth of costs, using an appropriate discount rate. The allocation of costs and benefits between the utility and its participants is not a consideration when applying the TRC test -- only the total of utility+customer costs and benefits. Most often, the appropriate discount rate for the TRC test is deemed to be the utility's discount rate.

In some jurisdictions, the TRC test is referred to as the "All Ratepayers Test." This test identifies if and how the same end-use results can be provided

Economic Evaluation

more efficiently with some integrated resource plan, considering the costs and benefits of the *program to the utility and to ratepayers, taken together.*

Program Benefits in an IRP are the benefits directly accruing to both electric utility and its customers, including avoided capital and operating savings. Indirect benefits, such as decreased environmental impact or conservation of resources accruing to society; or tax incentives accruing to the utility, are not included.

Participant Benefits are the incremental costs avoided by the electric utility customers due to the IRP. As included in cost effectiveness evaluation, these benefits do not include the electricity purchase costs avoided, nor do they give credit for any improvement in comfort levels.

Program Costs include both the electric utility's costs related to the program and the total of all participant costs -- if a utility pays only half of the cost of energy efficiency upgrades, the customers' half must be included in the All Ratepayers Test.

Participant Costs include those incurred by the electric utility's customers and which costs would not have been incurred but for program participation, and as normally defined include only those costs incurred to save electricity.

Basically, the TRC or All-Ratepayers test looks at the *sum* of both utility and customer-side costs for achieving the desired end-uses of electricity -- warm homes in winter, lighting, etc. It assesses as "good" any IRP that reduces this total cost to achieve those ends, even if that program re-distributes costs so much that one side (e.g, the customer, or alternately the utility) sees its costs increase. Such situations are viewed merely as cases for pricing or rate adjustments that will re-distribute the savings equitably. For example, if a new energy efficiency program means that customers have to shoulder a considerable cost even though the utility's costs go down substantially, then the utility may have to subsidize those customer costs, through incentives and rebates, to shift the economics to the point that both utility and customer see a portion of the savings.

A variant of the TRC test is the Societal Cost Test which includes in the evaluation the effects of externalities -- costs or issues outside of the utility/customer combination considered by the All Ratepayer's test. This is based on analysis of the economic efficiency of the use of society's resources, as a whole, to produce the desired products of energy usage (end-uses) as a whole.

Usually, regulatory guidelines will give very specific definition to what is included in this analysis and how benefits and costs are to be computed. An integrated resource program satisfies the Societal Test if the total cost of the program to society is less than the total value of the resources saved by avoiding electricity production. Costs include all resources used to produce energy savings, regardless of who pays such costs. Benefits include all resources saved by avoiding production of electricity, including environmental benefits, and any other social benefits external to the transaction between the utility and its customers. Transfer payments, such as tax incentives, are not included in this calculation. The Societal Test may use a discount rate different from the utility's to the extent of reflecting an interest rate more applicable to public projects.

Rate impact measure test (RIM)

Unlike the TRC test, the RIM test considers net lost revenues caused by reduced sales of electricity as part of the "costs" in its evaluation. As a result, this test quantifies the expected change in rates or customer bills that may result from an integrated resource plan. A DSM program that reduces sales of electricity (e.g., conservation) reduces the revenues the utility receives to cover its costs. As a result, the utility may have to raise its price per kWHr for the remaining sales, so that it can collect enough revenue to cover its costs. This means that while the DSM program reduced the use of electricity, and perhaps reduced the total electricity bills (amount times rate) for those customers who participated in the program, it may have raised the bills of those who did not participate, because no change was made in their energy usage while their cost/kWHr went up.

Basically, the RIM test is an analysis of the extent to which the average rate for electricity is altered by implementation of a demand-side management or integrated resource program. A program will satisfy the Rate Impact Test if the present worth revenue requirements per kWH after implementation of the program are equal to or less than the present value revenue requirements per KWH before implementation of the program, assuming that the end-use remains unchanged. This basically means that the rate charged for electricity cannot increase.

Comparison of TRC and RIM tests

The TRC test focuses on minimizing the cost of electricity services while the RIM test focuses more on minimizing electricity prices. Occasionally, these two goals can be in conflict -- as mentioned above, a program that lowers overall energy usage cost can raise electricity prices in some cases. In general, integrated

Economic Evaluation

resources plans that satisfy both tests are considered positive. Those that meet only one or the other are often rated as marginal and unacceptable.

A comparison of the costs considered under the TRC and the RIM cost-effectiveness testing methodologies is shown in Table 6.7. As can be seen, the TRC and the RIM cost-effectiveness test methods utilize some of the same cost information but, in essence, have different cost input data. As a result, the choice of evaluation technique is likely to produce different B/C ratios (particularly since the utility's avoided costs serve as the benefits portion in both methods). For example, a program that results in a large reduction in KWH sales or requires significant financial incentives to achieve a target market penetration may fail the RIM test while passing the TRC test. On the other hand, a program that requires a major investment by the program participants may result in the opposite conditions. It may not be accurate to attempt a complete generalization in this regard. Programs that focus less on energy conservation, and more on peak demand reductions, may tend to favor the RIM test. Conversely, programs that focus more on energy conservation, and less on peak demand reduction, may tend to favor the TRC test.

The various methods of conducting cost-effectiveness testing should be considered as different perspectives or various means of assessing any plan or integrated resource program. Each such measure provides some alternative insight to the feasibility of a program. Thus, the final selection of DSM programs is perhaps best made by including the results from different cost-effectiveness measures in the decision-making process.

Table 6.7 Components of TRC and RIM Cost-Effectiveness Tests

Item	TRC	RIM
Program Costs Paid by the Utility	X	X
Program Costs Paid by the Participant	X	
Program Administrative Costs	X	
Incentives Paid to Participants by the Utility		X
Lost Revenues Due to Sales	X	
Increased Fuel Costs (Fuel Substitution Programs)	X	X

6.5 VARIABILITY OF COSTS

Given that T&D planners have sufficient lead time and flexibility of selection in equipment, site, and routing, and can optimize T&D arrangements without severe limitation, the cost of moving power as perceived in their planning is usually fairly linear with respect to both the amount and the distance power is to be moved as illustrated in Figure 6.10: double the amount of power moved and the cost will double, move it twice as far and the cost will double. Thus, from a long-term or "Green Field" perspective, the marginal and incremental costs of delivering electric power are largely unconstrained and linear.

However, once a system is in place, its structure is defined, its fixed costs are sunk, and its embedded cost structure cannot be avoided. Its design affects the capability and costs of local power delivery. Its limitations constrain how much and where power can be delivered. Its capacity limits impose an exponential relationship between the amount of power moved and the losses costs. The existing T&D system creates constraints with respect to expansion and modification plans: marginal and incremental costs may no longer be so well behaved and linear. Cost of delivery varies within a T&D system depending on the time of delivery, the level of the system through which power is delivered, the location of delivery, and on any system or equipment contingencies.

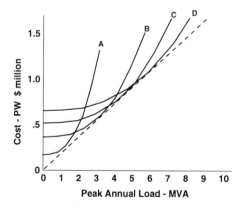

Figure 6.10 Present worth of building and operating one mile of 12.47 kV feeder with any of four different line types, as a function of peak load carried over the next 30 years (annual load factor of .60). Given that a planner has freedom to select which of several line types to build, as shown here, the locus of "minimum cost solutions" is fairly linear over a wide range, approximating a linear power versus cost relationship (dotted line).

Economic Evaluation

Temporal Variation

The cost per kilowatt for both power and power delivery varies with season, day of week and time of day. The cost of the power, itself, rises during peak periods because of increasing production cost (expensive generation is used only when all less expensive generation has already been committed). Delivery cost rises because of increased losses.

While the embedded fixed costs of equipment and operations do not change during peak conditions, the delivery cost rises, nonetheless, because of the non-linearity in losses costs and the increased production cost at peak times. A typical residential load curve may have an annual load factor of .55 and a minimum annual load that is only 20% of the annual peak. In such a situation, I^2R losses at peak are 25 times their value at minimum load conditions, and roughly five times their average value over the year. Second, the cost of energy to make up losses is greater during peak, because the cost per kWHr is higher. Table 6.8 gives computation of losses cost on a feeder under average, minimum, and peak conditions. Figure 6.11 illustrates the behavior of losses cost over a typical day.

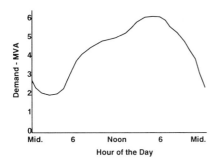 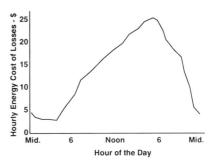

Figure 6.11 Coincident demand for a feeder (left) and losses cost (right) on an hourly basis for its annual peak day. While load varies by a factor of 3:1 during the day, losses vary by nearly an order of magnitude.

Table 6.8 Cost of Losses Per Kilowatt Delivered on a Feeder As a Function of the Operating Conditions

System Situation	Load kW	Losses kW	Price of Power (cents)	Value of Losses	Losses Cost per KW delivered
Minimum (April, Sun., 3 AM)	1200	7.9	5.0	$.40	.3
Average load (8760 hours)	3300	59.4	6.8	$4.03	1.2
Peak load (July, Tues., 5 PM)	6000	196.4	7.5	$14.33	2.5
Average, contingency	4900	130.9	6.8	$8.87	1.8
Peak, contingency	7600	315.1	7.5	$23.63	3.1

Contingency Variations

Economy of operation is normally a concern in the design of T&D systems only with respect to "normal" conditions, and no attempt is made to assure economy during contingency or emergency situations. During contingencies, when equipment is out of service and temporary switching or other measures have been implemented to maintain service, load levels, losses, and loss-of-life costs are quite high. Table 6.8 shows that during contingency operation of a feeder (When it picks up half of the load on another feeder) losses costs rise substantially.

Level of the System

Power delivered to customers at normal utilization voltage -- 120/240 volts -- travels through all levels of the T&D system on its way to the customer. If delivered at a higher voltage level, some levels are avoided and no cost is associated with these lower levels. Therefore, power is typically priced less if delivered in bulk at higher voltage, and even lower if the consumer owns or pays for the receiving-end power system equipment (substation, control, relaying and breakers, and metering, etc.). As an example, in Figure 6.12, power delivered at subtransmission voltage (69 kV) to a large industrial customer would not travel through a distribution substation, feeder, service transformer, or secondary system. As a result, its cost would be 6.4 cents per kilowatt hour, not the full 8.2 cents per kilowatt hour charged at the 120/240 volt level. Distribution costs were avoided and thus, the power is priced in accordance with just the transmission-only costs that were incurred.

Economic Evaluation

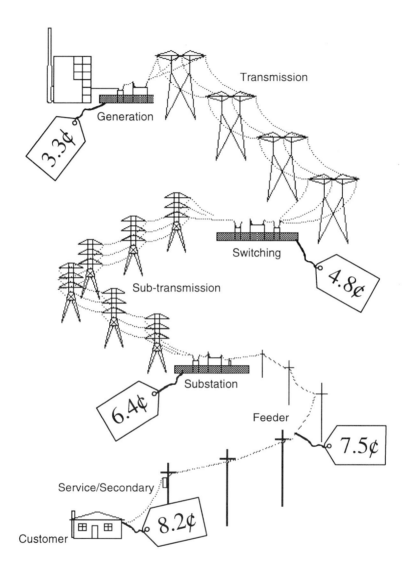

Figure 6.12 The cost of power depends on the effort and facilities used to deliver it. Thus, price varies depending on the level of the system from which it is delivered, as shown here. Prices indicate the average cost/kWHr.

Spatial Variation of T&D Cost

Cost to deliver power varies depending on location within a power system, both because there are simply some locations (isolated, long distances from other areas) where it is difficult to deliver power, and because once a system has been put in place, there will be some locations close to a substation (low losses, etc.) and others that happen to be far away from it. Figure 6.13 shows the cost of power delivery for a small port town, evaluated on a high resolution spatial basis. Here, as seems typical, embedded cost of power delivery varies by a factor of three among locations. Cost varies for a number of reasons: equipment in one part of the city is highly loaded, in other parts, lightly loaded; areas for the substation incur higher losses; load density/fixed cost ratio varies from area to area.

Marginal and incremental costs of power delivery or T&D expansion also vary on a spatial basis, often by a factor of ten to one, as shown in Figure 6.14. High local cost levels are usually due to tight constraints on existing equipment capacities and land availability -- for example a lack of any more room to expand a substation regardless of what is done.

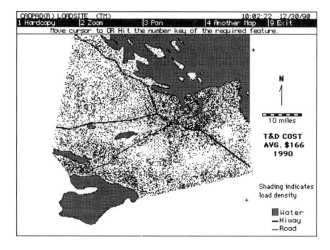

Figure 6.13 Shading indicates annual delivery cost of electric power on a location-dependent basis for a small coastal city.

Economic Evaluation

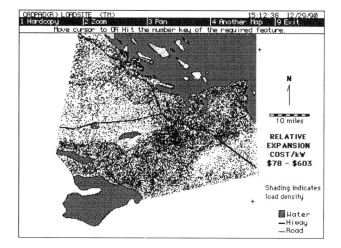

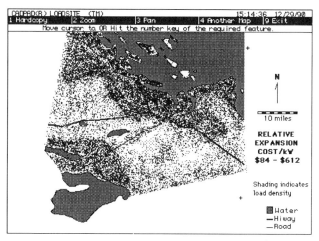

Figure 6.14 Maps of incremental T&D delivery cost for the region in Figure 6.13 under two different plans. Top, a proposed new power line is routed along a ROW near the coast. Bottom, it is not built because of environmental concerns. Either way, subsequent expansion costs vary spatially (a characteristic of all T&D systems), *but the locations of high and low-cost areas work out differently.* The specific locations of where power delivery is most expensive are often a function of planning and engineering decisions, not inherent difficulties associated with the terrain or the area's isolation etc.

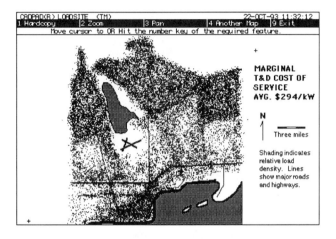

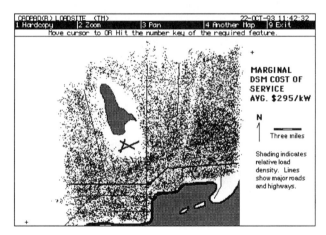

Figure 6.15 Both T&D and DSM costs vary spatially, depending on a variety of factors. Shown here are maps of the cost of expansion of T&D and DSM within a small city. Quite obviously, the cost-effectiveness of DSM (the ratio of these two) also varies spatially.

Economic Evaluation

Spatial Cost Variations and Integrated Resource Planning

Many aspects of integrated resource cost and benefit analysis vary geographically, whether the IRP is done on a resource value basis (traditional IRP) or a customer value basis (retail energy services marketing). The high local costs that occur in some areas of a power system are one reason why "geographically targeted" IR programs are often stressed by regulators, for even if energy efficiency, renewable resources, and other integrated resources are not cost-effective with respect to average T&D costs, they may have high benefit/cost ratio in the high cost areas.

The marginal and incremental costs of energy efficiency and renewable resources vary geographically, too, in much the same way as T&D costs. Usually DSM costs vary over a slightly narrower range than T&D costs, but as shown in Figure 6.15, the variations can be significant nonetheless. One reason DSM costs vary geographically is that the location of the target customer types varies. In addition, the market penetration of target appliances will vary -- for example some residential neighborhoods have high proportions of electric water heaters, others do not.

Climatological differences within a utility's service territory can have an impact on the distribution of weather-related appliances loads. For example, a very high market penetration of residential air conditioning can be found in many low-lying, coastal urban areas. Since air conditioners in these homes run at near 100% duty cycle during peak periods, every load controller will produce a very high reduction in coincident load. As a result, a reasonable market penetration of A/C load control can make a very noticeable impact on local distribution system peak loads. Yet an hour or two's drive away, at higher elevations, only a small proportion of homeowners may have A/C units and those may not produce a significant contribution to local distribution peaks. As a result, it would cost a great deal (many installations, many customers signed up with corresponding incentive payments) to "buy" any significant load control reduction.

6.6 CONCLUSION

All cost evaluation for power delivery planning should be based upon economic evaluation that is consistent over the range of all alternatives. Usually, this means that present and future costs must be compared. Present worth analysis, levelized costing, or some similar method of putting present and future value on a comparable based will be used. While methods and standards vary from one company to another, the important point is that planners apply their company's method consistently and correctly.

REFERENCES AND FURTHER READING

T. W. Berrie, *Electricity Economics and Planning,* Peter Peregrinus Ltd., London, 1992.

F. Schweppe, et al., *Spot Pricing of Electricity,* Kluwer Press, Cambridge, MA, 1988.

H. L. Willis and J. E. D. Northcote-Green, "Comparison of Fourteen Distribution Load Forecasting Methods," *IEEE Transactions on Power Apparatus and Systems,* June 1984, p. 1190.

H. L. Willis and R. W. Powell, "Load Forecasting for Transmission Planning," *IEEE Transactions on Power Apparatus and Systems,* August 1985, p. 2550.

World Bank, *Guidelines for Marginal Cost Analysis of Power Systems,* Energy Department paper number 18, 1984.

7
Line Segment and Transformer Economics and Set Design

7.1 INTRODUCTION

This is the first of six chapters addressing distribution system layout. The goal of distribution layout planning is to arrange the overall design of the distribution system so that it does its job in the best manner possible. The distribution feeder system is "assembled" by putting together a number of line segments (supported by ancillary equipment such as shunt capacitors, line regulators, etc.) into a working whole. "Layout" includes all aspects of how and why this is done, including selecting equipment, determining how to connect it to the rest of the system, and how to arrange and locate all of the pieces.

The key element in this layout is the distribution line -- the mechanism by which power is moved. The planner's goal is to build a feeder system from various line segments, arranging them so that the distribution system reaches every customer with sufficient capacity and proper voltage level to serve his load, while satisfying often conflicting goals of minimizing cost and maximizing service quality and reliability.

This chapter addresses the fundamental building block of distribution system layout -- the line segment. It begins by reviewing various types of lines and other aspects of their application. It then looks at the economics of their utilization, and how line capacity (conductor size) is selected for each segment based on economy. It concludes by considering "conductor inventory design basics" a fundamental, often overlooked, first step in distribution planning -- selecting a good *set* of line types (conductor sizes), appropriate to the planning requirements, so that the planner has the correct sizes and types of lines from which to build the system.

7.2 DISTRIBUTION LINES

The most primary function of a power delivery system is the movement of power from sources of supply to points of consumption. Power is moved on transmission or distribution *line segments*. In addition, it is often raised or reduced in voltage by *transformers*, in order to facilitate its efficient movement, or to render it of a more suitable voltage for consumption. Line segments and transformers are the basic "atomic elements" out of which the power distribution system is assembled. Other equipment, such as line regulators and capacitors, exist to support them, or to protect and control their functions, as in the case of relays and breakers. Both lines and transformers are available in a wide variety of types, capacities, and voltage ranges, over which both cost and electrical performance vary widely.

Line segments transmit electric power from one end of the other: except for a small percentage of losses, the power entering a line segment is the same as the power leaving the segment, and except for voltage drop the voltage along the line segment remains constant from one end to the other. A distribution system can be viewed as composed of numerous line segments, connected together so as to route power to customers as needed, interspersed with transformers, which change voltage-level as needed.

Line Types

Distribution lines are available in a variety of types suited to different situations and requirements, many of which are obvious at the beginning in any planning situation. Lines can be built as underground or overhead, and within the overhead category several special line designs exist for narrow rights of way and other special circumstances (Figure 7.1). The number of specialized conductor, cable and line equipment types available as standard products from suppliers is astounding -- not just dozens, but hundreds of different designs, many of the which possess small variations in design, which are important in unusual circumstances.

Many of the distinctions among these various sub-sub types become significant in the detailed engineering of the distribution system, and they are often the "solution" to vexing special situations where standard design cannot handle a particular constraint (a long span, severe weather stress, etc.).

Line Segment and Transformer Economics

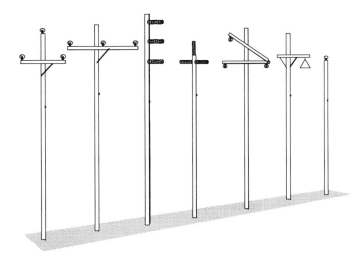

Figure 7.1 Distribution feeder lines can be built in a variety of styles, as illustrated by the variety shown here, by no means exhaustive. But what interests the planner most is not the exact style built, but the line capacity and cost, which are more a function of selection of conductor or cable size, as shown in Table 7.1.

Most utilities have chosen a dozen or fewer standard possibilities, as illustrated in Table 7.1. These are based on capacity (conductor size, cable size) within any one voltage class and type. In addition, both lines and transformers can be arranged in three-, two-, or single- phase configuration, and in either delta or wye configuration, as will be discussed later in this section.

Table 7.1 Standard Line Types Available for Distribution Layout at a Utility

14 kV OH	14 kV UG
795 MCM ACSR	1000 MCM XLP tri
500 MCM AA	500 MCM XLP
266 MCM AA	#1 XLP sheathed
3/0 AA	
#2 AA	
#6 CU	

Underground Distribution Lines

Underground feeder construction is used both in dense urban areas as well as suburban applications, both for esthetic and reliability reasons. In urban areas, a number of reasons proscribe anything but UG construction, beginning with the fact that there is simply not enough overhead space available for the number of feeders required to meet the very high load density. In many residential suburban applications, underground construction is desired because it rids neighborhoods of "unsightly" overhead lines and improves reliability. Most lengthy outages of overhead lines are caused by trees and/or adverse weather. Most momentary outages are caused by trees brushing against conductor during high winds. Underground residential distribution significantly reduces the incidence of both types of interruptions.

Underground feeders use cable, which generally consists of conductor, wrapped with one or more types of insulation, a neutral/ground conductor path of some type, and often a sheath to provide protection during installation, and to prevent penetration while in service (Figure 7.2). Cable is available in three-phase or single-phase form, in various sizes of conductor cross-section and various voltage levels (e.g., various types and thicknesses of insulation).

Underground cable is often installed inside buried *duct banks,* as shown in Figure 7.2. Ducts are usually made of concrete (but can be different forms of fiberglass, resin, and plastic) and are available with various numbers and sizes of cable positions. Occasionally metal, concrete, or pipe is used as a "single space" duct.[1] They are quite expensive to install but provide superb mechanical and electrical protection for the cables. Vaults -- underground rooms for cable pulling, repairs, and terminations -- are required at intervals. Ducts and vaults are installed first and the cable is pulled through them for installation. Routing is restricted to the routes of the ducts.

Duct bank/cable construction of distribution has very few advantages, beyond the fact that it fits the constraints found in densely populated urban cores where no other type of distribution construction can do the job. Duct bank construction is very expensive initially. While relatively immune to outages, when failures occur, they can take a long time to repair. Layout, planning, and operation are subject to numerous restrictions and constraints that limit design freedom and increase costs. (Chapter 8 includes a discussion of UG urban distribution layout).

Direct burial of cable is another option. Cable is buried in the soil with no duct bank or pipe protection (often a flexible plastic sheath or vinyl tube will be

---

[1] When UG cable must be routed over waterways, etc., in urban settings, it is encased in "duct pipes" routed on the underside of bridges and overpasses.

Line Segment and Transformer Economics

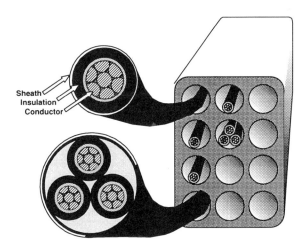

Figure 7.2 Distribution cable consists of conductor (either solid or stranded) wrapped with one of more types of insulation, and a sheath. Shown here are cross-sections of typical single-phase and three-phase cable, contained in a 3 × 4 concrete duct bank.

inserted around it). In particular, single-phase or small three-phase cable can be inserted in a streamlined operation in which a narrow trench is dug and the cable reeled in and covered -- all with one specialized machine in one operation, at a cost per mile comparable to that of overhead construction. Policy on direct burial of cable varies from one utility to another. Some limit its use to single-phase laterals. Others permit direct burial of all types and sizes of cable.

The advantages of direct buried distribution cable are low cost and fast installation speed (no type of distribution can be built more quickly), and esthetic improvement -- the lines are out of sight. Direct buried cable is also immune to many causes of outages that afflict overhead lines.[2]

Disadvantages of underground cable are that it "wears out" much more quickly and sooner than overhead conductor. It also has its own type of outage causes (dig-ins from construction, rodents, etc.), and generally repairs take much longer than on overhead. Finding the point of failure can present a major problem, and the cable often has to be dug out and repaired in a lengthy and expensive splicing process, requiring particular skill and care.

---

[2] Underground distribution is not susceptible to trees falling or brushing against them, or damage from ice and high winds. On the other hand, evidence suggests that it is susceptible to lighting strikes -- ground strikes can destroy cable.

Three-, Two-, and One-Phase System Components

Electric power systems and their major elements, such as line segments and transformers, normally consist of three-phases, and power is generated in three distinct phases, each of identical frequency and voltage, but with voltage displaced by 120 degrees between any two phases. Although a preponderance of equipment in power systems is three-phase, the vast majority of loads are single-phase. As a result, most systems are a mixture of three-phase and single-phase circuitry and equipment.

Delta and wye configuration

Three-phase equipment (line segments, transformers, or otherwise) can be configured in either delta or wye (star) configuration, both of which are shown in Figure 7.3. Delta configurations use only three-phase conductors. Wye-connected systems carry a fourth, neutral conductor, often grounded but sometimes left "floating" (ungrounded wye system). Regardless, the expectation is that normally the neutral voltage and current are both fairly close to zero.

Most power systems are a combination of delta- and wye-connected equipment, with high voltage transmission built as three-phase delta and distribution being predominantly wye-connected. Change from delta to wye or vice versa is effected simultaneously with change in voltage-current at a transformer (there are delta-delta, wye-wye, and delta-wye transformers).

In either case, the line rating of most equipment is usually based on the phase-to-phase voltage. For example, a distribution feeder is called 12.47 kV if

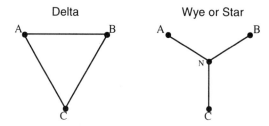

Figure 7.3 Three-phase equipment can be connected and operated in either a "delta" configuration (left) or wye (star) configuration (right). In delta systems loads are served and shunt equipment installed by connecting them between phases. In wye-connected systems, loads, and shunt equipment are connected between phase and neutral.

Line Segment and Transformer Economics

that is its phase to phase voltage, whether it is delta or wye connected, even though if operated as a wye circuit, its loads are actually connected between a phase and neutral (7.2 kV). Even a single-phase lateral which consists of one phase and a neutral with 7.2 kV potential between them, branching from the 12.47 kV phase-to phase line is referred to as "a single-phase 12.47 kV lateral."

Delta versus wye -- which is better?

If the only issue is moving power, a delta configuration is usually less costly. It requires only three conductors, not four, making equipment and construction cost lower. Losses and voltage drop are usually equivalent (at least if power flow is balanced). However, *wye-connected lines lead to a less expensive distribution system*, because transformers, reclosers, and lightning arresters for wye-connected systems cost less than their equivalents in delta systems. For example, single-phase service transformers require only one high-voltage bushing for wye-connected application; they require two for delta application (as well as slightly more expensive internal insulation and construction). They will also require two lightning arresters per transformer in delta applications, but only one on a wye-connected system.

There are service-reliability differences, too. A single-phase recloser can be installed on each phase of a wye-connected feeder, providing single-phase reclosing. On a delta circuit, the simple "recloser solution" is to use a three-phase unit, blinking customers on all phases every time the wind blows a tree limb against any one. Sensors and two-phase reclosers (more expensive and complicated) are required to provide single-phase reclosing on the delta circuit.

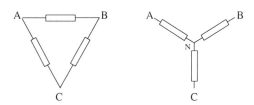

Figure 7.4 Left, a delta-connected system with only three phase conductors must have at least two phases present to provide even single-phase service to any one load (represented by rectangles). Either two- or three-phase service will require that all three phases be present. Right, a wye-connected system serves loads connected between phase conductors and the neutral. It requires the neutral everywhere service is provided, but can have one, two, or three-phase conductors.

Thus, while delta is the preferred style for transmission lines when power transmission cost is the major concern, wye-connected design is much preferred for distribution, because it provides both lower overall cost and better/simpler service reliability. In addition, historically many delta distribution systems built in the early part of the century were converted to wye-connected systems, in order to increase capacity to keep pace with growing customer demands. Delta to wye conversion is a relatively inexpensive way to upgrade primary-system capacity, because it can be done without replacing the service transformers.[3]

For example, a delta-connected, 13.8 kV (P-P) feeder system can be converted to 23.9 kV (which is 13.8 kV phase-to-ground), increasing capacity by 73% and reducing both voltage drop and losses significantly. The existing service transformers can be left in place, merely being re-connected to the primary in wye instead of delta fashion (see Figure 7.4). The cost of new transformers is avoided, as is the cost, mess, and customer inconvenience of making the replacements. The conversion is still expensive, and the logistics of conversion are complicated: on overhead lines the conductor can be left in place but new insulators, hardware, and crossarms (wider spacing) are required; on underground lines new cable must be pulled.

Such "once delta now wye" systems are easy to identify after the conversion. They have a neutral conductor and two-bushing single-phase transformers.

Two- and single-phase circuit elements

Current flow can be induced through a single-phase load by connecting it between any two phases (in a delta- or a wye-connected system), or between one phase and the neutral in a wye-connected system. Given that the majority of loads are single phase, the layout of most power systems tends to change from three-phase to single-phase circuitry at the extreme customer end.

In delta systems, single-phase service is provided by extending at least two phases to the customer and connecting the load across them (Figure 7.4). Two-phase service (rare) is provided by connecting loads between two pairs of the three conductors, in what is called an *open delta* load. Delta systems can thus have circuit elements consisting of two-phase conductors, or three.

In wye-connected systems, single-phase service is extended by connecting the load from any of the phases to the neutral as shown in Figure 7.4. Lines, transformers and their support equipment can be single-phase (one phase conductor and the neutral), two-phase (two-phase conductors and the neutral), or three-phase (all phases and the neutral).

---

[3] The author thanks Mr. Dan Ward, of Virginia Power, for his very cogent comments and observations about delta-versus-wye comparisons.

Line Segment and Transformer Economics 237

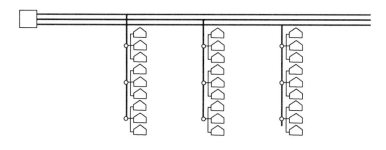

"American" Layout -- single phase laterals provide power to many small service transformers near the customers.

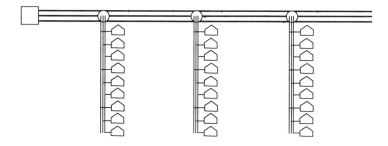

"European" Layout -- a few large service transformers feed single-phase customers with three--phase service-level lines. No laterals.

Figure 7.5 "American" (top) and "European" (bottom) layout methods for the service level to a neighborhood of 27 homes. American systems use single-phase laterals and relatively small, single-phase service transformers. Relatively low voltage (120 volt) service level (secondary) lines provide the same single phase of service to all customers in the immediate vicinity of each transformer. By contrast, European layout uses few if any laterals or single-phase primary circuit elements. Service transformers are much larger and often three-phase, as is the secondary circuits. Individual customers are provided with single-phase service from the three-phase secondary circuit.

Using Three-Phase Line Segments in Distribution Layout

From the distribution planner's perspective, three-phase line segments have a tremendous advantage: the current only has to be sent "out" to the load -- there are no losses or voltage drop associated with bringing the current "back" to the source in a completely balanced circuit. By contrast, if serving a load from a single-phase wye circuit (one phase and neutral) or a two-phase delta circuit, current flow goes out the load and then must return, and there are voltage drop and losses associated with each direction. Thus, use of *balanced three-phase components provides a two-to-one advantage* in terms of losses and voltage drop over completely unbalanced delivery.

Against this operating advantage is lower initial cost -- in a wye circuit there are four wires, and in a delta three, which must be installed, instead of only two for single-phase service.

"American" and "European" Service Level Layout

Exactly how and where the transition from three-phase to single phase circuitry takes place in a distribution system depends greatly on the specific situation, the design standards and practices in effect, local conditions, and even the individual planner's preferences. Generally, a utility will provide three-phase service to large customer loads -- for example at a factory or in a high-rise office building. The "split" to individual phases will be done within the wiring or plant distribution system inside the customer's facilities. However, in areas where individual customer loads are small, as for example in most residential neighborhoods, most customers will be provided with only one phase of service. Figure 7.5 illustrates two popular layout methods for accomplishing the multi- to single-phase transition in the layout of the system: what are referred to as "American" and "European" layouts.

"American" systems, typical throughout the United States, Canada, Mexico, and many others places, use a good deal of single-phase, primary-voltage laterals for delivery of power into various neighborhoods, and as a result, the majority of service-level equipment and circuitry are single-phase. The chief advantage of this approach is low capital cost when load density is low, as it was (and still is) in many parts of the United States when electric systems were first developed, and this style of distribution design was institutionalized as a preferred method.

"European" systems more commonly carry all three-phases into most of the service level, typically utilizing .416 kV three-phase secondary circuitry, with each customer provided with 250-volt service (phase to neutral). This layout is quite appropriate for higher load densities as are often found in urban areas. The

Line Segment and Transformer Economics

use of three-phase secondary provides fairly well-balanced loading on the service level, decreasing losses and increasing utilization.

While both the "American" and "European" styles were largely institutionalized in the early third of the 20th century, their continued use is due largely to the different utilization voltage. While the difference between 250 European service voltage and 120 volts of the American system may not seem substantial, it is a factor of two -- for this reason European circuits can "reach" four times as far given any equivalent load and voltage drop limitation. Add to this the fact that a balanced three-phase circuit can reach twice as far as a single-phase circuit, and the simple fact is that European systems do not need primary voltage laterals -- the secondary can carry roughly eight times the burden, replacing the role of laterals in the "American" system.

The foregoing discussion does not mean that there are not widespread areas of three-phase secondary service in many urban parts of America, nor that there are not areas of single-phase distribution and secondary circuitry in rural parts of Europe. Rather, the terms "American" and "European" systems are labels applied to two distinctly different ways of laying out the lower voltage portions of the distribution system, and of accomplishing the three-phase to single-phase transition and design of the "customer level" in a power system. "American" implies 120/240 volt single-phase service along with the conversion to single-phase circuitry at the primary level, using single phase laterals in many areas. "European" means 416 volt (phase to phase) three-phase service-level secondary from which 250 volt single-phase service is provided to customers by connecting them phase to neutral.

A third design type, applied only in very sparsely populated regions is single phase "earth return" distribution. Here, the earth is used as the return conductor and only one wire -- the phase conductor -- need be run in laterals to provide single-phase service. Earth return systems require augering of grounding rods deep into the soil at each customer site in order to establish a sufficiently low impedance to ground -- often costing up to $500 (US, 1995) per customer. However, the circuitry cost is very low, and in sparsely populated regions where the distance between customers is measured in kilometers the requirement for only a single wire can significantly reduce capital cost of the primary voltage system. Earth-return distribution with primary voltages as high as 66 kV and single-phase primary runs exceeding 120 miles has been used in very sparsely populated regions such as northern Canada and in parts of India and China.

Line Performance and Economy

From the distribution planning perspective, a line type has five attributes that determine how well it matches the particular needs of any distribution application. These are listed below in the order that most planners apply in evaluating what type of line will be best for a specific situation:

Does it fit? Only underground lines can be built in "UG-only" areas. In other cases constraints on esthetics, narrow widths of rights of way, clearance problems, or poor soil conditions, and other factors may limit the possible line types that can be built.

Capacity. Every line has a thermal capacity limit set by the maximum current its conductors can carry. Are all required loadings (normal, peak, contingency) within this limit for the line being considered?

Voltage drop. Every line experiences a voltage drop when transmitting power, due to its impedance. When delivering the load over the distance required, will the voltage drop be within proscribed limits?

Reliability. Line type can influence reliability. Generally, underground lines provide slightly more reliable service. Lines built with ACSR (aluminum clad steel reinforced) conductor rather than AA are less prone to parting and falling during snowstorms, etc.

Cost. Is the candidate line type the lowest cost alternative that meets requirements in the four categories above? Cost includes initial equipment and construction cost, continuing costs like property taxes, inspection, maintenance and repairs, and the costs of electrical losses.

The basic type of line built for any application -- overhead or underground, wood pole or metal pole, etc., has an impact on all five attributes, but mostly on only "fit," cost, and reliability. A wide range of voltage drop performance and capacity, as well as losses cost reduction, are generally available within any line type category -- UG or OH, crossarm or post insulator, wood or steel pole, etc., by selecting the appropriate "size" conductor or cable. To the planner, the chief aspect of line type selection in distribution planning is sizing -- determining the conductor or cable size in order to do the job required at minimum cost.

Impedance, Voltage Drop, and Losses

Both voltage drop, which determines how far power can be shipped along any line before unacceptable voltages are reached, and losses, which often create the highest portion of total PW cost for moving power at the distribution level, depend on the impedance of the line, as well as its loading. Line impedance is a function of the phasing (three-phase, two-phase, etc.), conductor resistance, and the conductor spacing (Figure 7.6). Generally, when additional capacity, lower voltage drop, or lower losses costs are desired, larger conductor can be specified,

Line Segment and Transformer Economics 241

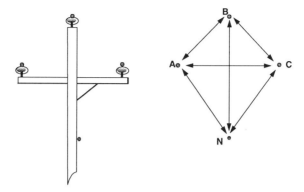

Figure 7.6 Impedance for a particular line (left) depends on the conductor's resistance (mostly a function of its size) and the reactance due to spacing of the conductors (right). Use of large conductor can reduce resistance but not reactance -- as a result beyond a certain point there are diminishing returns from using big conductor to reduce voltage drop and losses costs. Accurate assessment of the line impedances and coupling between all four wires is essential for quality distribution planning.

the additional cost of the larger size line "buying" improvement in all three categories. However, use of conductor above a certain size yields diminishing returns in terms of voltage drop and losses. Impedance (Z=R+X) does not drop much when R is already small and further reduced. Reactance, X, is a function of conductor spacing, which does not change with conductor size.

Little in distribution engineering is as important as accurately assessing and applying line impedances in the evaluation of distribution feeders. Any inaccuracy creates error in the estimated performance in both electrical service quality (voltage drop) and economic (losses) terms. The most accurate method of impedance computation is based upon Carson's equations (Carson, 1926). A somewhat less accurate method, but much easier to apply, is to use symmetrical components to compute the positive, negative, and zero sequence impedances. A thorough review of impedance computations and their application in feeder analysis is given in Kersting, 1992.

7.3 BASIC LINE TYPE ECONOMICS

At some point in the distribution process, the details of line segment design become important, but only the "big picture" aspects of a line segment economy and voltage-drop performance bear on distribution system planning and layout. At a later point in the planning process, precise design of feeders, down to and including details of pole placement, becomes important (engineering). But with respect to distribution planning, where the challenge is to determine the layout of the system and the requirements for each of its pieces, only basic electrical and economic performance of the line are of primary importance. The "model" -- the representation of performance -- used for a line in these phases of the planning process need only represent these aspects of a line.

Any particular type of transmission or distribution line (e.g. 12.47 kV P-P wye-connected, 336 MCM phase conductor with a 4/0 neutral) can be used to move any amount of power up to its maximum capability, from point A and point B, with certain consequences:

- power is moved from A to B
- voltage drop occurs from point A to point B
- losses in power are incurred, creating a cost
- the equipment and labor for the line create a cost
- maintaining the line in service creates a cost

Figure 7.7 shows the total 30-year PW cost to move power across one mile of this line as a function of the annual peak load (assumed 8760 hours, 90% power factor, 60% load factor, 46% loss factor). This computation was very easily performed using an electronic spreadsheet. Capital equipment and labor costs for the line have been added to the PW of annual fixed operating and ownership costs (evaluated over a 30-year period) to form the fixed cost (Y-intercept). This is the annual cost the utility incurs in just having the line in place. In addition, there is a variable cost, that of electrical losses, which depends on the amount of power shipped over the line, again evaluated over a 30-year period. This varies from zero to over $32,000 annually ($250,000 PW, 30 years) at this lines maximum 10 MW (thermal) limit. The resulting curve shows the total PW cost of serving any amount of peak load with this type of line.

A second line type can be compared to the first to determine if and under what conditions it might be more economical. Figure 7.8 shows the 336 MCM phase conductor line from Figure 7.8 and a similar line built with 636 MCM conductor. This line's fixed cost is higher -- the heavier conductor and its hardware costs more and requires more labor to construct -- but the line's impedance has dropped so that the value of losses at any level of power transmission has decreased. The 636 line also has a much higher thermal capacity -- 14.9 MW instead of 10.3 MW.

Line Segment and Transformer Economics

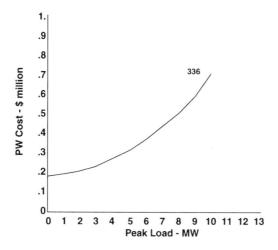

Figure 7.7 Annual cost of moving power one mile along a three-phase 12.47 kV distribution feeder built with 336 MCM conductor, with annual load measured in terms of its peak load over all 8760 hours (the load curve has an annual losses factor of .46).

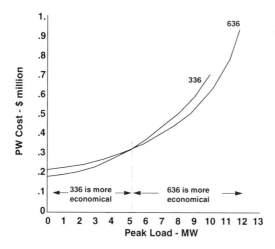

Figure 7.8 Computation of the cost curve for a second conductor identifies the ranges over which each conductor is the more economical of the two.

The two lines' cost curves cross at 5.1 MW. Below that value, the smaller line's lower capital cost more than compensates for its higher lifetime losses costs, and it has the lower PW. Above that peak load level, the larger conductor line is less expensive overall.

Effect of Load Growth on Conductor Selection

Figure 7.9A shows the evaluation from Figure 7.8 for a situation where load growth is assumed to grow by .5% annually from its original (plotted) peak value over the 30-year period of PW analysis. The value of future losses increases slightly with the higher load levels, shifting all points of both curves, except their Y-intercepts, upward slightly, and causing the crossover point between the two to shift slightly to the left. As a result the crossover point drops to 4.7 MW, not a large shift, but illustrative of the fact that load growth provides an economic incentive to install larger conductor.

Analysis such as shown here comparing any two lines types for economic applicability is easy to do using a PC-based electronic spreadsheet, applying PW losses analysis to any specific schedule of load growth expected over the 30-year

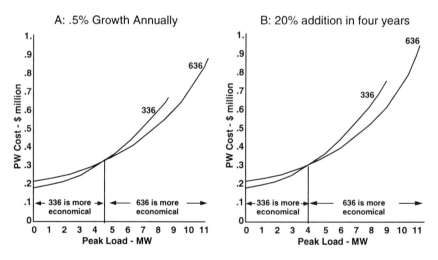

Figure 7.9 Anticipation of future load growth will shift the break point between conductors in favor of the larger (lower losses) conductor. (A) A steady .5% load growth over the next 30 years shifts the point from 5.1 to 4.7 MW. (B) The same evaluation with an assumption of an addition 20% MW of load growth added in four years.

Line Segment and Transformer Economics

period of PW analysis. Figure 7.9B compares the two conductors' applicability assuming that the peak load level shown sees no load growth for four years and then jumps by an addition 20% due to new customer additions.

In general, anticipation of load growth will move the breakpoint between two conductors to the left, favoring the larger conductor. Similarly, anticipation in any drop in load (for example due to conservation) will shift the break point for a particular analysis to the right.

Effect of Imbalance

Most line segments in a distribution system serve at best a slightly imbalanced load. The effect of anything other than perfect balance in load is to increase line impedance (now there will be a neutral current), increasing the losses and hence the losses cost, and increasing the value that larger conductor conveys. Assuming a 10% imbalance among loads in the analysis for Figure 7.8 shifts the breakpoint from 5.1 to 4.5 MW.

Effect of Shorter Evaluation Period

As they face de-regulation and the cost-cutting measures forced by competitive pressures, many distribution utilities are choosing to evaluate costs used in planning decisions over a shorter planning period than the traditional 30-year

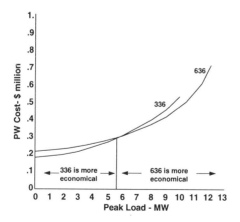

Figure 7.10 Evaluated over only a ten-year period instead of 30 years, the break point in Figure 7.8 moves to the right, to 5.7 MW.

cost evaluation period. Figure 7.10 shows the impact of using only a ten-year period of analysis (compare to Figure 7.10). The breakpoint between the two conductors has moved to the right, to 5.7 MW peak load. The impact is noticeable but not as dramatic as one might expect -- while the planning period has been shortened by a factor of three (from 30 to 10 years), the PW of losses drops by only 32% -- most of the PW for losses over the 30-year period are those that occur in its first decade.

Comparison of Changeout of Conductor for an Existing Line

Usually the cost of changing out an existing line type for a new, larger capacity line exceeds the cost of building the new line type itself. For example, a value of $145,000/mile is used in this chapter as the cost of building one mile of 636 MCM OH 3-phase 12,47 kV feeder. Converting an existing 336 MCM line to this larger conductor is estimated at $165,000.

To compare an existing 336 MCM to its possible changeout to "upgraded" 636 MCM conductor, only the Y-intercepts (fixed costs) of the two curves must be altered to reflect the changes in their respective costs. That for the 336 MCM drops -- the line already exists and thus there is no capital construction cost. Its fixed cost is only the value of the PW for future O&M, (which, if it is a very old line, might be higher than the future O&M for a new 336 MCM line).

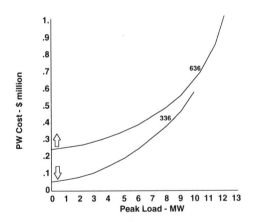

Figure 7.11 Analysis of an existing 336 MCM line against the same line upgraded to 636 MCM, as explained in the text.

Line Segment and Transformer Economics

The Y-intercept fixed cost for 636 MCM increases by $20,000, reflecting the higher changeout cost of installing the conductor when the line is already in place. Both curve shapes stay the same. Figure 7.11 shows the result (assuming no load growth -- compare to Figure 7.8). In this case the two curves never meet (computation shows they would cross at 10.9 MW, a value above the 10.4 MW thermal capacity limit of the 336 MCM conductor. Thus, in this case, once the 336 MCM is in place it is best to leave it even if its losses cost become quite high. More generally, changeout evaluation always moves the breakpoint to the right in comparison to cases where new line types are being evaluated. Figure 7.11 also demonstrates a very real and important aspect of distribution planning: once conductors are in place, replacement is generally not justifiable on the basis of economics. If a mistake is made, the system is "stuck" with it.

Evaluation of Multiple Conductors

In a manner similar to the comparisons done in Figures 7.7 through 7.11, other conductors can be added to the comparison. Shown in Figure 7.12 are lines of #2, 4/0, 795 MCM and 1113 MCM added to the conductors from Figure 7.7.

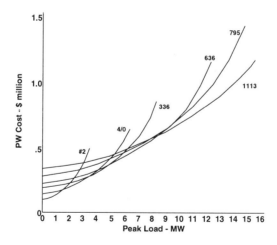

Figure 7.12 Analysis of six line types for overhead 12.47 kV crossarm type feeder, can be used to identify the range over which each conductor is most economical. The cost behavior of the entire set of possible line types at any voltage level can be characterized with an approximation that represents the lowest costs from among the conductor set (dotted line).

Table 7.2 Economical Ranges of 12.47 kV OH Three-Phase Line Types as Evaluated for Capital and Losses Economics -- Annual Peak MW

Line Conductor	Low	High
#2	0	1.6
4/0	1.6	3.7
336	3.7	5.1
636	5.1	8.5
795		
1113	8.5	12.4

Economical loading ranges for conductors

The range of loadings for which a particular conductor offers lower cost than any other available conductor is called its *economical loading range*. Table 7.2 shows economical ranges for the six conductors evaluated in Figure 7.12. In the range indicated, each is the most economical of the six evaluated. Note that 795 MCM conductor has no range over which it is the most economical -- it is a redundant line type. Redundant line types will be discussed later in this chapter.

The economical loading ranges for conductors, as well as the shape of the individual and group cost-vs.-capacity curves (e.g., Figure 7.12) are a function of power factor. VAR flow has a disproportionately high impact on conductors with high X/R ratios. Tables and plots in this Chapter assume a 90% power factor, about the best that can be achieved in practice (see Chapter 9, Section 9.6).

Evaluation of one- and two-phase lines

The preceding discussion focused on only balanced three-phase circuits. Similar analysis can be carried out on one- and two-phase line types to determine their economics and identify when and if they offer more economy than three-phase lines can provide. Figure 7.14 shows Figure 7.12 with cost curves for single-phase #2, 4/0, and 336 MCM conductor plotted on top of the approximate "best that can be done" relationship developed in Figure 7.12.

In general, single-phase circuits have a lower initial cost and a higher operating losses cost, than three-phase construction. They are suited only to moving small amounts of power short distances. However, single-phase lines are vitally important, for they make up the *vast majority* of a primary distribution system, a fact that discussed later in this Chapter (Section 7.5) and whose economic and design implications are explored in detail in Chapters 9 and 11.

Line Segment and Transformer Economics

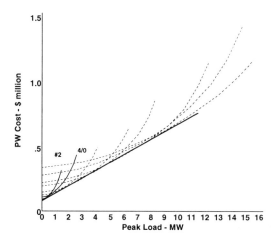

Figure 7.13 Single-phase lines, here of #2 and 4/0 conductor types (solid lines) can be added to the analysis. To a very good degree, "the best that can be done" by selecting the lowest cost line type at every peak load level is approximated by the heavy line shown, equal to $88,000 plus $45,250/MW.

Linear variable cost

If one takes the portion of each conductor's cost curve in which it offers the lowest overall cost, the set of these portions of the curves forms very nearly a straight line, as easily seen in Figure 7.12. Generally, resistance, R, of a line varies approximately in inverse proportion to conductor ampacity -- double the conductor's ability to carry current, and the resistance is cut in half. As a result, the conductor economics for the "conductor set" conform to nearly a straight line, as shown in Figure 7.13. Even though losses cost for any one conductor are proportional to I^2R, given that the planner has a choice of appropriate conductors, losses cost is linear. In this particular case, the "best that can be done" in Figure 7.13 is approximated very well by

$$\text{cost/mile} = \$88{,}000 + \$45{,}250 \times \text{MW}$$

Economics of Underground Cable

Underground cable systems in the core of large cities must be built in duct banks (direct burial of cable in a dense urban area is simply not an option). These ducts must be buried under streets (which means that construction disrupts traffic), and they require relatively wide trenching, considerable preparation and installation, labor. The utility must do significant clean-up work, including restoration and repaving of the street when done. All this work must be carried out in a crowded environment where permits, labor, and compliance costs are often quite high. The underground environment is crowded with other utilities and care must be taken to do no damage to those. Cable pulling and splicing vaults must be installed at intervals. The cable itself requires care and special skills in installation and splicing, etc.

For all these reasons, UG circuit cost is much higher than for overhead lines of equivalent voltage and capacity. Moreover, the very high initial cost of the duct installation changes the economic decision-making framework to one dominated by the fixed cost. From a planning standpoint, the high fixed cost changes the planning decision from "What size should be built?" to "Should anything be built?" Figure 7.14 shows PW costs for UG duct circuits in an urban area of the western United States.

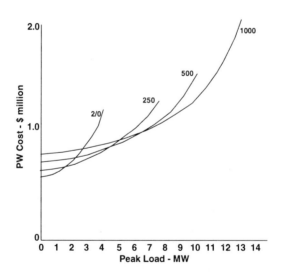

Figure 7.14 Cost curves for 15 kV underground distribution in ducts, as used for planning in the downtown area of a large city in the western United States. Cost is dominated by the high cost of building ducts. Compare to Figure 7.12.

Line Segment and Transformer Economics

7.4 VOLTAGE DROP AND LOAD REACH

The fundamental purpose of a distribution system is to move power. The evaluation that led to Table 7.2 looked only at the economic efficiency of moving power on a per mile basis, not at whether the feeders can move power far enough to meet the system's needs. The system's "needs" depend on a host of other factors: load density, subtransmission voltages, substation sites, substation capacity constraints, and other aspects of design which will be discussed in Chapter 11. But regardless of those needs, the process of evaluating conductor must be able to evaluate line performance against those needs.

Load Reach

The distance that an electrical line can move power before encountering limits in its performance due to engineering criteria, is called its *load reach*. If voltage drop criteria is set at a maximum of 7.5%, and a distribution line type has 2.5% voltage drop/mile at a recommended loading level, then its load reach is three miles. If the system requires power to be moved three miles or less, then the line is capable of meeting these needs at its recommended loading. But if the requirement is for power to be moved four miles, then either the line must be upgraded to one with 2% or less voltage drop per mile, or the recommended loading must be reduced by 25%.

Thermal load reach

The distance a particular line type can move power at its thermal loading limit is its *thermal load reach*. For example, the 12.47 kV (P-P) 336 MCM overhead line type analyzed earlier in this chapter has a voltage drop of 298 volts/mile, or 4.1% per mile, at its thermal limit of 530 amps. Thus, its thermal load reach against a 7.5% voltage drop criteria is

$$\text{Thermal load reach} = .075/.041 = 1.8 \text{ miles}$$

Thermal load reaches for various line types are shown in Table 7.3. Those for smaller conductor sizes are similar, because resistance (which dominates impedance for smaller conductor sizes) and thermal capacity vary roughly in reciprocal proportion. However, increasing a line's conductor size reduces its reactance only slightly. As a result, the impedance of large conductor lines does not drop in proportion to capacity, and the thermal load reach of lines larger than about 500 MCM is less than that of smaller lines. Table 7.3 gives the thermal load reach of three-phase 12.47 kV overhead lines made with nine different size conductors, including the six from Section 7.3, all evaluated on the same basis.

Economical load reach

A much more useful value for distribution planning is the distance power can be moved when the line's loadings are kept in its economical loading range. For example, Table 7.2 shows that a 336 MCM three-phase line has an economical loading range of from 3.7 to 5.1 MW, at which load reach against a 7.5% criteria is 5.1 and 3.6 miles respectively. The smaller value -- 3.6 miles -- is its *economic load reach* -- a planner can depend upon a 336 MCM line to deliver power this far, and within service standards, whenever utilized according to Table 7.2. Table 7.4 shows economic load reaches for various size lines.

Table 7.3 Thermal Load Reach Distances for 12.47 kV OH 3-P Line Type - Miles

Conductor	Thermal Reach
#6	1.8
#2	1.8
4/0	1.8
336	1.8
636	1.6
795	1.5
1113	1.4
1354	1.2
2000	1.1

Table 7.4 Economic Load Reach for 12.47 kV OH 3-P Line Type - Miles

Conductor	Economic Reach
#6	3.6
#2	3.5
4/0	3.6
336	3.6
636	3.5
795	3.3
1113	2.7
1354	2.4
2000	2.2

Line Segment and Transformer Economics

Comparing load reach and K-factor

The K-factor is a traditional way of assessing distribution performance, related in a tangential way to load reach. Developed prior to WW II, when primary voltages were predominately 4 kV or less and demand was far below today's standards, the K-factor (kVA factor) measures the percent voltage drop per kVA-mile of power delivery. For modern applications a factor expressed in terms of percent voltage drop per MVA-mile is more immediately useful. While this should perhaps be termed the "M-factor," it is normally termed K_m.

K_m = the percent voltage drop per MVA-mile

Load reach at a particular load level and the K-factor are related by

$$\text{Load reach (miles)} = C_v / (K_m \times L) \tag{7.1}$$

where C_v = voltage drop limit and L = load in MVA

For example, if the C_v is 7.5%, K_m is .5%/MVA-mile, and load 5 MVA, then the load reach is three miles.

When computed accurately and used in a knowledgeable manner, the K factor is a powerful *engineering* tool for estimating voltages and their sensitivity to load. In the hands of a master, it can be an amazingly versatile and quick tool for assessing feeder capability as well as evaluating design changes like conductor upgrades, capacitors, and regulators and LTCs (Kersting).

However, for distribution *planning* purposes the author prefers load reach, particularly economic load reach, for three reasons. First, load reach directly addresses the mission of a distribution system -- moving power -- with a factor measured in terms of distance. It is particularly powerful when used as a design guideline, as discussed in Chapters 8, 9, and 11. By contrast, K-factor addresses distance only tangentially.

Second, load reach is measured in a single dimension -- distance -- instead of two -- load and distance -- like the K-factor. This makes load reach easy to apply as something close to a planning criterion.

Third, economic load reach embodies both the desired electrical and economic design criteria. It assumes and is compatible with planning in which line segments are picked based on economics -- a step toward assuring that overall system design is at *minimum cost*. In addition, if used appropriately, it assures that maximum voltage drop is reached, but never exceeded. Voltage drop is *not* something to be minimized. It is a resource, never to be squandered, but always to be used prudently, and *fully*. A feeder that has less voltage drop than permitted is one that is very likely over-built, and hence probably more expensive than it needs to be. *Feeders built to the economic load reach guideline use all the voltage drop available to achieve low cost.*

Voltage drop in the economic range

The larger conductor sizes shown in Table 7.4 have shorter economic reach distances, because voltage drop does not drop in inverse proportion to increasing conductor size, as do losses. Losses are a function of resistance, which varies roughly in inverse proportion to conductor ampacity. But voltage drop is a function of impedance (resistance and reactance) not just resistance, and reactance depends mostly on phase spacing (Figure 7.6). As a rough rule of thumb, doubling a line's ampacity will cut resistance in half, but reduce reactance by only 10%.

For example, upgrading #2 ACSR ($Z = 1.7 + j.9$ ohms) to 4/0 ACSR ($.6 + j.8$ ohms) nearly doubles ampacity and nearly halves resistance. The small conductor has a relatively high resistance, and thus cutting that in half has a large impact on impedance and voltage drop. Both losses *and* the voltage drop are nearly halved.

But while upgrading 336 MCM ($.3 + j.66$) to 1113 MCM ($.09 + j.6$ ohms) doubles ampacity and cuts losses (by a factor of three in this case), the magnitude of its impedance is decreased by only 18%. Impact on voltage drop will depend on power factor, but in most practical situations (80-90% power factor) voltage drop will be reduced by only about half the amount that losses are reduced, whenever conductor size is above 500 MCM.

Figure 7.15 shows the voltage drop per mile of the various conductors from Figure 7.12 as a function of their loading at 90% power factor. Voltage drop in their economic loading ranges is highlighted (heavier portions of the lines), and the impact of the increasingly high X/R ratios for larger conductor can clearly be seen.

Adjusting Loading Ranges for Equivalent Reach

As shown in Figure 7.15, if 636 and 1113 MCM line types are limited to 7.8 and 9.3 MW respectively, then they, too, achieve a voltage drop sufficient to move power 3.6 miles while staying within criteria, essentially matching the performance of the rest of the conductors in the set. Table 7.5 shows the economic loading ranges of Table 7.2 adjusted so that each conductor has an economic load reach of 3.6 miles.

The loading ranges shown in Table 7.5 provide *equivalent capability for all the conductors in the set.* Those shown in Table 7.2 looked only at cost per mile and neglected the need to move power. Table 7.5 provides a more consistent set of guidelines.

Many distribution planners are familiar with the economic selection process outlined in the discussion accompanying Figures 7.7 through 7.12, which resulted

Line Segment and Transformer Economics

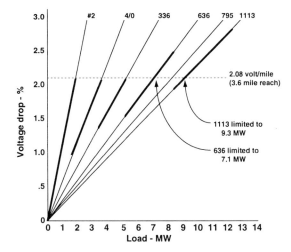

Figure 7.15 Voltage drop per mile of the conductors in the example as a function of loading. Heavy portion of each line is its economic loading range. The dotted line shows the voltage drop (2.08 volts/mile) corresponding to a load reach of 3.6 mile against a 7.5% voltage drop criteria. Due to their high X/R ratios, both 636 and 1113 MCM type lines have higher voltage drops for corresponding amounts of economic loading, and thus a much shorter load reach than the smaller conductor. Table 7.3 shows the economic loading table as revised, to limit their recommended loading ranges to less than 2.08 volts/mile, so they match the load reach of the smaller conductors in the set.

Table 7.5 Loading Ranges of the Conductors from Table 7.2 as Evaluated for Capital, Losses, *and* Load Reach -- Annual Peak MW

Line Conductor	Low	High
#2	0	1.6
4/0	1.6	3.7
336	3.7	5.1
636	5.1	7.8
795	not used (redundant)	
1113	7.8	9.3

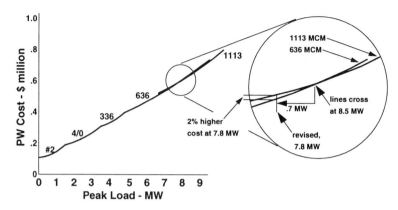

Figure 7.16 Only the portion of the cost versus loading curves in each conductor's economic loading range are shown here. Those for 636 and 1113 MCM cross at 8.5 MW, but moving the switchout load to 7.8 MW increases cost only a slight amount because the curves for the larger conductors are very tangential at that point.

in the loading ranges in Table 7.2. However, they are unfamiliar with the idea of de-rating large conductor as recommended in Table 7.5. It is important to understand why this should be done as a general recommendation.

First, planners should realize that only a very small amount of potential savings is given up in de-rating the large conductors (as illustrated in Figure 7.16). The cost curves for the larger conductors are quite parallel to one another in the range where they cross. Thus, deviations in loading range among the larger conductor sizes mean only slight changes in cost (the same is *not* true if similar adjustments are made in the smaller conductors).

Adjustment leads to maximum system economy

Whether the adjustment is warranted depends on whether the distribution planners *need* a reach of 3.6 miles to satisfy their system requirements. However, if planners do not arrange their system so they use all its equipment capabilities, they are forfeiting performance they could apply to further lower cost. The smaller lines in this conductor set *can* move power at least 3.6 miles within their economic loading ranges. Maximum overall economy will occur when the power system is laid out so that the feeder system uses all this potential, which means having all feeder contribute to moving power as far as the primary voltage can move it. This is discussed in more detail in Chapter 11.

Line Segment and Transformer Economics

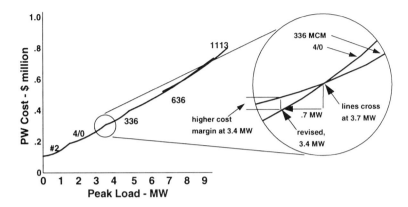

Figure 7.17 The cost curves for small conductor are not as parallel near their intersection points as they are for larger conductor. A .3 MW shift in the economic loading point between 4/0 and 336 can also "buy" a voltage drop equal to that obtained by moving the 636 to 1113 MCM transition point in Figure 7.16, but at a higher cost.

The structure typical of most radial feeders includes a single trunk (usually of large conductor) upstream of many branches and laterals (usually of small conductor). If the trunk's large conductor is loaded to the top of its economical range it will create an undue amount of voltage drop, forcing the planner to reduce loadings or accept a shorter reach.

As an example, 1113 MCM, the largest conductor in the set evaluated earlier, creates 2.8% volts/mile at its maximum economic loading. Suppose that it is used for one mile of express trunk at this load level. The remaining voltage drop available to the planner is only 4.7%. Any of the smaller conductors in the set -- those suitable for the more lightly loaded branches and segments downstream, can move power 2.25 miles at maximum economical loading before reaching a 4.7% voltage drop, all that is left within the 7.5% criteria. Total feeder length is thus limited to 3.25 miles -- or 10% *less* than the economic potential of most of its conductors. The planner has three options:

1. Live with this restriction, laying out the system so it does not need to move power more than about 3.25 miles. Overall, this is quite expensive, as will be discussed in Chapter 11.

2. Upgrade the line size of the various segments downstream of the trunk. Usually, this is more expensive than de-rating the large

conductors, or upgrading among larger conductor sizes, because the cost curves of smaller conductor are not as parallel near their intersection points (Figure 7.17).

3. De-rate the smaller conductor used in the segments downstream. Usually, this leads to having to upgrade some or all segments, and is essentially (2) above.

In general, if large-conductor feeder trunks are heavily loaded, they "use up" an undue amount of the permitted voltage drop, and cumulatively, cost is almost certain to rise. (This concept is discussed further later in this chapter, and in great detail throughout Chapter 11.)

Finally, every planner should understand that loading tables are only guidelines on usage. Exceptions will abound in actual application. Then, judgment and evaluation on a case-by-case basis must be used to determine, if, when, and how the exceptions will be handled. Regardless, the fundamental purpose of "loading range" tables is to guide planners in how to use conductors most as well as possible. "As well as possible" means both most economically and most effectively at doing their job (which is moving power). As much as any one table can provide a set of general guidelines for how to make the most of the conductors in the example given, Table 7.5 is that set of guidelines. Moreover, the method described here is the recommended way to determine such loading ranges.

The fact that planners will encounter many exceptions to any loading range guideline table is testimony to how much comprehensive planning methods, such as the automated circuit optimization methods discussed in Chapter 17, are needed to hone designs if the goal is the reduce cost to a bare minimum. Modern optimization methods can balance the conflicting constraints and requirements of voltage, reach requirements, conductor sizing, and losses costs, in ways that no table or simplified engineering analysis can ever accommodate.

Therefore, when all is considered, the recommended loading guideline for the 12.47 kV conductor set evaluated in this chapter limits it to 9.3 MW, less than 40% of the thermal capacity of its largest conductor and only 75% of its largest conductor's maximum economic loading (i.e., based only on cost/mile). Given the very small savings that 1113 MCM renders over 636 MCM at 9 MW (less than 3%), it is very likely many electric utilities would chose not to include it in their conductor inventory, and use 636 MCM as the largest standard conductor, limiting the recommended loadings on the primary feeder system to about 7.5 MW. Alternatively, they might delete 636 MCM and substitute 795 MCM. As mentioned earlier this "redundant conductor" essentially duplicates the costs of 636 MCM, but its larger wire size permits economic reach up to 8.5 miles.

Line Segment and Transformer Economics   259

7.5 PERFORMANCE OF THE CONDUCTOR SET

Table 7.5 is thus the recommended set of loading guidelines for this case; conductors evaluated on the basis of cost per mile, with large line types "de-rated" so their voltage drop performance is equivalent to that of the smaller conductors. Figure 7.18 shows an approximation to the overall conductor sets economics, when used as recommended in a table, such as Table 7.5, that includes assessment of both cost and reach. The approximation shown represents the essential elements of the behavior of this particular conductor set in the example case examined here, and qualitatively shows the type of behavior exhibited by most sets of conductors when evaluated on the basis of economics and power movement capability against a constant economic reach requirement (in this case 3.6 miles).

There is a minimum fixed cost to move power (in this case $88,000), a "linear range" in which power and variable cost vary in roughly constant proportion, and a "exponential range" where the planner who wants greater capacity must resort of every increasing sizes of (ever more de-rated due to high X/R ratios) conductor.

For purposes of planning and evaluation costs of the economic performance of the entire *set of conductors* available to the planner can be approximated with a function that summarizes the "best cost possible" of the entire. Figure 7.18 shows such an approximation. In the particular case given here, the economic performance of the conductor set consists of:

> *Minimum cost* to move power (the Y-intercept) in this case $88,000, is the approximate minimum cost of building *any* three- phase line.

> *Linear economical range,* between 0 and about 8.5 MW peak load, assuming the lowest cost conductor is selected for any load level, cost is approximately $88,000 plus $45,250 per MW.

> *Exponential "expensive" range.* Above about 8.5 MW, costs rise as losses follow an ever steeper curve.

In general, any well-selected set of conductors at this voltage and evaluated under these same criteria will have the same overall economic performance. For example, the set including #3, 3/0, and 288, 500, 795, and 1354 MCM line types also has a fixed cost very close to $88,000, a linear slope of $45,250/MW, and an exponential cost range that begins at 8.5 MW. The "best that can be done" curve (right side, Figure 7.18) is independent of the exact conductors used -- any good set of conductors will do the job.

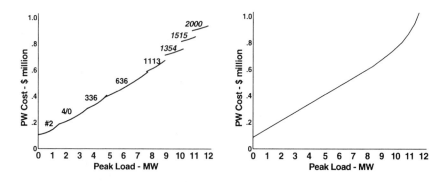

Figure 7.18 At the left, the portions of each conductor's cost curve in its loading range (Table 7.4). Three very large conductor sizes (1354, 1515, and 2000 MCM) have been added to illustrate how costs escalate exponentially beyond a certain point if equivalent reach is expected (ever larger conductor sizes must be de-rated in ever larger percentages to achieve the required reach). Right, the simple function described in the text represents the composite "best that can be done" of the conductor set quite well.

Effect of Different Voltage Levels

Similar cost versus peak load computations using these same six conductors as in Figure 7.12 can be carried out for other distribution voltages, and an approximate relation for "the best that can be done" found for each voltage level. Figure 7.19 compares the 12.47 kV cost behavior curve shown in Figure 7.18 with those based on analysis of the same six conductors for 4.16 kV, 25 kV, and 34.5 kV.[4] This displays the general behavior of cost versus peak load as voltage level of the distribution system is raised.

> The minimum fixed cost per mile changes roughly in proportion to something between the square and the cube root of the relative voltage level -- 25 kV costs about 27% more to build than 12.47 kV, while 34.5 kV costs about 60% more, and so forth.
>
> The width of the economical linear range is approximately proportional to voltage (i.e., it is between 0-8.5 MW for 12.47 kV, 0-2.8 MW for 4.16 kV, 0-17 MW for 25 kV, 0-23.5 MV for 34.5 kV).

---

[4] Crossarm distances used in impedance calculations were seven, nine, ten, and twelve feet, respectively.

Line Segment and Transformer Economics

The slope of both the economical linear range and the expensive exponential ranges varies roughly in proportion to the reciprocal of the voltage (linear cost is $45,250 MW for 12.47 kV, $22,600/ MW at 25 kV).

This last result is interesting and somewhat counter-intuitive. At any specified level of power (e.g., 6 MW) doubling voltage halves the required current, cutting the losses (and losses costs) by a factor of four. However, the reduction in losses costs wrought by doubling the voltage changes the economics of conductor selection -- the losses costs that can justify larger conductor are less at higher voltage. The net result is that, to a very close approximation, the slope (losses versus load) is inversely proportional to voltage, not inversely proportional to voltage squared.

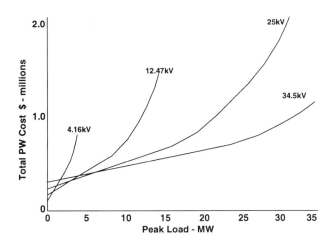

Figure 7.19 Behavior of overall cost versus peak load for various phase-to-phase voltages, based on urban/suburban construction costs and 30-year PW costs of O&M and losses for a utility in the mid United States, for conductor utilization with loading ranges adjusted in all cases to the "natural" economic load reach of the smaller conductors in the set. Losses are computed based on an 8760-hour loss factor in terms of the peak load plotted, and no load growth was assumed. Costs are reported here in relative terms because the author does not want readers to over interpret the results -- actual values from such an analysis are not general to all situations.

Figure 7.13 shows that every distribution voltage has some range of loading in which it is most economical. However, this comparison alone *does not* indicate (in fact, it does not even come close to indicating) what voltage is best for any given system situation. The fact that 25 kV offers lower cost per mile than 12.47 kV for all loads above 2.3 MW is almost irrelevant in determining primary voltage selection -- many other aspects of distribution performance apply in determining the "best" voltage for a particular utility distribution system. Voltage selection based on least-cost principles is discussed in Chapter 11.

Economic Load Reach as a Function of Primary Voltage

Table 7.6 gives the economic load reach (distance at maximum economical loading) for the set of six conductor line types analyzed earlier in this chapter, operating at voltages of 4.16, 12.47, 25, and 34.5 kV. Distances shown are not in exact proportion to voltage level because values at each voltage were based on impedances computed for the voltages' crossarm spacing (e.g., 12 feet for 34.5 kV versus 7 feet for 4.16 kV). Reactance is a function of spacing.

Doubling voltage doubles both the power delivered in the economical range, and *the economic load reach.* The 25 kV reach of 7.2 miles for 336 conductor in Table 7.4 is nearly twice that at 12.47kV, but at this distance the 25 kV is delivering 10.2 MW, the 12.47 kV only 5.1 MW (see Table 7.3).

The values in Table 7.5 *do not* indicate by themselves which voltage is best for any specific distribution system application, although they contribute greatly to such analysis. They indicate only the distance that "optimally-designed" systems of each voltage range can distribute power before costs begin to escalate because of voltage problems. If a planner knows that his system routinely has to deliver power over distances much greater than 1.3 miles, but seldom greater than 3.6 miles, then based on this table, 12.47 kV is probably the best voltage level for the distribution system as a whole, but there are *many* other factors that could change or influence this decision.

Table 7.6 Economical Load Reach Distances for OH 3-P Line Types as a Function of Voltage -- Miles

Primary Voltage -- kV	Economic Reach -- Miles
4.2	1.3
12.5	3.6
13.8	4.0
24.9	7.1
34.5	9.7

Line Segment and Transformer Economics

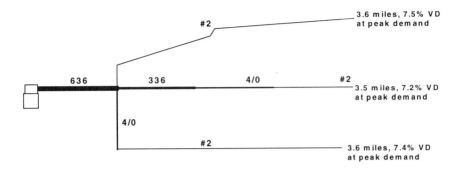

Figure 7.20 A planner can depend on economic selection criteria to produce a feeder with a load reach equal to that of the conductor set. If all line segments in this 12.47 kV feeder have had their conductor size selected on an economical basis from Table 7.5, then every path in the feeder will stay within voltage drop criteria out to at least 3.6 miles.

Characteristics of Feeders Laid Out Using "Economy and Reach" Tables

When used as a guideline for "putting together" a feeder or a feeder system, the conductor selection recommendations from Table 7.5, or any table similarly constructed, result in feeders which are nearly "optimal," or at least as optimal as any can be when designed on the basis of a general guideline table.

Feeders laid out based on economical conductor selection have the same economic load reach as their conductor set.

Very often, a feeder will be laid out as shown in Figure 7.20, with the conductor size "tapered" farther from the substation, as loadings decrease at line locations farther along the circuit run. When each segment is selected based on the adjusted, the resulting set of segments (i.e., the whole feeder) will have a composite economic load reach of about 3.6 miles.

> Distribution systems planned and utilized based on economic conductor selection can be depended upon to have a reach equal to the minimum economical load reach distance of their conductor set.

The ideal feeder length is *the economic load reach.*

Feeders significantly shorter than the economic load reach do not fully utilize the power movement potential of the primary voltage and conductors being used. Feeders longer than that limit have performance problems or must have their loadings reduced (in which case they are no longer economical).

> *Maximum economy and sufficient performance are both achieved when a feeder is laid out so that its length(s) are* just *at the economical load reach, with all segments sized by economic selection.*

Reach is important in determining the distribution voltage level.

If the substation spacing in a system is expected to be such that feeders will routinely have circuit runs of six miles, then a distribution voltage that provides economical load reach for the conductor set at that distance is desired -- a lower voltage could be used but will entail higher overall costs.

These and other aspects of voltage selection, feeder configuration, and the involvement of economic reach as a key design factor in distribution layout, are explored in much more detail in Chapters 8, 9, and 11.

Using a Conductor Set Beyond its Economical Reach

Often, a distribution planner has no choice but to use the distribution system to deliver power over distances well beyond its economic reach. For example, perhaps because no substation is planned in a particular sub-region, a planner knows that power will have to be distributed up to nine miles using 12.47 kV.

Table 7.7 Maximum Load at a Reach of Nine Miles for OH 12.47 kV 3-P Line Types -- MW

Line Conductor	Load
#2	.4
4/0	1.2
336	1.7
636	2.6
795	2.8
1113	3.7

Line Segment and Transformer Economics

That distance is well beyond the distances shown for 12.47 kV's economic reach in Table 7.4. One way to accomplish this task is to rely on voltage regulators and switched capacitors to boost voltage as required. This is expensive and has a relatively high continuing maintenance cost. Another way is to use larger conductor than specified for economical loading, or put another way, to employ line types at less than their most economical rating in order to gain additional load reach.

Table 7.7 gives the loadings that result in a voltage drop of .83% per mile for each of the six conductors shown in previous tables -- this voltage drop is equivalent to a reach of nine miles against a 7.5% voltage drop criterion. Figure 7.21 shows the portion of each conductor's cost curve for loads less than this limit. Linearization of "the best that can be done" in this case produces a 30% higher linear slope (compared to the normal economical linear range slope, also shown) -- the planner can expect to pay about 50% more compared to normal distribution costs to move this power. Since this margin is greater than the difference in capital costs between 25 kV and 12.47 kV (about 40%) but less than that between 12.47 kV and 34.5 kV (about 60-70%), the planner might want to look at 25 kV for this case.

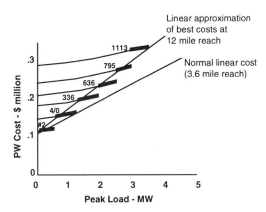

Figure 7.21 The portion of the six conductor cost curves from Figure 7.12 that fall within the nine mile reach criteria. Broad lines indicate the portion of each conductor's curve where it is the best choice of the six. Linearized, these choices result in slope more than 50% greater than the linearized cost for economical utilization (see Figure 7.12).

7.6 CONDUCTOR SET DESIGN

Although there are over one hundred types and sizes of overhead conductor and a similarly wide variety of underground cable types available for use, most distribution utilities standardize on between three and seven conductor types, and a similar number of cable types, which they keep in inventory and from which they build their system. The reason is that while a large number of available conductor and cable types would provide "a conductor for every situation," maintaining more than five or six different types as a standard increases cost. There is more to standardizing on a particular conductor -- for example 336 MCM ACSR -- than just deciding to use it and asking the Stores department to always maintain it in stock. A complete set of brackets, splices, terminators, clamps, and fittings must be kept in stock for this particular conductor, all of these probably different from those required for #2 or 1113 MCM conductor. Drawings and engineering specifications for lines built from each conductor must be developed and maintained. In addition, line crews must be provided with tools compatible with each conductor type (for example, the grips and the splice crimp jaws required to work 336 MCM conductor are slightly different than those needed for 1113 MCM), and line crews must be trained in the installation and procedures for each conductor type (in some cases stringing and tensioning procedures vary slightly from one type of conductor to another).

For all these reasons, maintaining more than six conductors or cables as standard, approved design-inventory is generally considered too costly, and in fact a limit of four has been found to be optimal by many utilities. On the other hand, using only one type is too limiting. As a result, nearly all utilities settle on using a set of between three and seven line types. The vast majority use four or five as their standard set.

What set of four or five conductors ought to be selected by a utility for use in putting together its system? Ideally, the planners need a set that provides very linear cost performance over the entire range where that is possible (0-8.5 MW for 12.47 kV) with relatively even spacing within that range covered by each conductor, as shown in Figure 7.22. These conductors will have similar thermal and economic reaches, too.

Note that conductor selection cannot widen the permissible economical linear band, due to the physics of line reactance as discussed earlier. Thus, the best that can be done is to pick a set that covers the range permissible within X/R limitations at the distribution voltage being used. Recommended practice is to select conductors to divide the possible economic range into N equal increments, where N is the number of conductors to be chosen, as shown in Figure 7.20. The largest conductor in this set is used to handle all large loads (those above the linear economical range).

Line Segment and Transformer Economics

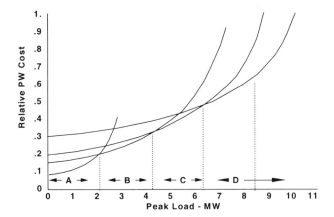

Figure 7.22 An ideal set of four conductors for 12.47kV application offers a very linear cost from 0 to 8.5 MW of peak load -- the full width of the possible economic linear cost range at 12.47. The economical range of each conductor type is roughly equal, at about 2.1 MW each, covering the 8.5 MW range in four equal increments with the "large conductor" having plenty of thermal capacity to deal with cases of very high load.

Common Problems in Conductor Inventory Design

Figure 7.23 illustrates two common problems that can occur with conductor sets. The first is a redundant line type -- a conductor without some range of loadings in which it provides the lowest cost. Unless there is some reason to maintain this conductor in stock, sufficient to justify the cost of doing so, it should be deleted from the standard inventory.

There can be valid reasons for maintaining a "redundant" conductor or line type. For example, some utilities apply AA (all aluminum conductor) to most of their design, but keep one or two sizes of ACSR (aluminum clad, steel reinforced) for applications where long span length or severe weather conditions require a particularly strong conductor. The ACSR, which has a higher cost and slightly higher losses compared to a similar AA conductor, will appear redundant when evaluated in the manner outlined here. From an economics perspective it is -- its existence in the utility's inventory should be justified on the extra reliability it provides in those situations for which it is intended.

The second problem shown in Figure 7.23 is a "gap" in economical application (shaded area), a range of loading where linear performance is not

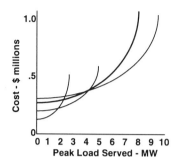 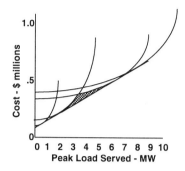

Figure 7.23 Common problems with conductor inventory selection are revealed through evaluation of the peak versus cost curves of two sets of four conductors. At the left, a redundant conductor (broad line). At the right, a gap (shaded area).

achieved, because none of the conductors in the set offers outstanding economics in that range of loadings. At its worst (around 3.5 MW) the difference between the "should be attainable" linear performance and the actual represents a 30% increase in cost -- this set of conductors leaves the distribution planner with no choice but to pay 30% more than is really necessary to move 3.5 MW of power.

The solution is to add a conductor to cover the gap, deleting one of the existing ones if necessary to keep within a four-conductor limit. While there are rare situations where it is impossible to find a set of conductors that offer linear cost-performance over a wide range of possible loadings, close to linear performance usually is attainable from a well chosen set of conductors and line designs.

Review Conductor Economics Every Five Years

Conductor economics should be reviewed every five years, or whenever any change in construction practices, costs, or planning method is made. Figure 7.24 illustrates what can happen when such review is not done. In this case, a utility in the United States standardized on a set of five conductors for 12.47 kV application in 1965, selecting a set that offered good, linear cost performance based on the costs and economic evaluation method used at that time (left diagram in Figure 7.24).

Three decades later this set does not provide good conductor economics, in spite of the fact that there are five conductors from which to choose. The reasons

Line Segment and Transformer Economics

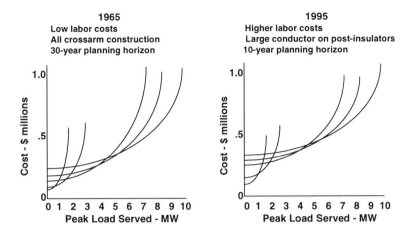

Figure 7.24 Example of how conductor economics can change, in this case for a utility in the northeastern United States. When originally standardized in 1965, the set of five conductors resulted in very linear economic performance over the entire range of loadings from 0-8.5 MW. 30 years later, costs and construction standards have shifted the economics so that these five conductors no longer provide good economical choices for new construction. All costs shown here are evaluated in 1995 dollars.

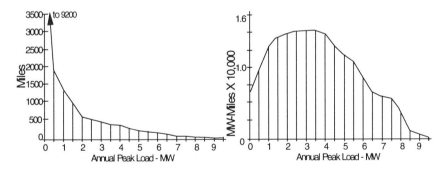

Figure 7.25 (Left) The 24,800 miles of primary voltage line in the system whose conductor economics are shown in Figure 7.24, plotted as a function of annual peak load. The majority of primary voltage segments carry less than 2 MW peak. (Right) same data with each segment length weighted by its MW loading. Diagrams like this surprise even experienced planners -- while large conductor lines carrying heavy loads get most of the planner's attention, single-phase laterals carrying less than 1 MW comprise a majority of length of most distribution systems. Chapter 11 will explore this in much more detail.

behind the change in cost performance provide a lesson in why conductor inventory should be reviewed periodically:

> Labor costs (in constant 1995 dollars) increased slightly over the 30-year period. Construction and maintenance costs, and hence the Y-intercept for every conductor type, are higher in 1995 than in 1965.
>
> Tax rate increased by 15% over the period, and while slight, this also contributed to a rise in the PW O&M and hence the Y-intercept fixed cost.
>
> Construction standards changed. In 1976 the utility revised its distribution standards to require post-insulator construction in a delta configuration (one phase on top and two laterals from each side of the pole) for all major feeder trunks. This was done for esthetic reasons, since large feeder trunks generally run along major streets and highways and post insulator is considered to have a somewhat neater appearance). Post insulator construction reduced line spacing (and hence reactance) by about 4%, thus increasing the cost of losses for the larger conductor.
>
> Losses costs changed. They actually decreased slightly (by about 3%) over the period, when measured in constant dollars.
>
> Planning period changed -- the utility shifted to a ten-year planning evaluation period for losses, from the previous 30-year period, effectively reducing both the cost of O&M (part of the fixed Y-intercept cost, and the cost of losses).

The net impact of all these changes is that fixed costs increased slightly, more so for larger conductor types than small, and variable (losses) costs went down, again more so for large conductor than small. This re-arrangement created the gap shown in Figure 7.24.

The reason why this gap is important is shown in Figure 7.25. Analysis of this utility's existing system on a segment-by-segment basis shows that fully half of all three-phase power shipped on its distribution system is moved in range of between 2-4 MW annual peak load. While planners often focus on only the large-current feeder segments, most systems have a characteristic like that shown in Figure 7.25, with a majority of the distribution system application lying in the lower levels of maximum possible feeder segment capability. Therefore, any gap left by the conductor set in this range is serious, for it means the planners have no tool with which to move these amounts of power as economically as possible.

Line Segment and Transformer Economics

Recommended Method for Conductor Set Design

The purpose of "conductor set design" is to determine the best set of line types (conductor and cable sizes, line construction types) for new construction, to determine the recommended loading levels for both new and existing (already in place) lines by type, and to determine the loading points at which reinforcement or reconstruction to upgrade existing lines to new construction is justified. Many aspects of the distribution business will be involved in this decision, including planning, engineering, operations, stores, customer service, and others. While the distribution planner must be aware of all of these perspectives, as steward of the planning function his chief priority is to assure that the resulting conductor inventory permits economical and satisfactory service to be achieved in the system.

From the planner's perspective, the primary goal is to achieve a good linear economy over the full range of loadings needed in distribution, at the load reaches expected to be required.

Determine *switchout loads* and *switchout conductors* for lines already in place

Planners cannot "undo" mistakes made in the past, but in some cases due to load growth or poor planning, lines already put in place are loaded far beyond the peak loads expected when they were originally planned. Planners should know at what load level it becomes economical to upgrade each type of line, and which conductor is most economical as an upgrade. Figure 7.26 shows the basic method of analysis, and illustrates two important aspects of determining upgrades. The four curves shown represent line types taken from Figure 7.22, and have had their Y-intercepts adjusted, as explained in the figure caption, in keeping with the discussion about conductor changeout economics given earlier in this chapter (see Figure 7.10 and accompanying text).

Note that while conductor C had an economic loading range (in Figure 7.22) for new design of 4.3 to 6.4 MW, replacement of this type when it is already in place is not economically justifiable until its peak load reaches 8.7 MW. More interesting is the case for conductor B. Its changeout point is similarly much higher than its maximum new construction loading (6.7 MW versus 4.2 MW), but if it needs to be changed out, it should be replaced with conductor type D, not the next largest size.

> *Very often the recommended upgrade for an existing line is not the next largest size, but a conductor two or even three steps up the "capacity ladder."*

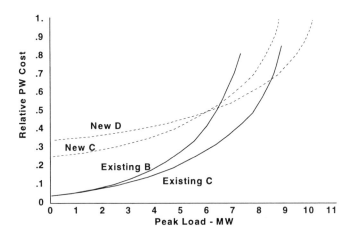

Figure 7.26 Conductors B and C (solid lines) from Figure 7.22 evaluated on an "already installed" basis -- they both have their Y-intercept adjusted to reflect only their future O&M PW costs, but no initial construction cost. The dashed lines show conductor C and D with higher Y-intercept costs than in Figure 7.22 (reflecting the fact that upgrades in place cost more than original construction). All four curves have the same shape (losses cost curve) as used in Figure 7.22.

Table 7.8 Data Used in Analysis of Conductor Sizing

Primary voltage (P-P)
Conductor spacing table
Conductor ohms/mile, GMR, and thermal limits
Conductor initial installed cost
Property tax rate
Annual line inspection cost
Annual maintenance cost expectation
The demand cost of losses
The energy cost of losses
Annual load loss factor
Load power factor
Annual load growth rate
PW evaluation period
PW factor

Include all costs

Line Segment and Transformer Economics

Include all costs

It is vital that all costs be included in the analysis and that the PW analysis include future costs over whatever period is applied for decision-making. Costs should include those shown in Table 7.8.

Number of conductors

Most utilities have found that between three and six overhead conductor types are needed, and between two and four cable types are necessary at each distribution voltage level. However, the proper answer to this question is "the minimum number that can provide linear cost with no significant gaps" over the range or required application which is normally the full range of loadings from zero up to the point where X/R ratio makes linearity impossible with any size conductor.

No incentive for an "extra large size"

Some utilities essentially set aside one conductor in their inventory as an "extra large" size, leaving N-1 to cover the range of normally expected loadings. This is usually not economical: selecting a set of N conductors that provides linear coverage generally leaves the largest conductor with plenty of capacity to handle extreme cases.

Single- and Two-Phase versus Three-Phase Evaluation

Studies of economic application should look at whether one- or two-phase circuits are more economical than three-phase, which can be determined in a similar manner to that illustrated earlier in Figure 7.13.

Method for Determining Economic Application and Conductor Sets

All of the data and criteria applicable to conductor inventory selection can be evaluated with an electronic spreadsheet on a personal computer. The results shown in all tables and drawings in this chapter was prepared by the author using a spreadsheet that took roughly two hours to put together and test. Table 7.6 lists the data used as input. Following these steps, the spreadsheet:

  A. Computes the positive, negative and zero sequence impedance for each conductor from conductor and spacing data (see Kersting, 1992).

  B. Uses the impedances to determine the losses and voltage drop caused by 8760 hours of load, at the input load factor, with a peak of 1 MW transmitted 1 mile along each conductor. It also computes the power delivered at the conductor's thermal limit.

Table 7.9 Economic Utilization Ranges for Conductors at 12.47 kV -- MW

Conductor	Phases	Normal (3 mile reach)		Rural (10.5 mile reach)*	
		Low	High	Low	High
#2	1	0	.25	0	.08
4/0	1	.25	.60	.08	.16
#2	3	.60	1.66	.16	.48
4./0	3	1.66	3.70	.48	1.06
336	3	3.70	5.10	1.06	1.45
636	3	5.10	8.75	1.45	2.50
1113	3	8.75	12.84	2.50	3.68

* for a 15 mile substation spacing (see discussion of spacing and reach in Chapter 11).

C. Computes two PW multipliers. 1) W, a simple PW multiplier (e.g., PW of a dollar over the next 30 years at a .9 PW factor equals $8.61, so W is 8.61 if period = 30 and PW factor =.9), and 2) a multiplier, U, that represents PW and escalation of losses costs due to load growth.

D. Fixed cost for each conductor is determined as initial cost plus W times its annual O&M costs (the sum of taxes, inspection, etc.).

E. Based on (B) and the demand and energy losses costs, it computes the annual cost of losses for all values of peak load from 2 MW to the conductor's thermal limit, and multiplies this times U to determine the PW losses cost, adding the fixed cost to get the total cost.

F. The author then would manually select sets of four to five curves to plot. The most difficult manual effort is accommodating a common reach (see "Adjusting Loading Ranges for Equivalent Reach earlier in this chapter). However, in numerous cases tested, it proved easy to find four to five conductors that always gave linear performance.

It is certainly feasible to develop a computer program with a more complicated algorithm that would optimize the selection against a performance index of economics weighted by distribution loading histogram (Figure 7.25), but this method seems more than suitable for developing the necessary decisions.

Line Segment and Transformer Economics

Comprehensive Line Sizing Planning Tables

Once a suitable set of conductors has been identified, a table of line type application ranges for planning can be developed, as illustrated by the values shown in Table 7.9, developed in 1992 for a utility in New England. These provide a guideline for planners in laying out their system and in determining if and when conductors should be upgraded.

7.7 TRANSFORMER SELECTION ECONOMICS

Transformers do not move power, they re-fashion its voltage-current combination, changing the effective economy-of-scale for the power's transmission and use. For a particular transformer, with a transformer ratio of α, the current leaving it will be α times the current entering it and the voltage (except for a slight amount of voltage drop) $1/\alpha$ times the voltage on the input side. Except for electrical losses, the total power (voltage times current) remains constant.

Transformers Change the Economy of Scale for Transporting Power

If the economics and economy of scale for moving and using power were the same at 500 kV as they are at 500V -- for example if it were economical to move small or large amounts of power at 500 kV at 500V, and it were just as feasible to build a 120kV toaster as a 120 toaster -- transformers would not be needed. However, this is not the case -- large amounts of power are most economically moved long distances at high voltage; small loads are most economically satisfied with service at low voltage.

Transformers are a cornerstone of distribution system layouts and thus of distribution system planning. While this seems obvious considering the number included in any distribution system, many planners consider "layout" to include only the routing and connecting of line segments, with transformer siting and selection an ancillary step after layout has been determined. Quite the contrary, transformers exist in order to lower or raise voltage so that the economics of moving power match those for the situation. They are used whenever

> A. it is more economical to change the voltage in order to move it,
>
> B. the voltage level must be changed to match the needs of the consumption equipment and appliances.

Transformer Types

Like transmission and distribution lines, transformers are available in a wide variety of applications types -- pole mount for overhead application and pad mount or vault-type for underground systems. Within each category are various types -- live front and dead front pad-mounts, CSP (completely self-protected), etc., with differences that are important in some cases.

Transformers are available as three-phase, or single-phase units. Three single-phase units can be connected together so they transform all three phases of power flow, but generally a purpose-built, three-phase transformer will be more efficient (low capital cost, lower losses). However, one advantage of connecting three single-phase units together is that contingency backup is less expensive: many small substations in rural areas are built with four single-phase units installed -- three provide service and one is a stand-by in case of failure.

Special Types of Transformers

There are really no "standard" transformer designs -- transformers vary so widely in specification, types, styles, and internal arrangements that no generalization can be made. Very large transformers (over 100 MVA and over 230 kV) and large transformers (5 to 100 MVA) are typically available in an infinite range of styles and characteristics, because the manufacturers are willing to design and build a transformer to whatever characteristics are specified by the utility. However, there are several common types of "uncommon" transformer.

Low loss transformers

Some types of transformers employ designs and materials that reduce electrical losses. Usually "low loss" means the design is focused on reducing core (no-load) losses. The metal for the core will be specially selected to reduce magnetic losses -- it may be an amorphous metal or simply a metal carefully picked to be relatively low loss. Winding materials can be larger for lower impedance. Design can pay special attention to reducing losses throughout. The result is a unit that: a) costs more than a "normal" transformer of equivalent capacity rating, b) has much lower losses.

Three winding transformers

Occasionally a transformer will be designed to exchange power at three voltages, as for example a 230 kV/66 kV/12.47 kV transformer that accepts power at 230 kV and outputs it at both 66 kV and 12.47 kV. In all such cases, these units, their

Line Segment and Transformer Economics

ratings, and their application are rather specific -- a particular 230 kV/66 kV/12.47kV unit might be intended to accept 100 MVA on the high side, deliver 75 MVA out at 66 kV, and up to 25 at 12.47 kV. The unit might work in an emergency as a 66 kV/25 kV transformer, but at what rating and for how long and with what losses, etc. would have to be determined by detailed study.

Low footprint transformer

Often the barrier to upgrading the installed transformer capacity (size) of a distribution substation will be the room available in the substation yard for transformers. Small footprint transformers are designed to occupy as little space as possible, so that a 40 MVA small footprint unit can replace a 25 MVA unit where there is no room for a normal 40 MVA unit. Most often these units are "tall and narrow" so that substation buswork must be re-arranged, but occasionally they are simply a design compromised heavily in favor of small size in all three dimensions at the expense of other characteristics, in which case they are typically relatively high loss and perhaps prone to overheat more quickly if loaded to emergency levels for even brief periods (they are smaller and hence rise in temperature more quickly).

High impedance transformers

Often, a high impedance transformer is desired for a specific application, even though this may mean higher than normal electrical losses, in order to reduce fault current.

High phase number transformers

Specialized transformers are available to produce six, nine, twelve, and fifteen-phase power for certain applications, generally as the first step towards rectification of power where little ripple can be tolerated on the DC output.

Transformer Ratings and Characteristics

Transformers are rated in terms of several attributes that are important to their performance. Foremost is their voltage class rating. For example, a very popular substation transformer is 138 kV to 12.47 kV, which means the transformer has a turns ratio of 11 (138/12.47) and is insulated for 138 kV on its high side, and 12.47 kV on its low side. Three-phase units are also specified by the connection type; they can be built as delta-wye or delta-delta. A 138 kV/12.47 kV transformer could be built as either type, but most often one will find distribution substation transformers of this voltage delta-connected on the high side (transmission is almost invariably delta) and wye-connected on the low side (12.47 kV distribution is most often wye connected).

As with line ratings, voltages are almost always given as phase-to-phase voltage, even though power is often applied phase-to-ground in wye-connected systems. Thus, this delta-wye transformer actually accepts 138 kV p-p voltage across phases on the high side, and produces 7.2 kV phase-to-neutral voltage for each phase on its low side, which, if loading is balanced, results in 12.47 kV p-p. Regardless, this transformer is normally referred to as a 138 kV/ 12.47 kV transformer, not a 138 kV to 7.2 kV transformer. Despite this, most planners would understand what was meant if one of their colleagues mentioned a "138 kV delta to 7.2 kV wye" transformer.

Transformer nomenclature becomes less definite when referring to single-phase units. For example, a single-phase transformer, built so three such units can be connected into a 69 kV (delta) to 12.47 kV (wye) bank might very well be referred to as either a 69 kV/12.47kV single-phase or a 69 kV/7.2 kV transformer, depending on the nomenclature and customer at a particular utility.

Transformer voltage class nomenclature can be confusing at the service level, where custom and usage differ widely among utilities as well as countries. In "American" systems, service transformers are usually referenced with their utilization voltage given as phase-to-neutral voltage -- as in 12.47kV/120 volt, which is understood to mean that the transformer takes 7.2 kV phase-to-neutral (12.47 kV p-p) voltage on the high side and produces 120 volt (phase to neutral) on the low side. However, three-phase service transformers are occasionally referred to in terms of phase-to-phase on both sides, as in 12.47kV/208V (208V phase-to-phase is 120 V phase-to-neutral), rather than 12.47kV/120V. In "European" systems, service transformers are very often referenced by the phase-to-phase voltage rating on both sides -- an 11 kV to .416 kV service transformer accepts 11 kV on the high side and produces 416 volts phase-to-phase for utilization at 250 volts phase-to-neutral in European electric systems.

Capacity of a transformer is given in kVA or MVA, for example 25 MVA meaning that the transformer can handle the flow of 25 MVA. In planning, this is also often called the "size" of the transformer, as in "sizing" studies, where it is understood that the issue is determination of the capacity to be installed, not anything to do with physical size or shape of the unit. After voltage class rating (which determines if the transformer will fit the situation) capacity is the most important characteristic to the planner, and the one over which the most evaluation and study is expended.

Capacity rating is always determined with respect to very specific conditions. Unfortunately, there is no standard for these conditions and great variation occurs in practice from one utility to another. For example, the manufacturer of a particular substation transformer might refer to it as a 25 MVA unit. This rating

Line Segment and Transformer Economics 279

might be based on one of several industry "standards" which lay out very exact conditions, loading definitions, and periods. For example this transformer might be capable of handling 25 MVA on a continuous basis out (in other words this does not include the losses, and might entail 25.2 MVA in), when the air around the unit is at 25°C and 65% humidity and still, when there are no harmonics at all (pure 60 Hz or 50 Hz power, as the case may be), with a steady-state rise in internal temperature (measured at the top of the oil above the core) of 55°C.[5] However, upon purchasing this unit and putting it in service, one particular utility might refer to it as a "37 MVA unit," on the basis that under certain conditions it has decided to use as "rating conditions" the unit can handle 37 MVA.[6] Another utility might refer to it as a "23.2 MVA" unit.[7] (This situation is a real example). Regardless, the manufacturer and both utilities would all understand the transformer's capabilities -- engineering data and performance tables are available from which its specific performance under any set of conditions could be determined. For more details on transformer engineering see Gonen or Grainger and Stevenson in the References.

Three-Rating Nomenclature

Very often, transformer ratings are given as three rather than just one capacity rating -- for example, 25/30/36 MVA. Again, there is no uniformity from one utility to another in what and how these ratings are determined, for the reasons cited above. Further, the meanings of these three numbers vary from one utility to another, as will be discussed above. But what is important is that the planner understand what they mean with respect to his company's rating system and the applications for which he is the planner. There are two very common "tri-rating" systems, each with slight individual variations as applied within utilities.

---

[5] There are numerous, slightly different standards for rating transformers. A thorough discussion of these and their implications is beyond the scope of this book. The important point for a planner to keep in mind is that a transformer rating is meaningless unless the conditions under which the rating applies are fully known. All that is important in planning is that all transformers being considered are rated consistently and that the ratings apply to the load and conditions appropriate to the planning scenarios.

[6] The unit can handle 37 MVA for two hours without exceeding an emergency temperature rise of 65°C, if not loaded above 20 MVA for the previous two hours, and if the ambient temperature is not above 85°F.

[7] This utility expects ambient temperatures that exceed those in the manufacturer's rating conditions during its summer peak. This unit is expected to handle a 23.2 MVA load on a continuous basis, without exceeding a 55°C rise, when the ambient temperature is above 105°F, 65% humidity, with still air.

"FOFA" ratings

A transformer's ability to handle load is constrained by the internal temperature rise it experiences (due to electrical losses). Whether rated at 65°C, or 55°C, or some other temperature rise, ultimately it is temperature rise of the hottest spot on the transformer core (usually at the top) that sets a bound on capacity. Many manufacturers offer optional oil pumps that circulate the transformer oil for more even distribution of heat, and radiators with fans to assist in removing heat, much as the radiator in an automobile engine does. Thus, very often a "tri-rating" for a transformer will refer to the "FOFA" convention of labeling capacity, so that a 25/30/36 MVA transformer would have a rating of:

> 25 MVA with only convection cooling: no oil pumps or radiator fans operating.
>
> 30 MVA if its oil pumps are operating to circulate its oil internally to distribute temperature more evenly (FO, for "Forced Oil").
>
> 36 MVA when the oil pumps are working and the fans are blowing air over its radiators to accelerate the rate of heat removal (FOFA, for "Forced Oil and Forced Air").

In general, these three ratings will refer to otherwise identical conditions of internal hot-spot temperature rise, load period, ambient conditions, etc. (However, there are exceptions). A transformer might be rated in this way, even if the oil pumps and radiators/fans are not installed. Often a utility will buy a 25/30/36 unit for service, without pumps and fans, knowing that if need be, the unit can be upgraded to 30 or 36 MVA peak capacity in the future.

Normal, contingency, and emergency rating

Many utilities apply an alternative tri-rating system, in which a transformer is rated with three numbers that refer to different conditions or application, so that a 25/30/36 MVA transformer could mean

> 25 MVA under normal conditions: contiguous load of 25 MVA at 55°C under standard ambient conditions.
>
> 30 MVA under "contingency" conditions defined as a 60°C rise for six hours.
>
> 36 MVA under "emergency" conditions defined as a 65°C rise for four hours.

Line Segment and Transformer Economics

Table 7.10A Standard Single-Phase Service Transformers

14 kV OH	14 kV UG
10 kVA	25 kVA
25 kVA	37 kVA
50 kVA	50 kVA
75 kVA	75 kVA
100 kVA	100 kVA
150 kVA	300 kVA (3-ph)
	500 kVA (3 ph)

Table 7.10B Standard Substation Transformers

69 kV/13.8 kV	138 kV/13.8 kV
5 MVA	
10 MVA	12.5 MVA
15 MVA	22 MVA
22 MVA	32 MVA
	40 MVA
	53 MVA

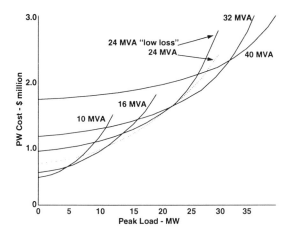

Figure 7.27 Transformer cost as a function of peak load can be analyzed exactly as line cost was in section 7.3.

Standard Transformer "Sizes" and Types

Usually, in a manner analogous to the selection and use of distribution line-type inventory sets, a utility will standardize on a set of transformers that are normally used as the building blocks for its system, as shown in Table 7.10 for a utility from the central United States. However, it is much more common for special transformers to be used on a "custom-specified" basis than non-standard line types to be used.[8,9]

Transformer Costs, Economics, and Sets

Transformers have initial costs (equipment and installation), ongoing operating costs (taxes, inspection, and maintenance), and electric losses costs, all exactly analogous to those discussed earlier in this chapter with respect to distribution line types. Unlike the case with lines, transformer losses consist of no-load losses (also called core losses) which occur because the unit is energized and are constant regardless of load, and load-related losses (also sometimes called copper losses) which are proportional to load squared.

The total lifetime cost of a transformer can be plotted as a function of MW (at a specified power factor) or MVA in exactly the manner as line segment costs were evaluated in Section 7.3. Several similar transformer types can be plotted together to determine how their costs compare as a function of loading, as shown in Figure 7.27. Note that the same linearity of application applies here as it did in lines, although the fixed cost (Y-intercept) is higher.

The fixed cost shown here is based on the author's recommended method of computing transformer costs for planning: it includes not only the cost of equipment, labor and PW of lifetime O&M for the transformer itself, but also for all the ancillary equipment such as buswork, breakers and measurement, and control machinery required for this unit to be put in service. Fixed cost also includes the future no-load losses.

---

[8] In the utility system from which Table 7.10 is taken, less than .6% of distribution lines are non-standard, in the sense that unusually structures or conductor sizes and types are used due to special circumstances. Similarly service transformers are almost all from standard inventory. However, 7% of the utility's substation transformers are "special design" having three windings, high impedance, small footprint, or other similar characteristics due to a desire to tailor them to the specific situations.

[9] While application of a transformer precisely tailored to a specific situation is always desired, planners must weigh the benefits against the disadvantages: both cost and delivery time will be higher for customer-designed units.

Line Segment and Transformer Economics

The dotted line in Figure 7.27 shows the economics of a "low loss" transformer, in this case a particular design with low core losses. This has been included to emphasize the complex nature of transformer specification and selection. This particular low-loss unit has a higher equipment cost, but that is more than offset by its much lower no-load losses, which means its PW of future no-load losses more than makes up for its higher initial cost, hence it has a lower Y-intercept that a "standard" 24 MVA transformer. However, this particular "low loss" transformer (but by no means all low loss transformers) has a higher than typical load related impedance, so that the slope of its curve as a function of loading has a higher slope. The two curves cross at 22 MVA, meaning that in very high loading scenarios, the "low-loss" unit is actually the more expensive of the two on a lifetime basis.[10]

Standard Transformer Inventory Sets

Recommended practice for distribution planners is to establish a standard transformer set -- a group of transformer types and sizes covering the required ranges of loadings, all selected similar to the conductor sizing/set selection covered earlier in this chapter, so that they provide a linear range of application. This should be done for both distribution substation transformers (in which case the "cost" should include the transformer, all equipment needed to support its use, and their installation and O&M and losses) and for service transformers. Service transformer selection is complicated by coincidence effects of the load behavior, so that the methods described here are not appropriate to evaluate losses economics completely. Service transformer selection is discussed in Chapter 13.

7.8 SUBSTATION TRANSFORMER SIZING AND LOADING

Selection of size and optimal loading for substation transformers (e.g., 69 kV to 12.47 kV) can be addressed with the same approach discussed in Section 7.7 for service transformers. Typically, a utility will standardize on between one and three sizes of substation transformer, building substations out of multiple units as discussed in Chapters 10 and 12. However, in recognition of the significantly

---

[10] Some low-loss core designs, including several amorphous-core types, have much lower core losses costs, but slightly higher winding losses than standard units. In a few cases, the total lifetime cost curves (e.g., Figure 7.27) will cross at a point less than the transformers' maximum ratings in which case the issue is quite confusing about which is actually the "low-loss" design. Usually, the curves cross above the transformer rating and the low-loss unit is less expensive at all loadings that it can handle.

greater cost of substation transformers as compared to service transformers, and their greater impact on customer reliability, evaluations at this level generally include a more comprehensive model of load curves and other factors unique to each individual situation. Such "tailored" evaluation is justified based on both cost and reliability considerations. Typically, both equipment reliability and customer reliability needs (perhaps on a value-basis), O&M costs, as well as a detailed lifetime evaluation of the transformer, are included in the process of selecting its size. Loss-of-life, contingency requirements, and "durability impact" are often included in loading evaluations (see Chapter 17, pages 766 and 767).

7.9 CONCLUSION

Line segments and transformers are the building blocks of the distribution system. They are available in a variety of styles, voltage levels, conductor types, phases, and sizes, with capacities and reaches that vary from very small to quite large. Distribution planners interested in achieving overall economical distribution line want to make certain they have a good set of conductors from which to choose, and that they apply correct economic evaluation to select the best line type for each segment in their system. This alone does not guarantee sound and economical distribution system design, but it is a necessary start -- without the right parts, the whole apparatus can never be assembled.

REFERENCES

J. R. Carson, "Wave Propagation in Overhead Wires with Ground Return," *Bell System Technical Journal,* New York, Vol. 5, 1926.

J. K. Dillard, editor, *T&D Engineering Reference Book,* Westinghouse Electric Corporation, Pittsburgh, 1965.

W. H. Kersting, "Feeder Analysis," Chapter 6 *of IEEE Distribution Planning Tutorial,* IEEE text 92 WHO 361-6 PWR, Institute of Electrical and Electronics Engineers, New York, 1992.

T. Gonen, *Electric Power Distribution System Engineering,* McGraw Hill, New York, 1986.

J. J. Grainger and W. D. Stevenson, *Power System Analysis,* McGraw Hill, New York, 1994.

8
Distribution Feeder Layout

8.1 INTRODUCTION

This is the second of six chapters addressing distribution system layout. The distribution system is composed largely of feeders -- "neighborhood size" delivery circuits which route power from distribution substations to the vicinity of homes and businesses. These circuits perform the bulk of the power distribution function, in the sense that they route power from a few utility company sources (substations) to many points (service transformers) each of which are always no more than a short distances from the customer. Feeders are the basic building blocks of distribution systems, composed themselves of the line segments discussed in Chapter 7, assembled according to layout guidelines and principles which will be discussed in this chapter. Groups of feeders compose substation service area sets, and cumulatively make up the distribution system as a whole. The decisions on how many, what type, and how to arrange these feeders is the essence of distribution planning and the subject of this chapter.

Section 8.2 begins with a look at feeder systems, their mission, characteristics, and the constraints that shape their design. In section 8.3, the layout of radial and loop feeders is examined, including styles of design, overall approach to feeder routing, use of single or dual voltages, and other matters in planning. Section 8.4 then looks at concerns about alternate sources and routes that must be arranged as contingency backup, and how this is generally accomplished. This material provides the foundation for examination of sound feeder system planning, given in Chapter 9. Chapter 10 will build further on this chapter, examining how feeders, as the building blocks of distribution systems, are used to lay out a system to effect the greatest economy.

8.2 THE FEEDER SYSTEM

Mission and Goals

The mission of the feeder system is to distribute power from a few system sources (substations) to many service transformers, that are scattered throughout the service territory, always in close proximity to the customer. It must accomplish this mission while achieving adequate performance in three categories:

> *Economy* - While meeting other goals, the total cost must be kept as low as possible.
>
> *Electrical* - The system must be able to deliver the power required by all customers.
>
> *Service* - Reliability of service must be very high, and voltage and quality of the power delivered must be satisfactory.

Substations and Feeders

Generally, this mission is accomplished while meeting these goals by distributing power from a number of substations strategically located throughout the utility service territory, as shown in Figure 8.1. Power is brought to these substations by the transmission/subtransmission system (not shown), at transmission voltages somewhere between 34.5 kV to 230 kV. In turn, that power is lowered in voltage to a *primary distribution voltage* (generally somewhere between 2.2 kV and 35 kV) selected as appropriate for the service area and load density, and typically routed onto between two and twelve feeders that serve the area surrounding the substation.

A feeder is an electrical distribution circuit fed from a single source point (breaker or fuse) at the substation. It operates at the primary distribution voltage and disseminates power through a portion of the substation's assigned service

Distribution Feeder Layout

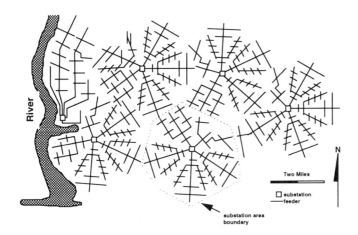

Figure 8.1 Distribution of electric power is accomplished from a set of substations (squares) which, together, must cover distribution needs for the service territory. Feeders (lines) emanate from each substation, distributing power within their own service areas. Although planners attempt to locate a substation near the center of its territory and load, occasionally circumstances force situations like that for the substation at the far left in this diagram, which sits of the edge of the area it must serve.

area which is its *feeder service area.* There are several basic types of feeders and feeder systems, including radial, loop, and network, the characteristics, advantages and disadvantages of which will be discussed later in this chapter. Together, the feeders emanating from a substation form that substation's *feeder set,* the circuit system that must serve all the load and cover all the territory assigned to that particular substation.

Central location for the substation

Usually, planners try to locate a substation near the center of the load or service area it will service, or put another way, they try to arrange the feeder system so that the substation serves the distribution needs of the area all around it. Thus, the feeders emanate from that central site in all directions, as is shown for most of the substations in Figure 8.1. There is considerable electrical, economic, and reliability incentive for a location near the center of the load served, as will be discussed in Chapter 12. However, there are unusual situations forced on planners by constraints of geography, unusual circumstances, or just plain poor planning done in the past, which cause exceptions to this rule, as illustrated by the substation on the extreme left in Figure 8.1, which is on the edge of its service territory due to geographic constraints.

Feeders must reach between substations

The feeder line types and loading criteria used in the design of the feeder system must be able to move power reliably, economically, and within engineering criteria (loading, voltage drop) to all locations between the substations. Thus, the fewer and farther apart the substations planned for a system, the greater the load reach needed by feeders to accomplish their part of the delivery function. (The interaction of substation spacing and feeder reach economics is examined in Chapters 11 and 12).

Contiguous, exclusive service areas

In most types of distribution systems, both the feeders and the substations have contiguous and exclusive service areas -- each feeder and each substation serves a single connected area for which it is the sole source of electric power (except for DG) in that area. There are only rare exceptions to this rule, and they fall most often into the category of "planning mistakes" -- situations forced on planners by circumstances which where not foreseen, and which, in hindsight, they wish could have been corrected. Thus, by design the goal for most distribution system planning is to lay out substations and feeders so that all have exclusive, contiguous service areas.

The exceptions are cable feeder systems built in downtown areas, where convenience occasionally leads to alternating customer pickup, and distribution networks, either primary voltage feeder networks, or interlaced feeder/secondary networks, that require overlapping feeders. Such systems are generally more expensive but more reliable than standard types of loop or radial distribution design.

Most of the load is far from the substation

Substation service areas tend to be vaguely circular in shape (as illustrated in Figure 8.1) with a substation's feeders each serving roughly triangular "slices" within that. This geometry means that a majority of the substation's load, and of each feeder's load, is located closer to the service area boundary than the substation, as shown in Figure 8.2. The feeders for any particular substation will have to carry more than half of that substation's load more than half of the distance to its boundary with other substation service areas.

Distribution Feeder Layout

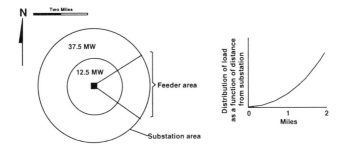

Figure 8.2. Most of the load in a substation or feeder area is "more than halfway" from the substation to the service area boundary. At the left is a theoretical substation area with a radius of two miles (equivalent to four mile substation spacing), and the area served by one of six feeders, typical of 12.47 kV systems in many urban utilities. Given a load density of four MW per square mile, the substation would serve 50 MW, only one quarter of which is within one mile of the substation.

Power must be delivered to the proximity of the customer

A feeder system routes power to various locations where it feeds the service transformers that reduce power to utilization voltage and then route that power to individual customers. Depending on the utilization voltage, equipment specifications, and type of layout used, the service level is limited to a distance of between 50 and 200 yards over which it can effectively and economically deliver power. Therefore, there must be a service transformers within 50 to 200 yards of every customer, and the feeders must reach each of these along their routes, as shown in Figure 8.3. (Service level equipment capability, layout, and planning will be discussed more fully in Chapter 13).

Branching and splitting structure

To cover its service area so that feeder-level delivery reaches sufficiently close to all customers, feeders typically split their routes many times, in what is often called a *dendrilic configuration,* as illustrated by the feeder drawn in Figure 8.4. There are various design guidelines on how branching is best accomplished, which will be covered later in this chapter. Regardless, a feeder consists of a single route leaving the substation, which branches and re-branches, gradually splitting the power flow into more but smaller-capacity routes for delivery as power moves from the substation to the customer.

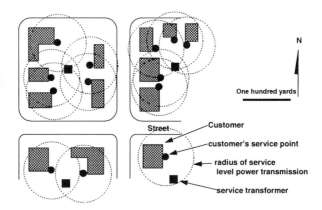

Figure 8.3 The feeder system must deliver power to the service transformers (black squares), which are each within a short distance, in this case 75 yards (dotted radii), of the customers' service points. Therefore, whatever feeder is to serve this small group of customers must be routed to reach all four service transformers shown above.

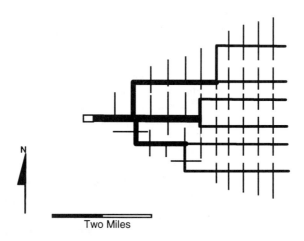

Figure 8.4 Feeders typically branch repeatedly on their way from the substation (a single source) to the customers (many loads). Line capacity is typically "tapered" as shown (line width indicates line size).

Distribution Feeder Layout

$D = (|\Delta X| + |\Delta Y|)$ works better than $D = \sqrt{(\Delta X^2 + \Delta Y^2)}$

Feeders generally follow roads, highways, and property boundaries. This means their routing usually is restricted to a rectangular grid, as shown in Figure 8.4, where the feeder's line segments are built along a rectangular pattern of roads and streets (not shown). As a result, the straight-line distance between two locations, computed by the Euclidean metric Distance $=\sqrt{(\Delta X^2 + \Delta Y^2)}$ is seldom a good estimate of the length of feeder needed. A much more useful distance measure for distribution planning is the *Lebesque 1 metric,* Distance = $|\Delta X|$ + $|\Delta Y|$ (often called the *taxicab travel* distance measure), which gives a reliable, quick estimate of feeder routing length. This metric measures the distance through a grid of streets which a taxicab, or a feeder, must take from one point to another when restricted to moving along the streets.

One important aspect of distribution planning that this distance measure highlights is illustrated in Figure 8.5 -- *there are usually many different routes, all the same shortest distance, between a substation and a particular customer or service point.* When feeder routes are restricted to a grid as they usually are,

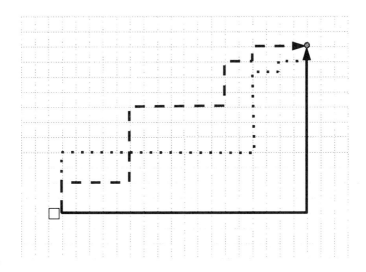

Figure 8.5 When restricted to routing feeders along a grid of streets or property lines, distribution planners find many "shortest routes" from one point to another. The three routes shown above, plus many more that could be arranged in this grid, have the same total length, and hence very similar cost.

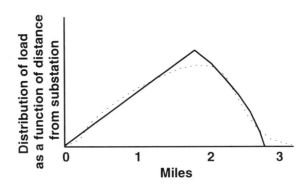

Figure 8.6 Distribution of distances is affected by the realities of rectangular routing for feeders. Shown here are the theoretical distribution of minimum feeder run lengths in the pie-slice shaped feeder area from Figure 8.2 (solid line), as well as the average of 36 actual feeders in a 12.47 kV distribution system with an average 3.93 miles between substations, 6.1 feeders per substation, and 48.6 MW/substation peak load (dotted line).

any route that takes the same cumulative X and Y total will have the same length (and hence probably roughly the same cost). This creates both a planning problem (because there are so many paths from which to choose), and a planning opportunity (one or the other of these paths may help achieve other objectives such as improved esthetics, reliability, etc.).

Recognition of this reality of feeder routing changes the actual feeder route length distribution shown in Figure 8.2 to that given in Figure 8.6. As shown, when interpreted with this distance metric, the maximum feeder-level delivery distance is now .75 times the distance between substations, and the average feeder-level delivery distance is very nearly equal to half the distance between substations. This is a general rule of thumb that applies to nearly all feeder systems.

Most feeders are the same "size"

Most feeders are planned by *starting* with the premise that the main trunk (the initial segment out of the substation, through which all of the power is routed) will the largest economical conductor in the conductor set. The feeder layout is arranged so this segment picks up enough load for its peak load to fall somewhere in the middle or upper half of that largest conductor's economical range. Using the six conductors listed in Chapter 7, Table 7.2 as a conductor set

Distribution Feeder Layout

would mean selecting 636 MCM conductor, with a target loading in the vicinity of 6.8 MW, halfway between 5.1 and 8.5 MW. Thus, all feeders in a power system are somewhat the same "size" in terms of capacity and loading.

Targeting peak loads at the upper end of the linear economical range, and using the largest economical conductor in the set as the main trunk assures not only the economical performance of the main feeder trunk, but also that there will be smaller conductor available to maintain that economy with "conductor tapering" as branches split off and load levels fall as the routes extend far from the substation (see Figure 8.4). Theoretically, the same overall feeder costs should be obtainable with more but smaller feeders, since the costs discussed in Chapter 7 are linear. However, in practice the greater number of feeders would require more routes and rights of way (and hence have a higher esthetic impact), and they would require more substation buswork and breakers, adding to cost.

Radial, Loop, and Network Feeder Arrangements

Radial feeders -- a single path

More than 80% of all distribution worldwide is accomplished using *radial feeder* systems, in which there is only one path between any customer and the substation (Figure 8.7, left). In some cases radial feeders are designed and built as fixed radial circuits, but in a majority of cases the feeder system is physically constructed as a network (many paths from many sources), but is operated radially by opening switches at strategic points to impose a radial flow pattern.[1] Radial circuits are both the least expensive type of distribution system and the easiest to analyze and operate. Both low cost and simplicity of analysis and operation made radial systems popular in the beginning of the electric era, before computerization made analysis of complex circuit behavior reliable and inexpensive. This early popularity helped institutionalize radial circuit design as *the* way to build distribution. Although simplicity of analysis is no longer a major concern, low cost continues to make radial circuits the choice for more than 90% of all new distribution construction.

The major drawback to radial circuit arrangement is reliability. Any equipment failure will interrupt service to all customers downstream from it. On average, failure of a segment on a feeder will interrupt service to about half of the customers it serves.

---

[1] In Y-connected radial systems, the neutral conductor is connected through all open switch points, thus forming a network connecting feeders and substations.

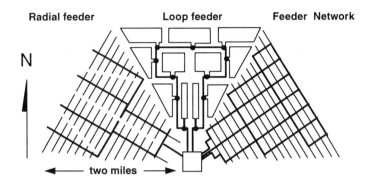

Figure 8.7 Three types of feeder, each serving an area of roughly three square miles. At the left, a radial feeder has only one path from the substation to any load point. Line thickness represents relative line segment capacity. Middle, a dual-voltage loop feeder, broad lines indicating the high voltage loop, circles the high-to-low voltage transformers, and thinner lines the lower voltage loops, all operated as closed loops. Right, three feeders connected as a network so that loss of any one causes no interruption of service.

Loop feeders -- two paths

Distribution can also be built and operated as *loop feeder* circuits in which the power flows into each "end" of a feeder and moves outward to customers, there being a "null point" somewhere on the loop where no power passes (Figure 8.7, middle). This is basically a "dynamic" radial circuit, with the open point (null point) shifting as loads change. When built and protected properly, it can provide very high levels of customer reliability -- any equipment failure causes interruption to only a small group of customers and on average two simultaneous failures result in interruption of service to only 1/4 of the customers. Generally, loop feeder systems cost about 20% (UG) to 50% more (OH) than radial systems. They are slightly more complicated to analyze and operate than radial circuits.[2]

Sometimes loop feeder systems are operated as open loop systems, with an open switch near the middle of the loop, in which case they are basically radial circuits.

---

[2] At the time of this writing (early 1997) computer programs to analyze voltage drop, loading, and fault behavior in closed loop and network systems have become widely available as options to commercial "radial distribution" analysis packages.

Distribution Feeder Layout

Feeder networks -- many paths

Feeder networks consist of groups of feeders interconnected so that there is always more than one path between any two points in the feeder network (Figure 8.7, right). If designed with sufficient capacity and protection throughout, a feeder network can provide very high levels of customer reliability: the loss of any segment or source will not interrupt the flow of power to any customers, and multiple failures can occur with little or no interruption. Among their disadvantages, feeder networks cost considerably more than radial systems, usually 33 to 50% more in UG construction and 100 to 150% more in overhead construction, and they require much more complicated analysis and operating procedures. They also require more expensive protective devices and coordination schemes.

"Normal," Urban, and Rural Distribution Feeder Categories

The essence of the distribution system perspective outlined in Chapters 7 through 12 can be summarized in this manner: If line and equipment types and sizes, primary voltage level, and feeder layout styles are selected properly, and if design standards have taken into account the electrical and economic interaction of distribution with other levels of the power system; then it is possible to lay out a "naturally compatible" feeder system in which lines and equipment can be defined on the basis of maximum overall economy alone, with few capacity or voltage drop limitations interfering with achieving this goal.

It is possible to specify distribution system line type conductor sets, select primary voltage, determine the substation spacing, number of feeders and their layouts, and define the other design variables in a distribution system so that over a very wide range of situations, neither capacity nor voltage drop limitations inhibit the economical design of the system. The author characterizes this very wide range of situations as constituting the "normal" category of distribution planning circumstances. But in dense urban areas, distribution systems are dominated by capacity limitations. Here, even the highest primary voltage with the largest possible conductor (1354 MCM, 34.5 kV) may not have enough capacity to serve less than two square miles, even though it can move its full thermal rating ($\approx$ 70 MW) nearly 12 miles before encountering range A voltage drop limits. No amount of care in selection, no amount of skill and innovation in design, can overcome the fact that in "urban areas" capacity limitation dominates the design of the distribution system. On the other hand, voltage drop is seldom if ever an issue in such planning.

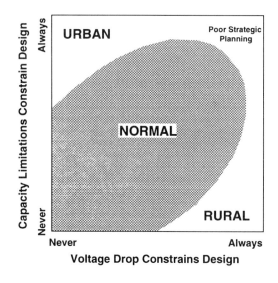

Figure 8.8 Planning situations with respect to feeder layout fall into one of three categories depending on the constraints dominating the distribution planner's freedom of design. In urban areas capacity versus load limitations drive design considerations and prohibit many standard practices. In rural areas capacity is seldom the limitation in achieving further cost reduction, but voltage drop is. "Normal" situations are those where careful selection of distribution equipment, voltage, and sites can balance design tradeoffs so that neither unduly limits the planner in achieving economy of design. There should be no situations where both capacity and voltage drop constraints cannot be overcome. Such situations develop due to poor strategic distribution planning.

At the other end of the distribution scale, in sparsely populated rural areas, voltage drop dominates the considerations which the planner must overcome. Here, load density is orders of magnitude lower than in urban areas, but the distances between nearest customers are often dozens of miles. A single feeder may have to distribute power over more than 1,000 square miles. Even so, few feeders ever run up against capacity constraints, but few fail to encounter severe voltage drop as a result of moving small amounts of power dozens of miles.

Thus, these categorizations have *nothing* to do with geography, load density, or type of system built *per se.* Instead, they are based upon differing relationships and constraints that affect distribution system layout, economics, and decision-making in each category -- of what type of situation confronts the planner in laying out the distribution system and achieving satisfactory

Distribution Feeder Layout

performance from both cost and customer standpoints. Normal, urban, and rural categories each present the planner with a uniquely different framework of economic and electrical interactions, natural physical limitations, and design tradeoffs, as illustrated in Figure 8.8. The "rules" are different in each of these three categories.

Underground feeder layout in urban areas

The highest load density, and the most restrictive limitations on feeder layout, occur simultaneously in the central cores of major cities. For a variety of reasons, distribution and distribution planning in these areas are quite different than in other types of situations. Load densities routinely exceed 60 MW per square mile and occasionally reach 100 MW per square mile, and all electrical equipment must be kept underground. In many cases, utilities will choose to serve these areas using feeder networks, of which there are many types and variations of design. However, radial circuits are used extensively in urban areas, and the author's experience is that radial, loop, and network feeder systems all provide satisfactory performance and economy if well designed, constructed and operated, even at load densities exceeding 100 MW/square mile.

Underground feeder system design is generally much more expensive than typical overhead construction, particularly in the core of large cities, where feeder circuits including laterals must always be enclosed in duct banks routed under streets. Direct burial of cable in a dense urban area is simply not an option, for a number of reasons. To begin with, underground is surprisingly densely populated with pedestrian walkways, as well as water, sewer, storm-drain, phone, steam, data-com, and other utility systems in addition to electrical. There really isn't a lot of room. The electrical utility must stake its claim to the routes and space allocated to it with its own duct banks. Second, duct banks are needed to protect underground cable from the constant dig-ins of other utilities, stress from settling, and heat and moisture which abound in this environment.[3] Third, digging into the street in an urban area, as would be required occasionally for either new additions or repairs of direct buried cable, is very expensive. Municipal governments discourage such activity, for it disrupts traffic and generally creates an esthetic nuisance. Such work requires permits (which are often not immediately forthcoming), and tight control of construction schedules. Traffic control and other requirements add to cost, as does digging around all

---

[3] In several downtown areas, there have been recorded failures of underground cable due to overheating caused by running too close to steam pipes.

those other utilities. Even routine maintenance and repair often take much longer than for overhead lines in less restricted areas.[4]

Thus, electric utilities have no choice but to use duct banks and cable vaults to create their own "cable tunnel system" under the streets in dense urban areas. Duct bank construction costs, in addition to other ones unique to urban UG construction, creates a unique economic situation, dominated by construction cost of the duct bank (see Chapter 7, Figure 7.18). While there is still a linearity of scale to the economics of UG feeder planning, as there in OH feeder planning, it is minor compared to the fixed cost of building the duct bank. On the other hand, given duct banks are in place and space is available, cost is much lower, and the economic decisions to build and size a feeder are roughly similar to those involved in overhead distribution planning.

Other characteristics in underground urban areas also heavily influence design and planning. To begin with, many more feeders can be routed down a street than in an overhead area -- since the lines are out of sight, if duct space can be arranged, four, six, or even eight circuits can run parallel along one route (see Figure 8.9).

Second, the individual loads are very large -- there are portions of some larger cities, where no load served off the primary system is less than 1 MW. Almost all customers are served with three-phase power -- single-phase loads and construction are the exception in these downtown areas. As a result, a feeder may have only a dozen or fewer loads, but all of them very large.

In addition, feeder lengths are short compared to distribution in other areas. At 60 MW per square mile, even 34.5 kV feeders can serve only slightly more than 1/2 square mile. Feeders at less than 15 kV often have a total length of less than 1/2 mile (see Figure 8.9), and rarely go more than a mile. Substations serving such areas generally are of very large capacity compared to the average substation in the utility system as a whole, and typically this means there are a lot of feeders emanating from this urban-core substation. Urban distribution substations can have as much as 475 MW capacity, with a peak load of 400 MW served by more than 40 feeders.

Finally, branching of underground cable (providing for "T" ox "X" connections of paths), while possible, is not as simple or as inexpensive as in overhead lines. In overhead construction, the branch conductors are simply clamped to the trunk conductors using inexpensive and reliable hardware. "T" or "X" connections in underground systems require a cable vault, special terminators, equipment, and splicing which not only costs money, takes time and

---

[4] One utility on the US Atlantic coast has experienced an *average* 73 hours time-until-repair on its urban UG feeder system, roughly 60% of which is due to the arrangements and scheduling that must be done for permits, arranging for traffic control, etc.

Distribution Feeder Layout

care, but often are perceived as a reliability weakness. Thus, distribution layout in an underground urban area has the following characteristics:

Capacity limits design. Voltage drop and losses cost are seldom a major concern -- feeder runs are too short to produce substantial voltage drops and losses costs. Capacity/load dominates design.

Layout is restricted to the street grid. Laterals and paths through blocks are not available. Feeders must follow streets. Service entrances (from under the street) are through the block face.

Loads are large and invariably three-phase. In extreme cases, a single load can absorb the entire capacity of a feeder.

Fixed cost is very high. Doubling the capacity of a feeder may increase cost by only 5%.

The cost of capacity shortfall is extremely high. Often new ducts and cables simply cannot be added. Other times, municipal restrictions or

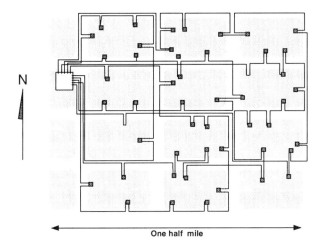

Figure 8.9 Underground distribution in many downtown areas has the characteristics shown here: short feeder runs; only a few service points per feeder; loop configuration (in this illustration, closed loops); many feeders in parallel under a street; winding routes for individual feeders; adjacent customers served by different feeders. Lightly shaded rectangles represent city blocks, small squares, customer service points, the large square, the substation, and dark lines the various loop circuits. There are a total of five.

codes may limit the utility to major construction in an area only once every five years, etc. As a result, making sudden changes to accommodate unexpected load growth is simply not possible.

Reliability requirements are above average. The generally accepted notion is that large commercial buildings have a higher than average demand for reliability of service. Value-based planning studies tend to support this conclusion.

Repairs can require a day or more. Even with duct banks in place, repairs to underground cables can proceed slowly. Outages, when they occur, tend to last a minimum of several hours.

For these reasons, UG practice at many distribution utilities includes several common adaptations to work within these design constraints.

Loop feeders are the rule, rather than the exception, for three reasons. First, the number of customers per feeder is generally small. Second, branching is relatively more expensive to arrange, and with distances short, a "winding route" through the few customer service points, as shown by several of the circuits in Figure 8.9, is preferred. Third, the loop feeder, operated as an open loop, provides contingency backup with quick restoration. Operated as a closed loop, it provides better reliability that any radial circuit.

Maximum size cable is often installed everywhere. The cost to build a small duct bank -- with six spaces -- is nearly the same as to build a larger one with 12 or more spaces -- both require the same permits, tearing up of the street, trenching, backfilling, etc. In addition, the cost to purchase and install large cable is not relatively more than for small cable. On the other hand, the cost of running short of capacity in an underground system is significant -- additions and new construction are often simply out of the question. As a result, policy at many utilities is always to build the greatest capacity possible whenever underground construction is performed.

Very grid-like planning. The layout and planning of the UG downtown feeder system itself is very grid-like in many respects, even if it is a radial or loop system. Expansion is often done by reinforcing only selected segments and re-arranging connections to reconfigure the network, as shown in Figure 8.10, which has more in common with the way actual networks are planned and expanded than with overhead radial systems.

Distribution Feeder Layout

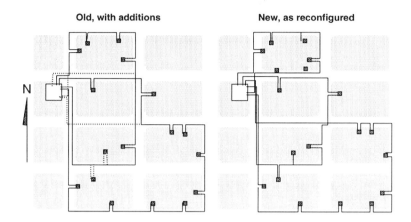

Figure 8.10 A portion of the urban area shown in Figure 8.9 with only two of the original feeder loops shown. Left, the original form of the two loop circuits, along with new cable segments added by the distribution planners (dashed lines). The two loops and the new segments are reconfigured as shown at the right, into three loop circuits.

In many cases, all of the reasons cited above motivate a utility to use a feeder network. The major cost increase in moving to network design is in protection, and the major gains are both improved reliability and improved planning robustness -- networks are expanded by making continual small (often simultaneous) additions, not feeder additions. Regardless, underground urban systems are the most expensive type of power distribution, fulfilling the most difficult distribution task -- delivering very high load densities at high reliability while keeping completely out of sight.

Feeder layout in rural areas

Sparsely populated areas present special challenges to distribution planners. Customers are far apart, load density is low, distances are great. Generally, routing is as restricted as it is in urban and suburban areas: to either a grid of roads in agricultural areas (plains and similar terrain), or to following valley, canyon, river, or lakeside roads in areas dominated by mountains, water, or other similar geographic features. Either way a feeder in a rural area is dominated not by capacity but by load reach considerations dictated by voltage drop.

Even a "densely populated" rural area may have a load density far less than that of a "sparse" suburban area. Farming country with 160 acre (one fourth square mile) farms would have a load density of about 40-75 kW per square mile. This compares to the low end of suburban load densities of 2,000 kW per square mile.[5] In addition, there would be only a handful of customer connections -- no more than a half dozen -- per square mile, as opposed to roughly 500 per square mile in suburban/urban areas. Even more sparsely populated regions, particularly ranch and farm country, in regions where rainfall is limited (and thus relatively many acres are required for any type of productive agricultural activity) can have load densities less than *one-tenth watt per acre* (this is equivalent to one farm household every ten miles).

Rural distribution design is dominated by the requirement to move power many times farther than the economic reach of most voltage/conductor combinations, and to deliver it to customer locations that are widely separated. Other aspects of rural distribution worth noting are:

> *Rural distribution systems often are not profitable.* In many rural planning situations, a normal business case could never be made for the required investment based on the expected revenue (hence the reason many governments provided electric service or special financial arrangements, such as the Rural Electric Authority in the United States.) Rural service in very sparsely populated areas is the "loss leader" part of a utility's obligation to serve its franchise area. Strict cost control and some deviation from preferred levels of voltage drop and regulation are the order of the day.
>
> *Voltage drop limits design.* Voltage drop is the major limitation on which planners must focus. "Running out of conductor capacity" seldom is a concern and is relatively easy to fix.
>
> *Losses costs are high.* Moving relatively small amounts of power over long distances results in losses which are high in proportion to the amount of power delivered.
>
> *Layout and customers are restricted to the road grid.* Laterals and paths off road are often available, but can present repair access problems and usually serve little purpose anyway. Most customers are located some distance off the road, and a utility will run primary across their property to a transformer located near their demand(s).
>
> *Loads vary from very small single-phase to medium sized three-phase.* Some loads are so small they require no more than a 3 kVA

---

[5] Load densities in suburban and fully developed residential/light commercial areas near large cities run from about 2-7 MW/square mile.

Distribution Feeder Layout

service transformer. Occasionally, oil pump or water pump motors may require substantial amounts of three-phase power.

Distances are tremendous. The longest rural feeder the author has personally seen is 115 miles long (25 kV), while the longest single-phase "lateral" was 86 miles (66 kV).

The cost of construction and capacity upgrade is relatively low. Costs for new construction, upgrades, and maintenance of distribution are invariably lower for rural utilities. There are a number of reasons, but in general costs are less than half of those in suburban/urban areas, and as little as one tenth of those in downtown urban cores.

Reliability requirements are below average. Generally, homeowners and businesses alike accept that locating in a remote or sparsely populated area means they will have to accept electric service that is less dependable than in urban and suburban settings.

Generally, best practice for rural distribution follows two principles. First, application of a higher voltage than is typically used in urban and suburban distribution, in order to buy the very long load reach higher voltages provide. As was discussed in Chapter 7 (Table 7.4), at any constant load level a 25 kV feeder can move power four times as far as a 12.5 kV feeder. Voltages of 25 kV, 34.5kV, and in an extreme case, 66 kV, have been used in rural distribution where extreme distances must be covered with relatively small loadings.

Second, use of single-phase "feeders." It is not uncommon for distribution in a rural area to follow the scheme shown in Figure 8.11. The substation shown consists of 4.5 MVA transformer capacity (consisting of four 1.5 MVA single-phase transformers, three connected for service and one spare as contingency reserve against the failure of any of the other three) and serves only one feeder. This feeder's three phases split within a short distance of leaving the substation, with each phase following a completely different route, covering a separate contiguous service territory, the three serving over one hundred square miles.

Using single rather than three-phase lines reducesboth the amount of power and the reach that a particular voltage/conductor can deliver. However, in most rural feeder cases the amount of power is seldom a constraint and long load reach is obtained via relatively light loading and high primary voltage. The resulting single-phase construction has a lower capital cost than three-phase (using large conductor in combination with single-phase, crossarm-less construction usually is much less expensive than building three-phase crossarm structures and using small conductor).

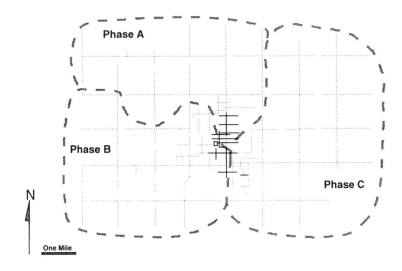

Figure 8.11 A typical "rural feeder" served by a one-feeder substation provides distribution to a small town (two hundred customers) and then splits by phase to distribute single-phase power throughout the surrounding one hundred miles.

Table 8.1 shows an economic conductor sizing table developed for rural application at 25 kV, following the principles and logic and using the conductors discussed in Chapter 7, for a designed load reach of 21 miles (sufficient when working within a grid (Lebesque distance measure) for a system with up to a 30 mile substation spacing). This table shows loading levels for both three-phase and single-phase application. The table assumes the most economic application of each conductor in this situation, and if all segments in the feeder are sized according to this table, any point on the feeder within a 21-mile distance from the substation will have a voltage drop at peak of no more than the class A 7.5% criterion. The author stresses that Table 8.1 has been computed specifically from the data and example system in Chapter 7 and is not a general recommendation on rural conductor types, economical loading levels, or design practices. It is an example of the result of the recommended feeder layout procedures covered here and in Chapter 7. In general, rural layout using such conductor practices, without installation of line regulators, is recommended -- line regulators can be added later, as needed, as a way of handling load growth.

Distribution Feeder Layout

Table 8.1 Conductor Sizing for Rural Application at 25 kV with a 21-Mile Load Reach - MW at 90% PF

Phases	Conductor	Minimum	Maximum
3	#2	0	1.1
3	4/0	1.1	2.3
3	336	2.3	3.4
3	636	3.4	4.7
1	#2	0	.50
1	4/0	.50	1.1
1	336	1.1	1.7
1	636	1.7	2.4

Rural feeder planning is otherwise similar to other feeder planning, except that contingency support switching in rural feeder design is often not feasible due to the distances involved and the fact that often there is no "other feeder" to provide support (outages are restored only by repair, not temporary switching). Thus, distribution layout in a rural or sparsely populated area has the following characteristics:

High primary voltages are favored. Primary distribution voltages higher than 13 kV work best in rural applications, and often 25 kV and above are economically the best choice. In extreme cases voltages as high as 66 kV have been used for distribution (including "single-phase" feeders).

"Single-phase feeders" are common. They have the capacity to meet the load, particularly if built with relatively large conductor, and can be built without crossarm construction, so that they are less expensive than three-phase lines of equivalent capacity or load reach.

Extreme and innovative measures are often used. These include using high voltages (up to 66 kV), earth return (normal single-phase construction requires two wires -- earth-return requires only one), and unusual construction (400 foot spans with all steel conductor).[6]

---

[6] In some parts of northern Canada, rural distribution is built with steel wire as the conductor (what most utilities would use only as guy-wire). Resistance is much higher than with aluminum or ACSR conductor, but it does carry current and it can be strung with high tension on very long spans (hence fewer poles per mile) and its superior strength/weight ratio means it does not fall from ice loadings during winter nearly as often as other conductor.

No provision is made for contingency backup of feeders. Installation of switches to provide contingency support to restore power during outages. Generally there are no other circuits in the area from which to restore power.

Very branch-like planning. New loads and changes tend to be performed by adding branches as needed in what can often be an unplanned sequence. The relatively low cost of reinforcement or modification of the feeder, coupled with the usually slow growth of rural load, means that many rural feeder systems are not planned well from a strategic or engineering sense. This is not necessarily an incorrect practice -- the effort required is often not justifiable.

8.3 RADIAL AND LOOP FEEDER LAYOUT

Figure 8.12 shows two very different ways that a radial feeder can be laid out to serve a set of 162 load points (service transformers) positioned on a rectangular grid of 1/8 mile spacing (typical of city blocks), within a roughly triangular service area. Although slightly idealized, the diagram represents the major aspects of typical 12.47 kV feeders in suburban and urban situations well: At 50 kVA per service transformer, the total load sums to 8.1 MW; feeder routes must be routed through a grid; the area covered is about two square miles.

The two designs shown in Figure 8.12 represent two very different schools of thought in power distribution feeder layout, both widely used. In each case, the basic design concept -- big trunk or multi-feeder -- is interpreted and applied to the realities of actual geography and layout requirements, as illustrated by several examples in Figure 8.12. Roughly equal numbers of utilities apply either the "multi-branch" or the "big trunk" schema to their distribution feeder layout, through the use of engineering guidelines and design procedures developed in house, and through institutionalized tradition that basically boils down to, "We've always done it that way."

It is possible to find distribution planners who vehemently insist that "their" preferred style of the two is best, but in truth there is little to choose between the two feeders shown from the standpoint of electrical and economic performance alone. Both feeders have an identical total length (162 segments of 1/8 mile for a total of 20.25 miles), and an identical maximum feeder run (24 segments, or 3.00 miles). Built at 12.47 kV with from the conductor set given in Table 6.2, with every segment sized according to the economic principles developed in Chapter 7, either layout would provide more than satisfactory electrical performance

Distribution Feeder Layout

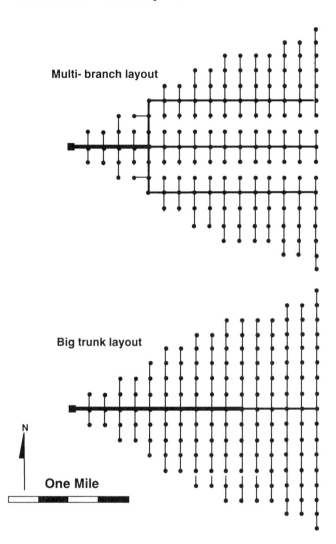

Figure 8.12 Two different ways of routing line segments to a set of 162 service transformers (dots), in a triangular feeder service area. Line thickness indicates segment capacity. Top, a multi-branch scheme, a feeder layout where medium size conductor is used in several fairly evenly loaded and sized branches fed from a short initial trunk of larger conductor. Short laterals branching off the branches provide service. This scheme shows three branches but a multi-branch feeder with between two and six branches are routinely built. Bottom, a feeder consisting of a single large trunk from which lateral segments (the shorter might be single-phase, the longer, three) run to the load.

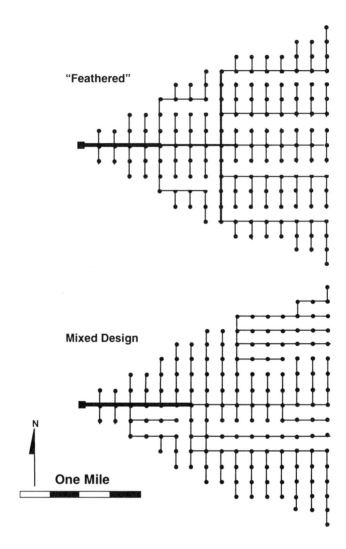

Figure 8.13 Two other ways a feeder could be laid out to reach the 162 service points depicted in Figure 8.12 illustrate the range of feasible alternative configurations that are possible -- the great flexibility that typically exists prior to the beginning of the distribution layout phase. Top, a configuration often referred to as a "feathered" feeder. Bottom, a feeder arranged according to no particular single design approach: it still accomplishes the mission and has very near optimal cost.

Distribution Feeder Layout

in the instance shown, and the resulting costs of the two feeders would be virtually identical.[7]

If feeder costs over the entire economic range were *perfectly* linear, and if every segment in both schemes could be protected with coordinated fusing to reliably isolate outages downstream from affecting the rest of the feeder, then the "large trunk" design would have an advantage in both cost and interruption rate of roughly 1.3%. In practice, neither of those two statements is completely valid, and the answer to "which is best?" depends on a number of factors that will be unique to each utility and even to each situation.[8]

Thus, as a general guideline, the two feeder styles shown in Figure 8.12, and variations on them, should be considered *identical* in terms of cost, electrical performance, and potential outage rate -- a distribution system that achieves maximum economy can be built by following either philosophy. Moreover, numerous other possible alternative feeder layouts, variations or mixtures of those two approaches, also have all these qualities.

Flexibility of Design

The major point illustrated by Figure 8.12 is that the requirements for satisfactory electrical and economical performance of a feeder often leave the distribution planner with tremendous flexibility in the exact design of the feeder. This freedom of design means there are nearly *one thousand* workable feeder layouts, built from the conductors shown and according to the line sizing recommendations of Chapter 7 (Table 7.2), which are capable of meeting criteria and achieving very nearly "optimal cost" while accomplishing the distribution task illustrated in Figure 8.12. Many of these alternatives are variations on either the multi-branch or large-trunk design, such as the "feathered" or heirarchical branched configuration shown at the top of Figure 8.13, but many others really are not identifiable as composed of any one style (bottom of Figure 8.13). However, all do the job, and among them are significantly different "ways to lay out" the feeder.

---

[7] The two configurations in Figure 8.12, and the two additional alternatives shown in Figure 8.13, all have an expected total PW cost including capital, losses and O&M&T over 30 years that is within ±1/2% of $1.8 million PW, or about $222 kW of peak load -- roughly an annualized cost of about $26/year/kW peak.

[8] With respect to protection, the alternatives work out virtually identically in any practical sense. They have the same total lengths of feeder run, so their inherent outage rates can be expected to be identical. The multi-branch has a longer run of "main" trunk, contributing to a slightly higher customer interruption rate, but it compensates by having shorter laterals, reducing customers interrupted by many outages, compensating for that.

Three aspects of distribution planning combine to create this tremendous range of alternative possibilities:

> *Trans-shipment and branching.* As highlighted earlier, feeders consist of repeated branching and splitting of power -- power being routed to two different locations can be combined along one route for much of its journey. Given that there are many locations and many potential route segments (in Figure 8.12, nearly every element of the street grid could be a route segment) this means the planner faces a myriad -- often literally millions -- of possible ways to combine segments and branching into feeder plans.
>
> *Grid routing.* Restriction to routing along a grid, with its consequent impact on route-length comparison (see Figure 8.5), means that usually there are many routes of similar length between any two points. Thus, when those myriad plans discussed above are examined on a length-of-route basis (length is usually the major determinant of both cost and reliability), there is often little difference --all the plans are similar.
>
> *Linearity of cost.* A good line type set provides linearity of cost over a wide range of loads, as was discussed in Chapter 7. This means that in many situations there is no large economic incentive to "put all your eggs in one basket" or to do the opposite and distribute power through a set of more but smaller lines.

As a result, from a cost standpoint alone, there is usually a very wide range of design freedom permitted to the distribution planner in laying out new feeders or making a major revision through new construction to existing ones in normal categories of distribution planning. In most distribution layout situations, many plans, some only slightly different, but others varying significantly in the type of feeders to be built, are possible with little if any difference in total cost.

> In most feeder planning situations, there are many workable plans with similar lowest cost.

The author is aware that many experienced distribution planners will dispute this statement, by pointing out that in most design situations it is often difficult enough to find even one workable feeder plan. But invariably, those planners are thinking of situations where *other criteria* beyond basic electrical performance and cost have been added to the requirements for the feeder plan. In such cases,

Distribution Feeder Layout 311

those other criteria and standards reduce the number of acceptable feeder configurations from thousands to only a few. The next section will discuss "spending" the freedom of design to obtain other goals or meet other criteria, but the fact is that viewed as what could be done and evaluated only on the basis of cost, the process of laying out feeders nearly always presents planners with a plethora of choices. This means that the process of feeder planning has two common aspects:

> *There are usually many "least cost" feeder alternatives. These will vary in many aspects, including routing, philosophy of design, and so forth, but cost will not be one of them.*

Finally, it is worth noting that this multiplicity of possible least-cost plans does not make feeder planning any easier. In fact, it makes it very difficult. There are close to one hundred thousand *apparently workable* feeder layouts that can be hypothesized for Figure 8.12, configurations that "look like a feeder" and deliver power to each load, meet voltage drop, and keep within all criteria. Of these perhaps fifteen thousand are not obviously non-optimal. But less than a thousand are truly "least cost." A planner must find those alternatives, and more, determine which of those alternatives he wants because it maximizes other attributes that are important to him.

"Spending" design flexibility on attributes other than cost

There *are* significant differences between the two layouts shown in Figure 8.12, as well as among the many other plans that are possible in that situation, but those differences are in areas other than total cost. Depending on which of those differences are most important to the planners, they can chose one of the two designs shown in Figure 8.12, or having identified what is important to them, they could go farther by optimizing their selection of a feeder plan from among all of those least-cost possibilities, selecting that least-cost alternative which maximizes their other criteria of importance.

Among the areas where the two feeders and other alternatives will differ are esthetics, expandability for future load growth, protection and expected reliability of service, and contingency support requirements and capability.

Esthetics and "beautification" costs

The big-trunk plan has fewer miles of "main feeder" (1.5 miles versus 4.0) and a greater length of laterals (11.25 versus 8.75), which makes it somewhat between slightly and much less esthetically impacting than the multi-branch feeder, depending on the utility's standards and construction practices. Often, a utility's design guidelines will call for all "main feeder trunks" to be located on the street so that they can be maintained by bucket trucks, while laterals are permitted to be built away from the street (e.g., down the middle of a block along property lines). Under such conditions, the large trunk style, with only 38% as much "main feeder" to run along a street, would have less esthetic impact.

The large trunk plan would also be less costly whenever applied by utilities who have standardized on "esthetic" distribution construction standards. Such standards typically specify steel poles or other attractive structures such as post insulator construction for all feeder along streets, and call for direct burial of all laterals. In such cases the large trunk plan shown in Figure 8.12 is much less expensive than the multi-branch, because it has less mileage of "main" feeder requiring those expensive structures, it has more mileage of lateral which can be direct buried, and it requires fewer riser poles where OH and UG meet.

Future load growth

The multi-branch plan will generally prove the most convenient and least costly of the two to expand for future load growth, particularly if the growth is unexpected in the sense that it is not included in the original planning. The multi-branch layout permits more design flexibility. It has more "main" feeder routes than the large trunk design, which means a main feeder branch is more likely to be near any new load. If need be most of these branch routes have less than largest conductor on them, so they can be expanded.[9] By contrast, the large trunk design requires *adding* major branches (converted from laterals) to expand capability should a large load unexpectedly develop in the area or loads exceed the capability of the laterals serving it. (Allowing for growth and expansion of feeders will be discussed further in Chapters 9, 11, and 12).

Contingency support

The single-path-to-any-customer characteristic of radial systems makes them quite susceptible to widespread interruptions whenever any equipment failure occurs. To mitigate this, radial systems are usually designed so that an alternate source/pathway can be switched into service to pick up the interrupted load, after only a small delay, until the repairs can be effected. The large trunk layout shown in Figure 8.12 is ideally suited to what is called a "single zone" or loop approach to providing such single contingency support. By contrast, the multi-branch layout best fits into what is called a multi-zone contingency support

---

[9] Such upgrades are expensive, but usually less expensive than having to augment the large trunk plan in a similar manner.

Distribution Feeder Layout 313

approach. Therefore, their use as a design guidelines at a utility often corresponds to the type of contingency support design policy used by that utility. Contingency support, single and multi-zone contingency policy, and contingency design are discussed in section 10.4.

Dual-Voltage Feeders

Many utilities have two or more distribution voltages in service, usually serving different areas of the service territory. Planning and layout of such multi-voltage systems will be covered in Chapter 11. However, it some cases, either by design or accident, a system can end up with feeders that are composed of two voltages (and the transformation between them).

Dual-voltage loop feeders

Figure 8.14 illustrates an idealized layout of a dual-voltage loop feeder system, in this case a 33 kV/11 kV feeder of the type employed in many locations throughout Europe (in Figure 8.7, the middle feeder shown is a less idealized diagram of a dual loop feeder). The particular circuit in Figure 8.14 consists of a 33 kV loop, usually operated as a closed loop, feeding several packaged substations from which a total of twelve 11 kV loops emanate -- these are usually operated as open loops. Such "feeders" are part of a very *hierarchical system structure* widely used in Europe, particularly in underground applications, consisting of 33 kV loops, each of which feed several packaged substations, each of which substations serve several 11 kV loops, each of which 11 kV loops feeds several large ($\approx$ 1,000 kVA) service ransformers, each of which feeds several .416 V service loops that feed the customers.

Whether the circuit shown in Figure 8.14 is a single feeder or not depends on one's perspective. However, it represents a popular design type. The hierarchical loop feeder layout is simple in concept and very structured in design -- it makes use of standardized unit 33/11 kV substations and loops -- and therefore this type of circuit proves very easy to lay out, engineer, and build. However, it has the cost and contingency advantages and disadvantages inherent to loop/single zone feeders, as discussed earlier in this chapter. The author's experience is that these types of circuits work well but cost 15-25% more than radial feeders designed to a multi-zone/multi-branch standard.

On the other hand, a hierarchical loop system is often built to operate as closed loops at both 33 and 11 kV levels (requiring only protective equipment and proper coordination for fault isolation of segments from both directions). If

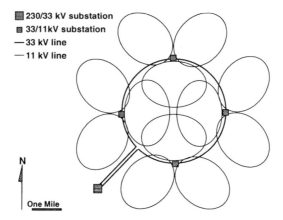

Figure 8.14 Hierarchical voltage loop feeder structure often applied in many areas of Europe, shown here in idealized form with 33 kV loop feeding a series of 11 kV loops. The substation would feed from 3 to 9 other, similar 33 kV loops. Cost generally exceeds that of multi-branch/multi-zone feeders by about 20%, but when operated as closed loops this "feeder" can deliver higher levels of delivery availability (customer reliability).

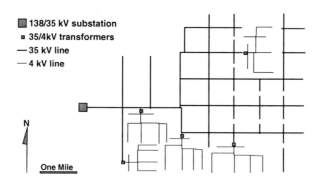

Figure 8.15 A dual-voltage feeder in a utility in the northeastern United States is vaguely analogous to the European dual loop system. Here, 34.5 kV (broad lines), once used only as subtransmission, is now used as direct primary while still feeding five 4.16 kV feeders (thinner lines). Unlike the European layout, this feeder provides power to service transformers from both its high and low voltages, and was never an intended design but resulted as a contrivance to lower cost -- the traditional sub-transmission (34.5 kV) was used as primary when distribution equipment became available for that voltage.

Distribution Feeder Layout 315

so operated, availability from the customer standpoint is substantially better than with any other type of radial system design. This type of hierarchical-loop layout, operated as closed loops, is intermediate, between the standard multi-branch/multi-zone radial layout and the interlaced primary/network secondary layout in terms of both cost and reliability.

Dual-voltage radial feeders

Somewhat analogous to the European style dual-voltage loop structure (with one major difference) is the concept of using feeders of a relatively high primary voltage to feed those with a much smaller voltage level, a situation often found throughout North America. For example, the 34.5 kV circuit shown in Figure 8.15 directly feeds 273 service transformers (1960 customers with 10.6 MW demand at peak) and provides power to five 4.16 kV feeders, which in total serve another 315 transformers (1600 customers and 6.4 MW). Many examples of such feeders can be found scattered throughout the United States. In general, these were not explicitly planned as dual-voltage feeders in the manner that the European systems were intended from the beginning of the planning staage as such. Instead, they resulted from a conversion of 34.5 kV, once used exclusively as sub-transmission, to a direct primary feed. Despite their unplanned nature, these types of feeders provide good service and it makes little sense to plan the retirement of the older, low voltage equipment as long as it is reliable.

Very few dual-voltage feeders are deliberately planned in "American" type systems. Selection of voltage and consideration of dual voltages depend more on "system," rather than individual, feeder economics, (as covered in Chapters 11 and 12).

Individual feeders that employ two voltages in series can occasionally be found, as diagrammed in Figure 8.16. Most such feeders occur because of unusual circumstances. The particular feeder shown in figure 8.16 resulted from the merger of two utilities with different primary distribution voltages. Shortly thereafter, a decision was made to retire (dismantle and remove) an older 22.9 kV substation in the first utility system (it was less than two miles from a substation in the second of the merged systems). However, it was decided not to re-build one of its lengthy rural circuits. Instead, a rack-mounted transformer set was installed, along with a short span of new wire, so that this 22.9 kV rural circuit could be fed from an existing *12.47kV* feeder in the second utility system. Such an "inverted" dual-voltage feeder makes no engineering sense unless the unusual circumstances leading to it are understood -- and such design is not a recommended practice for new construction.

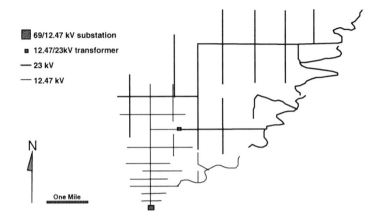

Figure 8.16 A dual voltage feeder serving 4.5 MW in a largely rural area in the northwestern United States consists of a 12.47kV portion leaving the substation and feeding a 22.9 kV portion, an arrangement that makes little sense unless one understands the history that led to it. The 22.9 and 12.47 kV portions intersect (with no connection) at several points and even parallel one another for nearly 1/2 mile. Portions of the feeder that have a twisting, erratic routing (ight-hand side) follow the bank of a meandering river.

Economics of dual-voltage feeders

Selection and application of various voltage levels is much more a factor of system rather than an individual feeder considerations, in that the choices of voltage(s) interrelate with the electrical and economic performance of more than feeder economics. These interelationships are explored in Chapter 11. Within a somewhat narrower scope, however, the impact of voltage selection at the feeder level can be planned where appropriate to select dual voltages, their impact estimated using the linearized cost versus load models for feeder segments and transformers as discussed in Chapter 7.

The idealized feeder shown in Figure 8.17 will serve as an example. The distribution task is to move 24 MW 15 miles and then split it onto 24 segments of 1 MW and move each an additional three miles to their load points. Table 8.2 shows the cost of this distribution, developed from the data used in Figure 7.13.

Distribution Feeder Layout

Table 8.2 Cost of Distributing Power in Figure 8.17

Voltage Used - kV	Trunk	Cost $ × 1000 Transformer	Branches	Total
4.16	$49,640		$13,723	$63,363
12.47	$17,610		$9,594	$27,204
25.00	$9,801		$9,619	$19,420
34.5	$8,046		$11,545	$19,591
34.5/12.47	$8,046	$1,001	$9,594	$18,631

Of the four distribution voltages shown in Table 8.2, 25 kV has the lowest total PW cost for doing the job in a single-voltage application, only 1% less than the next lowest cost (34.5 kV), but nearly 30% lower than 12.47 kV.[10]

Note that the lowest cost for the trunk segment alone is $8 million at 34.5 kV, and for the branches alone, $9.6 million at 12.47 kV, which sums to $17.6 million, about $2 million less than the cost of building both trunk and branch with 25 kV construction. Therefore, if the PW cost of a 20 MW 34.5/12.4 kV transformer and its associated switchgear and control equipment is less than $2 million, a dual voltage 34.5/12.47 kV feeder will be more economical for this situation. Using a cost model of $660,000 PW fixed cost and $14,190/MW slope (PW) as a model of transformer cost in this case gives $1million PW, for a total cost of $18.6 million for a 34.5 kV trunk leading to the transformation to 12.47 kV to feed the 24 branches, a savings of nearly one million dollars present worth.

This example lumped all of the transformation into one location so that an economy of scale could be obtained in the transformer cost. However, this basic economic advantage is behind the decision to use dual voltages in European type 33 kV/11 kV loop feeder systems, where a typical dual-voltage feeder has characteristics similar to the example in Figure 8.18 -- a fifteen mile 33 kV loop supports twenty-four, three-mile long, 11 kV loops through six 5.0 MW packaged transformers.

---

[10] Neither 4.16 nor 12.47 kV can move 24 MW 10 miles on a single circuit. Costs of multiple circuits were assumed for this exercise.

Figure 8.17 Hypothetical distribution situation where dual-voltage application makes economic sense. A total of 24 MW (at peak, with 8760 hour load factor of .5) is to be moved ten miles, then routed in 1 MW portions along 24 segments each three miles long. Although hypothetical, this feeder's economics are analogous to a ten-mile feeder loop supporting 24 loops, similar to that shown in Figure 8.18.

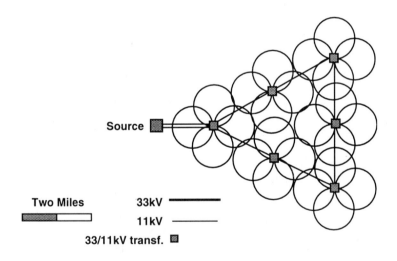

Figure 8.18 Analogous to the feeder in Figure 8.17 is a dual-voltage loop feeder consisting of a 15 mile loop at 33 kV delivering power to 24 loops of 11 kV, grouped into sets of four fed from one transformer, each three-mile loop with a load of 1 MVA.

Distribution Feeder Layout

8.4 CONTINGENCY SUPPORT CONSIDERATIONS

Occasionally, equipment in a distribution system fails due to damage from weather, vandalism, or other causes. In addition, it is recommended practice to have some way in which maintenance or replacement of every element in a system can be performed without causing lengthy interruption of electrical service to the customers it feeds. Thus, alternate sources, paths, and configurations of service must be planned so that both failures and maintenance do not affect customer service beyond a reasonable amount. In some cases, planning for alternate routes of service during equipment outages or emergencies -- will be the major aspect influencing selection of a feeder's capacity, type of route, or layout.

At most distribution utilities, nominal design criteria for voltage drop and equipment loading are relaxed slightly during contingency and emergency operation. While practices vary, many utilities use ANSI C84.1-1989 range A voltages as design criteria for normal conditions on their system, and range B for contingencies scenarios. Range A specifies a 7.5% maximum voltage drop on the primary feeder system (9 volts on a 120 volt scale, maximum). However, Range B voltage profiles permit up to 13 volts on a 120 volt scale (10.8% voltage drop).[11]

Similarly, loading criteria during emergencies typically permit excursions past the limitations set on expected loading during normal service. A particular conductor might be rated at a maximum of 500 amps, a substation transformer at 24 MVA for normal service -- whenever everything is operating as planned. Under emergency or contingency operation, however, the conductor might be permitted a loading of 666 amps (133%) for six hours, and the transformer up to 30 MVA (125%) for four hours. While not all utilities follow these specific standards, most follow something similar. The establishment and application of emergency loading and voltage standards is a recommended practice, so that while the same rigid requirements of normal service need not be met during contingencys, some minimally-acceptable standards do have to be met, to provide a uniform target for contingency planning and operations.

---

[11] See the IEEE Red Book -- *Recommended Practice for Electric Power Distribution for Industrial Plants*. Range A specifies a maximum of 125 volts and a minimum of 117 volts (on a 120 volt scale) for any service point of the primary distribution system. Range B specifies a range of 127 to 114 volts. Range A voltages are the design standard for normal operation. Range B voltages are permitted due to "operating conditions," but corrective action should be taken to restore voltage to range A, within "a reasonable time." When voltage falls outside of range B, "prompt corrective action" is recommended.

Most urban and suburban feeder systems are laid out so that every feeder has complete contingency backup through re-switching of its loads to other sources. Generally, the worst-case contingency for a feeder is the outage of its substation getaway -- the first segment out of the substation -- which leaves all the customers served by that feeder without power. In the event of the outage of this or any other major segment of the feeder, service can be *restored* by:

   a) opening switches to isolate the portion which is out of service,

   b) closing switches to connect the rest of the feeder to other source(s).

Such switching can usually be done within a hour, and leaves only a few customers (those with service from the outaged segment) without power while repairs are made.

Support from the Same Substation

Typically, planning for support of a feeder during the outage of its getaway or a major trunk will arrange to switch it onto adjacent feeder(s) served by the same substation, as opposed to feeders from another substation. There are three reasons why this is desirable:

Substation load balance. Transferring the feeder to other feeders from the substation means that no change is made to the distribution of loadings on the substation level and above.

Feeder load reach. In almost all cases, supporting a feeder from others within its substation (as opposed to supporting it from feeders served out of adjacent substations) results in shorter contingency power flow distances. This lowers the requirement for load reach under contingency conditions.

Make before break switching is more feasible under a wider range of circumstances. In cases where outages are planned (as for maintenance), even momentary interruption of service to customers can be avoided if the tie switch to the new source is closed before the tie switch from the present source is opened.

At times, depending on operating conditions, such "hot feeder switching" between substations can be risky -- it means paralleling the distribution feeder system with the transmission system, if only for a few moments. Slight differences in voltage angle among the substations at the transmission level can cause large circular flows through the connected feeders, which produce currents exceeding the

Distribution Feeder Layout

line tie switch interruption capability, so that the "break" back to radial service cannot be accomplished. In rare cases hot switching between substations has led to circuit overloads sufficient to damage equipment, and in one case, to catastrophic substation transformer failure. By contrast, hot switching of two feeders emanating from the same substation is much less apt to run into such operational problems.

Single zone or loop contingency backup

The simplest approach to feeder contingency backup is to arrange for each feeder to be backfed from a single other source, usually another feeder as shown in Figure 8.19 This is the typical arrangement used in what is often called "European" distribution layout, in which feeders are operated as open loops (or with additional protective equipment in place, as closed loops). When operated as closed loops, protective equipment is set to open at both ends of a segment upon sensing a fault on the segment. The remainder of the feeder stays in service and customers on all other segments notice at most a slight change in voltage.

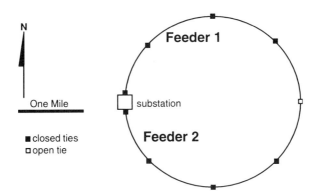

Figure 8.19 The simplest possible contingency backup, shown here implemented as it often is in a loop feeder layout, involves building feeders in pairs (or if one wishes to view it differently, as two halves of a loop) and operating them with an open tie between their ends. Additional switches located along the feeder routes permit isolation of outaged segments.

Single-zone operation makes for easy operations and quick restoration -- in the case of closed loop design an outage of a segment causes no interruption except to customers on that segment. In the case of open loop or "full rollover," service to most customers on the outaged feeder is restored with between two to four switching operations -- open one or two switches to isolate the outaged segment and close the open tie (and feeder breaker to pick up the remaining load.

Single-zone contingency support is the only viable option for certain types of distribution design, mainly underground loop cable systems, where it predominates.

However, such a simple contingency backup scheme requires that the feeder expected to pick up the outaged feeder have the capacity to pick up the load of an entire additional feeder (against the contingency where it is outaged at the substation) and the load reach to move power over the much longer distance required during contingency operation. For a loop, this means it must be designed to satisfy all the load when fed only from either end. Of course, during contingencies voltage drop and loading limitations are relaxed substantially, but even so in practice this means that each feeder faces *four times* the total burden (it must move twice its normal load, on average twice as far) and that it will cost nearly double what it would otherwise cost.

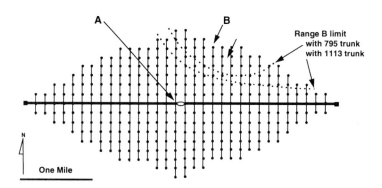

Figure 8.20 Two large-trunk feeders (see Figure 8.12) tied together for contingency support, with the feeder on the left supporting both sets of loads. Points A and B are referred to in the text and are respectively the tie point and the point of worst voltage drop under contingency operation. Dotted lines show the points where voltage drop reaches range B limit with 795 conductor as the trunk (voltages are symmetrical about the trunk, but the profiles are shown only for the northern side).

Distribution Feeder Layout

More than just the tie trunk may have to be reinforced

A point not often appreciated is that reinforcement of the feeder trunk alone (or whatever electrical route the additional current during a contingency takes) may not be sufficient to provide contingency support and meet range B voltage requirements during a contingency. Instead, regardless of the trunk size, some or all of the other conductors in a circuit -- elements that carry none of the extra current during the contingency -- may have to be reinforced as well.

Figure 8.20 shows two 12.47 kV feeders of the large-trunk style from Figure 8.12, tied together in a single-zone contingency form, so that during a contingency one will pick up the other. In the contingency scenario shown, the feeder on the left is energized and providing power to the entire double-feeder circuit. Every segment in both feeders has been economically sized according to the line type tables in Chapter 7, except the trunks, which for tie support purposes are not tapered as economic sizing would prescribe (if designed only for normal circumstances the feeder trunk would utilize tapered conductor, beginning with 636 out of the substation and reducing in size as load on the trunk decreased moving out toward the end). For tie purposes, the entire length of both trunks has been upgraded to 795 MCM, for two reasons. First, contingency load at peak is 16.2 MW, slightly above the thermal rating of the 636 MCM. Second, as will be seen, large conductor is needed for voltage drop reasons. This upgrade increased the original capital cost of each feeder by about $110,000 but because the larger conductor reduces losses costs over the lifetime of the feeder, the total PW cost increases by only $55,000.

Under normal (non-contingency) peak circumstances voltage drop at the end of the 795 MCM trunk on either feeder (the open tie point) will be 3.3%. Another 2.7% voltage drop occurs on the one mile of lateral leading from the end of the trunk to point A, resulting in a maximum (normal) voltage drop on the feeder of 6.0%, well under the 7.5% range A voltage drop limitations.[12]

Under contingency situations, feeder A will pick up the feeder B at the tie end, resulting in a 16.2 MW load, comfortably under the 795 MCM thermal limit of over 18 MW. Voltage drop at the tie point is now 8.4%, and at point A, 11.3%, a full volt worse than the maximum permitted under range B (contingency) standards. Worst still, the voltage drop at the ends of some laterals on feeder B is as low as 11.7 volts, nearly one and one half volts more than permitted.

---

[12] If designed with the tapered trunk (i.e., 636 MCM trunk up to a point where load drops below 5.1 MW, 336 MCM until loading drops below 3.7 MW, then 4/0) the circuit would have 7.1% voltage drop at the end point. The larger trunk reduces voltage drop by more than one volt.

Table 8.3 Upgrade Cost per Feeder for Contingency Support Using a Single Zone Plan

Upgrade	Capital	Total PW
Trunk from 636 to 795 MCM	$110,000	$55,000
Further increase to 1113 MCM	$100,000	$60,000
Upgrade five miles of lateral	$320,000	$160,000
Total cost of conductor upgrades	$530,000	$275,000
Percent of Total Basic Feeder Cost	29 %	5 %
Total cost including switches, etc.	$572,000	$330,000
Percent of Total Basic Feeder Cost	19 %	10 %

Increasing the conductor size of the trunk will not provide much improvement in voltage. As was noted in Chapter 7, beyond a certain point increasing the conductor size adds more ampacity (thermal capacity), but makes only a small improvement in impedance, since it reduces only R, not X. Thus, converting the entire length of the trunk of each feeder to the largest conductor in the set, 1113 MCM, (at an additional $100,000 capital, or $60,000 PW, per feeder[13]) brings down the voltage drop at the tie point to 7.8% and at point A to 10.5%. The worst voltage drop (on feeder B) is 10.7%, still .4 volt too much.

Therefore, to comply with range B requirements under this scenario, twenty-two laterals -- the last two on feeder A and the twenty nearest the tie point on feeder B, need to be upgraded with larger conductor to reduce their voltage drop by .4 volt. Since reciprocal contingency support (B feeding A) is also desired, this really means the end twenty laterals on each feeder must be reinforced. Between one and three segments on each lateral -- about five miles of conductor in each feeder -- must be upgraded in order to reduce voltage drop at their end by a sufficient amount. This increases the capital cost by $320,000. PW increases by only $160,000 because losses do down somewhat, due to the larger conductor). Total contingency upgrade cost for this feeder is summarized in Table 8.3.

---

[13] PW cost impact is less than capital cost impact because the larger conductor reduces losses costs, somewhat counteracting its increased capital cost in the PW total.

Distribution Feeder Layout 325

This example is not a contrived or overly sensitive case. Quite the contrary -- it is typical of feeder contingency planning in a number of important ways:

Voltage drop issues and upgrade requirements for areas of the feeder outside the trunk actually carrying the tie current, necessitate upgrades to a significant portion of the feeder.

Contingency support requirements increase initial cost more than PW cost -- upgrades increase the capital cost ($572,000, or 19%) by much more than the PW cost ($330,000, only 10% of PW cost).

Although not shown in the example, if the loads in the example grow by 5%, the cost of providing contingency support nearly doubles. Typically, contingency requirements are very sensitive to load growth.

Switched Contingency Zones

An alternate arrangement for contingency support is to make provision to split a feeder into several *switched zones* which are distributed among neighboring feeders during a contingency. This produces two advantages.

The additional load transferred to any neighboring feeder is only a fraction of a full feeder load, meaning that the percentage increase in load it sees during an emergency is much less than in a loop or full-trunk scheme.

Generally, the maximum distance (load reach) increase for power flow during an emergency is reduced, too.

As a result, usually only a few strategic segments need be sized above their economical size (as determined for normal operations) in order to achieve contingency capability for every feeder using this approach. A disadvantage, however, is that more switches have to be operated in order to restore service.

Figure 8.21 illustrates this concept, where a feeder area is split into three switchable zones. In the event of an outage at the substation, these three zones are each transferred to different feeders from this same substation, as shown. Each of the three picks up an additional 33% load. This scheme uses a popular contingency support method which is to provide switches near the substation to merge two feeders within a short distance of the substation (Figure 8.22).

The dotted line in Figure 8.21 shows that the maximum expected reach (distance as electricity flows using the Lebesque 1 measure) in this "theoretical" example is no longer than the longest in normal feeder service. However, as a rough rule of thumb, in practice the limited locations at which switches can be arranged means that feeder runs up to 1/8 longer (+12.5%) than under normal conditions may have to be accepted as part of such a three-zone scheme.

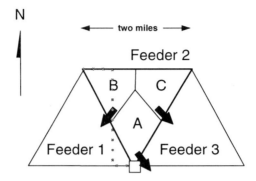

Figure 8.21 Shown here are the service areas for three of six feeders in the upper half of an idealized, hexagonal substation area. Feeder 2's area is shown split into three contingency "zones" -- A, B, and C -- two of which are transferred to adjacent feeders, while that nearest substation (A) is "switched across the substation" to a feeder not shown on the other side of the substation. The dotted line shows the greatest "theoretical" feeder run under the contingency switched configuration, a distance no longer than the normal longest reach needed by this feeder system.

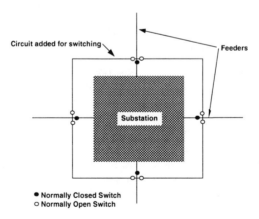

Figure 8.22 Switching flexibility "around the substation" is often provided by installing lines and switches near the substation as illustrated above for a substation with four feeders. Any one feeder can be isolated near the substation and switched onto either of two neighbors. Usually all the switching is outside the substation, sometimes located several blocks away.

Distribution Feeder Layout

Therefore, voltage drop on some portions of the feeder system during contingency switched operation could be as much as 33% + 12.5% = 45.5% greater than normal. Range B voltage tolerance limits permit 13/9 = 144.4% greater voltage drop during contingency operation, very nearly the same increase.

In practice, a zonal scheme with between three and five zones of this nature will usually provide complete contingency backup for all feeders, with no *additional cost* for line segment reinforcements. However, line switches have both capital and maintenance costs, the planning for multi-zonal schemes is considerably more difficult than for loop or single-zone systems, and the required switching operations required during contingencies take more time. However, these costs come to far less than the cost of additional capacity required for loop or single-zone contingency support, and therefore multi-zone switching is the preferred contingency approach for most feeder systems.

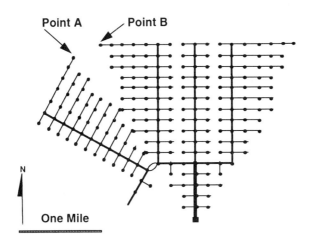

Figure 8.23 Two multi-branch feeders (see Figure 8.12) tied together for contingency support, with the feeder at the bottom supporting one of three zones from a feeder at its left (only the zone supported is shown. Points A and B are referred to in the text and are respectively the point of worst voltage drop on each feeder. Dotted lines show the points where voltage drop reaches range B limit with both 795 and 1113 MCM conductor. Voltages are symmetrical on both sides of the trunk, but the profiles are shown only for the northern side.

Low cost for a multi-zone contingency scheme

To illustrate the lower expected cost of providing contingency support with a multi-zone scheme, Figure 8.23 shows two 12.47 kV feeders of the multi-branch style from Figure 8.12, tied together in a three-zone contingency form, so that during a contingency, the feeder on the right will pick up a branch containing 63 of its neighbor's 162 load points, totaling 3.15 MW (39% of the load). Every segment in both feeders has been economically sized according to the line type table in Chapter 7 (Table 7.2), with no conductor upgrade. During normal (non-contingency) peak load conditions, voltage drop at the (open) tie point on either feeder is 2.7%, and at points A and B it reaches its maximum of 7.4 % (just within range A limits). This is essentially a "multi-branch" version of the example given earlier in Figure 8.20.

In this contingency scenario, the 636 MCM main trunk sees a load of 11.2 MW, comfortably below its thermal rating of over 15 MW. The perpendicular leg of this feeder uses 4/0 conductor, which sees a contingency load at peak conditions of 6.2 MW, 10% less than its thermal limit of 6.8 MW. All other segments are loaded as they are in normal circumstances. Voltage drop at the tie point on the supporting feeder increases to 4.3 volts, an increase of 1.6 volts. Voltage drop to points A and B increases to 9 %, well under the range B limit of 10.3 % voltage drop.

In fact, the feeder shown can actually pick up a second zone without suffering adverse overloads or voltage drop effects. Assuming that it picks up a second, similar zone on the other side of it, flow on its main trunk rises to 14.3 MW, near but still under its thermal limit, and voltage drop at the tie point rises by another .7 volts, to 5 volts, and thus total voltage drop at the end points A and B increases to 9.7%, still under range B voltages.

This multi-zone contingency plan obtained greater economy of contingency support than the case discussed in Figure 8.20 through two stratagems. First, it distributed the contingency load increase over more feeders and more paths. Second, it kept the contingency support paths short.[14] As a result, it can provide contingency support at very little additional cost (Table 8.4).

---

[14] In doing so, this particular case supported the outaged feeder from the same substation that served it, a recommended practice for feeder contingency support. However, that is irrelevant in this discussion. A single zone loop feeder is essentially equivalent in contingency regards to the case shown in Figure 8.20, and yet it provides contingency support out of the same substation that normally fed the outaged feeder it is supporting. In so doing, however, it uses a long path, one reason costs are high.

Distribution Feeder Layout

Table 8.4 Upgrade Cost per Feeder for Contingency Support with a Three-Zone Support Plan

Upgrade	Capital	Total PW
Total cost of conductor upgrades	none	none
Percent of Total Basic Feeder Cost	0 %	0 %
Total cost including switches, etc.	$82,000	$110,000
Percent of Total Basic Feeder Cost	4.5 %	2 %

Providing Switch Locations to Provide Complete Coverage

Failures and unexpected events do not always occur as and where expected. Therefore, a good contingency plan, multi-zone or otherwise, provides flexibility by locating switches at various strategic locations so that *parts of a feeder* can be picked up in the event of line outages at various places. Figure 8.24 shows the locations of various switches installed on the multi-branch feeder from Figures 8.12s and 8.23. Locating a switch at the end of each branch permits a portion of it to be picked up from the far end (assuming a source is arranged there) should a failure occur somewhere along the branch.

The purpose of this switch is not to pick up the entire branch from the tail end. To do so would create much the same problem encountered in Figure 8.20 with single-zone feeders picking one another up from the tail end, and would require the same solution -- conductor upgrade of various line segments. Provision here has already been made to support the entirety of each branch, in the manner shown in Figure 8.23. Rather, the far end switch is a feed for part of the feeder in the event of a failure along the branch trunk. The segment failure can be removed and the portion of the line restored to service, through the normally open tie at the feeder end. This provides contingency support to a portion of the branch (but generally not all of it without conductor upgrade), improving expected customer reliability. If the improvement is not sufficient, conductor upgrade can be considered if the cost is justified by the improvement.

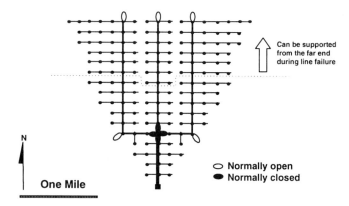

Figure 8.24 Three-branch/three-zone scheme with nine switches. The purpose of these open ties at the end of each branch is not to support the entire branch in the event of a failure (such support is done as shown in Figure 8.23), but to support only a portion of the load in the event of a failure in the branch path closer to the substation. Portions of the loads above the dotted line can be picked up in each branch without any upgrade.

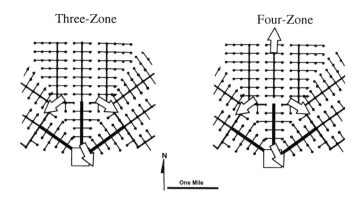

Figure 8.25 Three and four-zone contingency plans for a multi-branch style feeder. At the left, the multi-branch feeder from Figure 8.24 (above) and parts of two neighboring feeders. Left and right branches of the feeder have been isolated (open point at feeder trunk) and switched to a neighboring feeder (arrow), center section has been switched around the substation (as in Figure 8.22). In the four-zone plan the middle branch is isolated by an open switch and transferred to a feeder (not shown) from another substation.

Distribution Feeder Layout

More zones are needed to keep cost low when normal loadings are near thermal capacity limits

One reason that contingency support schemes work well is that there is usually a large difference between the economical and thermal load ratings for any particular line type. In the economic valuation of conductors given in Chapter 7 (e.g., Table 7.2), the upper end of a line type's economical range was generally about half of its thermal capacity limit. For example, the 336 MCM three-phase overhead line listed in Chapter 7, Table 7.2, can carry 10.3 MW at 12.47 kV, but has an economical range of application of from 3.7--5.1 MW. Thus, when built on minimum overall cost principles, it will have an emergency load-carrying capacity margin (thermal minus actual load) of at least 100%, which provides a cushion for it to accept both the higher loading levels and meet the longer load reach needed during contingency operation.

In systems where the PW evaluation period is very short, cost of future losses is weighted muuch less, and economic sizing methods as described in Chapter 7 will respond by recommending much higher loadings for each conductor type -- and consequently leave much less operating margin for contingencies. Further, there are always a few situations, either due to deliberate design or the effects of years of load growth, where feeder trunks may have no capacity margin at all. In such situations, subdividing feeders into more and smaller zones (Figure 8.25) can help keep the additional cost required for contingency support (as compared to the cost of what is required to provide only normal operation) as low as possible. However, increasing the number of zones adds greatly to the complexity of switching operations during contingencies. A general rule of thumb guideline on the number of zones required to meet feeder-level contingency requirements is given by:

$$\text{Zones required} = \text{Max} \left(1 + E/(T-E), 1/((D - 1.125))\right)/B^3 \qquad (8.1)$$

where T = the economical rating of lines used for tie support
E = the thermal rating of lines used for tie support
D = ratio of maximum voltage drops permitted under emergency and normal operation
B = ratio of capital cost, contingency plan over no contingencies planned (i.e., cost of providing contingency support is $B-1$)

This rule of thumb evaluates the two different types of constraints that define the number of switching zones required. The first is that set by the thermal limit of conductors, the second is defined by voltage drop limitations during contingency. One or the other of these will limit contingency switching. It also responds to the ratio B, the ratio of cost the planner is willing to spend versus the minimum (economic design, no contingency support at all), which experience has shown has a squared effect.[15]

As an example, consider a 12.47 kV OH system built with the line types discussed in Chapter 7. Of the four economical conductors shown in Table 6.2 (#2, 4/0, 336 MCM, and 636 MCM ACSR) the two larger sizes can be expected to be used for switchable trunks and feeder ties. These have economical loading points of 5.1 and 8.4 MW respectively, and thermal limits of 10.3 and 14.3 MW. Averaging these two sets of values gives E = 6.77 MW and T = 12.3 MW. If the contingency/normal voltage drop ratio is 1.43 (Range B over Range A voltage drops), and it is desired to keep budget ratio to 1.04 (about the best that can be done considering the need to add switches, etc., -- see Table 8.4), then

$$\text{Zones required} = \text{Max } (1 + 6.77/(12.3 - 6.77), 1/(1.43 - 1.125))/1.1^4$$

$$= \text{Max } (2.22, 3.27)/1.17$$

$$= 2.8$$

Effectively, this system will need to be planned with three zones per feeder. On the other hand, if the planner is willing to spend 1.33 times the minimum possible layout cost (i.e., Table 8.3) in order to gain contingency support, then

$$\text{Zones required} = \text{Max } (2.22, 3.27))/3.27$$

$$= 1.02$$

and the system can support contingency operation of feeders and transformers with single-zone/loop configurations.

Three zones per feeder with a capital cost premium of only about 4% above minimum, or single-zone operation with a premium of about 33%, are fairly typical values to find in traditional distribution planning, the type that uses cost of losses evaluated on a PW basis over 20 or more years into the future to size conductor. That period of evaluation leaves conductors loaded to roughly 50% of thermal rating when applied most economically (i.e., (T-E)/E is close to 100%). Shorter periods of evaluation, or very low values assigned to electrical losses costs will reduce this available margin. Equation 8.2 indicates that as (T-E)/E shrinks to 33%, then five zones per feeder are necessary if budget ratio is to be kept to a minimum of 1.05, and so forth.

---

[15] This is an empirical rule-of-thumb, for which the author offers no explanation, other than it is useful and works after a fashion.

Distribution Feeder Layout

Zone planning for contingencies is also sensitive to the ratio of contingency to normal voltage drops permitted. If D is reduced from 1.43 to 1.33, then the number of zones required if cost escalation is to be kept to a minimum rises to 4.4 (effectively a voltage ratio of 1.33 requires five zones per feeder or a cost premium of 10% and four zones). The budget premium required to maintain single-zone capability rises to 120% minimum no-contingency cost. If D is relaxed to 1.66, then the number of zones required drops to only two, and the cost premium to obtain single-zone capability falls to only 50%.

Zones do not have to be switched to separate feeders, only "separate paths"

The different zones in a feeder do not necessarily have to be switched onto different feeders in a contingency to stay within capacity and voltage drop constraints, but they must be switched onto *different paths,* so that the extra flows they cause share only a minimal common path. This was illustrated in the discussion of Figure 8.23 earlier in this chapter, where it was pointed out that without any conductor upgrades, the multi-branch feeder shown could pick up a branch on both sides of it during contingency situations. And while that feeder and its contingency support capability are very robust, if normal loadings were increased by about 10% (as they would be if the conductor tables used for line sizing were based on five to eight year future cost evaluation period, not 30 years) a feeder designed in accordance with the more heavily recommended loading tables would not have as much difference between E, the loadings designed for normal operation, and T, the maximum capability of its conductors, in an emergency. As a result, contingency plans may run into capacity, rather than voltage drop problems. [16] A design response is to divide a feeder into more

---

[16] There is a complication here with regard to the earlier example (Figures 8.20 and 8.23). There, a change in conductor economics brought about by a shorter PW period would *not* change the conductors selected -- increasing loading would decrease load reach, and the feeders shown require a reach near the maximum that the example conductor can provide. Those feeders need that load reach and thus need the conductor sizes/loadings in Table 7.2 just to perform within range A voltage drop requirements -- the sizes are required *other than just* economic reasons. In this sense, they represented optimized designs where their distance and layout were compatible with the capabilities of an economic conductor set (and vice versa) as determined along the lines of the stratagems to be discussed in Chapters 11 and 12. However, in cases where a utility decreases evaluation period, or otherwise increases the target conductor loadings, the result is that the ratio E/T increases, and *less* margin is left for contingency operation. Hence, the marginal cost for providing contingency capability may rise even as the base cost of moving power drops as a result of the re-rating of the line types.

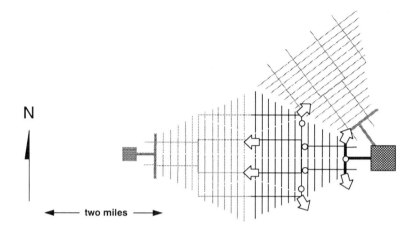

Figure 8.26 A feeder divided into six zones is transferred during contingency situations to three other feeders, each of those picking up two zones. Only small portions of each feeder (thick lines) will need reinforcement above economic rating in order to support this contingency scheme, even if normal loadings are up to 83% of thermal. Open circles indicate normally closed switches that are opened, arrows normally open switches that are closed. Contingency transfer requires a total of 12 switching operations.

zones and transfer each of these smaller zones onto different paths. Thus, the same amount of load may still be transferred to each neighboring feeder, but this can be done under a more restrictive contingency capacity margin (i.e., a smaller T-E) by distributing it along more paths.

Figure 8.26 shows a feeder with six zones (about the maximum that is generally ever thought to be practicable with manual switching), part of a distribution system that is planned to take advantage of this situation, in which two zones from this feeder are transferred to each of three neighboring feeders. Each of the two zones transferred to the same feeder is supported through a largely independent path. This arrangement can perform to range B voltage and loading limitations at about 15% higher normal loadings (i.e., with E a much larger portion of T) than a plan switched in only three zones or four zones. It is both the division of the outaged feeder into six separate load modules, and the use of two instead of one path in each support feeder, that contributes to lower voltage drop increases during contingency operation.

Distribution Feeder Layout

"Many-Zone" Switching Has a Cost of its Own

It is possible to create feeder system layouts that achieve remarkable contingency support economics, even as their normal peak loading levels approach thermal capacity, by utilizing six, seven, or even nine switchable zones per feeder. Such plans are of questionable practicality and effectiveness, because of the complexity and time required for the switching operation.

As an example, consider an extreme case, a feeder system laid out with nine switchable zones per feeder. Equation 8.1 predicts a need for nine zones when the contingency to normal voltage drop ratio is 1.43, lines are loaded by design to 90% of their thermal rating at normal peak conditions, and no more than a 5% premium is to be paid for achieving contingency support capability. It is possible (although difficult) to design feeders with nine switchable zones. But such a scheme would require a total of eighteen switching operations to complete each feeder transfer.

Each switching operation requires only 5 to 15 minutes including drive time, and contingencies are rare, so that the total lifetime PW cost of making switching operations for this nine-zone circuit would be small -- insignificant compared to the predicted savings effected by using such a high multi-zonal contingency plan rather than increase capacity so that a four- or five-zone scheme could work.

However, such a switching scheme is impractical, and will not improve customer reliability substantially. Switching operations often have to be done serially and usually only one or two switching crews are available. Eighteen switch operations could take up to four and one half hours to complete -- hardly "quick" restoration. Beyond that, complicated switching schemes provide potential for human error.[17] When operations must be done manually, the recommended maximum practical limit is six to eight switching operations (three to four zones), which when scheduled correctly should take about one half hour to complete. When switches are remote control or automatic, a dozen or more switching operations may be feasible in a switching series.

Feeder Layout to Support Substation Transformer Outages

Most major distribution substations consist of two or more transmission-to-primary voltage transformers, each feeding a low-side bus from which between two to six feeders draw power. When a substation transformer or its low-side bus is out of service, power flow to the group of feeders it serves is interrupted.

---

[17] Contingencies usually occur during storms, when resources are scattered and stretched to the limit, communication is imperfect; and crews have been working long periods of overtime -- conditions that make errors almost inevitable.

Some distribution systems are planned so that this whole group of feeders can be re-switched to alternate sources, thus avoiding extended customer interruptions when the transformer or its buses are out of service.

"Transformer-level" contingency support is almost always more expensive to arrange than contingency support for only individual feeder outages. More load must be supported during these contingency scenarios than when only a single feeder is out of service, meaning that either a greater capacity margin must be set aside for contingency purposes, or the outaged load must be switched over greater distances, or both.

Usually, in systems that adopt this contingency support policy, the feeder system is laid out so that when viewed in rotation around the substation, feeders alternate between transformers, as shown in Figure 8.27. This arrangement provides support on both sides for each of a transformer's feeders during the contingency -- whenever a transformer is out of service, feeders connected to transformers that are in service will be on both sides of it.

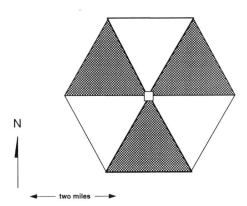

Figure 8.27 Alternating the feeders from different transformers by rotation as shown here makes contingency support for the transformer level via feeder switching somewhat easier to arrange. Here, shading indicates which of two transformers feed each of six feeders. If either transformer is lost, each of its feeders lies between feeders which are still in service, making the job of picking up its load from those remaining in service somewhat easier.

Distribution Feeder Layout 337

Transformer outage contingency support can be accomplished within either a single- or multi-zone framework. However, *a substation design policy calling for N transformers per substation is "naturally compatible" with a feeder-level contingency support plan built around it using N-1 switchable zones per feeder,* in that transformer outages can be supported with the same feeder design, and at basically no more cost, as feeder outages alone. For example, single-zone feeder contingency plans can usually accommodate transformer outage support in a two-transformer substation plan, with no additional feeder capacity and few changes in switch plans. When designed for feeder-level contingency support, each feeder has the capacity and reach to pick up another feeder. In a transformer outage, each feeder is being asked to do no more than that. For example, in Figure 8.27, if one of the two substation transformers failed, leaving three feeders without power, the three that remain in service could each individually pick up one neighboring feeder. Service would be restored (assuming the remaining transformer has sufficient capacity to handle the load). Little additional cost or complexity would be required in a single-zone or loop scheme in order to achieve contingency support of this type, satisfying both feeder and transformer failure scenarios.

However, a multi-zone contingency support scheme for this same two transformer substation (Figure 8.27) would present a far different situation. If the three-zone feeder transfer scheme shown in Figure 8.21 is used during transformer contingencies in Figure 8.27, each of the three remaining feeders would have to pick up 100% additional load, and the three-zone contingency planning gains nothing in terms of reduced capacity required over single-zone contingency support. Transfer of one or more zones to another substation, and clever design of the zones and "multiple path" switching, can reduce the additional capacity requirement substantially, at the cost of greatly increased operational complexity. However, a three-zone scheme would probably need significant reinforcement of conductor size in order to be able to provide transformer level contingency support for any two transformer substation.

On the other hand, a three-zone contingency support feeder system requires little modification in order to provide transformer contingency support, too, if four transformers have been specified per substation.[18] Used in a balanced design, four transformers would mean that three out of every four feeders would be left in service whenever any one transformer was out of service. Each feeder remaining in service could take one zone of a feeder out of service, giving it a 33% overload, exactly as is normally planned for feeder support. Potentially, no additional cost of capacity or load reach is required.

---

[18] Such a substation would very likely have 12 feeders -- three per transformer, as did the two-transformer example substation in Figure 8.22.

Layout to Support Substation Outages

In a few cases distribution feeder systems are designed so that the load served by an entire substation can be served during a substation outage through transfer of all loads to nearby substations via feeder re-switching. Only a small minority of distribution utilities build such levels of support into their system as a matter of policy at all substations. Usually full substation contingency support is provided only in special circumstances, for example when a substation is served by a radial transmission circuit and thus more prone to outages than average, or when the substation serves critical loads, or when repair or replacement of its equipment is expected to take much longer than average.

As was the case with transformer-level support, a system built with single-zone feeder contingency capability requires only modest additions to meet substation contingency support needs. By contrast, multi-zone approaches require considerable reinforcement to achieve this result. Thus, use of any more than two zones per feeder usually brings scant improvement in cost. Table 8.5 gives an indication of how costs vary for different types of substation/feeder systems built to provide varying levels of contingency support.

Choosing Between Single and Multi-Zone Contingency Schemes

As was shown in Table 8.6, multi-zone contingency support schemes can materially reduce the cost premium for providing contingency support at both feeder and transformer levels over single-zone or loop approaches. Multi-zone switching takes more time (which increases duration of service interruptions due to outages). In general, the decision on contingency support policy should be made at a systems level. The contingency support design of the substation and feeder levels should be selected so that overall customer reliability targets are met with the lowest possible cost. This means (as much as is practicably possible) selecting the number of transformers and type of substation design, and the style of feeder layout and number of switchable zones per feeder, so that the cost of reliability improvement has the same marginal cost at both the substation and feeder levels. (This will be covered in more detail in Chapter 11).

The large trunk layout can transfer its load with only two switching operations when its substation feed has failed, given that the trunk has been sized so that it can support feed from either end. If the trunk is sized accordingly, this feeder is similarly capable of picking up an additional feeder's load when that needs support, and thus providing contingency support as needed. These designs make automatic rollover switching quite easy -- just two automated, coordinated switches. They are also the only type of contingency support scheme that fits the layout of loop feeder systems, and are therefore required in those systems.

Distribution Feeder Layout

Table 8.5 Relative Capital Cost of Various Types of Feeder Layout When Providing Different Levels of Contingency Support

Transformers Per Substation	Switchable Zones Per Feeder	Providing Support for ...		
		Feeders	Transf.	Entire Sub
1	1	1.33	1.40	1.40
	2	1.15	1.25	1.33
	3	1.05	1.25	1.33
	4	1.05	1.25	1.25
2	1	1.33	1.40	1.40
	2	1.15	1.25	1.33
	3	1.10	1.15	1.33
	4	1.10	1.10	1.25
3	1	1.33	1.40	1.40
	2	1.15	1.15	1.33
	3	1.10	1.10	1.33
	4	1.05	1.10	1.25
4	1	1.33	1.40	1.40
	2	1.15	1.15	1.33
	3	1.05	1.10	1.33
	4	1.05	1.05	1.25

* Value of 1.0 is assigned to the cost of a feeder system planned only from the standpoint of meeting peak load at lowest cost using conductor tables as described in Chapter 7.

Table 8.6 Relative Cost and Expected Outage Duration Differences for Various Feeder Styles and Switched Zone Schemes Applied System-Wide

Feeder Layout	Zones	Relative Cost	Relative SAIDI	Load Growth
Large trunk	1	1.33	1.00	0
Large trunk	2	1.20	1.1	12%
Multi-branch	3	1.05	1.2	15%
Multi-branch	4	1.08	1.3	27%
Closed hierarchical loop	-	1.44	.2	10%
Interlaced feeder/network		1.83	.1	25%

Table 8.6 summarizes the various options discussed with respect to layout and conductor sizing and total cost (including switches) that meet contingency support requirements, and includes several layout types (loop, interlaced). SAIDI figures assume a 100 minute annual duration as average. Load growth shown represents the expected increase in peak load before reinforcement would be required.

Using Reliability-Based Rather Than Contingency Criteria

The foregoing discussion used a traditional "contingency coverage" perspective rather than an explicit, calculate-reliability-indices-throughout-the-feeder approach, which the author believes is actually superior for modern utility planning needs (see Chapters 3 and 5). This was done because it is easier to study and explain reliability and layout interaction on a configuration (contingency plan) basis rather than a "computed reliability" basis. In general, the qualitative results, as well as the recommendations, made through this section apply regardless of whether reliability-index or contingency coverage perspectives are applied. For example, the concept that "zones do not have to be switched to separate feeders, only separate pathes" (page 333) is a guiding rule regardless of which perspective, contingency or reliability index, is used.

8.5 CONCLUSION

Feeders are the major building block of power distribution systems. They are themselves composed of line segments, and in some cases transformers, which must be selected with care from a well-designed set of available components, in order to assure that the feeders are economical but have the capacity and load reach required. In some situations -- urban or rural areas -- conditions will heavily constrain the distribution planner with respect to configuration and usage of the feeder for power distribution, but for a wide variety of situations feeder layout includes considerable flexibility, if done correctly. This gives the designer freedom to meet other requirements besides just economy, and very often a major concern is contingency support planning to improve reliability.

REFERENCES

M. V. Engel et al., *Distribution Planning Tutorial,* IEEE text 92 WHO 361-6 PWR, Institute of Electrical and Electronics Engineers, New York, 1992.

H. L. Willis et al., "Load Forecasting for Transmission Planning," *Transactions on Power Apparatus and Systems,* Institute of Electrical and Electronics Engineers, New York, March 1988.

9
Multi - Feeder Layout and Volt-VAR Correction

9.1 INTRODUCTION

Most distribution systems consist of many feeders arranged into a "feeder system." While it is important that each feeder be well engineering and properly laid out, the arrangement of the feeders into a system is of paramount importance.

This chapter looks at several layout concerns that revolve around the arrangement of two or more feeders and their interaction, as well as volt-VAR control from a system (overall planning) aspect. In addition to the discussion given here, multi-feeder systems and their layout are also examined in Chapters 11 and 12 with regard to how the feeder system fits into the overall system.

The Systems Approach: Feeders Are Only Part of the System

As discussed in Chapter 1, a power delivery system consists of many levels of equipment, including sub-transmission, substations, feeders, and service. The recommended perspective for planning is to always view each level as part of the larger whole, and to plan it using *the system's approach*. This means that when laying out a particular feeder, the goal is not to minimize its cost, but to plan it so it contributes to achieving the lowest overall total system cost.

Similarly, decisions about the economy and reliability of feeder system as a whole -- voltage, number of feeders, and so forth, should all be made with this same perspective. In some cases addition construction or capacity installed at the feeder level will permit larger savings to be achieved at another level. Similarly, there are cases where "feeder level" performance issues are best solved by making alterations at another level in the system to change the situation at the feeder level. There is no hard and fast rule about how to do this, and thus the only consistent way to minimize cost is to *always* apply the systems approach -- feeders are only one part of the whole and their interaction (electrical and economic) with other parts of the system must always be kept in mind during all phases of the planning process.

Avoid "feeder-at-a-time" myopia

When working on the details of feeder system enhancement and design, is it common for distribution planners to focus on the study of one feeder at a time. A particular feeder will be studied in isolation from the rest of the system, as its specific routing, conductor selection for each segment, equipment selection, and a myriad of other details are developed. This feeder-at-a-time focus is essential in certain phases of planning, but it can lead to a kind of design myopia which is responsible for many missed opportunities for savings and service quality improvement.

Planners *must* remember that individual feeders are not rigidly defined units unto themselves. A "feeder" is all the downstream circuitry fed by a specific fed point (usually defined as a breaker position from the low side bus) at the substation, and the feeder's configuration at any moment is defined by the open points which terminate its electrical pathways. The route, service area, and load for any particular feeder can be changed by simply opening and closing several tie switches, or even more extreme changes rendered by constructing a few short new segments and adding open points so as to alter the feeder's boundaries quite significantly.

Very often, problems on a particular feeder (e.g., overloads, low voltages) that appear to require expensive reinforcement can be mitigated by re-allocating portions of that feeder's load to nearby feeders through such means of switching or light construction of a few segments to permit a different split of the load. In many cases the result is that the operational problems no longer exist, and in other cases, while a problem remains, its solution is now less costly.

While almost all distribution planners understand this concept in theory, many tend to forget it when evaluating and planning feeders, particularly when examining alternatives to upgrade existing feeders to handle new or growing load.

Multi-Feeder System Layout

9.2 HOW MANY FEEDERS IN A SUBSTATION SERVICE AREA?

How many feeders should a substation have? The best number of feeders depends on many factors, beginning with the concept of "feeder" being used. As stated above, perhaps the most unambiguous definition of a feeder is "a feeder is all primary voltage circuitry downstream of a low-side protective device at the substation."

Classical (and Inappropriate) Perspective on Feeders

The classical perspective on feeder layout and the recommended number of feeders to serve an area views feeders as each being a tree-like circuit serving a "slice" of a round or hexagonal service area (Figure 9.1). Increasing the number of feeders is seen as a way to increase the capacity of the substation, and also as a way "prevent an increase in feeder voltage drop" as load increases.[1] This perspective is valid -- in a very narrow sense -- but it is based upon three assumptions about feeders inappropriate to modern distribution planning.

Role rather than economic line-type selection. Feeders in classical distribution analysis are laid out based on line-type selection criteria defined by a segment's role, not its loading or load reach requirements. For example, the results in the Westinghouse "Green Book" are all based on feeder trunks being 4/0 conductor, branches either #2, 2/0, or 1/0 (depending on type of layout), and laterals being # 4, regardless of loading, reach requirements, or other factors.

By contrast, modern distribution planning lays out feeders with line types sized according to economic and performance requirements. Thus, the two layouts shown in Figure 9.2, serve the same load, have the same voltage drop, and consist in large part of the same segments with the same economically-selected wire sizes, even though the feeder layout on the left consists of two "feeders" and that on the right only one "feeder." Both layouts have virtually the same cost.[2]

No reliability considerations. Impact of layout on service reliability in the sense it is used at the present time is simply not assessed in classical distribution system analysis.

---

[1] See, for example, D. N. Reps, "Subtransmission and Distribution Substations," Chapter 3 of the *Westinghouse Electric Utility Distribution Systems Engineering Reference Book*, pages 80 and 81.

[2] As mentioned earlier there is a very slightly higher cost for the single-feeder layout, but this amounts to no more than .02% higher cost for several segments of "big wire" that represent the only different between the two layouts.

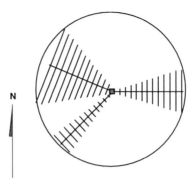

Figure 9.1 Three feeders depicted from a classical perspective on feeders, which viewed feeders as composed of standardized line segment roles and serving a defined "angular slice" area. The three feeders here all have trunks that head straight out from the substation, and all three would have trunks of identical wire size, even though they serve vastly different loads. Results interpreted from this perspective are often inappropriate for actual feeder system layout.

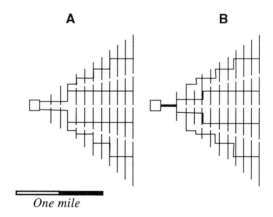

Figure 9.2 A more modern, and applicable, view on feeder layout and choosing the number of feeders. (A) Two feeders are used to serve an area of one and one half miles. (B) Feeders have been combined into one feeder. The only difference between the two configurations is the "simmering" of their first segments into one large line type. Otherwise, feeder layout is similar. *The optimal tree structure for getting power away from a substation is not really effected by the decision of how many feed points to use to leave the substation.*

Multi-Feeder System Layout

Euclidean shortest-distance rather than grid routing. This real-world factor and its impact on feeder layout and performance estimation was discussed in Chapter 7. The effect of using Euclidean distance in analysis of feeder system behavior and performance is to underestimate cost and voltage drop, as well as to lead to assumptions that are unrealistic about feeder routing and layout in general. The classical view of a feeder and all the computations based on that view assume that the feeder trunk heads away from the substation in a straight line in any direction needed (as shown in Figure 9.1), and that laterals can always branch at a ninety degree angle to the trunk). This is one reason why a few of the rules of thumb given in the Westinghouse "Green" Book (ibid.) are often inappropriate for actual distribution system layout.

At the heart of classical distribution engineering and planning is the concept of a feeder as a specific piece of equipment -- a set of lines assembled with 4/0 trunk, 2/0 branches, and #4 laterals, period. Shape and number of these units can be changed somewhat, but as the service area shape changes, their shape is changed, and as the load increases, more are added to increase capacity.

Recommended Perspective on Number of Feeders

A more realistic approach to feeder system planning is to realize that a "feeder" is a part of the distribution system tailored to the loads, routing, and needs of the system at its location. The definition that it is "all circuitry downstream from a protective device at the substation" identifies the portion of local distribution which must be analyzed as a unit (because it is electrically connected and therefore acts as a unit). However, planners have great discretion over the definition of where normally open and closed points are in a system, and thus they define what is identified as a feeder.

Figure 9.2 illustrates the definition of a "feeder" in actuality and underscore the fluidity of its definition as discussed elsewhere in this chapter. In Figure 9.2, two feeders are laid out with optimum economical conductor sizing and routing (left side). At the left, they are combined into "one feeder" at the substation using very large capacity conductor for the single common segment leaving the substation. Otherwise, they consist of identical elements except for the final hundred feet into the substation. *From electrical loading, voltage drop, power flow, power quality, and cost standpoints there is no significant difference between these two layouts.* Both configurations cost virtually the same amount of money, both initially and over time. Both have identical line equipment, open points, flow patterns, and voltages throughout, except at the substation. *The optimal circuit tree structure for getting power away from a substation into the surrounding area is not really effected by the decision of how many feed points to use to get power out of the substation.*

Reliability is the major difference

It is protection and reliability concerns that distinguish the two configurations shown in Figure 9.2. That on the left presents the customers it serves with expected service interruption exposure from only half as many miles of line as that on the right. Reliability, specifically frequency of service interruption, is the major difference in the two plans, and the consideration which can and should be used to decide if there will be one or two feeders.

Protection coordination is not a problem: each side of the "Siamese feeder" can be coordinated with the substation relay breaker as easily as either could have been coordinated individually. Combining the two feeders into one at the substation does means that whenever a fault cannot be cleared, a lock out of the entire feeder will effect twice as many customers. When siamesed, the configuration doubles every customer's exposure to such problems. However, such whole feeder lock outs are usually a small part of overall service interruptions, so this makes only a small incremental change-- usually only 10% to 15%. The real problem is momentary and instantaneous operations. Customers see twice as many with the feeders siamesed.

Deciding on the Number of Feeders

In laying out the feeder system, the only regard in choosing between Figure 9.2 A and B from the electrical standpoint is capacity limitation: if the load downstream of a point exceeds the available capacity of the conductor set, then that part of the system must be split into two paths -- which is done by starting with a definition of open points downstream of it, and then sizing of all segments. But, reliability considerations may require that the feeder be split into two or more portions in order to: a) provide adequately coordinated protection, and b) reduce the extent of service interruptions caused by any feeder lock-out sufficiently that the expected service interruption rate is within standards.

In a situation like that shown in Figure 9.2, the decision between building one or two feeders would generally be made early in the planning process, and would be based on evaluation of both reliability and economic guidelines. Figure 9.3, to be discussed below, shows the fundamental steps in feeder layout planning. As shown, determination of the number of feeders is in response to capacity and reliability constraints which are applied early in the process. Once that decision is made, and assuming that a good set of conductors are being used and that the line segments are subsequently sized appropriately (economically), layout of the system proceeds from there and will result in the same overall cost, performance, and probably very nearly the same layout except for a few key segments near the substation.

Multi-Feeder System Layout

9.3 PLANNING THE FEEDER SYSTEM

The multi-feeder planning process is only part of distribution planning, but a big part, for feeders are the building blocks of the distribution system. Generally, it is recommended that the planning of feeders be done in conjunction with the layout of the whole system, and as a step after the forecast and substation-level or "strategic" planning has been completed (see Chapter 12). This section discusses various aspects of feeder planning that will lead to better quality: more economical and reliable plans; a more orderly and quicker planning process; more credible and dependable procedures.

Table 9.1 identifies three "truths" about actual feeder system layout, which can be summarized as: the feeders will "reach everywhere" regardless of how they are laid out; branches and trunks are just larger versions of small lines; and open points are used to define service areas and loadings. Table 9.2 shows three constraints that effect what the distribution planner can do to lay out the feeder system: lines only have so much capacity; voltage drop must be considered; and the more customers served by a common circuit, the more will be out of service when that fails.

Finally, it is worth keeping in mind that the planner is juggling four goals in the layout of the distribution system (Table 9.3). Given the realities and constraints, the distribution feeder planning process and the layout of a feeder system boils down to something like the process shown in Figure 9.3, determining:

a) where open points will be put in the system,

b) which segments will be reinforced to act as branches and trunks, and

c) what part of the resulting "distribution grid" will be assigned to an individual protective device (fuse or breaker) at the substation.

The first two aspects of layout and planning are heavily interrelated, revolve around electrical performance, and have to be done jointly. The third aspect, which really boils down to answering the question "how many feeders are required?" is somewhat separate.

Figure 9.3 shows the major steps in planning the feeder system layout. The steps shown are simplified, and are actually interrelated -- for example steps 2 and 3 are somewhat mutually influencing -- exactly how and where the region is broken into feeder areas will depend somewhat on what routes are available for major branches, and vice versa.

Table 9.1 Three Key Observations about "Real" Feeder Systems

The feeder system is ubiquitous, reaching every service transformer with at least the smallest line type. Some type of line *must* run down nearly every street or block.

Branches and trunks, and all other application of "larger than smallest wire" are merely reinforcements of selected parts of the distribution grid so that power can be trans-shipped through it efficiently to reach required locations.

Open points (switches, double-dead ends, or missing segments in the grid) are "installed" where necessary to limit "load downstream" of a segment or location to a specific amount. A feeder's service area is defined and the load it serves determined by the open points in the grid downstream from it.

Table 9.2 Constraints That Define Feeder System Layout

Capacity is a limitation in all engineering applications. The lines in the conductor set only have so much capacity. For example, the 12.47 kV lines used in this chapter's case studies can move over 20 MVA (1,113 MCM), but are limited to 8.3 MVA as the upper end of their linear economical range.

Load reach is both a target and a constraint. For example, a 12.47 kV system built with economical conductor selection can move power up to about 3.3 miles while staying within range A voltage drop criteria. At this distance it achieves maximum economy as a system, not just as individual conductors.

Connectivity most be assured. All customers served by the same electrically-connected distribution assembly (a feeder) will experience service interruption if the assembly fails, or if the circuit "blinks" for any reason.

Table 9.3 Four Major Goals of Feeder System Layout

Economy should be maximized. The feeder system should be as inexpensive as possible while meeting other requirements.

Electrical performance must be satisfactory. The system must work within electrical and loading requirements.

Reliability (frequency of interruption) must be maintained within acceptable limits. The number of expected service interruptions should be kept low.

Contingency (duration) must also be contained. When an interruption occurs there should be a reasonable way to restore power in advance of repairs.

Multi-Feeder System Layout

1. Grid Must Reach Everywhere

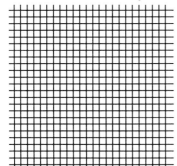

2. Capacity and Extent Constraints

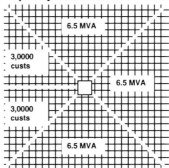

3. Identify Trunk and Branch Routes

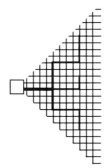

4. Size Segments & Radialize

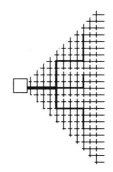

Figure 9.3 The fundamental feeder layout process begins by observing that the "distribution grid" must reach all load locations (step 1). Number of feeders is really determined by constraints of capacity (line rating) and extent (interruption spread) and must be applied on a multi-feeder basis, looking at all adjacent feeder areas (step 2). In the illustration here, presumably the constraints were to keep feeder load under 7 MVA (which defines three of the feeders), and the number of customers downstream from a breaker under 3,000 (which defined two others). Routes suitable for trunks and branches are identified in a step which can be done on a feeder-by-feeder basis (step 3). Then the actual feeder configuration is laid out, determining the open points and missing segments that define flow pattern and the line segment sizes that go along with them (step 4). Number of feeders has little to do with the routing or line sizing, but is an early part of the layout process and basically a response to capacity and reliability constraints, not engineering criteria or design goals.

Avoid "Feeder-at-a-time" Myopia

As stated in the introduction, the concept that a feeder is a single, immutable item and the resulting concentration on how to fix "it," is the single largest pitfall made by most distribution planners. The feeder *system* is the important entity -- feeders and feeder service areas are fluid -- they are easily changeable.

Coordinate Substation and Feeder Level Design Standards

Chapter 11 discusses the role of the feeder as a building block of the distribution system, and the improvement in economy, efficiency, and electrical performance that accrues from selecting voltage level, conductor economics, and layout styles to be compatible with the similar types of decision which must be made at the subtransmission and substation levels (e.g., how many transformers are used in substations, and what are the subtransmission voltages?). Good feeder planning should build on coordination of the entire subtransmission-substation-feeder level design standards and their equipment specifications and economies.

Always Start With A Good Economical Selection of Line Types

Chapter 7 discussed line segment economics, including economic selection of line type, design of an economical conductor set for laying out feeders, and the fact that in general all the line types in a conductor set will have roughly similar load reach, meaning that a feeder laid out by sizing line segments purely on economic principals will have a load reach equal to that of the conductor set from which it is built.

Economic selection of conductor in all cases, and the existence of a good conductor set designed to fit the needs of the distribution system, is assumed in all discussions of feeder layout and feeder system layout in this chapter (and in the this book for that matter). Good feeder planning starts by picking a good set of line types and sizes, so that economical linearity of cost is achieved over the entire permissible range from low load levels up to the point where high X/R ratios dictate non-linear costs with higher loadings (see Chapter 7). It means evaluating the requirements for and economics of load reach over the planning horizon being used (whether thirty years or thirty months) as outlined there and in this chapter. And it means adhering to the recommendations on line size and type that flow out of that analysis in all cases.

Multi-Feeder System Layout

Identify and Apply a Consistent Contingency Support Policy

Chapter 8 discussed both the degrees of contingency support which can be built into a feeder system and the approaches to providing that support. Planners should identify what level of contingency support strength will be part of their system (i.e., feeder-level, transformer level), as well as the general approaches they will use to obtain it (single zone, three zone, etc.) before beginning the planning process. There are two basic approaches to contingency policy determination and its application:

Contingency-scenario based planning. The traditional approach is to identify contingency support needs by designing the system to meet certain failure scenarios such as "the feeder is lost at the substation" by providing ways to pick up the load in every such case. A contingency support based planning would say something to the effect that "we always provide a way to pick up at feeder if it fails at the substation getaway, under peak conditions" and then apply this as a design constraint.

Reliability-index based planning. Newer, more effective, but more difficult than contingency-scenario planning are methods that apply contingency support planning selectively in proportion to need and results. This begins by picking reliability targets and then providing whatever contingency support capability is required to meet them. This is tantamount to saying something like "We design to provide an average frequency of 1 outage per year and duration of 2 hours per year on a feeder by feeder basis, and we provide whatever level of contingency backup -- multi- or single zone, feeder-level, transformer level, or substation level -- that is required to achieve this predicted level of reliability on each feeder."

This type of planning is very difficult without access to software and data that can compute reliability indices on a node or segment basis for any particular configuration.

"Backward-Trace" Layout Procedure Generally Produces the Best Results

Clearly, there is a powerful incentive for distribution planners to find a feeder configuration that has the shortest possible total length. Rarely (except in example problems in books like this) is this task easy to accomplish. Usually it takes study, and a bit of trial and error (even when using computer-aided design tools) to find a really effective configuration. The following four-step procedure

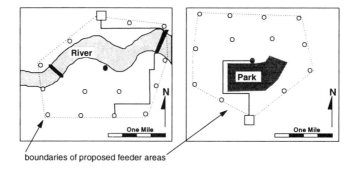

Figure 9.4 Backwards trace method of studying feeder configuration begins by identifying a set of load points for study (small circles). Included must be load points that are worst cases effected by any geographic constraints, such as a river (which can be crossed only at the bridges (dark lines) or a wilderness area/park (through which feeders are not permitted). Black dots indicate those points. A search for the a "total least distance feeder layout" starts by studying minimum length routes from these points back to the substation (solid line shown on each diagram).

works well in guiding planners toward finding the configuration with the least cost.

1. Identify all the geographic constraints -- areas that cannot be crossed, areas where construction must be underground, etc. These are areas to be avoided.
2. Identify all special opportunities -- diagonal routes, etc., which might contribute to lower cost.
3. Identify a set of load points on the periphery of the area to be served by the feeder, as well as a load point that is the "worst case" in terms of each constraint (Figure 9.4).
4. One by one for each of load points identified in step 3, *work backward from it toward the substation* (rather than the other way around) trying to find the shortest route(s) as shown by the solid lines in Figure 9.4.

This last step is the essence of the "backwards trace" method. Generally, it proves easiest to start with a particular load point and try to identify the shortest route(s) back to the substation, then do similarly for another, and then yet another load point, until a pattern or general trend among the routes starts to

Multi-Feeder System Layout

emerge, from which the planners can then build on to complete a configuration layout. Starting at the substation tends to blur the planners' vision -- they know that eventually they must go from that substation to many points -- and makes the job much more difficult. But by starting with individual load points and working back, one at a time, keeps focus on one part of the problem. As planners identify shortest backpaths from each point, they will begin to see commonalities among the pathes which can be used as major trunk and branch routes.

Try to Use Any Non-Rectangular Path That Is Close to the "Line of Sight" of the Needed Feeder Route(s)

One recommendation toward reducing total feeder length is to utilize any non-rectangular path through a grid of streets on property lines. Figure 9.5 illustrates an example. A diagonal street, ROW, or whatever through a street grid can give up to a 40% improvement in the length of feeder needed for elements of the system traveling along it. In laying out a feeder, planners should always look for such routes first, then try to work them into the plan, where possible.

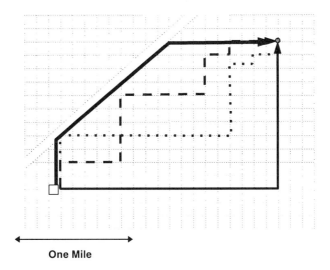

One Mile

Figure 9.5 The example rectangular street grid example from Figure 8.5, but with a multi-lane parkway (and available ROW alongside it) running diagonally through the area. The feeder route following it (broad line) saves 1/2 mile over any rectangularly routed line.

When there is such a right of way, it may change the planner's preference for large-trunk versus multi-branch layout. In the area shown in Figure 7.29, a large trunk design (whose one large trunk can fully take advantage of the diagonal route) would have an advantage over a multi-branch design (only one or two of whose branches could take an advantage of this special route) which would render it potentially more economical.

9.4 PLANNING FOR LOAD GROWTH

A good deal of distribution planning involves the planning and layout of distribution systems to accommodate future load growth. The cardinal rule of planning to accommodate such load growth is to plan for a way to handle long-term growth eventually. Distribution planners should take note of the words "plan," "way," and "eventually" in that statement -- distribution planning does not involve building initially to meet long-term needs. Instead, it involves finding ways to avoid spending money today, while leaving viable, economic options for tomorrow. The best way to accomplish this is:

Try to know what is going to happen. This means studying the possibilities of load growth, change, and shifts in conditions, and trying to forecast what is most likely. Planners should always keep in mind that any load forecast is probably wrong, at least to a small extent. But the fact that the future cannot be predicted perfectly is not an excuse for not forecasting at all.

Plan ahead. Often, planning for long term load growth means nothing more than leaving a way to add or make changes in the future. Spending anything today for long term growth should be avoided where possible. As mentioned Chapter 8, design flexibility, if identified early enough, is usually considerable and as a result it often costs nothing to choose a layout that leaves room for growth. The key is to look ahead, to identify what might happen and consider how that can fit into the plan in an orderly manner over the long term.

Load Forecast

The best preparation for load growth is to have a good forecast of: a) the load that is expected to develop, b) the load that might develop, but is uncertain, and c) an identification of the reasons for and conditions controlling that uncertainty. Good spatial forecasting can provide distribution planners with all three. Chapter 15 summarizes spatial load forecasting requirements, methods, and application. More detail on such methods can be found in available references (Engel et al., 1992; Willis, 1996).

Multi-Feeder System Layout 355

Recommended practice for distribution load forecasting is to apply some form of spatial forecasting, on a much larger-area basis than just individual feeder areas, as shown in Figure 9.6. Ideally, planners should start with spatial forecasts and perform a "top-down" strategic distribution plan that identifies how subtransmission and substation levels will handle the growth and looks at the feeder system as a whole, before focusing on individual feeders.

Spatial forecasting does not have to be a complicated or involved process -- although there are many cases where considerable effort is justified and quite exotic methods are most useful, often a simple assessment of possibilities and a basic forecasting method are all that is needed. The key step to is to do a forecast in some credible and documented manner. Often the greatest value added by such a spatial forecast will be in identifying the limits of possibilities -- i.e., essentially identifying scenarios for Multi-Scenario Planning -- what won't happen even if exactly what will happen is uncertain (Chapters 14 and 15).

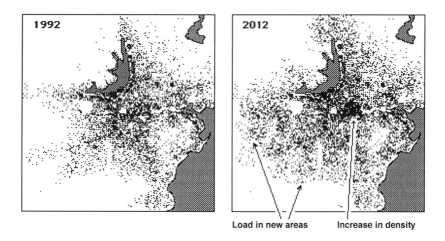

Figure 9.6 Maps of load (shading indicates amount of load) show the twenty-year projection for load growth in and around a large city. Load growth occurs both as new loads in previously vacant areas and as increases in the density of existing loads. Impact on feeder planning is different depending on which of the two cases occurs in an area.

Handling growth of existing loads in the feeder area

From a feeder planning perspective, there are two impacts of load growth, depending on the location of that growth relative to the feeder. Growth of the "original load" -- that native to the area already served by the feeder, requires that the feeder handle an increasing load over time. Growth in new areas, currently not served, requires additional sections or extensions of the feeder system to be built (Figure 9.6).

A general increase in most or all of the original load served by the feeder is best accommodated by including load growth in the economic conductor sizing and planning evaluation done when the feeder is originally planned. For example, if a forecast of future customer usage for a residential area projects a long-term growth in load that averages .5% annually, or if it projects an expected major new load in a few years, then the feeder planning should be done using

> *Conductor size based on loading tables adjusted for load growth.* The evaluation of present versus future cost must account for the increased future losses costs which load growth will produce (see Chapter 7, Figure 7.9, for a discussion of how load growth can be included in the conductor sizing economic evaluation).

> *Layout constrained by load reach for the final load level.* For example, .5% growth over twenty years results in a 10.5% growth in load, which makes a corresponding decrease in the load reach of feeder segments -- a feeder that could satisfy voltage drop constraints out to a distance X with original load Y will be able to do so only to a distance X/1.105 when the load has growth to 1.105Y.

Expanding to serve new spot loads in existing feeder areas

Often the load growth expected in the feeder area is not a general increase, but a specific amount at a certain site, such as in the planning case shown in Figure 9.7. Here, a new regional high school with a of 1.4 MW load is expected to be added about seven years after the construction of the feeder, which prior to that will serve a load of 7.0 MW. This 20% load increase of the feeder's peak load represents between a 20% and 100% increase in the line segments' loads along the feeder route leading to the new line (dotted line on left side of the figure). These segments should be sized using an economic evaluation of losses that shows a step function in their loads in year 7 and onward, because such large increases will make a big difference in the PW of losses on those segments. This will make only a modest change in the conductors size called for along the route, because the PW of the losses increase is discounted by being far into the future.

Multi-Feeder System Layout

However, this big of a load increase will make a noticeable increase in the voltage drop downstream of the new load (i.e., the increased loading will reduce the load reach of the feeder route shown by the dotted line). The slightly larger conductor called for on an economic basis cannot provide the required voltage drop performance. Voltage drop shown in the area circled by the dotted line on the left side of Figure 9.7 will drop below range A requirements when loads are at or near peak conditions. Thus, conductor along the route or elsewhere would have to be upgraded, more than is justified on the basis of PW economics alone, just to accommodate minimum voltage drop requirements.

Instead, the planners build the feeder initially as shown, but with a plan to convert it to the configuration shown at the right when and if the new load really develops. Only the small portion of feeder (dotted line) needs to be sized using conductor economic evaluation that shows a step jump in load after year ten. A new branch to serve the added load, and a few customers in its immediate vicinity, is planned to be built and cut into service as shown. The portion of the original feeder circled in Figure 9.7 is planned with economic conductor sizing evaluation of future PW losses costs that shows a slight *drop* in load after year ten (and thus tends to be specified as smaller than it would otherwise be).

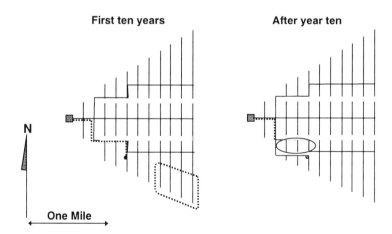

Figure 9.7 A 1.4 MW load (dark dot) is highly likely to develop ten years after this feeder is built (initial configuration on left). The most obvious plan for growth would be to plan the line segments along the feeder route leading to it (dotted line, left diagram). Instead, a lower cost option is as shown at the right. Dotted lines and areas circles are explained in the text.

Extension to Serve Load Growth in "New Areas"

As shown by the load maps in Figure 9.6, often load grows in previously vacant areas -- where there was no load, and no facilities, prior to the growth. The best way to plan for extension of a feeder into these areas of new load growth is to plan for such growth as part of a long-range strategic plan, in which feeder expansion is coordinated with the expansion planning and evaluation of options at the subtransmission, substation, *and* feeder levels, planning that balances their interaction and expansion costs to achieve the overall minimum cost. Assuming that such planning has been done, and a particular feeder has been identified as the one which will be expanded to pick up the load in a newly growing area, it is best to plan how it will be extended in advance.

Planning for load growth, and for new area growth in particular, often uses the approach of "splitting feeders" as shown in Figure 9.8. Most suburban feeder systems grow through a process of addition and reconfiguration that is a gradual, evolutionary change in configuration, involving splitting of feeders and branches as or more often than replacement and upgrading of conductor.

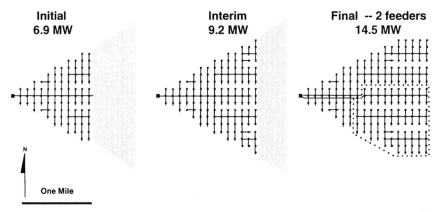

Figure 9.8 Extension of distribution into new areas (shaded) should be planned far in advance -- ideally when the original feeder is built. Here, what will eventually grow and be split into two feeders is initially built as one feeder, then expanded to where it serves a bit more than its optimum desired economic load for a brief period, then "split" into two feeders, as shown. The dotted line in the rightmost diagram encircles the area served by the second feeder created. Properly organized and planned, it is almost always possible to build toward the ultimate plan while never having to overbuild in the short run.

Multi-Feeder System Layout

Splitting Feeders

As an area's load grows, the feeders and branches serving it can be split, as shown in Figure 9.8. There are four reasons why "splitting" a feeder into two or more parts as the load grows is preferable to upgrading its conductor sizes in selected areas so that its capacity grows with the load.

"Feeder Economics." If the feeder was planned so it was a good investment prior to splitting, then it was probably planned so that its initial load was near the upper end of the economical linear range (see Section 8.2 -- particularly "Most feeders are the same size"). Continued growth will push load past the top end of the economical range, meaning there is no way to achieve really high levels of economy if it stays in a single-feeder configuration.

Upgrading an existing line costs more than building that larger line from new (see Chapter 9, Figure 6.23). Splitting a feeder, which means essentially building that capacity along another route, is less costly.

Load reach problems. Building a feeder so it can handle substantially larger amounts of load and service distances means that it will need oversized conductor purely to provide load-reach for future load growth.[3] This was the case in the example discussed in Figure 9.8 -- that planning situation was what one could term "splitting a branch" and it was done for all the same reasons as planners often want to split a feeder. Overseeing of conductor in order to meet load reach needs is not economical in the long run, and as costs in that period dominate PW cost, such oversizing is seldom justifiable.

Identify problems early. By planning in the long run to do so where possible, planners can foresee many of the situations where they will *not* be able to split feeders, and where reinforcement of feeders will be the only option. This allows them to make what provisions they can early on to reduce the eventual (high) cost of reinforcement.

The best way to achieve economy over time in the presence of load growth is to upgrade configuration, not segment capacity, of the feeder as the load grows.

---

[3] "Oversized" as used throughout this book means larger than whatever size conductor might be specified due to anticipation of growth and the higher losses costs it will bring. Losses economics alone will never specify that a conductor be upgraded for voltage drop constraint reasons.

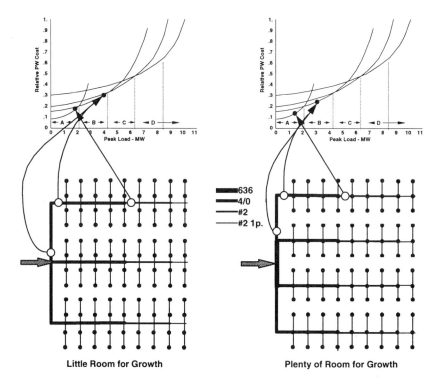

Figure 9.9 Artful selection of layout, particularly when using a multi-branch style, can result in a feeder where almost all segments are loaded economically but with loadings near the bottom or middle of their particular line-type's economical load range. At the left, when every segment of a three-branch layout to reach 108 is sized economically, critical points on the feeder are at or near the maximum loading for their conductor size's economical range (only points on one branch are indicated, but all branches are the same). If load growth occurs, this feeder will have to have some segments reinforced to larger conductor size.

In this case, a switch to a four-branch design (at the right) yields exactly the same total feeder length and virtually identical PW cost when all segments are sized economically, but results in trunk loadings that fall into the middle of the economical ranges. Thus, it has the same PW cost but capacity to handle expansion without upgrades or modification.

The great flexibility of layout discussed in Chapter 8 permits planners to search for, and often find, plans with identical PW costs that have differences like those illustrated here. Note: there is no hard and fast rule on how to find such plans. Increasing the number of branches does not always accomplish this "trick," but usually *some* such change does. This type of approach remains partly a "black art" and some planners are simply better at it than others.

Multi-Feeder System Layout

Risk Minimization for Uncertain Load Growth

Sometimes, great uncertainty exists about the forecast of future growth in an area. To minimize the cost impact of load growth uncertainty, planners can endeavor to lay out a feeder so that all segment loadings fall at the lower end of their economic ranges. There is as much art as science to this approach, but this "trick" can make a big difference at times, endowing a feeder with an almost magical robustness with respect to load growth -- economical if the load does grow; economical if it does.

In the examples given in Chapter 7, Table 7.2, a 12.47 kV three-phase line with 336 MCM phase conductor has an economical loading range of from 3.7 MW to 5.1 MW. Thus, a segment planned to have a peak load of 5.1 MW is economical, but has no room for load growth (beyond anything "built in" to the conductor sizing analysis). However, a 336 MCM segment planned with an original load of only 3.8 MW (the middle of the range) is as economical from a planning standpoint, and has room for 33% load growth before it is outside of its economical range for 336 MCM.

Given the very great flexibility of feeder layout as discussed in Chapter 8, it is usually possible to find a feeder configuration where a majority of line segments naturally fall into the lower part of their respective ranges. This can usually be accomplished, but requires artful arrangement of branches, lengths, and routes. The key is not to increase cost by changing configuration to achieve the desired "bottom of the economical range" loading characteristic, but rather to stay within the many minimum cost configuration- which normally exist. The author knows of no formal procedure to work toward this goal: It remains partly a "black art" and some planners are simply better at it than others. The additional planning effort required to do so is worthwhile in situations where the future load served by a feeder has uncertainty as to the eventual load growth. So arranged, the feeder will be economical whether the load grows or not, over a wide range -- cost will be minimum regardless of what is required.

Figure 9.9 shows a simple example, which illustrates both the concept and the way costs counterbalance as configuration is changed. A three-branch feeder layout to serve a total of 10.8 MW loads its branch trunks (4/0 conductor, 1.66 to 3.7 MW economical range) to 3.6 MW at peak, leaving less than 3% for growth within the economical range. A four-branch design results in branch loadings of only 2.7 MW, leaving 33% remaining capacity for growth. Total length of primary circuit remains unchanged -- while there is one more branch, there is less length of lateral. Capital cost of the four-branch design is slightly higher (the price planners must pay for load growth insensitivity in this case) but lower losses makes up for that in the long run and PW cost is identical, and the feeder can now handle consider growth with economy.

9.5 FORMULA TO ESTIMATE FEEDER SYSTEM COST AND ITS APPLICATION TO INDICATE DESIGN GOALS

The cost of a proposed feeder can be estimated by adding together the cost of all segments as individually estimated by the linearized cost model discussed in Chapter 7.

$$\text{feeder cost} = \sum (F_v (l_i \times d_i)) \tag{9.1}$$

where the summation is over all line segments, indexed by i
F_v is the function at voltage v that gives the cost approximation for l_i, the loading on segment i
and d_i, the length of segment i.

If all loadings are kept within the economical linear range -- definitely a desired goal -- then cost can be approximated even more simply. Since in this range

$$F_v (l_i \times d_i) = f_v \times d_i + s_v \times l_i \times d_i, \text{ for every segment, then}$$

$$\text{feeder cost} = \sum (f_v \times d_i) + s_v \times l_i \times d_i) \tag{9.2}$$

where the summation is over all line segments, indexed by i
f_v is the fixed cost at voltage v,
s_v is the slope of the cost curve v in the linear range
l_i is the loading on segment i
d_i is the length of i

This and Equation 9.1 both neglect any costs for the feeder due to reinforcement for contingency support.

Working with the fact that equations 9.1 and 9.2 are linear, and that distances are measured by the Lebesque metric, one can obtain the following equation which gives a very good estimate of the cost of the minimum-cost feeder that can be built

$$\text{feeder cost} = s_v \times \left(\sum (|x_s - x_j| + |y_s - y_j|) \times l_j \right) + f_v \times \left(\sum d_i \right) \tag{9.3}$$

where the first summation is over all loads, indexed by j
and the second is over all segments, indexed by i

Multi-Feeder System Layout 363

> f_v is the fixed cost at voltage v,
> s_v is the slope of the cost curve v in the linear range
> d_i is the length of segment i
> (x_s, y_s) is the location of the substation
> (x_j, y_j) is the location of the load, l_j, at point j

Equation 9.3 is an effective approximation: estimates are rarely off by more than 5% and usually are accurate to within 3%. Accuracy in ranking feeder costs (i.e., if evaluating which of two designs is less costly) is much better. If this equation is inexact to any significant degree, its estimates will be slightly low. It always computes a lower bound on the optimal cost of the feeder.

Reducing the Cost of Planned Feeders

Equation 9.3, while approximate, provides an interesting illumination into cost of feeders and ways that cost can possibly be reduced. Overall PW cost is the composite of two summations. The first summation is over all loads, and is called the *load service cost*. It is a function of the loads and their locations relative to the substation, and the term s_v, the linear cost from the conductor set evaluation. Essentially, this part of the equation predicts that PW cost of a distribution feeder will rise linearly as the load increases and as the Lesbeque metric distance between the loads and the substation increases. Neither is a surprising result, except perhaps that cost is linear with respect to both. I2R and Ohms Law not withstanding, given that a feeder is laid out well, total PW cost of distribution is linear with respect to distance and amount of power that must be served.

The second summation in Equation 9.3 is the *routing cost*. This is a summation over all line segments and is the product of the conductor set's y-intercept fixed cost, f_v (the cost to built the smallest type of line) times the total length of all lines built. It does not depend on the loading or types of lines built. As far as the equation is concerned, the value of this summation does not depend on the amounts or the locations of the loads.[4]

What Equation 9.3 indicates (assuming feeders are intelligently laid out) is that distribution feeder cost consists of two components: a *route cost* that is a function only of the total length of route taken by the feeders, not their sizes or the loads or their locations; and a *load service cost* that depends only on the loads and their distances from the substation, not the distances of feeder built of

---

[4] However, any acceptable feeder configuration must reach all the load points, so a bound on this part of the equation by that aspect of the loads is implicit.

the particular routes used within the feeder system to get power to any particular load.

The planners have control over none of the aspects of the loads -- those amounts and locations are set by the customers, not the planners. They have influence over only three aspects of this equation. First, they can try to reduce s_v, the slope of the conductor set's economical linear range (see Chapter 7). This is done by selecting a set of conductors well and making certain the lines and loads are as balanced as possible. Second, the planners can try to reduce f_v, the fixed cost (y-intercept), and third, they can work to minimize the total length of feeder that needs to be built through manipulation of the feeder layout.

Reducing f_v makes a big impact on reducing cost

The term f_v in Equation 9.3 is the y-axis intercept of the economical conductor set cost curve (see Figure 7.12). Equation 9.3 indicates that the cost of the feeder system is very dependent on this value -- which is essentially the capital cost of the smallest capacity primary feeder type in the conductor inventory. At least as far as Equation 9.3 is concerned, one way to reduce the total cost of building a distribution system a great deal is to reduce the cost of the smallest line that can be built. In fact, this is true and the reason is not entirely obvious.

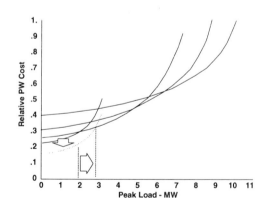

Figure 9.10 Reducing the y-intercept by cutting the cost of the smallest line type reduces the cost of a line type used most, resulting in a considerable direct savings. Beyond that, reducing the capital cost of the smallest line type moves its whole curve downward, moving the breakpoint with the next largest conductor outward, so that the smallest line type is used more often, too.

Multi-Feeder System Layout 365

The overall cost of a distribution system is much more sensitive to the cost per mile of the smallest line type than the largest. Part of the reason is that so much more of the small line type is built (this is discussed in detail in Chapter 11). However, equally important is that lowering the capital cost of the smallest type of line changes the economics of the conductor table itself, as shown in Figure 9.10.

Lowering the cost of the capital cost smallest line in the conductor set not only lowers the cost of many miles of a line type which must be built (which saves money), but shifts economics of line type selection so that even more of the system will be built with this smallest (and now even less expensive) line type (which saves even more).

This reduces overall PW cost, but it has a more dramatic positive impact on initial capital cost. For example, in Figure 9.10, reducing the fixed cost of the smallest line type shifts its breakpoint with the next size conductor (the load level where the next size should be built instead of it) from 1.9 MW to 2.8 MW. In that range of loads, using the smaller line type in lieu of the larger will mean increased losses over the long term, but the new, lower capital cost makes this preferable and economically the best choice Thus, reducing the capital cost of the smallest line type built has the following impact on the distribution budget:

- it reduces the capital cost of many small capacity lines that must be built.

- it trades future PW losses costs for a larger immediate savings by substituting a smaller line type for larger lines in many cases.

Finding ways to reduce the cost of larger line types will help bring down costs, too, but finding a way to reduce the capital cost of the smallest line type in the conductor set will have the biggest possible impact. (Strictly speaking, f_v and s_v are interrelated and contra-influencing, but from a practical standpoint this is not important). [5]

---

[5] Mathematically, reducing f_v must lead to a slight increase in the linear approximation slope, s_v (the reader may wish to refer to Figure 7.13), so the two goals -- reduce f_v and reduce s_v -- are slightly at odds. However, while reducing f_v leads to a slight increase in s_v, reducing it does lower overall cost, so it should be done, but only to reasonable levels.

A practical demonstration of this interaction is the following: s_v can always be reduced by increasing voltage level ($s_{25kv} < s_{12.47\ kV}$), while f_v can always be reduced by lowering voltage level ($f_{25kv} > f_{12.47\ kV}$). Thus, the two factors are interrelated (the proper way to balance them is discussed in the section on selecting primary voltage level in Chapter 11). However, once the primary voltage has been selected the focus should turn to reducing f_v as much as possible, and then minimizing s_v, given that f_v is minimized.

366   Chapter 9

Focus on minimizing the total *route length*

Equation 9.3 indicates that reducing the *total* length of lines in the overall feeder route is *all* that matters in reducing cost through manipulating routing. Approximation or not, what the equation indicates is true, except in very rare cases. Intuitively, it might seem that changing the design of a feeder so that the length of "big wire" needed would be reduced would be more important than reducing the length of "small wire" needed, but in fact that is not the case. The reason that there are constraints on what constitutes an acceptable answer to "what is the best feeder configuration?" which are: it must reach all the load points; and it must have sufficient capacity to serve them. Among configurations that meet those two criteria, and assuming loadings are kept within the economical linear range, all that really matters as a first step in designing a least-cost feeder is reducing the total length of line that is needed.

The two different feeder configurations -- multi-branch and large trunk -- discussed in Chapter 8 (Figure 8.12), demonstrate this fact. Both have the same total length of feeder (20.25 miles), and the same cost (capital $1.8 million, total PW over 30 years, $6.09 million), yet differ greatly in the number of miles of large line (large trunk, 2.0; multi-branch, 5.25) and small conductor (18.25 versus 15.0 miles). But the proportion of large and small conductor doesn't effect the final cost -- what a designer gives away in reducing one, he or she will equally make up somewhere in the other. Thus, while the large trunk layout has far less "really big wire" (only 2.0 miles of trunk), it has many long laterals, so many, all so long, that it needs a total of 10.5 miles of three phase #2 lateral. By contrast, the multi-branch layout has more trunk (5.25 miles), but it has no laterals with more than .25 MW load (the economic cut off for single phase #2 conductor size, and thus it needs no three phase laterals. There are other differences, but the point is that the sizing of lines all balances out so that in normal circumstances only the total length really controls the cost.[6]

Any combination of feeder segments that can reach all 162 load points in Figure 8.12 with less total length than the two feeders shown would work out to be less costly than either of those two plans.[7] Any feeder plan that calls for a greater total length will prove more expensive, regardless of whether or how it uses big, medium, and small conductor to get the job done.

---

[6] This assumes, of course, that after finding one of the (many) feeder plans with a "shortest possible total length" the planner follows up with good engineering to minimize the line size of each segment through careful application of economic sizing, etc.
[7] An impossibility, both configurations shown in that drawing have the minimum total length possible to reach all 162 load points, given feeders must be routed through a grid.

Multi-Feeder System Layout

Under only two conditions is this rule not true. First, whenever a feeder must be designed to serve a total load that is far greater than the maximum load in the economical linear range of the conductor set. Then, there may be an economy of scale in accepting a plan with more total distance but which splits routes (usually in such cases a multi-branch feeder layout, with more total distance but which splits the load onto two routes right out of the substation, will prove less costly than a large trunk plan that carries the whole load). The second situation is when there are severe geographical conditions or constraints that limit feeder routing (and usually in these cases, the rule holds).

9.6 VOLT-VAR CONTROL AND CORRECTION

Power factor correction and voltage regulation are important elements in the achievement of high efficiency in a power distribution system. Many aspects of capacitor and regulator application, particularly details of exact placement, sizing, switching, and protection, are part of the engineering function, not planning. This section, which discusses power factor, power factor correction, and voltage regulator application, could fit in many places in this book. It has been included in this chapter to emphasize that *while volt-VAR control is implemented on an engineering basis, its application is a strategic factor* to be included in the planning of the power system and to be specified on an overall basis in the distribution plan.

Power factor is the ratio of the effective to the total current on a line, the difference due to lagging (or leading) current phase with respect to voltage phase. Figure 9.11 illustrates a typical situation. Due to VARs causing an 80% power factor, a feeder with a 6 MVA capacity limit can deliver only 4.8 MW.

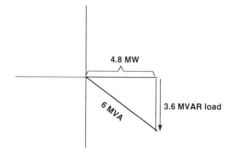

Figure 9.11 A line carries 6,000 kVA of power at 80% power factor, delivering 4,800 kW of effective power along with 3,600 in kVAR flow.

Why VAR Flow Occurs

The phase of an alternating current usually lags the phase of the alternating voltage, particularly through any "wound" device or other apparatus that depends on magnetic field interactions for its operation (e.g., transformers, motors). The concept of current "inertia," while theoretically unjustifiable, works well as an intuitive guide to VAR flow; current takes a while to move through any device that creates a magnetic field, even as the voltage (pressure) acts instantaneously. Thus, current lags behind the voltage rise and fall.

Actually, a type of inertia is involved in the magnetic fields which link the voltage and current flow inside devices such as transformers, reactors, and motors. Time is required to disturb the magnetic domains in core materials and magnetically switch them back and forth to create the alternating fields. These magnetic domains, though small, are physical entities and are immensely large (orders and orders of magnitude) compared to the quantum scale. Their "activation time" is often measurable in milliseconds, and being that a power cycle takes slightly less than 17 milliseconds, these magnetic transition times have a decided effect, separating current from voltage.

If a purely quantum-level magnetic material can ever be created, one in which magnetic fields are established only due to quantum effects, equipment such as motors and solenoids could potentially be built so they would cause no reactive load. (Such material, however, would be worthless for some applications, such as reactors, etc., where a reactive load is desired.)

Effect of VAR Flow on Feeder System Economics

Excessive VAR flow on a feeder "uses up" its capacity and increases both the voltage drop and percentage losses. For example, 4/0 ACSR overhead conductor has a nominal thermal capacity (Westinghouse "Blue" Book) of 340 amps. Therefore, a 12.47 kV feeder built with this conductor can deliver up to 7.34 MW at 100% power factor, but only 6.6 MW at 90% power factor, and only 5.14 MW if the power factor slips to 70%, (at which point reactive and real current flows are virtually identical). In all three cases, the electrical losses on the feeder are identical, equal to 340 amps squared times the line's resistance of .591 ohms per mile, or 68 kW/mile. However, since the amount of power delivered drops as power factor worsens, the percentage losses increase as power factor worsens. At 100% power factor, losses amount to only .92% per mile, but at 70% power factor, they have grown to 1.3% per mile.

The relative impact of power factor on capacity and losses is independent of line size or type. A line built with the much larger 636 MCM ACSR conductor instead of 4/0 can carry up to 770 amps, at which it is moving 16.6 MW with 96

Multi-Feeder System Layout

kW per mile of losses (.6% per mile) if power factor is 100%. Worsening power factor degrades the amount of power the line can carry, and increases the percentage losses on this line, in exactly the same proportions as it does for the smaller conductor.

Power Factor and X/R Ratio

While power factor's impact on capacity and losses does not vary as a function of line size, its impact on voltage drop and load reach does, because it depends greatly on the conductor's X/R ratio. The 4/0 circuit described above has an X/R ratio of 1.34 (Z = .590 + j.79 ohms per mile). At its thermal current limit of 340 amps, and at 100% PF, this circuit creates a voltage drop of 2.7% per mile, for a thermal load reach of 2.8 miles. Voltage drop increases to 4.1% per mile at 90% power factor (1.8 mile reach) and 4.6% per mile at 70% power factor (1.6 mile reach). Load reach (thermal or economic) drops by nearly a factor of two as power factor worsens from 100% to 70%.

A similar shift in power factor would degrade the performance of a larger conductor line much more. If built with 636 MCM conductor, this same line would have an X/R ratio of nearly 4.0 (Z = .16 + j.62 ohms per mile). At its thermal limit of 770 amps and at 100% power factor, it creates a voltage drop of only 1.7% per mile, for a potential load reach of 4.3 miles. But if power factor slips to only 90%, voltage drop more than doubles -- to 4.45% per mile -- and thermal load reach drops to only 1.6 miles, the same reach the smaller line reached only when power factor slipped to 70%. By the time power factor reaches 70% on this conductor, voltage drop is 6% per mile, and thermal load reach is only 1.25 miles. Load reach drops by a factor of four as power factor worsens from 100% to 70%. The larger line, with a relatively high X/R ratio, is twice as sensitive to shifts in power factor.

Shunt Capacitor Application

Shunt capacitors inject a reactive component of current flow into the circuit at their location (Figure 9.12). Given that the poor power factor is due to lagging current (it usually is) and assuming the capacitor is appropriately sized, this will reduce VAR flow, and consequently improve voltage, losses, and line capability. Figure 9.12 also illustrates an often overloaded detail of capacitor application: *Capacitors are impedance devices.* The 1000 kVAR bank in this example will inject 1,000 kVAR only if voltage across it is 1.0 PU, but due to voltage drop to this site, voltage is less and its injection is less. Such differences in injected versus rated capacity of capacitors are often important in detailed computations of their application or in decisions about exactly where to locate them.

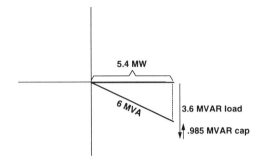

Figure 9.12 A capacitor bank injecting 985 kVAR has been added to the situation originally depicted in Figure 9.11, raising the power factor from 80% to 90%. This increases the real power that can be delivered to 5,400 kW, a 12.5% increase.

Decreasing marginal effectiveness of shunt capacitors as power factor improves

Only 985 kVAR was required to correct the power factor from 80% to 90% in Figures 9.11 and 9.12, effectively "buying" planners 600 kW of additional capacity on this line segment, along with corresponding improvements in losses and voltage drop. Doubling that injection to 1,970 kVAR would improve power factor to 96%, and would "buy" only another 375 kW of capacity. Tripling it, to 2955 kVAR, provides only a further 185 kW capacity increase. Due to the trigonometric relationships involved, capacitors are less effective at making improvements when power factor is already fairly good.

Capacitor Action

The impact of a capacitor on a feeder's VAR flow can be analyzed by any of several means. A very useful way, at least for conceptual study, is depicted graphically in Figure 9.13. Shown at the left in that figure is the VAR flow along a three-mile feeder with an even VAR loading of 2 MVAR/mile, for a total impact of 9 MVAR-miles (total shaded area). At the right, a 3,000 kVAR capacitor bank located one and one half miles from the substation corrects power factor at its location to 100%. Note that:

- VAR flow upstream of the capacitor is reduced by 3,000 kVAR.
- VAR flow downstream of its location is not affected at all.
- MVAR-miles are cut in half, to 4.5 MVAR-miles.

Multi-Feeder System Layout

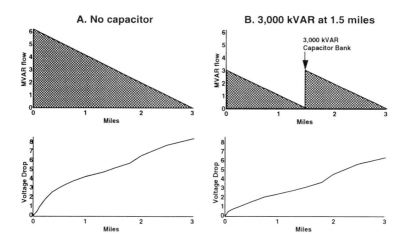

Figure 9.13 (A) A three-mile long feeder trunk has a power factor of 70.7% (real and reactive loads are equal) both evenly distributed at 2,000 kW and 2000 kVAR/mile. VAR load is shown at the top, and voltage drop profile at the bottom (the uneven slope is due to different size conductors). VAR-miles total nine. (B) A 3,000 kVAR bank located half way out the feeder corrects power factor at its location to 100% and power factor as seen at the substation to 89%. MVAR-miles are cut in half, and voltage drop at the end of the feeder improves by nearly two volts.

Estimating the optimal location and size for capacitors

It is the cumulative impact of VAR flow along a feeder (MVAR-miles) that is mostly responsible for degrading electrical performance. The graphical method depicted in Figure 9.13 measures this directly, and is thus useful for estimating how moving the capacitor bank to other locations might improve a feeder's electrical performance, as illustrated in Figure 9.14. Analysis of incremental changes in location can be used as a guide in determining an "optimal location" that minimizes MVAR-miles on the feeder. As shown in Figure 9.14, the best location for a 3,000 kVAR capacitor bank in this example is three-quarters of the way out the feeder. Similarly, analysis of incremental changes in size can help identify the best size for a capacitor at any particular site, as shown in Figure 9.15. Simultaneously, varying both the size and location identifies their best combination, in this case, a bank equal to 4,000 kVAR (two-thirds of the total VAR load) located two miles from the substation (two-thirds of the way from the substation to the end of the feeder -- the *two-thirds rule*.

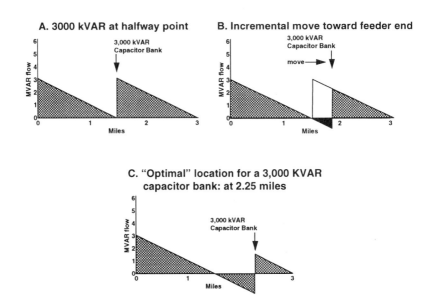

Figure 9.14 A shift in location from the halfway point (A) farther out the feeder (B) adds the MVAR-miles shown as darkly shaded, and removes the MVAR-miles shown unshaded. The net result is a reduction in overall MVAR-miles, and an improvement. The 3,000 kVAR capacitor in this analysis can be "moved" in small increments until the marginal gain and loss of any further movement are equal, at which point it is at the best location, in this case 2.25 miles from the substation (three-quarters of the way toward the end), as shown in (C).

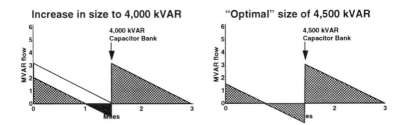

Figure 9.15 Similarly, the "optimal size" for a capacitor at the halfway point on the feeder can be found by comparing the MVAR-mile areas gained (dark) and given up (no shading) as the capacitor bank size is changed. The indicated best size for a capacitor at the halfway point is three-quarters of the VAR load, or 4,500 kVAR.

Multi-Feeder System Layout

The Two-Thirds Rule for Capacitor Application

A traditional rule-of-thumb for capacitor application to a feeder is to "place a capacitor equal to two-thirds the VAR load of the feeder at a point two-thirds of the distance from the substation." The graphical method of capacitor impact analysis, or an algebraic equivalent, can be used to confirm that this minimizes the total MVAR-miles, given that the VAR loading is uniform.[8] The left side of Figure 9.16 shows the MVAR profile for this application, which reduces MVAR-miles on the feeder to one-third of their uncorrected total.

Multiple capacitor banks and the general form of the two-thirds *rule*

The impact of two capacitor banks can be similarly represented graphically, as shown in Figure 9.16. Inspection of such graphs, or algebraic manipulation that accomplishes the same purpose, can establish that for a feeder uniformly loaded with a VAR load of Q, the optimal sizing and locations for N banks is

$$\text{Size of each of N banks} = 2/(2N+1) \times Q \qquad (9.4)$$

$$\text{Locations} = n \times 2/(2N+1) \times L \text{ from the substation, for } n = 1, N \qquad (9.5)$$

As a result, the MVAR-miles on the feeder are reduced from $(Q \times L)/2$ to

$$\text{Total MVAR miles} = 1/(2N+1) \times (Q \times L)/2 \qquad (9.6)$$

Thus, the best two-capacitor solution is two equally sized banks of 2/5 the VAR load of the feeder, located 2/5 and 4/5 of the way from the substation, which reduces MVAR-miles to 1/5 of their previous level. The best three-capacitor solution is banks of 2/7 the VAR load, at 2/7, 4/7, and 6/7 of the length of the feeder, reducing MVAR-miles to 1/7 of their uncorrected level, etc.

Note that correct *sizing* of N capacitor banks calls for a total VAR capacity equal to $2N/(2N+1)$ times the feeder's VAR load, thereby reducing the feeder's total VAR load to $1/(2N+1)$ of its uncorrected value. The correct *locations* for these capacitor banks will allow this amount of correction to make a similar reduction in feeder impact, but can do no better than an equivalent amount of reduction (i.e., optimal placement will reduce the MVAR-miles also by a factor of $2N/(2N+1)$, to $1/(2N+1)$ of their uncorrected total).

---

[8] See for example, *Fundamentals of Power Distribution Engineering* by J. J. Burke, Marcel Dekker, 1993, pages 93 through 95, for a basic algebraic derivation of a single-capacitor case.

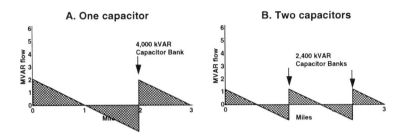

Figure 9.16 The generalized "two-thirds" rule for capacitor sizing and placement. (A) One capacitor, equal to 2/3 of the VAR load, located at 2/3 of the feeder length. (B) Two capacitors, each 2/5 of the VAR load, at 2/5 and 4/5 length, respectively.

Capacitor Application When Loading Is Uneven

The two-thirds rule applies only in situations in which the VAR load on a feeder is continuous and uniformly distributed. Most feeders have so many loads scattered throughout, that they can be modeled as having a continuous distribution of VAR load. However, most feeders do not have an evenly distributed VAR loading, and so the two-thirds rule is not completely applicable. While the two-thirds rule does not apply to these cases, the graphical method (or its algebraic equivalent) which applies the concept of MVAR-mile minimization, can be used to develop guidelines for typical situations of uneven loading.

Typical feeder

If any one feeder configuration can be used as "typical," it is most likely the large-trunk design serving a triangular service area, as discussed in Chapter 8 and depicted in the bottom of Figure 8.12. As shown in Figure 9.17, uniform area loading (a uniform VAR loading *per length of lateral*) results in an uneven distribution of VAR load on the feeder trunk, with a majority toward the end of the trunk. The capacitor application to minimize MVAR-miles in this case is a bank equal to 7/8 the VAR load of the feeder, located 3/4 of the distance out from the substation. This reduces MVAR-miles to 2/9 of their uncorrected total. This rule *is* generalizable to all trunks with "triangular" load distributions.

This rule minimizes only MVAR-mile flow on the feeder trunk. No consideration of any minimization of VARs for the laterals is given in this example, for two reasons: a) they are on laterals not influenced to any great degree by the capacitor, and b) they are very small lines whose impedance is mostly resistive so they are not overly sensitive to VAR flow.

Multi-Feeder System Layout

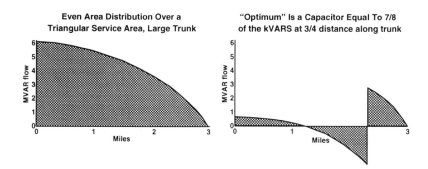

Figure 9.17 A large trunk feeder serving a triangular service has a VAR flow along the feeder trunk like that shown at the left. A capacitor equal to .87% (7/8) of the feeder's VAR load, situated 75% (3/4) of the way toward the end of the feeder is the best single-capacitor solution to reduce the MVAR-miles.

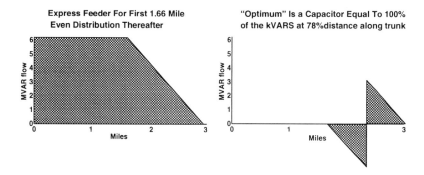

Figure 9.18 Many feeders have an "express" portion, a certain length out from the substation on which no load is placed. None of the load (and VAR load) will be on this portion. All will be on the end portion. Left, VAR flow on a three-mile feeder trunk when the first 1.66 miles is express and a 6 MVAR load is spread evenly over the final 1.33 miles. Among single-capacitor applications, a bank equal to 100% of the feeder's VAR load, located 77% (7/9) percent of the way out the feeder, reduces MVAR-miles to 1/7 their uncorrected value.

Express feeder

Often a feeder is deliberately designed with no load on the portion of the feeder trunk for the first mile or so out from the substation. This results in a very uneven loading for the feeder trunk, as shown in Figure 9.18. While the best size and location for a capacitor depend on the VAR load, length of express trunk, etc., if the express portion of the feeder represents a significant portion of its length, then the optimum capacitor location is usually at the half-way point of the remaining (non-express, loading) portion of the feeder. Note that in Figure 9.17, the capacitor is located in the middle of the 1.33 mile portion of line which has load. Situations with express trunks of one, one and one half, and two miles also conform to this rule. Generally, if the express portion is greater than one-third the length of the feeder, then the optimum capacitor bank size is usually 100% of the feeder VAR load.

Generally: More *than two-thirds the kVAR located at* more *than two-thirds distance*

Due to their area geometries and typical layouts, most feeders have a VAR loading biased toward the end of their trunks and branches (see Chapter 8, particularly Figure 8.2). As shown by the two unevenly loaded examples above, MVAR-miles in many actual situations are minimized by applying more kVAR capacity than the two-thirds rule recommends, at a distance farther away from the substation than that rule recommends.

General Guidelines Based on the Two-Thirds Rule

The MVAR-miles minimization method used above is basically a more flexible application of the two-thirds rule, which can accommodate uneven VAR loadings. Thus, the following guidelines can be thought of as corollaries to the two-thirds rule, applicable to situations distribution planners are more likely to face.

On typical feeders (large trunk, triangular service area), the best single capacitor solution is a bank size equal to 7/8 of the feeder VAR load, located 3/4 of the way out the feeder. The best two-capacitor solution is 45% of the VAR load at .53 the length and 50% load at .90 the length of the feeder.

In cases where an express feeder trunk is used, the best single-capacitor bank application is usually a bank size equal to the VAR load of the feeder, located at the halfway point of the load. The best two-capacitor solution is banks

Multi-Feeder System Layout

equal to half the VAR load at the 1/4th and 3/4th points on the loaded portion's length.

Correct any large VAR load at its location. In most cases, a large or special load will create a very large VAR load at a one point on the circuit. Analysis using the graphical method (or any other technique) will shown that the best strategy for minimizing its impact is to install capacitor equal to its VAR load at its location (i.e., cancel the VARS at their source).

Practical Guidelines on Capacitor Application

Application to feeders with branches

Most feeders have several major branches. The graphic method can be used to picture the VAR loading on such feeders and to study the impact of various combinations of application, as shown in Figure 9.19. However, the situation with almost any branched feeder is complicated to the extent that simple analysis of the type described here is not likely to yield a completely satisfactory result. Generalizations about results are not possible, although use of the 2/3 rule on each branch of a large feeder is a feasible way to begin manual application studies.

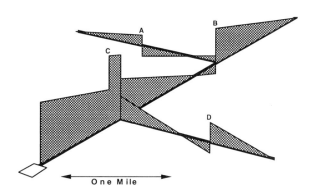

Figure 9.19 A three-dimensional perspective of the MVAR-miles impact on a feeder with two branches and capacitors at locations A, B, C, and D. Vertical axis is VAR load. Although the graphic method shown here is useful for picturing VAR flow on a branched feeder, most branched-feeder situations are too complicated to permit application of simple rules for its use as a planning tool, as illustrated earlier for single-trunk layouts.

Capacitors: never on a lateral

Capacitors are rarely installed on laterals, for four reasons. First, the load (VAR or otherwise) on most laterals is usually quite small, far below the minimum size of a capacitor bank. Thus, installing them on a lateral almost always grossly "overcorrects" power factor locally. While such capacitor placement may still improve overall feeder power factor, it is most often decidedly non-optimal. Second, laterals are most often single-phase. While capacitors can be utilized as separate phase installations, they are less expensive and their impact on imbalance is easier to anticipate and control if they are installed as three-phase units on three-phase portions of the system. Third, capacitors should be installed only where they can be reached with bucket trucks, etc. to perform maintenance. Many lateral locations are impossible to reach with such equipment. Finally, laterals usually have low X/R ratios. Reactive flow does not impact them relatively as much, and economics are not as much in favor of their utilization.

Incremental sizes and maximum sizes

Capacitors are usually available only in standard unit sizes (e.g., 100 kVAR/phase units at 12.47 kV), and there is a limit to the size of bank that can be installed at any one location -- typically no more than five to six units per phase. In addition, fault duty of capacitors must be considered. Large banks may have intolerably high fault currents. These practical realities often fashion the "optimal" solution to a capacitor application problem. In the example used here,

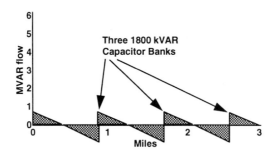

Figure 9.20 The best practical application to the evenly loaded trunk example (Figure 9.13, top left) is three capacitors. Any fewer and the required amount of kVAR/bank exceeds the maximum size permitted. They are located as specified by the two-thirds rule at 2/7, 4/7, and 6/7 of the feeder's length.

Multi-Feeder System Layout

the best practical application, based on the two-thirds rule, is to use three capacitors of 1800 kVAR (slightly larger than the rule's specification of 1714 kVAR each) located as specified by the rule at 2/7, 4/7, and 6/7 of the feeder length. Figure 9.20 shows this example, which will be the basis of the capacitor switching examples used later in this section.

How Good Is the Two-Third's Rule?

Like any rule of thumb, the two thirds rule and its various corollaries have limitations, and there are frequent exceptions where they are not applicable. However, as an overall guide for capacitor utilization, they are as dependable as any simply guidelines and superior to any other simple and easy-to-apply capacitor guidelines.

The most important point about the two-thirds rule, and the graphical-algebraic interpretation of it described in this section, is that it addresses the real issue at hand, which is *minimizing the impact of VAR flow on the feeder*, and not just correcting power factor. As was illustrated by Figures 9.14 and 9.15, the method tries to reduce the MVAR-miles to a minimum. Given a uniform VAR loading along the feeder trunk, this will result in a series of similarly sized and shaped triangular areas -- three in the case of a single capacitor, five in the case of two capacitors, seven for three capacitors, and so forth. (For non-uniform distributions, not such simple geometry exists.)

Table 9.4 shows how various "two-thirds" rule applications reduce MVAR-miles on both evenly loaded and "typical" feeders. The two-thirds rule and its variations also correct overall power factor (as seen at the substation) to a very high level in nearly all circumstances (Table 9.5).

Table 9.4 Expected Reduction in MVAR-mile Flows from Application of the Two-Thirds Rule as a Function of Number of Capacitors

Number of Capacitors	Percent Reduction in MVAR-mile Flow on Trunk of	
	An Evenly Loaded Trunk	A Typical Feeder
1	66	77
2	80	87
3	86	93
4	89	95
5	91	96
6	93	97

Table 9.5 Corrected Power Factor at the Substation After Application of the Two-Thirds Rule as a Function of Uncorrected Power Factor

Uncorrected Power Factor	Power Factor at the Substation After Application	
	With One Capacitor	With Two Capacitors
90	99	100
80	97	99
70	95	98
60	91	97
50	87	94
40	80	91

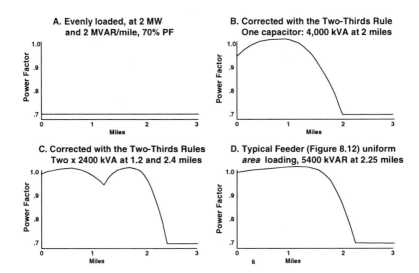

Figure 9.21 (A) Power factor profile for a distribution of 2 MW and 2 MVAR/mile along a three-mile feeder whose MVAR-miles were plotted in Figure 9.13 A. (B) The same after correction with one capacitor according to the two third's rule as shown in Figure 9.16 A. (C) The same after correction with two capacitors according to the two thirds rule as shown in Figure 9.16 B. (D) Profile of the large-trunk, "typical" feeder Figure 9.17 after correction.

Multi-Feeder System Layout

Shortcomings of the two-thirds rule

The graphical MVAR-mile minimization method used in the examples above is a very useful mechanism for illustrating the basics of VAR-capacitor interaction, and for deriving "first order" approximate guidelines, such as the two-thirds rule, for capacitor application. A number of important factors are not considered, however:

> *Complex power flow.* Actual power flow is complex. The MVA-mile analysis deals with only one dimension -- VARS -- without recognizing that its impact, or importance, is somewhat a function of the real power flow, too. Looking at the MVAR-miles of the corrected single-capacitor case (Figure 9.16, A), the rule has treated those MVAR-miles near the substation as just as important as those far from the substation (specifically for this example, with those in the farthest third of the feeder).
>
> In fact, in many situations, it is slightly less important to minimize those near the substation, because (in this example) they are being combined with three times as much real power, and the resulting trigonometric relationships mean the VARS contribute less impact to worsening flow. (The power factor profile in Figure 9.21 B illustrates the difference -- 95% PF at the substation and 70% at 2+ miles.)
>
> *Economics.* The value of VAR reduction depends on the cost of losses and the need for the additional capacity and reach "released" by the improvement in power factor. Capacitor application ought to be based on economic benefit versus cost analysis.
>
> *Line impedance.* Both the response of a feeder to changes in VAR flow, and the importance of reducing VAR flow vary depending on the impedance of the various line segments, whereas the approximate method essentially treats all portions of the feeder as equivalently important.
>
> *Discontinuous Load.* Actual kW and kVAR load on a feeder is discontinuous, whereas in all the representations given here, it is modeled as continuous.

Detailed analysis of capacitor interaction for each specific feeder, taking into account all of the above, is necessary to optimize capacitor application. Software to accomplish this task reliably and practically is available. It can often improve upon the economic savings rendered by the two-thirds rule and its derivative by useful margins -- 10 to 15%.

Power factor profiles

The example power factor profiles shown in Figure 9.21 demonstrate an important point; even feeders with good power factor correction at the substation have portions that are not corrected well. Thus, while power factor is corrected to better than 90% on a majority of the three corrected cases shown, there are places, particularly at the feeder's end, where power factor is much lower.

Power factor can be corrected to an average 90%

Either of the two lower examples in Figure 9.21 has a power factor, averaged over its length, of about 90%. Application of the two-thirds rule results in correction of to an average near 90%. This is the basis for the assumption used in Chapter 7's computation of conductor economics, that power factor is 90%.

Switched Capacitors

Shunt capacitor banks may be static (unswitched), in which case they are in service all the time, or switched, in which case they are cut in or out of service as needed by any of several means of switch control. There are two reasons for switching capacitors, both related to the fact that load varies over time, and thus the needed kVAR correction changes.

Voltage rise above permitted limits

In some cases, when the load is quite low, shunt capacitors can boost voltage above permitted levels. In such cases, they must be switched off when the load is low. The voltage boost (120 volt scale) at the end of a feeder, due to a capacitor, can be estimated as

$$\text{Voltage rise (120 volt scale)} = .12(\text{CkVA} \times \text{X})/(\text{kV}^2) \tag{9.7}$$

Where X is the line reactance to the capacitor location and CkVA is the capacitor's capacity. For example, 4,000 kVAR at three miles on a 12.47 kV feeder with X = .68/mile, would boost voltage about 6.31 volts.[9] At very light load conditions, voltage drop to the feeder's end might be only a volt, meaning that if the substation is maintained at 125 volts, the feeder's end would reach over 130 volts. Under light load conditions, this bank must be switched off.

---

[9] Normally, capacitor banks are not this large, something that is discussed elsewhere in this section, but this example illustrates the principle.

Multi-Feeder System Layout

Power factor correction, fixed capacitors, and load variations

But there are additional economic and electrical benefits that accrue from operating capacitors only during peak VAR periods, even if during off-peak periods voltages do not rise to unduly high levels due to static installation. The capacitor size needed to correct power factor during peak conditions may seriously overcompensate during off-peak conditions, increaseing losses.

Figure 9.22 shows real and reactive loads for a residential/commercial feeder in a city along the Gulf of Mexico in the southern United States. During summer peak, power factor drops to 68%, when peak reactive load (6.01 MVAR at 2 PM) actually exceeds real load (5.89 MW) for a brief period. While extreme, this situation is not unusual: Power factors approaching 70% occur on many feeders in the southern US during summer, due to the high reactive loads of air conditioning induction motors, which are at maximum mechanical load during very hot, humid periods, and operating at marginally low voltages during peak. Off season, both real and reactive loads are far below their summer values, but VAR load lessens more, so that power factor is never less than 80%.

Suppose that the example feeder used earlier in this chapter had its uniform distribution of real and reactive loads following the variations shown in Figure 9.22's load curves. Figure 9.23 shows the VAR flow on the feeder that will result at the time of minimum annual load, from the three capacitors specified by the two-thirds rule applied to peak conditions (Figure 9.20). The (leading) power factor at the substation is 50%, and as poor as 25% near the end of the feeder.

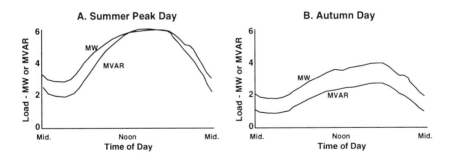

Figure 9.22 (A) Load curves measured at the head of a feeder serving 1350 large, single-family residences in an older area of a city on the US gulf coast, on a peak day when temperature reached 102 degrees F and humidity averaged 98%. (B) Load curves for the same feeder during an off-season day, with far less air conditioning load.

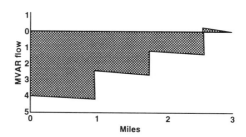

Figure 9.23 The feeder/capacitor combination from Figure 9.20, during annual off-peak hour (early morning in Figure 9.22 B). Due to severe overcorrection of VAR flow, power factor at the substation is only 50% (leading) and is as poor as 25% (at the capacitor location farthest from the substation).

Determining if capacitors should be switched

The situation shown in Figure 9.23 must be kept in perspective in order to understand *why* the distribution planners may wish to use switched capacitors. Unless there is some special aspect or constraint to this particular feeder not included in this example, this feeder *will* operate within loading and voltage specifications with this three-capacitor configuration operating during off-peak conditions. Even during the annual minimum, with the gross over-correction of power factor made by the three capacitors, the MVAR-miles are reduced from peak conditions (from 9 MVAR-miles to slightly less than 7), and because the load itself is greatly reduced, the magnitude of current flow during off-peak situations is much less. Thus, despite the poor power factor, current flow and voltage drop under these conditions will be less than at peak: Thus if the feeder can serve the peak conditions within loading and voltage drop specifications, it can do so at this time, too.

But the *power factor itself might be objectionable*. Near the end of the feeder, the capacitors have shifted power factor to .25 leading, a phase shift between current and voltage so great that it is likely that a good portion of electronic power supplies will not operate properly, and many other apparatus including motors and other devices will find the leading power factor not entirely to their liking.

But forgetting that objection to leaving the capacitors as a static installations, the planners in this particular case should focus on an important point: the annual cost of losses with these capacitors left in place (static) are higher than annual

Multi-Feeder System Layout

losses' costs when no capacitors are installed.[10] This static capacitor "solution" may solve intolerable operating problems during peak conditions, but its cost is an increase in annual losses' costs.

In this case, one or more of the capacitors should be switched, both to avoid the dramatically low power factors caused during off-peak periods, and to lower losses' costs.

Manual, automatic, or automated switching?

Capacitors can be switched in one of three ways: manually, automatically, or as part of an automated distribution system.

Manual switching. One option for switching the capacitor in the example above is for planners to call for it to be manually cut in and out of service on a seasonal basis. This requires a service crew to visit the capacitor site twice a year to cut the capacitor's disconnects in and out of service with a hot-stick, short-duration tasks that will probably incur an annual cost (depending on accounting procedures) of from $100 to $400. In this case, based on an actual engineering project, this was the author's recommendation. Manual switching of capacitors is often overlooked by distribution planners and engineers, particularly in a world full of automation options, but it has several virtues: there is no capital cost for switching equipment, the marginal cost of labor for the switching is often nil, and it is very flexible.

Automatic switching. Capacitor banks can be fitted with automatic switches controlled by any of several means of local (autonomous) control. Popular means of control are: voltage, temperature, current, time of day (sometimes done by photocell), or power factor. Each has advantages and disadvantages with respect to the others. All have two disadvantages: maintenance costs and inflexibility.

The author will admit to a strong prejudice against switched capacitor banks on the basis of reliability problems often encountered, a prejudice based upon his experience while working both on utility line crews and as a distribution planner early in his career. Capacitor switches under local control cycle several hundred times per year, and they often malfunction due to any of several possible failure modes.[11] If one of the three (phase) switches required at any bank fails, it

---

[10] Details on 8760 hour load curve shape sufficient to reach this conclusion have not been provided here. This example is based on an actual capacitor switching study in which the annual feeder losses increased by 210% with the capacitor in place.

[11] A survey the author made of one utility in the early 1980s indicated that slightly more than *one-third* of all switched capacitor banks were not switching properly due to mechanical failure, vandalism, or weather damage of one type or another.

creates severe imbalance which defeats much of the purpose of the bank and may contribute additional operating problems. Beyond this, if the switches are operating properly, the control, itself, may fail. As a result, in order to assure operation, switched capacitors must be visually inspected several times a year to guarantee they are operating properly. This annual inspection cost is roughly the same as the cost of switching them manually twice a year.

In addition, another frequent problem with automatic switching is encountered when feeder segments are switched. A capacitor bank may have been programmed (its voltage or current sensing control set to a certain limit) based on the original feeder configuration. When the feeder is re-configured through re-switching, these may be quite inappropriate. Few utilities re-compute and re-set capacitor switch controls after re-configuring feeder line switches.

Automated distribution. Remote control of capacitor switches from a central location using an automated distribution system provides results superior to anything possible with automatic switching. To begin with, the automated distribution system can sense whether the capacitors are being switched, identifying when failures occur. This both assures operation that is as expected on a continuing basis, and negates any requirement for manual inspection.

But more important, the capacitor switching can be coordinated with that of other capacitor banks, LTCs, and regulators on the feeder, so that the overall volt-VAR control scheme is optimized. Whether the automated control of capacitors is justifiable on the basis of cost is something that must be determined through detailed study. However, given that a communications system already exists or can be upgraded at low cost, automated control is less expensive than automatic control, and provides superior control (lower losses, tighter voltage control, and higher utilization factors).

Analysis of switching -- if and how often?

Using data on the 8760 hour behavior of the load curve, it is possible to evaluate the benefit of various capacitor applications and switching schemes, to determine if automatic switching, manual switching, or static capacitors are best. Such analysis should be done using the actual load curves, not on the basis of load duration curve analysis. Table 9.6 shows the result of a comprehensive analysis of several capacitor switching schemes for the example feeder. In this case, manual switching of two of the banks on a seasonal basis -- on for four and one half months, off for seven and one half -- proves effective and less expensive (this is not the case in all situations).

The manually switched case calls for the banks nearest and farthest from the substation to be switched out of service during all but the peak season. This is the best overall compromise, so that overcorrection during the off-peak season is

Multi-Feeder System Layout

tolerable. This results in: a 23% increase in MW capability of the feeder during peak conditions, as compared to the uncorrected case; correction of power factor at the substation during peak conditions to 98%; overcorrection off-peak to no worse than 85% PF (leading) at the substation; a worst PF on the feeder during off-peak periods of no more than about 70% (no worse than normal, but leading instead of lagging); a reduction in losses during the peak season worth $3250 annually (enough to cost justify the capacitors on this basis alone).

During off-peak season, the single capacitor in place will overcorrect during minimum load conditions, with the worst overcorrected power factor on the feeder (at the capacitors location) being an overcorrected 70%, and that at the substation being an uncorrected 89%. But during peak daily conditions during the off-peak season, this single capacitor corrects power factor from 83% to 96+% and lowers losses by noticeable amounts. Overall, its contribution during the eight off-peak months of the year is close to neutral -- it overcorrects as much as it corrects, but it is not worth the cost of switching it in and out on an annual basis.

In general, widespread application of automatic switched capacitors justifies itself only in cases where losses' costs are high and inspection and maintenance costs are low.

Table 9.6 Comparison of Several Capacitor Switching Alternatives for the three by 1,800 kVAR configuration (Figure 9.20)

Capacitor Switching Scheme	Times Switched per year	Change in Annual MVAR-miles (%)	Worst PF Due to Over-correction
Static	none	+42%	25%
Static, only half the VAR size*	none	+6%	45%
Manual, #s 1 and 3 (see text)	2	-18%	85%
Automatic			
current, # 3 only	260	-4	55%
current, #s 1 and 3	226/204	-26%	85%
current, all three	208/174/192	-27%	none
temperature, all three	190/160/140	-20	none
Automated, #s 1 and 3	240/182	-27%	none
Automated, all three	262/200/104	-28%	none

* Provides insufficient correction during peak.

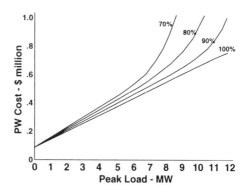

Figure 9.24 The linearized "cost curve" from Chapter 7's analysis of conductor economics (from Figure 7.18, based on an assumed 90% power factor), along with curves recomputed on the basis of assumed power factors of 70%, 80%, and 100%. This shows the importance of setting strategic targets for power factor correction throughout the feeder system. Higher power factor yields a lower slope, a higher limit to the top of the linear cost range, and a less exponential "high cost range." If 100% power factor could be achieved throughout, overall costs would be lower and large conductor lines would be the most economical rather than the least economical, possible. However, obtaining such high power factors over any feeder's length is practically impossible (see text).

Strategic Planning of Capacitors and Power Factor

In general, recommended practice for capacitors is to include them in the strategic planning of the distribution system, setting targets and policies for their inclusion in the layout of the system as part of the overall standards and planning guidelines. The goal ought to be to optimize the economic utilization of power factor correction, taking into account the reduction in losses, the improvement in load carrying capacity, and the improvement in load reach that result from correction of poor power factor. Details of how capacitors are utilized on each feeder can be left to engineering studies done prior to construction, but the interaction of the overall target in power factor correction and capacitor application must be considered in establishing conductor loading tables and economic loading ranges.

As shown in Figure 9.24, the degree of power factor correction that is expected influences the economics that can be depended upon by the planners. Generally, it is possible to depend reliably on correction of power factor to 90% during peak conditions, but no more. The benefits of very tight control to maintain power factor close to 1.0 are obvious from Figure 9.24: The slope of

Multi-Feeder System Layout

the power versus cost curve drops, and the end of the economical range rises, as power factor is improved. If power factor could be kept to exactly 1.0, larger conductor lines (those with high X/R) have a longer reach than smaller lines.[12]

Regardless, most important is the establishment of guidelines on reduction of MVAR-miles (power factor profile). Such correction to whatever level works out as optimally on an economic basis can then be depended upon in the rating of conductors and lay out equipment. The assumption of 90% power factor used in Chapter 7's economic evaluation of conductors is an example.

Voltage Regulators

A voltage regulator placed strategically can nearly double the reach of the particular path it lies upon. For example, 12.47 kV feeders as discussed in Chapters 7 and 8 can deliver power up to about 3.3 miles. A regulator placed at the end of this distance, where voltage drop reaches the maximum, can boost and control voltage, allowing another 3.3 miles of power flow. Losses and maintenance costs are higher, and capital cost lower than alternatives that handle the job with larger conductor, but in cases where voltage problems evolve due to growing load on a system already in place, regulators are the lowest cost option.

Often, it is better to reserve voltage regulators as *engineering tools* to be used when the system's capabilities are exceeded, rather than to plan for their use as a matter of policy as part of strategic planning, as is the case with capacitors: Regulators represent a low-cost, flexible way to accommodate the inevitable problems that occur unexpectedly.

9.7 SUMMARY OF KEY POINTS

A *feeder system* consists of many feeders -- radial electric circuits, usually with multiple branches and laterals, that include everything downstream of a feed point/protective device at the substation. The configuration, service area, and load of a feeder are defined by the connections and open points used to define its

---

[12] From a practical standpoint this is impossible. Maintaining control of power factor at the substation very near 1.0 -- say in the range of 98.5% or better, is difficult, but can be done, requiring tight monitoring and staged switching of capacitors. But this is not nearly enough. To achieve the conductor economics shown for a power factor of 1.0 in Figure 9.24, power factor must be kept within 98.5 or better everywhere along the feeder. Note the power factor profiles in Figure 9.21 to see how difficult this would be: it would require a multitude of very small banks scattered along the feeder, many of them switched in increments. From a practical standpoint, maintaining power factor at 90% or better is about the best target planners can depend upon.

connectivity. These can be changed with little effort, and often no cost, and thus the configuration of any "feeder" is quite fluid -- it can and will change over time. Proper planning should focus on designing the system, not the individual feeders in a feeder-at-a-time manner. Important points about multi-feeder system planning are:

- Classical geometric analysis of feeders and feeder layout is of little real use in laying out feeder systems, as it often provides results and analysis inappropriate for real-world systems.

- "Number of feeders" is a very over-rated design consideration. The optimal circuit design configuration for getting power away from a substation into the surrounding area is not really affected by the decision of how many feed points to use to get power out of the substation.

- The backwards trace method of laying out feeders -- starting at the loads and working back to the substation -- requires slightly more planning effort but generally provides much better results than planning methods that start at the substation and work out to the loads.

- Expansion to meet possible future loads should be planned when the system is first built, the range of future possibilities being defined by a load forecast. Expandability, particularly towards a certain target, can be built into a feeder system at little cost if that is a goal at the beginning.

- Load growth and changing needs over time are best handled by changing configuration, not by upgrading conductor size.

- Capacitor application is best planned strategically to identify expected performance and economic target, and set guidelines on capacitor application. Such policies are then assumed in subsequent planning and implemented in detailed engineering studies on a feeder by feeder basis.

REFERENCES

J. J. Burke, *Fundamentals of Distribution Engineering,* Marcel Dekker, New York, 1994.

J. K. Dillard, editor, *T&D Engineering Reference Book,* Westinghouse Electric Corporation, Pittsburgh, 1928 and 1958 (slightly revised, 1965).

M. V. Engel et al., editors, *Tutorial on Distribution Planning,* IEEE Course Text EHO 361-6-PWR, Institute of Electrical and Electronics Engineers, Hoes Lane, NJ, 1992.

H. L. Willis, *Spatial Electric Load Forecasting,* Marcel Dekker, New York, 1996.

10
Distribution Substations

10.1 INTRODUCTION

The purpose of a substation is to take power at high voltage from the transmission or sub-transmission level, reduce its voltage, and route it onto a number of primary voltage feeders for distribution in the area surrounding it. In addition, it performs operational and contingency switching and protection duties at both the transmission and feeder levels, and provides a convenient local site for additional equipment such as communications, storage of tools, etc.

Substations are somewhat more important to system performance and economy than their cost alone might indicate (they are the least expensive level in a power system) for they are the meeting place of the transmission and distribution systems, and therefore play an inordinately large role in influencing the cost and performance of both those levels as well as their own. From both a cost and reliability standpoint, their interaction with the T and D systems is often more important than they themselves, in the sense that their influence on T&D reliability and costs often outweighs their own costs and reliability contributions. Thus, in many ways, good planning of the substation level is the key to good distribution system planning. Certainly, poor substation-level planning forfeits any hope of achieving outstanding performance and economy at the distribution level.

Substations represent the end delivery points for the transmission system. They are the sites to which the transmission system must deliver power, and their cumulative demands (the sums of all feeder loads at each substation) are the loads used in transmission level planning (e.g, load flow analysis generally lumps all the feeder loads into a single demand at each site). Substations' performance and economy interact greatly with the transmission system. They define absolute constraints on the transmission system, which must reach them and deliver power to them, even in cases where that task is costly or unreliable because of other system constraints. Conversely, if the system can deliver only so much power to a substation, then the distribution planners must live with that restriction, costly as it may be.

On the low voltage side, substations are the starting points for the feeder system. A set of feeders emanates from each substation, serving the area around it, and in combination with those of other substations, serving all of the utility's load. The location and capacity of the substation materially affects the feeder

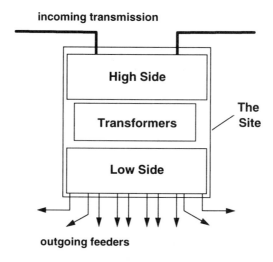

Figure 10.1 Substations consist of four fundamental parts: the site itself, the incoming transmission, the transformation, and the outgoing distribution. Design, cost, and reliability of a substation can be analyzed by examining each of the four parts and their interaction with one another. Configuration of each part heavily influences the others.

Distribution Substations

system attached to it. If the substation is in a poor location from the standpoint of the feeder system, it increases feeder cost and decreases reliability. If it has insufficient capacity, then the feeder system may not be provided with the power it needs at this site, and service or economy may suffer when service has to be provided by substations located at sites farther away.

Compatibility of Standards and Design is the Key

Thus, the specification, layout, and design of substations is an important element in the power system, and most critical is that they be compatible in size, location, and voltage level with the type of transmission and distribution systems built on either (electrical) side of them. Such compatibility interactions and the planning issues associated with them, are more fully explored in Chapter 9. This chapter will focus on the substations, themselves, and the "sub-transmission" level most often associated with their high-side design.

Four Fundamental Parts

Substations can be built in a wide variety of types and sizes, but all have four fundamental parts, as shown in Figure 10.1. These are: 1) the high-side buswork and protection, 2) the low-side buswork and protection, 3, the transformers, and 4) the site itself. The transformers are the most distinguishing characteristic of a distribution substation -- the feature for which it is built and which normally defines its capacity -- if a substation has three 16 MVA transformers, then typically it is referred to as having 48 MVA of capacity.[1]

This chapter will examine each of these four aspects of a substation in turn, looking at how cost and reliability vary as a function of different alternatives for each level and the way all four fit together into a complete substation. Sections 10.2 through 10.5 will look respectively at the high-side and the sub-transmission, the transformer level, the low-side, and the site, itself. Section 10.6 will examine how the four levels fit together and explore typical costs and performance at the substation level. Sections 10.7 and 10.8 address several planning issues that deal specifically with the substation level.

---

[1] In some cases however, other aspects of system performance constrain capacity. For example if the incoming transmission were limited to 45 MVA, or if the capacity of the low-side buswork were similarly limited to below 48 MVA, then the substation is, in effect, a lower capacity. Some utilities ignore these restrictions and refer to a substation by its transformer capacity. Others do not, using the minimum constraining capacity of any aspect of the substation as its maximum capacity.

10.2 HIGH-SIDE SUBSTATION EQUIPMENT AND LAYOUT

The high side of a substation provides the termination for incoming transmission or sub-transmission as well as the protection, switching, and monitoring and control. It consists of all buswork, breakers, relays, meters, and all other equipment needed to terminate the incoming transmission, as well as handle any transmission being switched at this location or otherwise passing through. This level incorporates everything, including the high-side buswork, required to provide the power to the substation's transformers. Typically, the high side represents from 1/4 to 1/3 of a substation's total cost.

This portion of the substation can take on any number of configurations, depending on the number of incoming transmission lines, the arrangements made for their switching and contingency backup, and the type of protection and isolation switching desired. Its arrangement depends somewhat on the transformer portion of the substation, particularly the number of transformers and the arrangements desired for their contingency backup.

Combined Transmission-Distribution Substations

Occasionally, a good deal of the high-side facilities at a "distribution" substation may be related to transmission-switching functions (i.e., functions associated more with the bulk power grid than with the delivery of power to substations). Given the great difficulties utilities have obtaining sites, the practice of combining transmission and distribution functions at one substation has become common. Thus, the "high side" at many substations may include equipment that has little if anything to do with the distribution function.

Generally, this complicates matters for both transmission and distribution planners, there being four issues worth noting:

a. Incoming transmission lines and outgoing feeders compete for the available route space into and out of the substation. There is only so much room overhead and underground.

b. The combined equipment requirements of both T and D create the need for a large site. This reduces planning flexibility by limiting the number of sites which are minimally acceptable.

c. Location of a combined T&D substation has to be a compromise between transmission and distribution needs, that means it may not be ideally located for either purpose.

d. The site, itself, may become quite congested, with minimal clearances and no available space for future expansion of either side.

Distribution Substations

Very often, organization procedures within a traditional large electric utility do not include frequent communication and close coordination between the transmission and distribution planning functions. Many of the problems that result revolve around combined use substations -- transmission and distribution planners each plan to use available space and rights of way at a combined substation for their own long-term requirements, but do not communicate that intention to the other department. By the time a conflict is detected and resolved, it may be too late to work out a really effective, low-cost solution. In particular, item (d) above suffers.

In a de-regulated utility industry, transmission and distribution functions are usually either functionally or organizationally disaggregated. One result will be a reduction in the construction of new combined-use substations. But more important, disaggregation will force both entities to formalize the relationship between T & D planning, identifying specifically what is and is not theirs and informing the other entity of their intentions.

Sub-Transmission Lines and Voltages

Sub-transmission voltages of between 25 kV up to above 230 kV are used to route power from sources within the power grid to the distribution substations. The same capacity and load reach concerns discussed for feeders in Chapter 7 apply to moving the "substation-sized" amounts of power required at the sub-transmission level, and often higher voltage is sought in order to move more power greater distances more efficiently. As at the feeder level, higher voltage brings greater capacity and reach, and a higher minimum cost, both in the sub-transmission lines, themselves, and at each substation, which must be built to the voltage level of the sub-transmission used.

In general, voltages of 25 kV to 46 kV are holdovers from sub-transmission voltages applied in the first half of the twentieth century, and usually are found in company with rather small (10 MVA peak load) substations with distribution voltages in the range of 4 kV to 11 kV. Higher voltage sub-transmission (in company with larger substations and higher distribution voltages) has been a general trend throughout the twentieth century, partly because of the need to meet higher load densities and distribute power over greater distances, but also because among other advantages higher voltage reduces the number of sub-transmission routes and substation sites required. New applications rarely use sub-transmission voltages below 69 kV. However, there is no "best" or preferred sub-transmission voltage -- application depends on a number of factors that will be discussed later in this chapter and in Chapters 11 and 12.

Table 10.1 Relative Line Cost, High-side Substation Cost, and Reliability as a Function of Sub-transmission Voltages

Subtr. Voltage kV	Typical Capacity MVA	Most Typically Feeds Number of substations	Each with peak MW	Typical Line Cost/mi.	Typical Sub. High-Side Cost	Cost/MW Trans. Line	Cost/MW Subs. H.Side	Single Line Reliability
25	25	2	5 - 12	0.06	0.11	3.00	2.00	2.38
34	40	2 - 3	5 - 15	0.12	0.20	2.80	1.50	1.66
46	60	2 - 3	5 - 20	0.20	0.30	2.70	1.37	1.48
69	90	2 - 4	10 - 30	0.30	0.47	2.00	1.33	1.43
87	125	2 - 4	15 - 30	0.44	0.59	1.30	1.25	1.32
115	175	2 - 4	20 - 40	0.74	0.84	1.00	1.05	1.06
138	225	2 - 4	22 - 45	1.00	1.00	1.00	1.00	1.00
161	275	2 - 4	25 - 55	1.31	1.33	0.85	0.93	0.92
230	325	2 - 3	75 - 150	1.71	2.05	0.75	0.85	0.81
277	500	1 - 2	120 - 240	3.17	2.77	0.66	0.70	0.64
345	800	1 - 2	200 - 400	5.92	5.05	0.55	0.60	0.53

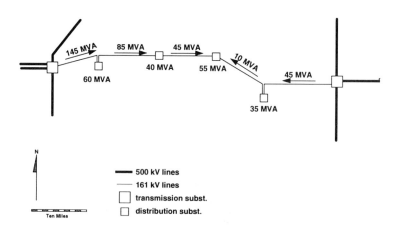

Figure 10.2 Sub-transmission lines are often "loops" that serve several substations. A "sub-transmission loop" isn't actually a loop circuit -- it terminates at different locations in the transmission grid, but loops through several substations, as shown. Configured appropriately, it will provide duplicate sources (from either direction) to each of the substations it serves, for greater service reliability than is available with radial transmission feed.

Distribution Substations

Table 10.1 compares cost, utilization and reliability of sub-transmission by voltage level, giving values typical of average applications at each level. In general, systems with very low sub-transmission voltages have only one or two substations per sub-transmission circuit (due to capacity limitations), as do systems that use transmission grid voltage (lines above 161 kV) to feed substations directly. (Such high-voltage lines are an important part of the overall power grid, to the point that they cannot be totally devoted to sub-transmission duties because some of their capacity is needed in order to move bulk power and maintain system integrity). Intermediate voltage lines (between 69 and 230 kV) typically have three to four and occasionally as many as six substations fed from one transmission ROW, in the manner depicted in Figure 10.2.

Cost per mile of line, as well as for the high side of each substation, increases as voltage level is increased, due both to the higher cost of providing insulation to the higher voltage level, and because higher voltage substations have a greater capacity. But the capacity of both the lines and high-side facilities increases greatly as voltage is raised. On a per-MW basis, higher voltage is usually less expensive.

In general, reliability of higher voltage lines is somewhat better than lower voltage lines (230 kV lines typically have an outage rate per mile of only about a third of that for 34.5 kV sub-transmission) due to the higher insulation and lightning strike withstand capability, as well as to the fact that higher voltage transmission is usually built on very robust towers with considerable clearance from trees and other items that can fall into the line during a storm.

Radial supply

The simplest arrangement for both the routing of power into a substation, and its high-side configuration, consists of only one incoming transmission line (Figure 10.3). This is a radial configuration -- the substation will be completely out of service if this only incoming line fails for any reason. Generally, the lowest possible alternative for the power feed to a substation is a single radial feed with no high-side breaker at the substation (Figure 10.3A): protection for the transmission line and the high-side of the substation is provided by a breaker at the other end of the transmission line. Given that such protection can be completely and dependably coordinated, there is no loss of reliability or safety in such a design (versus one with a high-side breaker, as shown in Figure 10.3B). Often a breaker is needed, or preferred, in order to provide protection which covers all required fault levels and coordinates with the low-side protection. But given that coordination is possible from the other end of the transmission line, a local high-side breaker provides no additional reliability of service.

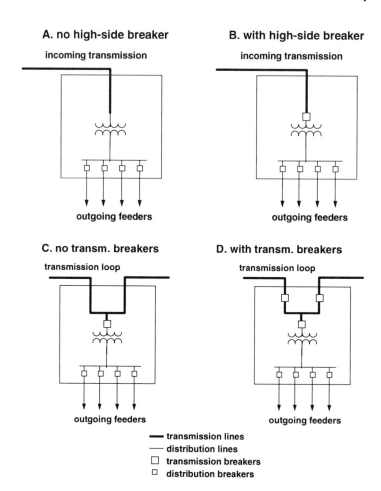

Figure 10.3 Top: radial transmission provides incoming power with no duplication of supply. In some cases a "breakerless" configuration can be built (A) where all high-side protection for the line, buswork, and transformer is covered by high-side protection at the other end of the line. In other cases (B), a local high-side breaker is preferable or required in order to coordinate protection completely. Bottom: dual incoming routes. Configuration C has no breakers to isolate a transmission line fault on one side or the other of the substation. It is essentially a radial feed -- a fault anywhere on the line on either side of the substation will outage all incoming transmission to the substation -- and it has the additional handicap of having a greater length of exposure than a radial line that terminates at the distribution substation. Configuration C has a breaker to isolate each incoming side of the transmission route, providing duplication of source and greater reliability as a consequence. In some cases, the transformer breaker can be left out of configuration C or D without undue reduction in protection coordination.

Distribution Substations

Loop or dual transmission feeds

To avoid interruption of service when a transmission line is outaged, most substations are fed by two sub-transmission circuits, each of which has sufficient capacity to serve the load (at least at its emergency rating, for brief periods). Very often this is accomplished by looping a transmission circuit through several substations, as was illustrated in Figure 10.2. Such sub-transmission "loops" (which generally start and end at different locations in the grid) typically serve between two and five substations. Given that the various transmission line segments are sized so that the load of all substations on the loop can be supported when fed from either end, such circuit arrangement provides duplicate sources of transmission-side service, and therefore gives immunity from service interruption due to any one transmission line outage alone.

Improved reliability from loop circuits requires sub-transmission sectionalizing breakers

Transmission feed from a loop circuit does not necessarily assure greater service reliability than that provided with radial service. A loop arrangement without breakers to sectionalize the feed on either side of the substation (Figure 10.3C) typically provides reliability *worse* than that of a single radial transmission line, because the outage exposure (line length) is greater for the loop circuit than for a single radial route, and if a fault happens anywhere along this longer line, it is cleared at both ends by other protection, and service to the substation is lost.

In order to gain the maximum potential reliability, the transmission loop must have both the capacity to serve the load from either direction *and* breakers that sectionalize it in the event of a transmission line fault on either side of the substation to isolate the fault. Thus, the most common practice in distribution substations is to provide two transmission feed lines, each with sufficient capacity and breaker protection so that they form a duplicate supply (Figure 10.3D).

In cases where each section of the sub-transmission loop does not have individual capacity sufficient to serve the entire peak load, both are required to meet the load at peak and perhaps many off-peak times. Service reliability is more difficult to evaluate in these situations, because it depends on the annual load curve shape, specifically how often the load exceeds the capacity of either incoming line. For example, if both parts of the loop shown in Figure 10.3D are limited to 40 MVA capacity, but the substation peak load exceeds 40 MVA 1,000 hours during the year, then complete service can be provided during those 1,000 hours only with *both* lines in service. For those 1,000 hours, the likelihood of interruption is similar to that provided by the configuration in Figure 10.3C.

For the other 7760 hours per year, interruption rate resembles that provided by Figure 10.3 D, when load is less than the capacity of either line.

Loop circuits on one structure

In Figure 10.2, the sub-transmission into two substations (the first and last in line) take a short dog-leg from the main circuit route into each substation. Such transmission routing is common and usually accomplished by putting both sides of the loop (into and out of) on the same double-circuit structures for the distance into the substation. Such paralleling of circuits into a substation reduces reliability slightly, as compared to a configuration where the routes are completely separate, for both are at jeopardy in the event of a failure of one of the structures or any other event that would disable the right-of-way (e.g., a tornado). In cases where the likelihood of damage to structures is high, exposure should be limited. For example, if both sub-transmission lines are on the same set of double-circuit wooden poles, placing these poles immediately alongside a busy street is sure to increase the likelihood of substation outage -- an automobile colliding with and knocking down a pole will take both circuits out of service.

But while paralleling of duplicate sources should be avoided where possible and economical, reliability is not unduly affected if the length of parallel route is limited and the probability of structures being damaged is not high.[2] Reliability of supply is still better than being provided with only radial service along the same route, because the availability of power is supported by two locations in the grid.

High-side Configurations and Reliability of Service

Reliability of service from two incoming transmission sources depends on more than *just* the high-side protection/segmentation of the transmission loop into the substation. Most substations include two or more transformers, one reason being to provide contingency support in the event of a transformer outage for improved reliability. The switching and protection/segmentation provided by the high-side configuration in company with these transformers has a large role in determining overall reliability of supply. Figure 10.4 shows seven arrangements for one pair

---

[2] However, in cases where the likelihood of damage is high, exposure should be limited. For example, if both sub-transmission lines are on the same set of double-circuit wooden poles placing this poles immediately alongside a busy street for any great length of route is sure to increase the likelihood of substation outage -- any vehicle collision with a pole is very likely to take both circuits out of service, and hence, the entire substation.

Distribution Substations

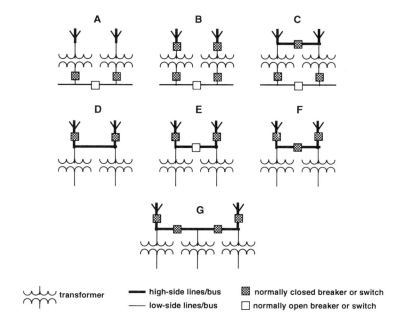

Figure 10.4 Alternate configurations for arrangement of two incoming transmission lines and multiple transformers.

of incoming lines serving two transformers. Reliability varies slightly, and operational flexibility varies greatly, depending on the arrangement. Configurations A, B, and C each pair a transformer with an incoming subtransmission line. Contingency support during high side or transformer outages depends on low-side switching (shown), and reliability depends greatly on the configuration and performance of equipment at that level. In the event of an outage of one incoming line, or an outage of one transformer, all load is thrown over to the remaining line-transformer pair's low-side bus -- the outage of either one transformer or one line takes *both* a line and a transformer out of service.

Configurations D, E, and F shows arrangements with contingency support of incoming service arranged on the high side. The low-side bus is not shown because it is not relevant to contingency support issues related to the high side.[3] The outage of either line leaves both transformers in service. The outage of either transformer leaves both lines in service. Configuration G shows this same concept extended to a substation with three transformers.

Despite these differences, configurations A through F all give similar expectations of service interruption when evaluated for line-substation reliability. Given that the substations' peak load can be supported entirely by either incoming line, or either transformer; that the low-side switching in A, B, and C functions well; and that there are no "common mode" failures (an incoming line and the opposite transformer failing simultaneously due to a common cause); then configurations A, B, and C offer service reliability similar to configurations D, E, or F. The only difference in expected outage rate is in situations when a line or transformer on the opposite side of the substation goes out of service unexpectedly while one transformer or line is already out of service, a contingency that configurations D, E, or F can support but A, B, or C cannot. Under typical circumstances, this might be expected to occur no more than once every ten years, and would contribute up to five minutes per year to expected annual outage duration.

Quality of service during contingencies would differ more significantly. During any line or transformer outage, configurations A, B, and C revert to radial transmission service with one transformer, with both line and transformer loaded to range B levels. Voltage drop and voltage regulation problems could result. Configurations E and F provide superior low-side voltage and voltage regulation in all cases because they have dual incoming feeds when a transformer is out of service, and two transformers (lower voltage drop, better regulation) when a line is out of service.

An equally important consideration is flexibility and robustness of contingency support. Configurations A, B, and C depend entirely on low-side switching for support during transmission line or transformer outages, and plans for protection and equipment performance during contingencies must work perfectly for those configurations to provide the target service reliability. Given that they have similar low-side switching arrangements, configurations E, D, and F can always use low-side switching to support contingencies if needed, but can rely on high-side/transformer level contingency support, too. They are more robust, for they depend less on the flawless performance of any one contingency support mechanism. They also leave greater flexibility to the system operator.

---

[3] This does not mean that configuration, protection, and switching at that level are not important, only that they do not play a role in determining contingency support when there are *high-side* outages.

Distribution Substations

Choosing a high-side configuration for a distribution substation

The advantages of the more complicated configurations must be weighed against their cost. In most situations configurations D, E, and F will cost significantly more than A, B, or C. The choice between the two sets of configurations is basically one of purchasing reductions in expected outage duration. Within A, B, and C or within D, E, and F, configuration selection will affect both the coordination of protection (some configurations shown in Figure 10.4 may not be feasible due to protection coordination reason in some situations), and the frequency of interruption (A, B, C, and E are all most likely to create discernible "blinks" or flicker during the clearing of a fault).

Selection depends on conditions and contingency requirements specific to each situation, as illustrated by Table 10.2. The table compares the expected cost, worst low-side voltage level under contingencies, and expected service interruption (annual values) for various configurations, for two similar substations in a T&D system in the central United States. The lowest cost alternative at each substation is one incoming line and a single 36 MV transformer. Other alternatives include two incoming lines and two 22 MVA transformers arranged in any one of configurations A through F.

For substation case 1, configuration A improves expected interruption duration to 8,100 customers by an expected 84 minutes per year at a cost premium of $2,800,000, giving an improvement cost of $4.11 per annual customer minute of reduction. Configuration E (preferred over F in this case because sub-transmission loop flow is *not* desired under normal conditions) costs an additional $700,000 for a further reduction of 12 minutes, or 97,200 customer minutes reduced, at $7.20 per customer minute.

In case 2, configuration A, at a premium of $465,000 over radial/one transformer service, will buy only 15 minutes of interruption duration reduction for 9,100 customers, a value of $3.40 per customer minute. Configuration B (selected in this case for transmission related reasons because it permits network flow on the sub-transmission loop when the breaker is closed) would provide no further improvement. Configuration E or F would buy a further 27,000 customer minutes at a cost of $500,000, working out to $18.31 per customer minute.

The present design threshold for customer minutes cost in this particular utility system is $10/customer minute of improvement. Typically, values used by other utilities vary from $3 to $15, so this is about average. Though additional reliability might be needed at this site, there may be other places in this system (e.g., feeder-level) where a reduction in duration can be bought at a lower cost than $18.31/minute. Therefore, the selection of substation configuration is best coordinated both with customer needs and design at other levels of the system.

Table 10.2 Comparison of Cost, Contingency Voltage, and Expected Service Reliability for Various Configurations at Two Substations Planned on a Value-based Basis -- Cost in $ × 1000, Voltage in per Unit, Duration in Minutes

Situation >>	Case 1 115 kV/25 kV 31 MVA peak avg. 16 mile exposure on both incoming transm. routes, high isochronic level. 8,300 custs.				Case 2 115 kV/13 kV 35 MVA peak avg. 2.3 mile exposure on both incoming transm. routes, med. isochronic level. 9,100 custs.		
Configuration	Cost	Voltage	Freq.	Dur. Cost	Voltage	Freq.	Dur.
radial/one trans.	$4,400	.96	.73	112	$1,735	1.00	.25	20
A	$7,200	.95	.18	28	$2,200	1.00	.07	5
B	$7,600	.95	.18	28	$2,500	1.00	.07	5
C	$7,400	.95	.18	28	$2,300	1.00	.07	5
D	$7,700	.98	.10	15	$2,600	1.03	.05	4
E	$7,900	98	.07	12	$2,800	1.03	.03	2
F	$7,900	98	.07	12	$2,800	1.03	.03	2

Multiple incoming transmission lines

Other configurations for transmission service into a distribution substation are possible, beyond the radial and dual-circuit/loop configurations shown in Figures 10.3 and 10.4. It is not uncommon for a large distribution substation (one having four or more transformers and a peak load exceeding 100 MW) to be fed by three or more transmission circuits, with its configuration designed so that it can fully meet its peak load with any one line out of service, while still within or very close to range A standards (with so many lines and transformers, situations where at least one line or transformer will be out of service are frequent, and it is best to make provisions to be able to provide *good* service during those times.)

Switching and protection configurations of such schemes play a critical role in determining reliability, and most often the high-side includes at least some transmission level switching and re-configuration ability, so that the substation is to some extent a transmission substation and switching point among the incoming lines.

Salient High-Side Features for the Planner

From the planner's perspective, the important issues with respect to alternative substation high-side configurations are the cost, capacity, and reliability of each

Distribution Substations

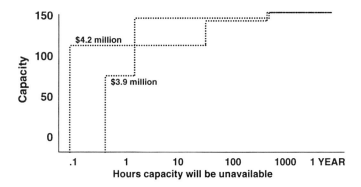

Figure 10.5 Which is better? Availability of capacity relationships developed from probabilistic analysis of two alternate transmission arrangements into a planned distribution substation show the complexity a planner can face in determining high-side configuration for a substation. The first alternative, at $3.9 million capital cost, is expected to be able to meet the projected peak load of 135 MW all but 90 minutes per year, but will have complete power unavailability about 30 minutes. A slightly more expensive arrangement with a cost of $4.2 million provides at least 110 MW all but five minutes per year, but will fail to meet the peak load level of 135 MW more than 50 hours per year. "Which is better?" depends on cost (a $300,000 difference) and what is more important, serving all of the load, or serving a majority of it.

Table 10.3 Relative Capital Cost and Expected Hours of Outage Per Year for a 40 MVA Substation with Various Types of Transmission Configurations*

Type of Transmission Service	Relative Cost	Frequency Events/year	Duration Hours/year
Radial, 69 kV, no hi-side breaker at subst. (Fig. 8.2A)	1.00	1.1	6.0
Radial, 69 kV, one breaker at subst. (Figure 10.2B)	1.10	1.1	6.0
Loop, 69 kV, only one breaker at subst. (Figure 10.4A)	1.22	1.2	8.2
Loop, 69 kV, three breakers at subst. (Figure 10.4B)	1.35	.02	0.1
as above, but load exceeds indiv. line capacity 1000 hours/year	1.35	.15	.50
as above, but load exceeds indiv. line capacity 8760 hours/year	1.35	1.2	8.2
Three incoming 69 kV (6 mile) lines, any one is sufficient	1.66	.01	.05
Three incoming 69 kV (6 mile) lines, any two are sufficient	1.50	.25	2.0

* Computations for this table were based upon a six-mile distance from transmission substations for each of the transmission circuits, and typical outage rates for transmission lines and equipment.

of the various alternative designs. In addition, the expandability and flexibility of adaptation of an alternative may be an issue when future load growth is uncertain. The various factors may be related in complicated ways, as illustrated in Figure 10.5, to the point that merely identifying the "best" alternative once all the evaluation has been completed may be a challenge in itself.

Costs, as well as comparison and selection of high-side features for distribution substations, will be discussed in more detail later in this chapter. Table 10.3 compares relative cost and expected outage statistics for a distribution substation under various configurations of from one to three incoming transmission service, illustrating the variations that planners may have at their selection.[4]

10.3 TRANSFORMER PORTION OF A SUBSTATION

Transformers are not only the *raison de' etre* for distribution substations but often represent the largest portion of the cost, typically representing from 1/2 to 2/3 of the total substation cost. Distribution substations utilize from one to six transformers to convert incoming power from the sub-transmission voltage to the primary distribution voltage.

Generally, the sum of all transformer ratings defines a substation's capacity -- if a substation has two 22 MVA transformers and one 33 MVA transformer, then it is rated as "a 77 MVA substation." Variations in the way transformer "capacity" is defined, as covered in Chapter 6's discussion on transformer ratings, apply to total substation capacity, as well. Thus, a 77 MVA rating could mean that the substation can handle a load of 77 MW continuously (and much more for brief emergency periods), or it could mean that it is capable of handling 77 MW only at emergency load levels for brief periods, or it could mean something else.

Beyond this, a few utilities do not sum transformer capacities to determine substation capacity, but instead define a substation's capacity based on some variation of N-1 contingency evaluation of the number and type of transformers. Such a nomenclature is consistent with standards that call for contingency capable at the transmission-substation level. The substation cited above, with two 22 MVA transformers and one 33 MVA transformer, might be rated at 66 MVA with such a terminology: its worst-case contingency is the loss of the largest transformer (33 MVA), limiting it to the contingency capability (in this

---

[4] The many values and details for the system analysis required to compute this table are not given here. It is meant only to illustrate the typical differences that arise from variations in the transmission service provided to a substation.

Distribution Substations

case, a 4-hour overload of 50 %) for the two remaining transformers, giving a capacity of 2 × 150% × 22 MVA = 66 MVA.

Transformer Types

Distribution substations most often utilize three-phase transformers, which require less space than three single-phase units, and offer superior economy (a three-phase unit generally has slightly lower losses than three single-phase units of equivalent capacity) and lower inspection and maintenance cost. The only exceptions are rural substations (see below). Type of transformer utilized depends on the application, but typically autotransformers are used due to their lower cost. Depending on application, these may or may not be fitted with oil pumps, radiators, and fans to augment capacity to FO and FA ratings (see Chapter 6). In some cases where load growth is expected, each transformer may be installed initially without such equipment -- perhaps with a rating of 16 MVA -- the plan being to add pumps, radiators and fans at a later date to boost rating to 22 or 24 MVA, if a growing peak load creates the need.

Except for very rare exceptions, distribution substation transformers are invariably delta connected on the high side (transmission is almost always delta connected, but there are exceptions where wye-connected sub-transmission is in use), and either wye- or delta-connected on the low side, depending on whether the primary distribution system is wye- or delta-connected.

As mentioned in Chapter 7, numerous types of special transformers exist to accommodate special requirements. Chief among special transformer designs important to distribution planners are "low footprint transformers" designed to fit into substation locations originally built only for smaller units. For example, a 25 MVA low-footprint transformer may slip into the space originally occupied by a standard 16 MVA unit, its greater capacity fitting within the physical, clearance, and air circulation constraints originally designed to accommodate no more than the 16 MVA unit. The 25 MVA unit may have higher losses, lower contingency overload capability, or other shortcomings when compared to a standard 25 MVA design, but it will prove far superior to any other alternative when capacity is needed in a substation of limited expandability.

Another category for special application is high impedance transformers, those specifically designed to limit fault currents, may be required in some locations. These have higher costs than normal transformers, but reduce fault currents and breaker requirements to levels that fit within the capabilities of other equipment at the site.

Often distribution applications will require three winding transformers in which power is simultaneously converted to two voltages (e.g., 138 kV to 25 kV and 13.8 kV, as shown in Figure 10.6.

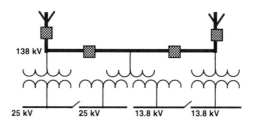

Figure 10.6 Substation with a three-winding transformer.

Multiple single-phase transformers in rural substations

It is common practice in rural or sparsely populated areas, where substation loadings are relatively small and distances between utility facilities quite long, to use single-phase transformers connected in the appropriate Y or delta configuration to convert sub-transmission to primary voltage, rather than a single three-phase unit. A 69 kV to 25 kV substation serving 4 MW of load might have four 1.66 MVA transformers installed -- three in service to convert each of the three phases, and a fourth as standby against the failure of any of the other three. In some situations, the standby transformer is provided with buswork and switching so that it can be cut in as a replacement with only a few minutes effort. In other cases several hours of work are involved in physically disconnecting the outaged unit and connecting the standby unit. Either way, contingency backup is provided at much less cost than if using a full three-phase unit.

Multiple Transformers

Most substations have two or more transformers, for a number of reasons, among the most important of which are:

> *Reliability and contingency support.* A substation with only one transformer will be unable to provide any service if that transformer fails. By contrast a substation with four similar transformers can still provide full service during the loss of any one by raising the loadings on its remaining transformers to 133% -- an overload most transformers can withstand for quite a few hours before overheating Of course a four-transformer substation will be roughly four times as likely to have a single

Distribution Substations 409

transformer out of service, but overall it is much less likely to experience a complete failure of service due to lose of transformers.

Size and transportability. Distribution substation transformers over about 60 MVA are difficult to transport by road, and those over 120 MVA are costly to move even by rail (and restrict the substation to being on a rail siding). Thus, even if there were economic or performance reasons for building a substation out of a large, single unit, practical considerations would discourage such a decision.

Expandability. If a substation is planned for multiple transformers, its capacity can be increased in stages, as needed. A new substation can be built initially with two 15 MVA transformers (each providing backup for the other), and nothing spent on future growth beyond allowing space for future transformers, should they be needed. If the load grows, a third and even a fourth 15 MVA transformer can be added, if and when needed. In cases of significant growth, the third or fourth unit added can be larger than the initial transformers. For example, the third unit could be a 30 MVA, with the two original transformers the same level of contingency coverage for it they provided for one another. The fourth could be somewhat larger still.

Transformer Configuration

Many different ways of laying out the transformer portion of a distribution substation have been tried and are in service, so many that there is a successful exception to nearly every rule or generalization. However,

All transformers are fed from a common source. Generally, all transformers at a substation are served from a common high-side bus, rather than being arranged so that each incoming line serves only a portion of the transformers (i.e., not as in Figure 10.4 A or B). The bus may be segmentable for contingency and maintenance reasons, but typically for contingency reasons it is operated so that all transformers take their power from a common source.

Transformers are applied as individual capacity units with no direct connection other than to a common source. Generally, paralleling of transformers (Figure 10.7A) is not recommended for several reasons: the internal impedance of the two transformers must exactly match in order for loading to be distributed evenly, and tap changing features are difficult to coordinate or not used. Similarly, direct serial application (Figure 10.7B) is rare. Much the same effect is often provided in substations where

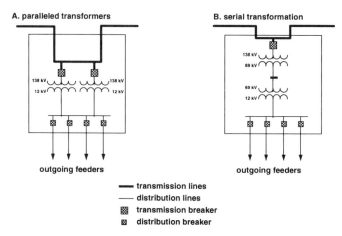

Figure 10.7 Two types of utilization rarely seen in distribution substations. At the left, two transformers are paralleled, which can lead to loading and voltage regulation problems. At the right, serial application in which two transformers -- in this case a 115 to 69 kV followed by a 69 kV to 12 kV transformer -- are used to effect a voltage reduction. This works, but has a higher cost than a single 115 kV to 12 kV transformer.

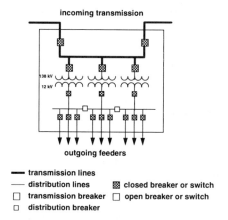

Figure 10.8 Characteristics used in most substations are shown here. First, multiple transformers are all served from a common source -- all three transformers are served from both incoming transmission lines. Two, exclusive feeder service by each transformer -- while the low-side buses of the three transformers can be connected this is only done during contingencies and normally the configuration is as shown, with open breakers (or switches) between each set of feeders.

Distribution Substations

the transmission and distribution functions are performed in different parts of the same substation but the overall purpose and application differ.[5]

Each transformer provides service to an exclusive set of from one to six feeders that it normally serves, these being fed from a low-side bus connected to and associated with that transformer alone (Figure 10.8). Each low-side bus is operated in isolation from the others (otherwise this would parallel the transformers). However, for contingency and maintenance purposes, the low-side buses of all the transformers are arranged so that they can be connected one to the other, so that service can be maintained whenever a transformer is out of service due to failure or maintenance.

Reliability of the Transformer Level

By far the most important reason for multiple transformers is to provide contingency support and improved reliability. Reliability of service for a substation is best measured by evaluating the likelihood of interruption from the feeder perspective, not by looking at the probability of having transformer capacity available. (The discussion of reliability of source given in section 10.2 (Figure 10.4 and Table 10.2) did not take this perspective because it was explicitly concerned with evaluating source (high-side) contingency matters).

Reliability evaluation of net transformer capacity as a guide in selecting substation configuration can be misleading. For example, analysis of a four-transformer substation might indicate an expected 50 hours per year when there will be at least one transformer out of service, but only one second in 30 years when all transformers can be expected to have failed simultaneously. However, due to the availability of contingency backup on a transformer basis combined with limitations on how flexible that backup can be switched (not every transformer in a multi-transformer substation can back up every other transformer), each feeder might have an expected occurrence of interruption due to insufficient transformer availability of once every six years, with an expected duration of two hours -- an annual average of 20 minutes.

---

[5] There are many cases, however, where serial flow through different parts of a combined transmission and distribution substation will lead to the same result. For instance, a substation might consist of a transmission-level facility on one side of the yard in which 138 kV power is reduced to 69 kV sub-transmission level. Power from the low-side 69 kV bus might be routed to the distribution side of the yard where it serves a 69 kV/12.47 kV distribution transformer. This is not serial application, but instead only the result of having the transmission and distribution "substations" at the same location.

Reliability of transformers

Basic transformers (those consisting on only cores and windings) are among the most reliable and maintenance-free of electric devices, because they contain no moving parts to wear out and are reasonably immune to damage by weather and the elements (except for lightning, from which they are usually well protected). A basic transformer -- most at the service level fall into this category -- is installed and forgotten until it needs replacement -- a period that *averages* between 40 and 60 years in most systems.

However, distribution substation transformers are seldom so basic. First, many contain tap-changers and perhaps line-drop compensator circuitry -- mechanical devices that have many moving parts and that carry a good deal of current (they are located on the low side of the transformer). No matter how well-designed and robust, these will need routine inspection and maintenance, and in spite of the best care and service, occasionally they will fail unexpectedly. As discussed earlier in this section, many transformers also have pumps to circulate oil through radiators, and radiator fans, all requiring further maintenance and providing more possibilities for failure.

In addition, distribution substation transformers are generally inspected annually or more often for signs of incipient failure (leaking cases, cracked bushings, internal vibration or electrical noise, gas in the oil), for several reasons. First, the unexpected outage of such a transformer would interrupt service to a large number of customers. Second, replacement and repair is costly and lengthy. Third, some failures can be catastrophic, the resulting fire destroying all or part of the substation. And finally, incipient failure can often be detected early enough that repairs can be made quickly and problems avoided.

Table 10.4 Typical Annual Outage Rate Expectation for Substation Transformers Applied within Generally Accepted Loading and Maintenance Guidelines

Distribution Substation Transoformer Size and Type	66 & 69 kV high side				115 & 138 kV			
	Scheduled		Unscheduled		Scheduled		Unscheduled	
	Events	Hours	Events	Hours	Events	Hours	Events	Hours
Smaller than 25 MVA nameplate								
Basic, no LTC, no pumps or fans	.06	16	.04	40	.2	50	.03	60
with pumps and fans	.08	24	.07	40	.22	65	.04	70
with pumps and fans and LTC	.11	24	.07	48	.27	70	.03	84
Larger than 25 MVA nameplate								
Basic, no LTC, no pumps or fans	.10	24	.03	63	.2	48	.02	152
with pumps and fans	.12	36	.05	70	.22	64	.03	144
with pumps and fans and LTC	.18	32	.06	72	.27	72	.02	168

Distribution Substations

As a result, transformers are routinely inspected and maintenance performed as needed, during which time the transformer may have to be taken out of service. Actual failures while in service are rare but do occur, and usually the resulting outage period is not measured in hours, but days. Data on outage rates for distribution transformers is sketchy at best, partly because little reliable data is available and also because definitions and recording practices differ among the utilities that do have data.

Table 10.4 gives information the author believes is representative of equipment in service, listing expected events per year and duration per event.[6] From the data shown, one can determine that planners can expect transformers to be out of service, for scheduled and unscheduled reasons, a good deal more often than customers will tolerate service interruptions. For example, large 138 kV transformers with pumps, fans, and LTC are expected to be out a total of nearly one day per year, although failures are expected only once every five decades, taking a week to repair (replacement) on each instance.

$$\begin{aligned}
\text{Annual Outage Hours} &= \text{Scheduled Outages} + \text{Unscheduled Outages} \quad (10.1) \\
&= .27 \times 72 \quad\quad\quad + .02 \times 108 \\
&= 22.8 \text{ hours/year}
\end{aligned}$$

The nearly one day of outage expected means that contingency support for this transformer is mandatory in order to provide acceptable service.

Voltage Regulation

Voltage regulation at the substation is often provided by use of load tap-changing transformers. Generally, load trap-changing transformers provide variation of plus and minus 10% in low-side voltage through variation among 32 tap settings built into the low-side windings.[7,8] Equipment is provided to sense

---

[6] This data was collected from eight utilities which have different types and ages of systems, whose operating, inspection, and maintenance practices differ considerably (one does *no* scheduled maintenance), and who serve different types of load patterns and experience different climate and isochronic behavior. No attempt was made to adjust various sources to consistency and thus these figures do not reveal expectations under any one type of system or patterns in operating/maintenance philosophy, but rather are only a representative sample of the industry as a whole.

[7] Some types of load tap-changing transformers provide a narrower range of variation and fewer tap settings, with ranges of only plus or minus 7.5, 6, or 5%.

[8] In addition, to account for expected differences in high-side voltage, some substation transformers have high-side taps (that must be manually adjusted while de-energized) that permit selection of plus and minus 2.5% in addition to the normal turns ratio.

the voltage on the low side and automatically vary the tap settings to adjust for variations in voltage from either or both of two causes:

High-side voltage variation. Voltage provided to a substation transformer by the transmission system may vary by up to ten percent during normal situations due to variations in voltage drop caused by to loading and system switching. Beyond this, during transmission contingencies the voltage may drop to lower than normal levels. An LTC transformer can adjust the transformer turns ratio to compensate, providing nearly constant low-side voltage as the incoming voltage varies over a 20% range.

Primary feeder voltage drop. Under full load, voltage drop on the primary distribution feeders will increase. A line drop compensator (LDC) circuit can be added to the LTC transformer control equipment, so that as load increases, the transformer boosts its turns ratio to compensate for the higher voltage drop expected downstream of it (Figure 10.9). The LDC can be set with R and X that mirror the feeder's impedance to a certain "regulating point" (usually a point about halfway out the feeder trunk), causing the transformer to adjust its taps to compensate for the anticipated voltage drop to that point, effectively keeping voltage at that point constant as loading increases.

In some cases LCD's will be set to provide a voltage boost with increasing load that is based on the anticipated average voltage drop of the several feeders it serves. However, most often in such situations voltage regulation is provided by separate voltage regulators installed for each feeder, which are part of the low-side portion of the substation.

Salient Transformer Features for the Planner

From the planner's perspective, the important issues with respect to alternative substation transformer configurations are the cost, capacity, and reliability of the various alternative designs. In addition, the expandability and flexibility of adaptation of an alternative may be an issue when future load growth is uncertain. Generally, reliability of a multi-transformer substation is a function of both the number and size of transformers, and the high-and low-side bus/switching configurations. Providing multiple transformers with capacity sufficient to support the load when any one is out of service is necessary in order to achieve high levels of service availability. However, only compatibility of the high and low-side designs will assure that the substation has the capability to utilize this potential. Therefore, selection of the number, type, and configuration of the transformer level is very much a function of the high- and low-side configurations.

Distribution Substations

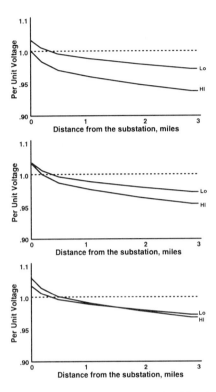

Figure 10.9 (Top) Voltage profile at high and low load levels along a three-mile long feeder trunk with no regulation. (Middle) With regulation to maintain constant low side voltage regardless of load or high-side voltage. (Bottom) Voltage with a line drop compensator set to a point 1.5 miles from the substation.

10.4 LOW-SIDE PORTION OF A SUBSTATION

The low-side portion of a substation operates at the voltage associated with the primary distribution feeders, and includes all buswork, switching and protection, voltage regulation, metering, monitoring, control, and other equipment required to provide service to the substation's feeders. Normally, any low-side breakers or fuses protecting the transformers, are also included in this level, not at the transformer level.

While the low-side portion of a substation operates at low voltage, it operates at much higher current levels than the high side, and has high fault current levels, too, so that the size of buswork and components like switches, and the structural integrity of some components must be as great or greater. However, its insulation requirements are considerably less. Typically, it is the least costly of the four parts of a substation, representing from 1/15 to 1/5 of a substation's total cost. Cost of this part is somewhat more linear with respect to substation capacity than either the high-side or transformer portions, with cost being largely proportional to the number of feeders built.

Typical low-side layout includes a low-side bus associated with each transformer, from which one to six feeders are serviced. Figure 10.10 shows several arrangements for the low side of a substation. Generally, provision is made so that adjacent transformer low-side buses can be switched together, so that in the event of a transformer being out of service, the bus it normally services can be fed from another transformer.

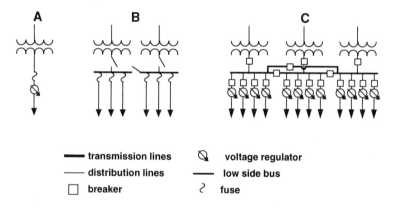

Figure 10.10 Several types of low-side substation arrangement (high-side portion of each substation is not shown). A: a single feeder with a fuse and voltage regulator, fed by a single transformer, as might be installed in a small rural location. B: a two-transformer substation with each transformer serving three feeders, a tie switch (normally open) between its two buses, and tie switches (normally closed) to isolate each transformer on the low side in the event of an outage, so that either transformer can serve all six feeders. C: a three-transformer substation with breakers connecting buses and transformers (each transformer-to-bus breaker is closed, each bus-to-bus breaker open, during normal operations) in a low-side ring-bus arrangement so that any transformer can support any other transformer's load during contingencies.

Distribution Substations   417

Protection -- either breaker or fuse -- is *always* provided for each feeder. In fact a workable definition for a "feeder" could be: "The circuit downstream of an individual low-side protective device at the substation." Additionally, some or all feeders may have voltage regulators installed at the substation. These can be either gang operated (all three phases move together and are set to the same tap) or individually operated, and set to regulate voltage at the substation, or using a line drop compensator, to a regulation point on each feeder.

In addition, there is considerable other equipment not generally important to distribution planners (which doesn't mean it is not important) nor relevant to the evaluation of alternatives in planning substations. This includes switches to allow the isolation of each bus and breaker, lightning arresters, potential and current transformers, etc. The cost of all such items should be included in cost estimates for budgeting. However, as this equipment is mandatory for every alternative, it is not germane to the issue of cost versus performance, and of no concern to the planners as they decide among alternative designs.

Low-side Reliability Considerations

Service reliability and the evaluation of alternatives based on differences in reliability is seldom a large planning concern in the layout and specification of the low-side portion of a substation, for several reasons. First, with the exception of the buswork, itself, and a few other similarly dependable, low maintenance elements, a substation's low-side equipment consists of feeder-related "sets," for example the majority of low-side equipment in a substation with eight feeders consists of the eight sets of breaker-regulator-switches-monitoring-getaway associated with serving the feeder system. A scheduled or unscheduled outage anywhere in one of these eight sets interrupts service to only a single feeder, and can be restored through feeder-level restoration procedures (i.e., switching to another feeder) within criteria and times considered acceptable for distribution-level restoration using feeder switching. Second, this equipment operates at a relatively low (primary) voltage, and is quite durable, reliable and straightforward to maintain. Thus, reliability considerations associated directly with providing contingency support for the low-side equipment are normally not a big concern in planning the layout of a substation.

Low-side compatibility with transformer contingency support plans is a major planning concern. The chief benefit derived from multiple substation transformers is the ability to provide contingency support during scheduled or unscheduled outages, but the low-side configuration and equipment must be able to support this if the potential reliability improvement is to be realized. Figure 10.10 B and C are two arrangements where switching is provided so that loads can be transferred among transformers. In B, switches provide the ability to isolate either of a pair of transformers and then tie their two buses together so that either transformer can support the load of both during an outage of the other. In C, breakers provide switching among three transformer low-side buses for the same purpose. Such bus re-switching can be accomplished in a number of ways, all essentially the same switching arrangement but implemented with different levels of speed and control characteristics.

Manual switching using load breaker switches for low-side bus ties is the least costly alternative, but the one that produces the longest interruption durations. In order to restore service, someone must travel to the substation and manually close and open appropriate switches to restore service, typically resulting in an interruption duration one-half to one-hour. Despite the delay in restoration time, use of manual switching is recommended in situations where the utility's remote monitoring of equipment and conditions at the substation (i.e, SCADA or automation) is insufficient to provide enough information to make certain that such switching is safe and will be secure. The person doing the switching can inspect the substation and its equipment to assure that conditions are appropriate, before making the transfer of load.

Motor-operated switches, either remote controlled or designed for automatic rollover, can be provided to effect both isolation of the transformer and closure of the tie to another bus, resulting in restoration within less than five minutes (perhaps within less than a minute). If done by remote control, it is sound practice if on-site monitoring is available to provide the operator making the remote switching with the status of high-side breakers, the switches, both transformers (monitoring of both temperature and load is best), and the low-side bus voltages and flows, in order to be able to ascertain:

a) The cause of the outage.

b) If conditions appear appropriate for switching.

c) That service is restored when the contingency switching is completed.

If done with automatic equipment, the "roll-over" must be controlled with equipment that affects switching in the proper sequence (i. check that the high-side breaker is open; ii. open low-side tie to transformer; iii. close low-side bus tie). Similarly, if restoration were affected by automated equipment under central control, the program logic would follow the same series of steps.

Distribution Substations 419

Breakers can be used instead of switches, providing much faster switching along with superior interruption isolation.[9] Protection can be arranged so that upon a fault or other condition in the transformer, both high and low-side breakers open to isolate the transformer, immediately followed by closure of the tie breaker. This reduces interruption time to a matter of seconds (it will still be very discernible to both digital equipment and human beings). Again, as with motor-operated switches, this function can either be implemented with an automatic (on-site) set of control-relaying equipment, or done by remote automation from a central control facility (in which case data communication back and forth adds slightly -- perhaps a second or more -- to the interruption time).

Thus, low-side reliability planning mostly consists mostly of selecting from among alternatives in switching speed and flexibility, and assessing their application to isolation and switching of the low-side buses to control outage duration -- more expensive equipment can provide shorter duration of outages.

Recommended planning procedure for this portion of the substation is to balance the cost of reducing customer-minutes of interruption here, against the cost of buying similar improvements elsewhere in the system. The working concept is to "buy" improvement were it costs the least. Frequently, this leads to application of low-side breakers and their automation, for while such equipment is relatively expensive, outages at the substation affect many customers and manual switching requires considerable time, so that on a "cost per customer-minute of improvement," money spent here has the largest impact in the entire sub-transmission/substation/feeder chain.

In addition, the actual arrangement of the buses and switching may present alternatives that need to be evaluated. In Figure 10.10 C, the low-side buswork in a three-transformer substation is configured as a ring-bus, so that the two end buses can be connected together, but kept isolated from the middle bus. Under some situations, such a bus arrangement may not be necessary, or justifiable based on cost, but there are two reasons why this may be advantageous.

---

[9] Protection would be no safer, but would be superior in terms of minimizing the extent of any interruptions caused by further problems. For example, in the configuration shown in Figure 10.10B, after contingency switching to operate with one transformer, if problems already exist, or occur on low-side bus (it may have been damaged by whatever event took the transformer out of service) then the only protection option left is to isolate the entire substation by opening the operating transformer's high-side breaker. On the other hand, if the tie switch is replaced with a breaker, it can immediately open to isolate the problem and cut the interruption extent in half.

1. Flexibility of transformer support. Without the ring-bus arrangement, the substation has full transformer contingency support for any single transformer outage, but the middle transformer must be used when either of the two end units is out of service. With the ring-bus arrangement, operators have the option of supporting the load of either end unit from the other end unit, rather than the middle transformer. There may be times or conditions in which this is preferred.

2. Unequal size transformers. The three transformers shown in Figure 10.10C may not be the same capacity. If the middle unit is the smallest of the three (perhaps they are rated at 32, 24, and 32 MVA respectively, left to right), then a ring bus is warranted so that the two larger units can provide support to one another.

Regardless, the important point in this discussion is not specifically about the ring-bus configuration, but rather that the low-side bus configuration must be compatible with and meet the contingency support needs of the substation as a whole. Figure 10.10 C is an example of such a need, but there are many other quite different situations. Planners need to ascertain what contingencies are to be supported on the high side and transformer portion and assure that appropriate low-side configuration is possible to support them.

10.5 THE SUBSTATION SITE

The substation site must be appropriate to contain the high-side equipment, the transformers and their support equipment, the low-side equipment, and anything else required, in whatever configurations are decided upon. Except for vault type packaged substations, and certain very rare types of basement or underground situations, substations are normally built above ground with the equipment exposed to the elements. Purchase and preparation of the actual site is a significant portion of substation cost -- from 1/10 to 2/3 of the site cost depending on the location and other requirements. It also varies greatly from site to site (Figure 10.11), making site cost a factor in siting and planning a substation. Elements of site cost are:

The land. Substations require anywhere from several hundred square feet (for a package one-transformer substation) to ten or more acres for a transmission-distribution substation. Both the land itself and the surveying/filing/legal process necessary to obtain a site are expensive.

Recommended procedure for site purchase is to identify siting needs far in advance through long range planning, and buy sites at whatever time provides the lowest present-worth cost, taking into account the expected escalation in price and the risk that current

Distribution Substations

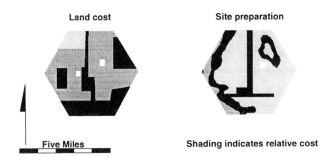

Figure 10.11 Relative site cost including land purchase cost (left) and preparation costs (right) for an 18.0 square-mile candidate substation service area on the outskirts of a major metropolitan area of the United States, developed based on assumptions about local land prices as a function of amenities (proximity to highway, etc.) and geological analysis of site substrate, etc. (This is from the substation siting example in Chapter 12).

forecasts and plans may not be perfect. This has the benefit of assuring the utility that it will obtain the site when needed -- too often waiting as long as possible means that all optimal and near-optimal, and even workable sites have been taken for development, and the utility must make do with a site that presents severe limitations or is in the wrong location.[10]

In cases where a site is bought far in advance, a further recommendation is to fence it immediately and store several large pieces of spare or surplus substation equipment there. A fenced yard with breakers and transformers inside makes it clear that the site is a utility substation, avoiding accusations of non-disclosure later on.

Civil/electrical/mechanical preparation includes providing a grounding grid, foundations for racks and equipment, and underground

---

[10] Some utilities are prohibited by municipal law or franchise regulation from buying sites in advance of need, a policy aimed at preventing a utility from benefiting from land price escalation and from any temptation to indulge in land-speculation. However, the author believes strongly that purchase of substation sites *in the open market* in advance of need, in order to obtain sites that lead to efficient T&D design, is a proper business practice for all utilities.

ductwork for control and communications cables as well as underground electric T&D cables, a control house, and other facilities such as oil spillage containment tanks, etc., to support the electrical function of the substation. These costs can depend greatly on the type of underlying soil and rock.

Feeder getaway. Routing a large number of feeders out of a substation can become a challenge in congested or otherwise restricted sites. Very often, particularly if not picked carefully far in advance, the sites available to a utility may have severe limitations on the routing of distribution rights of way and easements out of the site. Figure 10.12 illustrates this point. The substation in question was the only one available for a 138 kV/12.47 kV 3 transformer substation with 12 feeders. Wetland and other considerations limited the utility to less than 90 degrees of the compass for routing of feeders out of the substation. Even though all construction was intended to be overhead, eight feeders had to be routed up to 1/6 mile away in underground ducts before available space above ground was found for them. Total cost of meeting these special feeder getaway requirements totaled over $660,000.

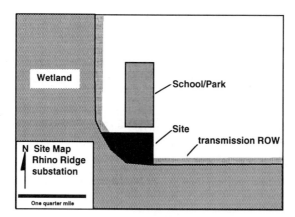

Figure 10.12 A particularly poor site for a large substation from the standpoint of feeder getaways. Three quarters of the directions that could normally be used are precluded due to a federal wetland area. A nearby school further limits feeder routing.

Distribution Substations

Table 10.5 Representative Substation Site Costs -- $ × 1000

Substation Type	Area Required	Cost
Rural 46 kV/22 kV 12 MVA capacity, wood pole constr.	8,000 sq. ft.	$11
Suburban, 161 kV/25 kV, two 60 MVA transformers, normal	1.3 acres	$490
Same as above, but site is steep hillside, rock subsurface	1.3 acres	$1,290
33 kV/11 kV packaged "European" UG vault, 15 MVA	2 x 4 x 2 m.	$21
Urban, 138 kV/34.5 kV, 4 x 25 MVA transf. brick wall	1.6 acres	$2,000
Urban, 230 kV/25 kV, 5 x 50 MVA, underground	1 acre	$3,500
Suburban, 345 kV/230 kV/115 kV/23 kV/13.8 kV, 340 MVA	8 acres	$4,200

Public safety and esthetic site preparation includes at a minimum a fence or wall around the site to secure it from public access. In problem areas, the fence or wall may need to be up to ten meters high. The utility may be required to landscape the site, with a green space including foliage to block view of the equipment. In some cases, the substation may be enclosed in what appears to be a building (often called a "cottage substation") to completely hide it.

A substation utilizing underground transmission and distribution can be built in the basement of a large building, but this is generally an option only in dense urban areas where land cost for an above-ground substation is absolutely prohibitive. Space and clearance limitations of basement substations often require GIS insulated equipment to be used throughout.

Taxes and permits. Investor-owned utilities usually have to pay initial building permit fees, etc. and continuing annual property taxes on the site and its installed equipment. Tax rates vary enough among different counties and municipalities that the location of a substation in one or another can noticeably affect its present worth cost.

Substation site cost varies greatly depending on size, type of substation, location and subsurface material. Table 10.5 gives costs for several substation sites. Generally, the specific site and the manner of preparation have little if any influence, on reliability.

10.6 SUBSTATION COSTS, CAPACITY, AND RELIABILITY

As discussed above, substations can be viewed as having four major components, each of which has a cost. Generally, cost increases with higher voltage, greater capacity, and higher reliability. *Very rough rules of thumb* about cost are:

> *Minimum site.* The minimum economical capacity for a substation is roughly equal to one fourth the high-side voltage. A "minimum" 69 kV/XX kV substation can serve about 17 MVA -- anything less costs little less than a 17 MVA substation. A "minimum" 230 kV/XX kV substation will serve about 57 MVA.
>
> *Linearity of cost with increasing capacity.* Site, high-side, transformer, and low-side costs are roughly proportional to capacity when it is above the minimum economical size for a substation.
>
> *Site cost* runs about one fifth of total cost. It tends to increase slightly as low or high-side voltage level is increased for substations that are otherwise identical in capacity and configuration, due to the larger clearances/equipment required at higher voltage and the greater preparation and site costs for the higher voltage -- a 230 kV/25 kV substation of 120 MVA capacity is about 12% more costly than a 161 kV/19 kV substation of the same capacity.
>
> *High-side cost* above 138 kV tends to increase as the 1.25 power of the voltage level -- 230 kV costs about 2.35 times that of a 115 kV. Costs at voltages below 87 kV and above 33 kV are generally linear as a function of voltage -- 87 kV costs about twice what 46 kV does, etc.
>
> *Low-side cost* tends to increase as the square root of primary voltage level -- increasing voltage from 4 kV to 34.5 kV (8.6 times) increases low-side cost by about a factor of three.
>
> *Transformer cost* varies about halfway between high- and low-side variation.
>
> *Customization is expensive.* Whether dictated by unusual circumstances or a desire to reduce cost, departure from standard design practices usually increases cost, often in unanticipated ways.

Table 10.6 lists representative costs (1997 dollars), capacities (maximum continuous), and service interruption rates (minutes per year) for several substations.

Distribution Substations

Table 10.6 Representative Costs of Distribution Substations - $ x 1000

Substation	Location	Capital	O&M
69 kV/25 kV, one incoming circuit (radial) with high-side fuses, 7.5 MVA transformer (no LTC) one feeder (fused) with voltage regulator. All pole and crossarm rack construction. Peak load 4.4 MVA. All OH. Built with used (rebuilt) transformer and regulator.	rural	$61	$3
33 kV/11 kV completely packaged 15 MVA vault type unit with high-side fuses, three low-side feeder connections (all fused). All UG. No voltage regulation. Typically serves 12 MVA.	urban	$100	$3
69 kV/13 kV, two incoming UG circuits with breakers two 33 MVA LTC transformers, two low-side buses each leading to three 12 MVA UG feeders each with a breaker and voltage regulator. Metal-clad switch. "Cottage structure" enclosed. Peak load of 45 MVA.	urban	$3,200	$135
161 kV/ 24 kV, two incoming OH circuits with breakers two 60 MVA LTC transformers, two low-side buses each leading to three 22 MVA feeders each with a breaker and voltage regulator and UG getaway, all steel construction, enclosed control house. Extensive landscaping. Serves a peak load of 96 MVA.	suburban	$4,100	$135
230 kV/25 kV to 12.47 kV, five incoming OH circuits with ring bus, three 75 MVA 230/25 kV and two 50 MVA 230/12.47 kV LTC transformers, low-side bus breakers, twelve 25 kV and eight 12.47 kV OH feeders, all with breakers, voltage regulators on 12.47 kV, 200 yard UG getaways. Steel outdoor construction. Control house. Peak load 288 MVA.	suburban	$11,200	$320
230 kV/25 kV, three incoming UG circuits with ring bus, five 75 MVA 230/25 kV LTC transformers, (four in service and a fifth energized as standby), breakers, fifteen 25 kV UG feeders with breakers. Underground (basement) facility. Metal clad switchgear. Peak load 225 MVA.	urban core	$19,200	$350

10.7 OTHER SUBSTATION PLANNING CONSIDERATIONS

Unit, or Completely Packaged, Substations

Unit substations include all parts of a substation packaged in a single assembly, usually a metal-clad weatherproof housing, with separate portions for each of the three parts of a substation, and include, usually in separately accessible cabinets: a high-side section with connectors for one (for radial service) or two (for loop service) incoming high-side circuits. Some unit substations have breakers for one or both sides, but more typically only fuses or no protection at all is provided. The transformer may or may not be a load trap changing unit with or without a line drop compensator.

Such units are available for overhead application, as well as for underground systems in either padmount or vault-type applications. Available combinations include high-side voltages as high as 115 kV and as low as 5 kV, and low-side voltages from .416 kV to 34.5 kV.

Unit substations are also sometimes called CPS (completely packaged substations) or CSP (completely protected substations). They are essentially modular, "completely protected" transformers. If used in "dual voltage" feeder systems (which can be viewed as a combined sub-transmission/primary voltage system) they can provide a very standardized, modular, and effective way to lay out economical and efficient T&D systems.

Mobile Substations

Occasionally, a catastrophic equipment failure may mean that an entire substation will be out of service for an extended period of time, regardless of design provisions for contingency backup. In such cases, mobile substations can be used, their equipment being mounted on one or more large tractor trailer assemblies that can be moved to the site to provide temporary service.

Mobile substations are generally modular in design and limited to about 40 MVA maximum capacity, due to size and weight limitations. They are not much more expensive than equivalent permanent facilities, but typically take three to six hours to connect and energize once moved to a site, in addition to whatever time is required to actually move them to the site.

In situations where several substations in a system do not have feasible contingency support for major failures (buswork/transformer), mobile substation provide a prudent contingency backup plan. One mobile substation can provide "support" to a number of substations.

Distribution Substations

Planning with "Transformer Units"

Often, the planning and design of substations for a particular system is done using transformer "units," a unit including all equipment associated with a transformer and its standard application as part of the system. This equipment is usually not contained in a pre-packaged assembly as is the case with unit substations. Rather, the utility's engineering department will have laid out and specified a series of standard designs (units) each of which includes a transformer of a certain capacity, along with the site preparation (foundation, grounding, ductwork, etc.) necessary to install it, as well as appropriately selected and sized racks and buswork, switches, breakers, monitoring and control equipment, oil spill containment and fire prevention systems, and any cooling and noise abatement equipment.

Everything in a unit is compatible in size, type, and installation with the utilization of the transformer. The buswork specified will be capable of supporting the full load of the transformer. The breakers specified will be the most economical to match the short circuit duty requirements of the transformer, etc. They form an economical "unit" that fully utilizes the transformer's capacity while meeting corporate criteria on safety, protection, loading, etc.

Each standard unit design will include a standard load rating (usually the rating of the transformer) that indicates its capacity application, as well as an estimated cost, both of which can be used in system planning. For example, while a 24 MVA, 138/12.47 kV transformer might cost $166,000, a complete 24 MVA unit installation with breakers on the high and low sides, appropriate switches, racks buswork, and control equipment, and all site preparation and labor might have an estimated cost of $380,000. A "transformer unit" might also include all the low-side equipment associated with the feeders it is intended to serve. For example, planning evaluation may have determined that a 24 MVA transformer unit should most efficiently provide power to four feeders. Including the switches, breakers, voltage regulators, monitoring and control equipment, in-yard ductwork and riser poles (if any) for feeder getaways might raise the price of this unit to $640,000.

Unit designs will typically include several sizes/configurations (Table 10.7 with all equipment in each unit selected to be compatible as to size and to all applicable corporate safety, protection, and engineering criteria (Figure 10.13). A few utilities have standardized on only one unit size-- which they apply to all new substations -- if more capacity is needed, more units are built at a particular site. Typically, this size will be slightly greater than the "minimum economical size" for the high-side voltage, and utilize a "commodity" transformer size -- one that is common and thus available at low cost due to competition among suppliers. For example, this is about 22 to 25 MVA FOFA at 115 to 138 kV.

Table 10.7 Typical "Unit" Costs as Used for Substation Planning of 138 kV/25 kV Overhead Distribution Substations -- $ × 1000

Item	Capital	O&M
Site (land, preparation, grounding, control building, fence, etc.)	$770	$7.0
High-side cost including station towers, buswork, switches, site work		
radial service (up to 40 MVA) including one breaker	$510	$2.5
loop service (up to 120 MVA) including two breakers	$860	$4.0
loop service (up to 240 MVA) including two breakers	$1,180	$5.5
Transformer units inclusive		
22/29/37 MVA with two 19 MVA feeders	$820	$12.0
33/44/56 MVA with three 19 MVA feeders	$1,090	$15.5
50/67/83 MVA with four 19 MVA feeders	$1,430	$20.0

Example: substation with	Capital	O&M
basic site	$770	$7.0
120 MVA incoming loop capacity	$860	$4.0
3 × (33/44/56 MVA units + 3 × 19 MVA feeder)	$3,270	$46.5
Estimated cost	$4,900	$57.5

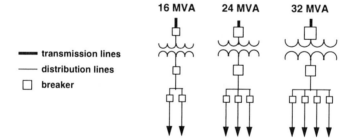

Figure 10.13 Concept of "transformer units" includes specifying all equipment for the complete installation and application of a particular size transformer as a unit, with all support equipment including buswork, switching and breakers, monitoring and control, etc., sized and selected to be compatible, and to be the most economical that matches the transformer's capacity. Illustrated here are three units encompassing 138/12 kV transformers of 16, 24, and 32 MVA respectively, along with all equipment to support the high-side and low side -- including equipment for two, three, and four 9 MVA feeder bays, respectively. Size of symbols gives an indication of capacity.

Distribution Substations

10.8 PLANNING WITH "TRANSFORMER UNITS"

Substation planning, engineering, layout, and design based on a series *of unit transformers is a recommended practice,* for several reasons. First, developing sets of compatible, least-cost equipment of varying capacity levels helps focus attention on standardization and economy, and assures compatibility of design among all parts of a substation.

Even in cases where the utility has not engineered such unit designs in detail, recommended practice is for distribution planners to define a set of transformer units and estimated costs for use in planning, as shown in Figure 10.13. With such a system, a substation is reduced to a series of modules including a site cost, a high-side cost, and one or more transformer units that include the transformer and the low-side facilities associated with it. While such standardization is theoretically less accurate than explicit analysis on a case-by-case basis, in practice it is better, for it streamlines procedure, helps avoid omissions, and focuses attention on how they whole plan fits together.

10.9 CONCLUSION

Planning: More Is Involved Than Just The Substation

Substations are key elements of the T&D chain, and proper selection of equipment and configuration is crucial to successful planning of the delivery system. However, much of a substation's impact on system performance and cost goes beyond the design, capacity, or reliability of equipment inside the substation fence. Rather, it is a function of the substation's location and its role in the system as a whole. As mentioned at the beginning of this chapter, because substations are the meeting point of T and D, they tend to have an influence on system design all out of proportion to their cost. This influence, and planning to accommodate it to advantage into the overall least-cost system design, is discussed in Chapters 11 and 12. A critical element of T&D planning is assuring that the substation equipment and configuration fits this overall system design.

REFERENCES AND BIBLIOGRAPHY

J. K. Dillard, *Distribution Systems, Electric Utility Engineering Reference Book,* Westinghouse Electric Company, Pittsburgh, 1959.

M. V. Engel, *IEEE Distribution Planning Tutorial,* IEEE text 92 WHO 361-6 PWR, Institute of Electrical and Electronics Engineers, New York, 1992.

11
Distribution System Layout

11.1 INTRODUCTION

A distribution system is composed of many parts, and although the distribution planner has great discretion over the selection of these parts and many aspects of their siting and installation, ultimately they must be assembled into a complete system leading from the power grid to the customers. A combination of good economy and efficient performance comes only from joining the myriad levels and types of equipment involved in a T&D system into a compatible whole, and that happens only by design. "The system" is more important than any of its parts, and specification and design to fit each part into the system is the critical step in effective delivery planning -- *it is more important that the parts fit together well than that any of them be individually ideal.*

This chapter, the fourth of four that examine the elements and layout of distribution systems, focuses on how all the parts of a T&D system act in concert and how electrical performance, reliability, and most of all, cost, "add up" to define the character of a power delivery system. This examination begins in section 11.2, with a look at a typical delivery system -- an example used throughout this and subsequent chapters -- and the contribution each of its parts makes to performance and cost. An example T&D system is then examined in detail in section 11.3, looking at how its cost and performance change as various

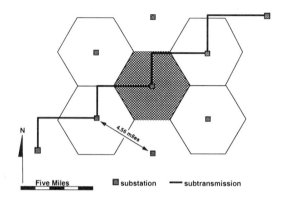

Figure 11.1 Slightly idealized concept of a delivery system to be used in analysis of costs and performance in this chapter. Sub-transmission at 138 kV delivers power to 81 MVA capacity substations for distribution on a 12.47 kV feeder system. Substations are 4.56 miles apart. Each substation service area (shaded area) is 18 square miles, giving a peak load of 58.5 MW (65 MVA) at 3.25 MW/mile^2 (3.61 MVA/mile^2 at 90% PF).

Table 11.1 Characteristics of the Example T&D System

Peak demand density of 3.25 MW/ mile^2 (1.5 MW/sq. km)
  525 customers/mile^2 6.19 kW/customer average peak
90% power factor, for a peak load of 3.61 MVA/mile^2
Peak occurs annually as 42 periods of 3 hours each.

Annual value of	load factor	loss factor
sub-transmission	.63	.45
substation	.61	.39
feeder	.57	.34
service	.45	.25

Roads & property lines on a 440-foot (1/12 mile) grid

138 kV sub-transmission
  600 MCM ACSR, 220 MVA thermal capacity

Substations hexagonally 4.56 miles apart (service areas of 18 sq. miles)
  peak load = 58.5 MW or 65 MVA
  3 × 28 MVA LTC, 138/12.47 kV autotransformers
  high-side ring bus
  low-side bus tie breakers

12.47 kV primary distribution, all OH
  conductor: 636 MCM, 336 MCM, 4/0, #2 ACSR
  1113 MCM for heavy getaways

Distribution System Layout 433

aspects of its design -- such as substation spacing, primary voltage level, and similar design factors -- are changed. Similarly, Section 11.4 examines an example rural delivery system, one dominated by voltage drop and load reach considerations instead of capacity limitations as in typical urban/suburban systems. The chapter concludes with a list of general traits about the structure and performance-cost tradeoffs involved in T&D system layout.

11.2 THE WHOLE T&D SYSTEM

Figure 11.1 shows a six-substation area of an example power delivery system, a slightly idealized layout in which every unit at every level is identical (i.e., every substation service area is the same size and shape, every feeder is the same area and peak load), but nonetheless this system is a very real, representative form of a T&D system. This example, one of several to be used in this chapter, is based on a large metropolitan suburban/urban system in the central United States.

Important characteristics of this example system are given in Table 11.1 Peak demand density is 3.25 MW per square mile, typical of the demand density one finds in the fully-developed sprawl that lies between the growing, peripheral regions of a large city and the dense downtown core. Power factor is 90%, giving a load of 3.61 MVA/ mile^2.

Substations are located 4.56 miles apart, arranged in a hexagonal pattern. Each has six neighboring substations, and serves 18 sq. mi., with a peak load of 58.5 MW (65 MVA). Substation capacity is 81 MVA (FOFA) utilizing three 17/22/27 MVA LTC transformers, giving an 80% utilization factor against FOFA rating at peak conditions. High-side configuration uses a ring-bus arrangement (Figure 11.2). Transformer and low-side portions of each substation have been laid out with transformer units consisting of a transformer, breaker and low-side bus, and three feeder breakers, as shown in Figure 11.2. There are no individual feeder voltage regulators in this system, but the transformer LTCs are used to maintain low-side bus voltage at constant levels.

Sub-transmission circuits operating at 138 kV with H-frame, 2 x 477 MCM ACSR bundled conductor, and rated at 220 MVA, serve three substations each, for a peak load of 195 MVA, as shown in Figure 11.3. Each circuit "loops" an average of 25 miles between end points that feed from the high-voltage grid.[1]

---

[1] Although substations are located an average of 4.56 miles apart (Euclidean distance), sub-transmission lines cannot always follow shortest-distance routes, and in some cases "dog-leg" into substations, as discussed in Chapter 10. In the system upon which this example is based, sub-transmission routes average 37% longer than line-of-sight distance between substations, about the best the author believes can be done, and the ratio used here.

434   **Chapter 11**

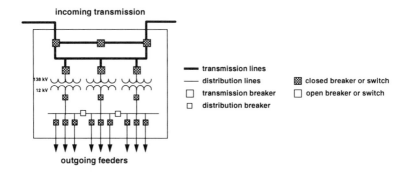

Figure 11.2 Substation configuration in the example system includes a ring-bus on the high side, and three 27 MVA "transformer units," each feeding three 12.47 kV feeders.

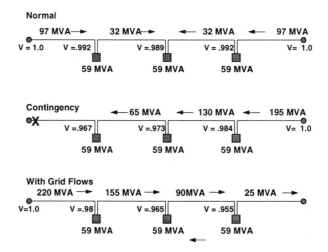

Figure 11.3 Sub-transmission circuits in the example serve three substations each, with a peak demand of 175 MW (195 MVA) -- (top). The worst-case contingency is the loss of either end segment (middle), in which case the three substations must be fed radially from the other end, with additional voltage drop reaching 2.5% above normal and line loadings reaching 90%. However, the worst voltage drop and highest line loadings will occur during times when the grid is using the line to transship power (bottom).

Distribution System Layout

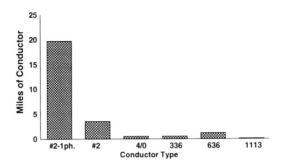

Figure 11.4 Miles of line by type for a feeder in the example system. Of the 25.6 miles of 12.47 kV overhead route, 19.7 miles -- over three quarters -- are single-phase #2 lateral. Only slightly less than two miles are "big conductor" -- 336, 636, or 1113 MCM.

Served with power from both ends, this is equivalent to two 12.5 mile radial lines serving loads of 65 and 32 MVA at distances of 6.25 and 12.5 miles.

Primary distribution in this example is all overhead, built on wood pole with crossarms (nine-foot spacing) where needed, with a conductor set including #2, 4/0, 336 MCM, and 636 MCM conductor, along with 1113 MCM available for large ties.[2] Single-phase laterals are built with #2, and occasionally 4/0. Feeder conductor economics are as computed in Chapters 7 and 9, with a linear economical range from 0 to 8.3 MW, an f_v of $88,000 and an s_v of $45,250/MW-mile.[3]

The nine feeders in each substation serve an average area of 2.0 square miles each, with an average peak load of 6.5 MW, almost exactly in the center of the economical range (5.1 to 8.3 MW) for 636 MCM conductor as determined in Chapter 7 (and applicable to this example). The feeders run through a typical suburban/urban street grid, with a block spacing of 1/12 mile (440 feet, about 133 meters). The longest feeder run (assuming all feeders are routed the minimum possible distance through the street grid) is 3.4 miles, just above the

---

[2] The 1113 MCM conductor is used solely for its high (30 MW) thermal capacity, and only for feeder getaway and line segments near the substation, where that high capacity provides switching flexibility -- the ability to support three feeders during contingencies and to tolerate limited loop flows during "hot switching."

[3] The conductor economics in Chapter 7 were computed in MW, assuming a 90% power factor, as in this example.

minimum economical load reach for 12.47 kV, as computed in Chapter 7. Figure 11.4 shows the distribution of line types for a typical feeder.

Feeders are laid out in a multi-branch configuration with all line segment types selected from the tables developed in Chapter 7, with three switchable zones per feeder, along the lines of Figures 8.24 and 8.25. Layout includes the near-substation switching arrangements shown in Figure 8.22, with line segments built to support switching near the substation using 1113 MCM for added thermal contingency capacity. The three-zone scheme along with near-substation switching results in an average of 12 line-switches per feeder.

Each feeder serves an area of 2.0 square miles, with a total length of routing (including laterals) of 25.6 miles, for a feeder route-density of 12.8 miles per square mile, or 106 feet per acre served. Figure 11.4 shows the distribution of line types in every feeder. Of the 25.6 miles of primary-voltage routing in each feeder, 19.66 miles -- 77% -- are single-phase #2. Only slightly more than 1.6 miles (less than 7%) are "big" conductor -- 336, 636, or 1113 MCM.[4] Were feeders laid out with a large trunk design instead of a multi-branch design, the results would be inconsequentially different -- still over 19.5 miles of single-phase #2 and less than two miles of large conductor. (Feeder system layout in this system will be discussed at greater length later in this section.)

Performance of the Example System

Capacity utilization

Under circumstances where all equipment is functioning and there are no contingencies, sub-transmission segments are stressed by their substation delivery duties to a maximum of only 44% (97 MVA vs. 220 MVA capacity), with the average loading at peak running far less, at only 30% (65 MVA/220 MVA) (see Figure 11.3, top). However, such statistics give a deceptive view of loading and voltage drop at the substations, for two reasons. First, during sub-transmission contingencies (Figure 11.3 middle) the maximum loading on some sub-transmission segments will increase to nearly 90% of their thermal rating. Much more important, and expected to occur more often, are "loop" flows where these sub-transmission lines transship current for interconnected grid purposes (Figure 11.3, bottom). This is actually the "worst case" scenario from the standpoint of voltage at the high-side distribution substation busses, with worst-case voltage drop reaching 4.5%, one reason for the use of LTC transformers.

---

[4] Despite this, every feeder is fully capable of completely meeting contingency switching capabilities within range B criteria, as described in Chapter 8, and Figure 8.25.

Distribution System Layout

The high-side ring bus assures that all transformers in service at a substation will be fed from both incoming lines, regardless of any single outage. With all three transformers in service, loadings at peak load reach 80% of FOFA ratings. The voltage on the low side is maintained at 1.0 by load-tap changing, and energy losses are 3.3% (2 MVA). During contingencies at peak load with only two transformers in service, average transformer loading increases to 120% of FOFA. Low-side voltage still can be maintained at 1.0, but energy losses swell to 4.2%.

As mentioned above, all feeders throughout this system have been laid out in a three-zone multi-branch configuration (see Chapter 8) with all line segments sized according to Table 7. 2, and feeder getaway (from the substation to the first near-substation line switches) using 1113 MCM conductor. Segment loading at peak load, averaged over all line sections in the system, is only 36% of thermal rating. This low utilization is a result of the minimization of PW costs in conductor sizing (and fairly typical of systems where 30-year losses costs have been included in the PW evaluation). During contingencies, short sections of 636 MCM feeder getaways and branches may be loaded to as high as 87% of thermal rating at peak load, the highest feeder segment loading expected during normal and planned contingency situations.

Voltage and Losses

Table 11.2 gives per unit voltages at each level of the system under various normal and contingency conditions. Table 11.3 shows percent peak and energy losses by level of the system.

Table 11.2 Lowest Voltages in the Example System

| System | Normal | | | Contingency | | |
Level	Low	Med.	High	Low	Med.	High
High-side bus	.98	.967	.955	.986	.98	.967
Low-side bus	1.035	1.035	1.035	1.035	1.035	1.040
Feeder ends	1.00	.9	.965	.975	.960	.950

Table 11.3 Percent Losses in the Example System

System	At Peak	Annual Energy
Sub-transmission	1.0%	.71%
Transformers	3.3%	2.5%
Feeder system	4.3%	2.6%
TOTAL	8.6%	5.2%

Table 11.4 Costs of Major Equipment Types in the Example System - $ × 1000

Equipment	Initial	Annual
Sub-transmission, per mile	223	4
Substation site	1,355	14
basic high-side (2 transf.)	850	18
to support add'nl. transf.	275	6
per transformer "unit"	500	22
Feeder system		
per mile of line	45-185	1-3
switches, per feeder, (12)	185	7

Table 11.5 Cost of a Complete Substation of the Example System - $ × 1000

Equipment	Initial	O&M	PW	Losses	PW	Total
Sub-transmission, per sub. (8.3 mi..)	1848	43	2214	81	695	2909
Substation	3980	104	4876	352	3046	7923
Feeder system	18798	433	22526	358	3085	25611
Total =	24626	580	29615	791	6826	36443
Cost per kW peak, dollars	421	10	509	14	117	623

Table 11.6 Cost of the Example System by Level - Percent

Equipment	Initial	PW
Sub-transmission, per sub. (8.33 mi..)	8	8
Substation	16	22
Feeder system	76	70
Total =	100	100

Distribution System Layout

Cost of the example system

Table 11.4 gives the unit costs for various elements used to build this example system. Costs are, as much as the author can determine, representative of the true construction and maintenance cost of each item.[5] Table 11.5 gives the cost of the system on a per- "complete substation" basis, including: 8.3 miles of sub-transmission, the substation, and the entire nine-feeder system associated with it -- basically everything in the T&D system required to take power from the grid and deliver it to the service level.

Table 11.5 shows that the feeder system is by far the most expensive element of the "complete substation." Total PW cost works out to slightly more than $620/kW of delivered peak load or 1.3 cents per delivered kWhr. Interestingly, if the cost of the feeder system is estimated with an f_v of $88,000 and an s_v of $45,250/MW-mile as developed in Chapters 7, 8 and 9, its total PW cost estimate is $25.25 million, within 1.5 % of its cost as computed in detail.

Comments on the cost of the feeder system

The feeder system in this example, as in most distribution systems, represents the bulk of the system cost: whether viewed as initial capital investment or total present worth, it is roughly two-thirds of the total. As such, it and its costs are the crucial elements of distribution cost. A thorough examination of feeder system cost is warranted in any attempt to understand power delivery cost and how it might be reduced.

The present worth of losses shown for the feeder system in Table 11.4, $3.1 million, is only one sixth of the construction cost of the feeder system ($18.7 million) and only 12% of the total feeder system cost (which includes O&M, in addition to capital and losses). This small contribution by losses may seem surprising considering the discussion of conductor economics in Chapter 7, in which the PW of future electrical losses, at economical loading levels, was shown to be roughly equivalent to the construction cost of a feeder. However, the result obtained here, and in most systems, is actually not inconsistent with the conclusions of that chapter, *for although cost of PW losses dominates feeder line-type selection economics, losses cost will be only a small portion of the cost of the resulting feeder system.*

---

[5] As mentioned in Chapter 6, construction costs used in planning differ greatly among utilities, due to varying accounting practices (e.g., cost for a mile of new 12.47 kV three-phase OH feeder with 336 MCM conductor varies from $25,000 to $230,000). The value used here is $142,000, which is 93% (100% minus a profit margin) of what an outside contractor quoted to build such lines for the utility upon which this example is modeled.

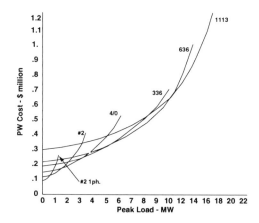

Figure 11.5 Conductor selection evaluation from Chapter 7 for the lines types used in the example system, evaluated at 90% power factor.

Figure 11.5 shows the conductor economics for the line types used in this example: basically those developed in Chapter 7. Losses costs, which are responsible for any increase in cost over the Y-intercept for a line type, appear to dominate this diagram. For example, lines built with 336 MCM conductor -- the middle of the five conductors in the #2-4/0-336-636-1113 set -- have a construction cost of $142,000/mile, which with the PW of future non-variable O&M costs added, gives a Y-intercept of $173,000/mile. The PW of future losses costs at 5.1 MW annual peak load (the upper limit of this line type's economical range) comes to $145,000 PW, roughly equal to its construction cost, and amounting to 44% of the total PW cost for that type of conductor. The largest cost plotted -- $1.2 million for delivering 17 MW with 1113 MCM conductor -- is composed of 75% losses cost.

But only a very small portion -- less than 7% -- of the example feeder system is built with heavily-loaded, large conductor. The vast majority of line segments are of #2 single- and #2 three-phase segments with loads *far less* than 1 MW -- in fact, there is twice as much length of single-phase #2 lateral carrying less than 25 kW (i.e., with only one service transformer downstream of it) than there is large conductor line carrying more than 1 MW. And for this smallest line type, losses costs represent only a small part of lifetime cost regardless of what its loading is. At its maximum recommended utilization -- about .5 MW -- only 20% of the total cost of #2 single-phase is due to losses.

Distribution System Layout

Table 11.7 Losses as a Percent of Total PW Cost; Percent Utilization in the System, and Percent Contribution to Total Feeder System Cost, by Line Type

Line Type	Losses - % of its Total PW Cost	Utilization - % of Feeder System	Contribution to Total Feeder Cost
#2, 1 ph.	8%	77%	6.2%
#2, 3 ph.	17%	15%	2.6%
4/0 3 ph.	29%	2%	0.6%
336 3 ph.	39%	2%	0.8%
636 3 ph.	51%	4%	1.9%
1113 3 ph.	30%	.3%	0.1%
	Total losses cost, as a percent of total feeder cost		12.05%

Table 11.7, first column, shows the proportion that losses contribute to a feeder line type's total PW cost, weighted by usage in the example system. For instance, as utilized in this system, #2 conductor single-phase lines average only 8% PW losses cost, because on average they are loaded to considerably less than their maximum economic limit.

The second column in Table 11.7 shows the utilization of each line type in the system -- for example #2 single-phase represents 77% of all feeder miles. Column three is the row product of the first two columns, and gives the contribution to total losses cost by each line type, showing where the 12% total losses cost comes from. Over half of the losses occur on the single-phase #2 laterals. Overall, this analysis highlights a very important aspect of distribution system layout in almost all systems:

> While it is important to minimize losses, they represent only a small fraction of the total cost of distributing power.

Table 11.8 shows total feeder system cost, again broken out by line type. Capital cost represents nearly three quarters of total PW cost. And the capital cost for the smallest conductor type represents almost half (47.5%) of the total PW cost of the feeder system. Since the feeder system itself represents 76% of the total capital cost of the T&D system, this means *that construction of the smallest line type cumulatively represents over one third of the capital cost of the entire power delivery system.* This is a typical result, not one unique to this example system.

Table 11.8 Breakdown of Feeder System Cost by Line Type and Cost Category - Percent

Line Type	Utilization	Capital	O&M	Losses	% of Total $
#2, 1 ph.	77%	47.5%	11.0%	6.2%	64.7%
#2, 3 ph.	15%	14.2%	1.7%	2.6%	18.5%
4/0 3 ph.	2%	2.4%	0.4%	0.6%	3.4%
336 3 ph.	2%	2.7%	0.4%	0.8%	3.9%
636 3 ph.	4%	6.1%	1.0 %	1.6%	9.0%
1113 3 ph.	.3%	.5%	0.1%	0.1%	0.6%
TOTAL	100%	73.4%	14.6%	12.0%	100.0%

Table 11.9 Annual Failure Data for the Example System Frequency in Events/year and Duration in Minutes/year

Equipment	Frequency	Duration
Grid (source)	.03	1
Sub-trans. per mile	.015	12
High-side bus	.02	36
Transformer	.033	72
Low-side bus	.038	36
Feeder, lateral per mile.	.48	3
Feeder re-switching time, avg.	.55	
Service	.15	3

Table 11.10 Annual Outage and Interruption Data for the Example System from Non-Catastrophic Causes Frequency in Events/year and Duration in Minutes/year

Equipment	Frequency	Duration
No power, high-side	.033	.03
Insufficient Transf. Cap.	.001	.01
No power for low-side bus	.045	.15
Feeder, at head of trunk	.075	.25
Feeder, at end	1.65	.65
Avg. Lateral	1.31	.90
Service transf.	1.45	1.50
Customer	1.52	1.53

Distribution System Layout 443

The capital cost of this smallest line type, at $80,000 per mile,[6] is the least the utility can spend to run primary voltage to a location. Nearly 20 miles of it are required in every feeder, which is what makes it expensive. This emphasizes the point made in Chapter 7: *lowering the initial cost of the smallest line type is the single biggest factor in reducing feeder cost*, and not just feeder cost, but total cost. If this cost could be reduced by 15% -- from $80,000/mile to $68,000 per mile -- the savings would exceed the cost of the entire sub-transmission system.

In fact, that reduction in capital cost for this example system can be effected by substituting #6 conductor for #2 in laterals and using slightly lighter poles and hardware throughout. Increased PW losses costs caused by the smaller conductor size's higher resistance equal almost exactly the capital savings, so overall cost changes very little. Despite this, most utilities would take the capital savings in company with the increase in long-term cost, since the two are equal. However, if the change is made, reliability worsens (experience has shown the lighter line hardware is less robust in service, leading to a 16% higher service interruption rate), and as this line type represents over three quarters of the system and is always in close proximity to the customers, this cost-reduction measure is considered unacceptable).[7]

But the overall message of this analysis is clear: *roughly a third of T&D delivery cost is due to the need to "run a wire" to every location.*

Outage and service reliability of the example system

Table 11.9 gives the inherent equipment reliabilities and switching times used in estimation of the service reliability for the example system. Table 11.10 gives the resulting expected outage statistics on an annual basis at various levels of the system -- the values shown in each row are for complete cessation of power delivery capability at a particular point in the system, and are not additive (example: if a low-side bus is out of service, its feeders can be switched to another bus so that although the bus is out of service, its feeders are not, etc.). Outages and the service interruptions that result shown in the tables, exclude any that might be caused by major storms (defined as weather expected less often than once every 20 years), such as a "direct hit" to the area by a hurricane, or a tornado, or natural catastrophes (earthquakes, floods).

The resulting service reliability level -- basically three interruptions and three hours interruption every two years, is typical of many North American utilities.

---

[6] With the PW cost of future O&M&T added, this cost is $88,000/mile, the f_v value.
[7] The utility upon whose system this example is based used #6 as a line type for overhead laterals until 1989, when reliability concerns dictated a change to #2.

Table 11.11 Comparison of the Example System to An Actual Utility System Built with the Same Standards

Measurement	Actual	Example
Total load - MW	354	351
Area covered - square miles	108	108
Number of substations	6	6
Number of subs. transformers	17	18
Installed subs. cap. MVA (FOFA)	523	486
Miles of sub-transmission	49.9	50
Number of feeders	55	54
Miles 3-ph. main trunk	108.8	101.5
Miles lateral	1,205	1,272
Number of line switches	661	648
Service interruptions, events/year	1.47	1.52
hours/year	2.05	1.53
Capital cost, per kW peak	$449	$421
Total PW cost, per kW peak	$656	$623
Feeder losses as a % of feeder PW	12.0	12.0

Comparison of the Example and Actual Systems

Table 11.11 compares various data on the example system with those for a six-substation area of the T&D system on which the example is based -- the western suburban region of a large metropolitan area in the central United States. The example system differs from the actual system in only four categories:

a) The actual system has 523 MVA instead of 486 MVA of transformers: due to a difference in past design standards from those used in this example, some substations have 35 MVA rather than 27 MVA units.

b) Actual routing of feeder trunks is 7% longer than in the example system. The real world puts barriers in the way of least-distance routing. Distribution must go around cemeteries, parks, etc.

c) Length of laterals in the actual system is 9.5% less than in the example system, because the distribution system does not have to reach into cemeteries, parks, etc.

d) Interruption duration time is nearly half an hour better in the example than the actual system. The example system assumes that a computerized distribution trouble call and restoration management system, which is expected to reduce response time by one half hour, is in place (it is not yet complete in the actual system).

Distribution System Layout

11.3 DESIGN INTERRELATIONSHIPS

Introduction

This section examines how the cost, electrical performance, and reliability of the example system change as the characteristics of its layout, equipment, and spacing are changed.

Substation Size and Spacing

The set of substation service areas for a T&D system must "tile" the utility service territory, covering all locations where there is any demand, and each substation must have sufficient capacity to serve the load in its service area. As the distance between substations -- the substation spacing -- is increased, fewer substations are needed, but the average substation service area becomes larger, and substations will need a greater individual capacity to serve their loads. For instance, the example system described in section 9.2 has six substations serving an area of 108 square miles, evenly spaced in a hexagonal pattern 4.56 miles apart. Each serves 18 square miles with a peak load of 58.5 MW (65 MVA), with 81 MVA capacity, for an 80% utilization rating. If the capacity of each substation were doubled, to 162 MVA, each could serve twice the area -- 36 square miles -- and only half as many substations would be needed.

Area, and load, increase with the square of the spacing, so doubling the capacity and area served will result in an increase of 41% in permissible substation spacing, in this case from 4.56 to 6.45 miles. Doubling the spacing, to 9.12 miles, would require construction of substations with four times the capacity, or 324 MVA each, but on average only one-fourth as many would be needed.

Changing substation size

One way to change the capacity of the substations in a system is to vary the number of transformers of a standard size used in each. For example, if each substation in the example system were composed of two 27 MVA transformers instead of three, each would have a capacity of 54 MVA and each would be able to serve a peak load of 43.2 MVA (39 MW) at 80% FOFA, equivalent to 10.8 square miles of service area. Average spacing would be 3.72 miles. Nine substations would be needed instead of the original six. If substations were composed of four transformers -- capacity of 108 MVA -- each would be able to serve a peak load of 86.5 MVA (78 MW) at 80% FOFA, equivalent to an area of 20.75 square miles, or a spacing of 5.26 miles.

Another option for varying the capacity and hence the potential service area of a substation is to vary the size of the transformers while keeping their number constant. For example, rather than increase the number of 27 MVA transformers at each substation from three to six in order to double capacity, every transformer could be doubled to 54 MVA, resulting in 162 MVA capacity, and the same ability to handle a peak load of 117 MVA at a spacing of 6.44 miles as a substation composed of six 27 MVA units.

Cost impacts of changing substation size

In general, cost for any capacity level is lower if larger, rather than more, transformers are used. Figure 11.6 compares total installed cost for substations of various capacities, all variations on the 3 × 27 MVA substation used in the example system. Cost per MVA drops in both cases as substation size is increased, but it decreases more if transformer size, rather than the number of transformers, is increased. When capacity of a 3 × 27 MVA substation is further augmented by adding one, two, or three transformers of a constant size, the incremental capacity cost for each additional increment is constant -- using the numbers shown in Table 11.4, each additional 27 MVA of capacity in this example costs an additional $775,000. Cost per MVA drops with each increment because these additions are spread over the fixed site cost of $1,730. Thus, while the 3 × 27 MVA substations in the example system cost $49,000 per MVA, a 6 × 27 MVA substation would cost only $39,000 per MVA.

But a three-transformer, 162 MVA substation would cost even less. The installed cost of a 54 MVA transformer is almost always substantially less than twice that of a 27 MVA transformer, so that even a larger savings is effected by increasing capacity to 162 MVA by using three 54 MVA transformers instead of six 27 MVA units. The installation "cost" of a transformer includes more than just the transformer itself -- it includes the foundations, spill containment, fire prevention, labor, breakers and buswork, and all other items associated with putting the unit into service, some of which have a cost that is almost proportional to transformer capacity, and others which change little if at all as capacity is changed. Some elements increase in cost roughly in proportion to capacity, others change very little. The value used here represents the author's experience that transformer cost varies as the .8 root of capacity; doubling capacity increases cost by about 74%.[8] Thus, a 54 MVA transformer, installed, would cost $1,350,000 rather than the $1,550,000 cost of two 27 MVA units.

---

[8] This rule-of-thumb refers to total installed capital cost for units of otherwise identical specification, and assumes no special requirements must be met such as high impedance, small footprint, etc.

Distribution System Layout

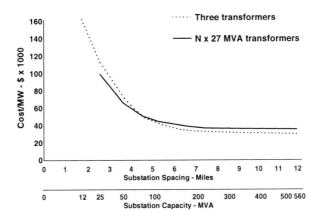

Figure 11.6 Cost per MW served for substations in the example system, as a function of capacity and spacing for three-transformer substations (dotted line) built with transformers of various size, and for substations built with units 27 MVA only (solid line).

As a result, 3 × 54 MVA substation would cost $5,705,000, or only $35,200/MVA, 11% less than a 6 × 27 MVA substation. Figure 11.6, left side, shows the cost for substations of various capacity, built from various numbers of 27 MVA units(solid line) and the cost for building them with three similar units (whatever size) at a substation.

Reliability impacts of changing substation size

Reliability of service is little changed if capacity and spacing is scaled up or down by changing the size of the transformers in a substation whose layout and configuration is otherwise the same. In such cases -- assuming reliability of the equipment itself does not change with size or capacity and that utilization levels are kept constant -- frequency and duration of outages as experienced by the customers remain unchanged.

Changing the number of transformers to vary capacity, however, has a more complex impact on reliability. As discussed in Chapter 10, the number of transformers installed at a substation and the configuration for their support have a great influence on the resulting reliability. Using the values in Table 11.9, if a substation had only one transformer (of whatever capacity was needed to serve its peak load), that transformer would be expected to fail unexpectedly about once every 30 years, requiring an average of 72 hours to repair or replace -- for an average *annual* expectation of 144 minutes (2.4 hours). In addition, service would have to be interrupted infrequently to do maintenance that could not be done while the transformer was energized, although conceivably this could be covered by switching feeders to another substation or use of a mobile substation during those brief times.

With two transformers at the substation, the expectation of both being out of service due to simultaneous and unrelated failures is very small -- less than a minute per year. The greatest risk of complete transformer outage is that one transformer fails unexpectedly while the other is out of service for scheduled maintenance, and that expectation depends on how carefully maintenance is scheduled and on how well the transformer to be left in service is diagnosed before the other is de-energized for maintenance.[9] Given the values shown in Table 11.9, a two-transformer substation would have an expectation of one transformer failing unexpectedly once every 15 years, and an annual expectation of about 4.8 hours per year with one transformer out of service.[10]

With the loading levels used in this example, transformer outages at a two-transformer substation during peak conditions put severe stress on the remaining transformer (loadings to 160% of FOFA). While such loadings are tolerable for *very* brief periods, they are not recommended as the major contingency support plan system-wide. Failures and inability to serve the load will result from widespread and prolonged use of this approach for contingency support. Assuming the likelihood of failure is evenly distributed throughout the year (it isn't, transformers are more likely to fail during summer and in peak periods, but this assumption is a great aid in computing expected reliability) and that peak periods occur only during one season, this 4.8 hours period represents about 1.2 hours (72 minutes) of expected inability to serve the load annually.

Adding a third transformer at a substation changes the situation further in two ways: a) there are more transformers, so that the likelihood of there being one out of service increases, b) each transformer represents less of the substation's capacity so that such situations present much less stress on the remaining units. As a result, as planned capacity of a substation is increased by addition of transformers, the expected availability of sufficient power available at the substation changes, too.

---

[9] The author has seen research data that suggest incipient transformer failure can be detected more than six months early by analysis of core sound, dissolved gases, etc.

[10] This is an *annual* expectation, and as such is subject to all the usual *caveats* about probabilistic analysis. Failed transformers are expected to fail once ever 30 years and require 72 hours to repair (Table 11.9). A simple but good way to interpret the 4.8 hour expectation per substation is: if a system has 15 substations of this type, one can expect one 72-hour incident per year (72 hours/15 substations = 4.8 hours per substation).

Distribution System Layout

Feeder switching used during transformer contingencies at peak

As noted in Chapter 10, paralleling substation transformers is not recommended (Figure 10.7A) unless considerable care has been taken and appropriate monitoring and control equipment is in place. When one transformer in a three-transformer substation fails, the substation has sufficient capacity in the remaining two transformers to serve peak load, but both can share in the burden only if they are paralleled (i.e., by tying all three low-side buses together). Since this is not a recommended step, the contingency plan in the example system is to immediately transfer the low-side bus of a failed transformer to one of its two neighbors. If the transformer failure occurred during off-peak conditions, when loads are far below peak levels, then there is no problem with this one transformer picking up all the load normally served by two.[11]

However, if the transformer failure occurred during a peak load period, the low-side bus is transferred to one of the two transformers remaining in service, and then feeder switching is used to immediately shift one of the outaged transformer's three feeders to the other transformer still in service. This shifts the burden of loading during the contingency from 160% on one transformer and 80% (i.e., normal) on the other, to a 134%/106% split. Although usually a line crew has to be dispatched to do such feeder switching manually, if contingency plans for such situations are prepared in advance, the switching can be done within a reasonable time -- one half hour -- long before serious transformer overheating occurs.

An assumption used throughout this section is that such feeder transfers are done with near-substation feeder switching and done "cold" -- the feeder to be transferred is first isolated from its current feed, before it is switched to the other feed -- causing another interruption event for the customer it serves.[12]

Reliability as a function of number of transformers

Table 11.12 shows the resulting service interruption statistics expected due to transformer failures in the example system. As a result of the feeder switching discussed above, the frequency of interruption due to transformer failures

---

[11] For this reason, maintenance is done at such off-peak times so that this can be done.

[12] Such switching is assumed done with near-substation switches, so that line impedance and load between the low-side buses and the switch locations will be small. Hot switching would mean tying one transformer that is very heavily overloaded to one that is not, via a feeder switch. Without special study to assure conditions will not cause high circulating currents, this is not a prudent operation to carry out.

increases slightly as the number of transformers at a substation is increased as shown. Were it not for that switching, frequency of interruption (due to transformer failures) would be constant. A single large transformer serving all the load might fail only once in 30 years, but its outage impacts all customers. Use of two transformers, each half the capacity but otherwise similar, will mean two transformer failures expected in the same period, but as each serves only half the customers, the interruptions per customer work out to be the same -- the expected interruptions per customer due to transformer failures remain constant as more transformers are introduced. The cold switching of feeders affects only a portion of customers and only has to be done under peak conditions. As a result it makes but a minor change in this otherwise constant value.

The duration of expected service interruptions drops as more transformers are added. With one transformer per substation, average expected duration is 2.4 hours/year. With two at each substation, the duration of service interruptions drops to about 1.2 hours/year. Three or more transformers mean that expected contingency loadings when any one is out of service are at acceptable levels even at peak, so that the only expected duration of interruption is from switching -- with low-side bus-tie breakers in place this is a matter of seconds. If feeder-switching must be done, it adds a very slight amount.

Table 11.12 was produced with a computerized reliability assessment program that takes into account several secondary and tertiary effects not discussed here, including both the likelihood of failures of one or more operating transformers during the scheduled maintenance outage of another at a substation, the switching time at the feeder level when that is necessary to restore service, the probability of failure of switches, and the actual 8760 hour peak load curve shape.

Table 11.12 Expected Service Interruptions Due to Transformers in the Example System as a Function of the Number of Transformers at an 81 MVA Substation

Number of Transformers	Interruptions per year	Duration hours per year
1	.033	2.5
2	.035	1.3
3	.045	.18
4	.045	.16
5	.046	.14
6	.047	.13

Distribution System Layout

Table 11.13 Marginal Cost of Reliability Gained or Lost by Using N Transformers of 27 MVA Capacity versus 3 Transformers of (N*27)/3) MVA, as a Function of the Capacity Installed at a Distribution Substation

Capacity MVA	Cost of Sub with Three transf.	N × 27 MVA	Num custs.	ΔCost/ cust.	Minutes Inter./Yr.	ΔCost/ cust.-min.
27	$2,935	$2,555	3,150	$121	153	$ .86
54	$3,600	$3,205	6,300	$63	80	$ .92
81	$3,980	$3,980	9,450	-	11	-
108	$4,555	$4,755	12,600	$16	9.6	$11
135	$5,130	$5,530	15,750	$25	8.4	$21
162	$5,705	$6,305	18,900	$32	7.5	$36

Table 11.13 shows the marginal cost of reliability as a function of number of transformers. Here, the cost of either 3 × X MVA or N × 27 MVA substations is shown, along with the difference in their reliabilities (duration of interruptions) and cost, and the incremental cost of reliability computed as $/customer/minute of improvement. The last column shows that building a two- or a three-transformer substation rather than a one-transformer substation yields a cost/customer-minute of less than $1, whereas adding additional transformers beyond three costs $11 or more per customer minute of improvement.

Summary of substation size and spacing interactions

The tables and figures given here illustrate the *qualitative* behavior of reliability versus cost versus number of substation transformers in most power systems. The *quantitative* results shown are *not* generalizable. For example, in with some types of substation configuration, the "break" in marginal reliability cost, where adding transformers improves reliability only by a very small amount -- occurs at two, or at four transformers, not three as in this example. As the size (MVA capacity) of the substations used in a power system is increased, few are needed and the spacing between them can be greater. In general:

Doubling substation size will increase the spacing by the $\sqrt{2}$, and only half as many substations will be required.

Substation cost/customer decreases roughly in proportion to the square root of size.

Substation reliability remains constant or improves slightly as size is increased, depending on how substations are configured.

Sub-transmission Cost versus Substation Size and Spacing

Suppose substations in the example system plan were halved in capacity, peak load, and service area. Then the sub-transmission system would have to deliver half as much power to each of twice as many points, spaced an average of only 71% as far apart. Figure 11.7 shows one example of how this could be accomplished. Basically, the same 138 kV circuit that previously fed three substations of 65 MVA peak (195 MVA) now serves four substations of 43.3 MVA peak (173.3 MVA).

In practice, route length for this application will be slightly greater than for delivery to three substations -- roughly 28 miles instead of 25, covering four instead of three substations. Cost per kW rises slightly as shown in Table 11.14, from $49.7/kW PW to $60.3/kW PW. Reliability would fall slightly, basically due to the fact that there is more exposure (longer line/MW served), and more substations, buswork, and breakers in line to cause potential problems.

Table 11.14 shows values computed for several size substation layouts. Figure 11.8 shows the economy (cost/MW) of sub-transmission as a function of substation spacing and capacity, for several popular sub-transmission voltage levels, as computed for this particular example system. The particular costs and curve shapes given in the figure are specific to this example, and in addition depend on the type of transmission grid and grid-level layout guidelines being used. However, the general qualitative behavior shown is characteristic of most systems.

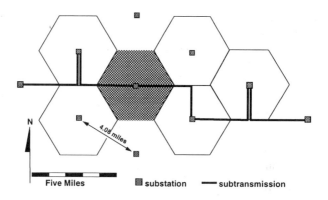

Figure 11.7 Sub-transmission line to four substations takes a somewhat more circuitous route and serves slightly fewer MW. As a result both cost and service interruption rate increase over the base three-substation per circuit case.

Distribution System Layout

Table 11.14 Comparison of Sub-transmission System Costs and Reliability on a per MW Basis for Substations of Various Size

Spacing Miles	Load MW	Cost ($ × 10⁶) Initial	PW	Cost/kW (Dollars) Init./MW	PW/MW	Serv. Inter. per Yr. Freq.	Duration
3.72	39.0	$1.50	$.034	$38	$60	.041	.04
4.56	58.5	$1.84	$.043	$31	$50	.033	.03
5.26	78.0	$2.07	$.048	$27	$42	.031	.03
8.84	220.	$4.88	$.113	$22	$35	.028	.03

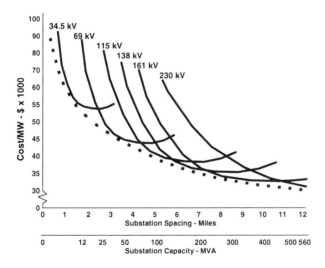

Figure 11.8 Cost per MW of the sub-transmission necessary to feed distribution substations, as a function of substation size in the example system, evaluated for several popular sub-transmission voltages, shows that each voltage level has a range over which is best suited to application. Assuming planners have selection of the most appropriate voltage to use in sub-transmission application (not always a good assumption), available cost versus size is represented by the heavy dotted line. Overall, sub-transmission cost per MW decreases as substation size is increased.

Substation Spacing and Feeder System Interaction

Suppose the substations in the example system plan are reduced in size, to 54 MVA (two transformers), each serving a peak load of 43.3 MVA. Each now serves only 66% as much area, and the feeder system emanating from it must distribute only 66% as much load. Figure 11.9 shows the result from the feeder perspective. At 81 MVA, substations each cover 15.6 square miles and the substation spacing is 4.56 miles, meaning the distribution system must reach 2.28 miles (Euclidean) between substations, which gives a maximum primary feeder reach requirement (through a street grid) of about 3.33 miles. At 54 MVA, substation spacing will be only 3.72 miles, and the maximum feeder reach required drops correspondingly to 2.7 miles.

Trunk feeders can be shorter. More important, the average distance that power has to be moved by the distribution system decreases by 19%. When substations were 81 MVA and spaced 4.56 miles apart, the feeder system had to move power an average of 1.8 miles to deliver it from substation to customer service transformer. At 54 MVA, each substation serves a smaller area, and the average distance power must be moved drops to 1.5 miles -- the feeder system has to perform fewer MW-miles of power transportation

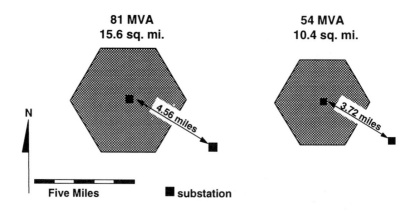

Figure 11.9 A reduction in substation size from 81 to 54 MVA, with all other aspects of the system held the same, results in a 33% reduction in substation service area size (to 12 square miles) and a 19% reduction in the required reach of the feeder system.

Distribution System Layout

Table 11.15 Comparison of Feeder System Costs on a per MW Basis for Substations of 4.56 and 3.72 Miles Spacing

Spacing Miles	Load MW	Costs $ × 1,000,0000				Cost/kW (Dollars)	
		Initial	O&M/Yr.	Losses/Yr.	PW	Init./MW	PW/MW
4.56	58.5	$18.8	$.43	$.36	$ 25.6	$321	$438
3.70	39.0	$12.2	$.28	$.22	$16.2	$311	$415

The feeder system associated with each 54 MVA substation has an initial cost of $12.2 million and a PW cost of $16.2 million, considerably less than that shown for the 81 MVA substations of the original example system layout (Table 11.5), but the feeder system represented there serves 58.5 MW, whereas this reduced substation serves only 39 MVA. Table 11.15 compares the two feeder systems on a per-MW basis. Initial cost drops by 3.1%, and present worth cost falls by 5.2% as substation size is reduced 33%. Present worth cost falls more than initial cost because losses/kW served, which are quite sensitive to the distance power is moved, drop by nearly 10%.

The savings in feeder costs amount to $10/kW served, not enough to overcome the increase in cost/kW for the substations involved: the 54 MVA substations ($3.21 million, $82/kW served) cost about $14/kW more than the 81 MVA substations ($3.98 million, $68/kW). Thus, although reducing substation size cuts the overall cost of the feeder system, the combined substation-feeder cost is higher than for three-transformer substations.

Interaction of Substation Spacing and Feeder Cost

The average distance power had to be moved dropped by 19% when substations were reduced in size from 81 MVA to 54 MVA (average distance power must be moved, like substation spacing, is related to the square root of substation size). Yet initial and present worth costs of the feeder system drop by only 3% and 5%, respectively. There are two useful ways to view this relationship.

Circuit basis

The first perspective for feeder system cost analysis is that regardless of substation sizes, locations, and capacities, the feeder system must "reach everywhere" -- the electric utility must run at least its smallest size lateral down nearly every street or block, reaching every service transformer location.

Reducing substation area size and feeder distance does nothing to reduce this requirement. Table 11.8 provides enough information to estimate the overall savings. Note that 77% of the feeder system consists of laterals of the smallest-size lines, representing 64.7% (47.5/73.4) of the initial cost. If 100% of the feeder were this line type, its cost would be only 64.7/77% = 84% of its actual cost. Taking this value, 84%, as the "cost of running a minimum wire to every location" leaves the remaining 16% as the only portion affected by the reduction in feeder size and distance. The closer spacing reduced the needs for this portion by about 19%, thus

$$\text{Predicted capital savings} = \text{reduction in required distance} \times \text{non-minimum portion of cost} \quad (11.1)$$

$$= .19 \times 16\% = 3.04\%$$

This estimate is in the range of total cost change shown in Table 11.15.

Area-approximation basis

The approximate formulae developed in Chapter 9 (equations 9.1-9.3) to estimate feeder system PW cost can be re-written in terms of the total length of feeder routes that must be built, F, and the average distance power must be moved from substation to customer, d_a

$$\text{Cost} = f_v \times F + s_v \times L \times d_a \quad (11.2)$$

The term, d_a, the average distance power is shipped in a triangular feeder service area with routing constrained to a rectangular street grid is .55 times the sum of the maximum X and Y distances from substation to the end of a feeder, which in a hexagonal substation service area are equal to .5 and .25 the substation spacing, d_{ss}, respectively (max X is equal to the "radius" of the hexagon --length of one side).[13] Thus,

$$\text{Cost} = f_v \times F + s_v \times L \times .4125 \times d_{ss} \quad (11.3)$$

---

[13] This approximation works fairly well for any service area shapes that are vaguely hexagonal, for circular service areas the ".55" term becomes .50, and for square service areas, the average distance is best computed as .55 × (max X + max Y).

Distribution System Layout

and given that, L = load = A × M, and F = total feeder length = A × r, then, A = L/M and F = L/M × r, thus,

$$\text{Cost} = L \times (f_v \times r / M + s_v \times .4125 d_{ss}) \qquad (11.4)$$

Setting L = 1 to put the results on a per MW basis gives

$$\text{Cost} = f_v \times r/M + s_v \times .4125 d_{ss} \qquad (11.5)$$

where:
  f_v is the fixed cost at voltage v ($88,000)
  s_v is the slope of the cost in the linear range ($45,250)
  F is the total length of feeder routing
  d_{ss} is the average distance between substations
  A = area served
  L = load served
  M = load density, 3.25 MW/mile2
  r = feeder route density (12.8 miles/mile2)

The terms f_v, s_v, r, and M are invariant among the two cases

Cost at 81 MVA (at 4.65 miles) = $88,000 × 14.8/3.25 + $18,550 × 4.56
= $346,585 + $84,588
= $431,170 per MW served

Cost at 54 MVA (at 3.72 miles) = $88,000 × 12.8/3.25 + $18,660 × 3.72
= $346,585 + $69,415
= $416,000 per MW served

for a calculated difference in PW cost of $15,170, or 3.5%. These values correspond well with those shown in the rightmost column of Table 11.14, too.

Generally applicable feeder cost approximations

Equations 11.1-11.5 serve as useful approximations for estimating the impact of other changes in the feeder system, and will be used throughout this chapter, and in several other parts of this book. They are applicable only when:

- All feeder loadings are within the voltage level's economical linear range.

- All line segments are sized for maximum economy in the manner discussed in Chapter 7.

- No power is shipped beyond the economical load reach of the voltage level.

- No significant geographic barriers to feeder routing exist in the area being studied.

These conditions mean that the formula provides an estimate of the lowest possible PW cost -- it provides an estimate of the lower bound on possible feeder system cost. Planners should bear this and the formula's approximate nature in mind -- wide dissemination of cost estimates based on this formula could put an unreasonable expectation in the minds of management and co-workers. On the other hand, the formula is very useful for quickly evaluating alternative plans.

The formula itself,

$$\text{Cost} = L \times (f_v \times r / M + s_v \times .4125 d_{ss})$$

has two basic terms, that related to fixed cost, and that related to the variable cost of the feeder system.

Fixed cost depends on the fixed cost, f_v, for the feeder voltage level, the feeder route density (how many miles of lateral the utility has to build, per square mile, to reach the customer locations) and inversely on the load density -- higher load density spreads that fixed cost over a larger base of usage, reducing it on a per-MW basis. Alternately, since L/M = area served, the equation establishes that the fixed cost is basically invariant with load -- as discussed earlier, the utility is "stuck" with the cost of having to reach every location, which means at least the f_v, fixed cost.

Variable cost, on a per MW basis, depends only on the slope of the cost curve for the voltage, system geometry (here embodied in the .4125 term applicable to most feeder systems), and the distance between substations. The important points here are that if loadings can be kept in the economical linear range, there is no economy or dis-economy of scale in moving power through a feeder system. If the amount of power moved is doubled, the variable cost doubles; if the distance it is moved is doubled, the variable cost doubles, too.

Distribution System Layout

Running out of economical reach

As substation spacing is further reduced from the 3.72 miles in the foregoing example, feeder system cost continues to drop slowly, toward a minimum of $347,000 per MW -- which is basically the minimum, fixed PW cost of building a feeder system to reach the customer locations. Cost estimates for substation spacings between 0 and 4.56 miles can be based on equation 11.4. However, for spacing beyond 4.56 miles, the equation is not applicable, because the required load reach exceeds that available for 12.47 kV. As noted in Chapter 7, under the conditions as evaluated here, 12.47 kV has a minimum economical load reach of about 3.3 miles. Since the required reach for feeders in a distribution system will be roughly .75 times the substation spacing (Figure 11.10), at 4.56 miles between substations, this requires

$$\text{Required reach} = \text{substation spacing} \times .75 \qquad (11.6)$$

$$= 3.4 \text{ miles}$$

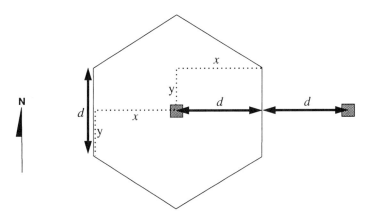

Figure 11.10 Maximum distance a feeder has to reach through a rectangular grid in a hexagonal substation service area is one and one half times the "hexagonal radius, d, or .75 times the substation spacing, as shown.

The example system, with a spacing of 4.56 miles, is actually slightly over this limit. It can "get by" with few operational problems and provide acceptable service quality because not all segments in a feeder system designed with the economical loading method presented in Chapter 7 will be loaded to the maximum end of their economical linear range. In particular, most of the #2 laterals in this or nearly any feeder system are loaded far below the upper end of their economical range, voltage drop is less than at the top of their economical range, and consequently the feeder system should be able to reach appreciably farther, *on average*. However, the analysis and conductor selection method developed in Chapter 7 assumes balanced loadings, something that never occurs completely on individual segments in a distribution system. Imbalance contributes to slightly higher voltage drop. Overall, however, planners can count on a small extra margin of reach (5% is a good rule of thumb) being available. Thus,

$$\text{"Rule of thumb maximum spacing"} = 1.05/.75 \times \text{feeder reach} \quad (11.7)$$
$$= 1.4 \times 3.3 \text{ miles}$$
$$= 4.62 \text{ miles}$$

This means that the example system, with 81 MVA substations and 4.56 mile spacing is at the extreme end of economical application of 12.47 kV. As a result, it will exhibit occasional, isolated voltage drop and voltage regulation problems, and any significant load growth will lead to voltage drop and regulation problems requiring line regulators and conductor reinforcement. If however the substation spacing is increased, then voltage drop and voltage regulation problems will no longer be occasional, or isolated -- they will become quite widespread.

When spacing goes beyond the maximum recommended for a voltage level by equation 9.6, line segments have to be used at less than their economical loading (see "Using a Conductor Set Beyond its Economical Reach" in Chapter 7). Variable cost of the conductor set rises, and a new s_v, computed on the basis of the increased reach required, can be found by working backward from the distance that must be met (see Figure 7.16). At six miles reach, s_v is $57/kW-mile, at nine miles, $68/kW-mile, versus $45/kW-mile at 3.3 miles or less

However, the increased slope is not the only factor that must be taken into account, or the only reason cost rises as spacing increases beyond the economical reach of the conductor set. At a reach of 9 miles, 636 MCM line can move only 2.83 MVA, and 1113 MCM only 3.45 MVA. Thus, any configuration that calls for more than 3.45 MVA to be routed along a single trunk or branch must be modified. Trunks and main branches have to be split into two (or at even longer reaches, three) paths. For example, one 6.5 MVA feeder trunk in the

Distribution System Layout

normal system would have to be split into two 3.25 MVA trunks. Even single-trunk layouts have to be modified to become multi-branch, and multi-branch configurations can evolve to five or six equal branches. Additional miles of line have to be built for these branches that split the routing duty of power -- sometimes branches can be double-circuited but usually this requires new routes/easements. The fixed cost of these lines is not included in any computation of equation 9.4, even if done with a revised sv. Basically, the term r in the equation expands as the load reach is greatly exceeded by the need to move power. Such requirements begin to happen at a reach that is about (8.3 MVA/6.5 MVA) = 1.28 times the maximum economical reach, equivalent to a substation spacing of about 5.82 miles. (The normal economical range at 12.47 kV extends to 8.3 MVA at peak. Feeder peak loads are expected to be 6.5 MVA. Voltage drop varies with loading, thus the ratio 8.3/6.5.) As reach requirements grow beyond that, more and more distance of routing must be added (r increases more) further adding to cost.

Figure 11.11 shows the overall cost per MW of a feeder system as a function of substation spacing and substation capacity, taking into account both the increase in s_v beginning at 4.6 miles, and the additional route length needed at 5.8 miles and beyond. Overall, cost per MW rises linearly up to 4.6 miles spacing, at which point it begins to increase exponentially.

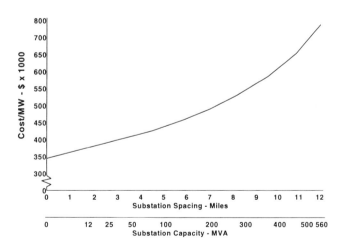

Figure 11.11 Cost of the feeder system on a per MW basis increases linearly up to the maximum economical reach of the primary voltage being used, then increases exponentially. Shown here is the cost per MW of a 12.47 kV feeder system built from the conductor set covered in Chapter 7 and used in the example system in this chapter.

Feeder system service reliability and substation spacing

The net service reliability provided by the feeder system varies as a function of substation spacing. How it changes depends on how the feeder system layout is changed as substation service area is changed. If the size and area served of feeders is kept the same, and the feeder system simply scaled up in number of feeders as substations are made larger, then little it any impact results. For example, if a 162 MVA substation is created with six (27 MVA transformers + 3 feeder) "units" -- basically a substation with twice the equipment shown in Figure 11.2, then feeders will serve about 1.73 square miles and 6.5 MVA peak load each, as in the base case. Reliability will be similar to that given for the base system (Table 11.10) and service interruption rate (due to causes on the feeder system) does not vary with substation size).

However, if the feeders also are scaled up as substation size is increased, then expected service interruptions do change. For example, if the 162 MVA substation is composed of only nine feeders, each served 3.47 square miles and 13 MVA peak load, then service interruption levels rise, because every customer is now served by more miles of exposure.

In general, service reliability increases proportional to the square root of feeder service area size. Thus, if the number of feeders is kept constant -- they are scaled in size as the substation is -- then service interruptions are proportional to substation spacing -- double the spacing; double the interruptions. Table 11.16 shows the resultant feeder level service interruption statistics.

Table 11.16 Annual Expected Interruption Rates Due to Feeder-Level Causes for the Example System as a Function of Substation Size

Spacing miles	Sub. Cap MVA	Number of customers	Nine feeders		Feeders = 6.5 MW	
			Freq.	Hours	Freq.	Hours
2.63	27	3,150	0.35	0.43	0.61	0.75
3.72	54	6,300	0.50	0.61	0.61	0.75
4.56	81	9,450	0.61	0.75	0.61	0.75
5.27	108	12,600	0.70	0.87	0.61	0.75
5.89	135	15,750	0.79	0.97	0.61	0.75
6.45	162	18,900	0.86	1.06	0.61	0.75

Distribution System Layout

Table 11.17 Present Worth Cost per MW Served of the Example T&D System as a Function of Substation Size and Spacing

Spacing miles	Sub. Cap MVA	Peak MW	Number of custs.	PW Cost per MW Served			
				Subtr.	Subst.	Feeder	Total
2.63	27	19.5	3,150	48	200	397	645
3.72	54	39.0	6,300	45	152	417	614
4.56	81	58.5	9,450	42	135	434	611
5.27	108	78.0	12,600	40	127	448	615
5.89	135	97.5	15,750	39	122	464	625
6.45	162	117.0	18,900	38	119	479	636

Table 11.18 Expected Service Interruption for the Example System as a Function of Substation Size

Spacing miles	Sub. Cap MVA	Number of customers	Annual Interruption	
			Freq.	Hours
2.63	27	3,150	1.45	3.75
3.72	54	6,300	1.33	2.25
4.56	81	9,450	1.50	1.50
5.27	108	12,600	1.50	1.48
5.89	135	15,750	1.51	1.45
6.45	162	18,900	1.52	1.45

Composite Cost and Reliability of the Entire T&D System as a Function of Substation Size and Spacing

Table 11.17 and Figure 11.12 summarize the cost of each of the levels of the system -- sub-transmission, substation, and feeders. Table 11.18 and Figure 11.13 summarize service reliability. Costs are on a per MW basis as computed and discussed earlier in this section, and include:

> Sub-transmission -- the lowest cost possible (best sub-transmission voltage) is used in every instance (dotted line in Figure 11.8)

> Substations -- the lower of the costs for either a three transformer or an $N \times 27$ MVA transformer substation from Table 11.13.

> Feeders -- PW of the feeder system is estimated using equation 11.4.

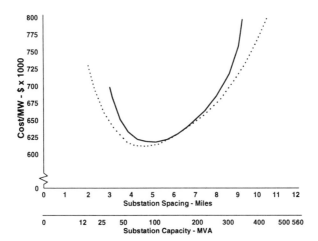

Figure 11.12 Cost per MW of the combined sub-transmission-substation-feeder system, as a function of substation size and spacing. Dotted line represents the cost, assuming that planners can choose the best sub-transmission voltage (dotted line in Figure 11.8) appropriate to the spacing. Solid line represents the cost with 138 kV sub-transmission.

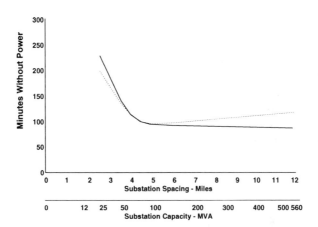

Figure 11.13 Expected annual customer service interruption duration as a function of substation size and spacing. Solid line assumes substations are built with feeders of a standard size and peak load (6.5 MVA). Dotted line assumes nine feeders (each 1/9 of substation load) per substation.

Distribution System Layout 465

Figures 11.12 and 11.13 show a minimum for both cost and service interruptions at about 4.56 miles spacing (81 MVA substations), indicating that among the options considered and within the design constraints outlined for the example system, substation spacing and size is very close to best from both an economy and reliability standpoint.[14]

Design tradeoffs in T&D layout

The optimal spacing of 4.56 miles results because it is the best compromise between "economies of scale" at the various levels of the interconnected system. At the sub-transmission and substation levels, there is a positive economy of scale with respect to substation spacing -- increasing the size and spacing of substations brings a lower cost per MW of demand, as well as better reliability, because larger substations and sub-transmission lines cost less per MVA and have slightly better abilities to provide their own contingency support. The major contributor to this economy of scale is the substation level. Substations represent about 20% of the cost of the T&D system (Table 11.6), and their cost per MW served drops by half as substation size is increased from 27 MW (2.63 mile spacing) to 162 MVA (6.45 mile spacing) -- an amount that represents about 10% of the entire T&D system's cost.

The feeder system has an opposite economy of scale. Larger substations, and the greater area that their feeder system most serve, put a greater MW-mile burden on the power distribution system, which increases cost. This negative economy of scale behaves in a relatively mild manner compared to the cost behavior at the substation level -- feeder cost per MW served increases by only 26% as substation spacing is increased from 2.63 miles to 6.45 miles, whereas substation cost drops by half in that same increment. In addition, the greater feeder route distances required to accommodate wider substation spacing means more exposure of distribution to outages -- so that reliability suffers slightly.

But the feeder level represents 70% -- nearly three times -- the cost of the substation level, so that as substation spacing is increased beyond 4.56 miles, the dollar amount of cost increase at the feeder level exceeds that saved at the substation level. Feeder costs tend to dominate design tradeoffs in T&D system.

---

[14] This, of course, is no accident. The author's thinking is that if close to 50 pages are to be devoted to analysis of a T&D system, it might as well be an "optimal" one, at that. More important, many of the rules of thumb developed in the chapter (such as "feeders represent about two-thirds to three quarters of the total cost of a T&D system" and "feeder losses represent about 1/8 of the total PW cost of the feeder system" are based on economies that change as design is changed. It makes sense to develop examples and infer such generalities from good examples of T&D system design, as done here.

Optimum system layout

The overall result for cost behavior of the T&D system (Figure 11.12) is that there is a definite economy of scale in substation spacing up to an optimum -- in this case in the vicinity of 4.56 miles spacing. However, cost behavior about this optimum is rather mild -- cost per MW varies by only 2% over the range from 3.5 miles to 5.9 miles spacing (dotted line, Figure 11.12). This observation is not made to suggest that selecting and working to achieve the optimal spacing and layout is not important -- 2% of the PW cost of one complete substation (over $36 million) is nearly three quarters of a million dollars -- sufficient to cover a distribution planner's salary for many years.

Rather, this low rate change in economy with respect to spacing means that a T&D system, like the example system in this chapter (one whose layout guidelines have been well-selected, and whose equipment types and design guidelines at all levels are appropriate to one another) will be relatively robust with respect to real-world practicalities. For example, "perfect" substation sites -- just the right amount of room, at just the right location, with just the needed access rights of way -- simply don't exist in the real world. A good T&D system layout standard -- like that of the example system in this chapter -- can accommodate "sub-optimal" sites and rights of way or other similar constraints without exacting a heavy cost penalty. By selecting the optimal spacing of 4.56 miles and everything that it implies as the target layout for their system design, T&D planners know that they can vary their substation spacing by from 3.5 to 5.9 miles (a *52% range* about 4.56 miles) in accommodating such practicalities, with little impact on the overall economy of the resulting system. *That* is no accident -- a T&D system whose layout is fixed in the manner done here both achieves overall economy *and* has flexibility.

"Optimality" should be recognized for what it is, and what it isn't. The economy and reliability of the system described here will be best if substation spacing -- and all that implies -- is kept at about 4.56 miles. Practical considerations will intervene to reduce economy slightly, but a well-selected design target, as shown here, is robust and can achieve near-optimal results over a reasonably wide range of situations.

It is important to note that the results given here apply *only* to the example system for which they have been computed. As will be shown later in this chapter, although such "optimality" is relatively insensitive to a host of factors such as the load density, it does change as conditions change and the figures and tables in this chapter *cannot* be used as general recommendations for all T&D systems. However, the methods used to analyze and determine the optimal structure of the system *can* be applied to all T&D systems.

Distribution System Layout

Effect of Changing Load Density on Cost and Substation Spacing

Changes in the load density of a region have a complicated interaction with the cost of a proposed T&D system plan, an interaction best understood by examining how load density interacts with each of the components of the T&D cost. Suppose the load density in the example system were doubled, to 6.5 MW/mile^2 (7.22 MVA/mile^2), while all other aspects of the distribution planning -- the grid constraint on feeder routing, the local geography, the equipment to be used and its costs, remained constant.

Effect of changing load density on substation cost

When density is doubled, a substation whose capacity was previously sufficient to serve an X mile radius around it now can serve only $X/\sqrt{2}$ radius. The same radius as before can be served by doubling substation capacity, either by doubling transformer size or installing twice as many transformers and associated equipment. As shown in Figure 11.14, the impact of the increase in load density on the "cost/MW versus spacing" curve is to move it proportionally to the left -- every point on the curve moves to an X coordinate that is $1/\sqrt{2}$ times its original X coordinate, as shown.

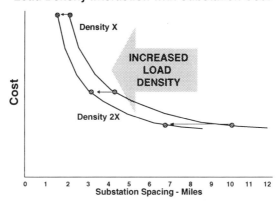

Figure 11.14 Substation cost curve shift to the left when load density doubles: every point on the curve moves 29.3% closer to the Y-axis (i.e., the point's X component is reduced to $(1-\sqrt{2})$ of its previous value).

Cost per MW at every substation spacing distance drops. For example, at a 4.56 mile spacing, which originally required 81 MVA substations, 162 MVA substations are now required. These larger substations have a better economy of scale -- they are about 13% less costly on a per MW basis, and substation cost per MW falls from the original $135/MW PW for 81 MVA substations to $119/MW. Due to the shape of the cost/MW versus spacing cost curve for substations, cost at lesser spacings drops more: at a 2.63 mile spacing, it drops by nearly 25%, from $200/MW to $152/MW.

Effect of changing load density on sub-transmission costs

Ideally, a higher sub-transmission voltage should be used whenever load density of the plan is raised; a lower one when density is reduced. When density is doubled, optimum sub-transmission voltage will rise to $\sqrt{2}$: in this case it rises from 115 kV (the true optimum sub-transmission voltage for the example system -- see Figure 11.8) to 161 kV. Making this increase gains the capability to serve the same number (three) of (now larger) substations per transmission circuit (in this case, three 162 MVA substations per circuit with a total peak load of 393 MVA. While these higher voltage lines cost more per mile, they have twice the capacity, overall cost per MW drop and behavior of the cost/MW versus spacing curve in general is qualitatively similar to that for substations.

If selecting a higher voltage is unfeasible (sub-transmission voltage often is constrained by available grid voltages, etc.), then twice as much sub-transmission will have to be built, by either doubling the conductor sizes of lines at the existing voltage, or using twice as many lines. Either way, cost will increase -- it may double, but since load has double, cost/MW remains constant for this segment of the system.

Since sub-transmission is a minor part of total T&D cost (see Table 11.6), the changes in its cost behavior make little impact on how overall cost changes -- they are overshadowed by the changes at the substation and sub-transmission level.

Effect of changing load density on feeder costs

The feeder system undergoes the most complicated cost behavior impact when load density is changed, and one difficult to master completely and use effectively in planning. As shown in Figure 11.15, fixed-cost per MW drops in inverse proportion to load density (if load density is doubled it drops to half its former value), but the slope of the curve increases proportionally to load density (doubling if density doubles). Thermal and economical load reach distances are both unaffected.

Distribution System Layout

The increasing load density does nothing to alter the minimum fixed cost of the distribution system, or the requirement that it reach everywhere. The smallest line type is still the lowest-cost option, and the same number of miles of "at least minimum size wire" must be run through the service area being planned. But the doubling of load spreads that fixed cost over twice as many MW, cutting its cost per MW in half. With respect to the variable costs, at any substation spacing, the feeder system must move twice as much power out from the substation to the surrounding area, which means that the PW cost of losses will quadruple. Thus, four times the cost is divided by two times the power being delivered, resulting in twice the cost per MW delivered.

As a result, the response of feeder cost curves to changes in load density is as shown in Figure 11.15. The two curves for high and low density will always cross, at a point where the savings wrought by the halving of fixed cost equals the doubled variable cost for moving power out through the distribution network. For the example system, this occurs at a spacing of nine miles. For spacings beyond this distance, cost per MW delivered is greater that the higher density. But at spacings less than that, cost per MW is less: serving the doubled load density requires a more expensive feeder system, but not one that costs twice as much. Importantly, in the range of substation spacings of most interest to the planner -- those around the optimum 4.56 mile spacing -- cost decreases.

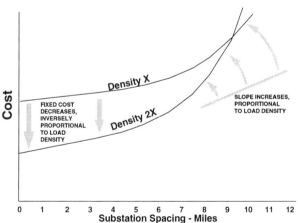

Figure 11.15 Feeder cost behavior is complicated: the curve moves down and slope increases: when load density doubles, the fixed cost (Y-intercept) drops to one half its previous value, and the slope of the curve doubles.

470   Chapter 11

Conductor limitations

The higher load density will very likely require a re-design of the distribution feeder system to effect these economies -- simply scaling up the existing plan by doubling conductor size in every segment will usually not do the job. For one reason, such large capacity line types are often unavailable. Secondly, as discussed in Chapters 7 and 8, even if larger conductor is available, the linear range is really limited in width to 8.3 MW at 12.47 kV due to X/R ratio reasons, which means extensive use of larger conductor in merely "doubling up" capacity would not be economical.

One way to accommodate a doubling of load density from 3.25 to 6.5 MW/mile^2 within the design limitations of the conductor set used in the example system would be to build more feeders. In the example system, when density is doubled to 6.5 MW/mile^2, eighteen rather than nine feeders could be built out of each substation. Each feeder would serve the same 6.5 MW demand as before in a narrower "slice" of the service area -- 20° rather than 40° of circular area in the substation's service territory, only 1.0 rather than 2.0 square miles. Narrowing the "slice" allocated to every feeder will slightly affect economy and reliability, and increase load reach minutely, but such effects are of secondary importance compared to the factors being discussed here, and would make little difference. Such interactions will be discussed later in this chapter

Considering that eighteen feeders often present a feeder getaway problem, an alternative would be to change the design to have fewer feeders, for example, by combining pairs as shown in Figure 11.16 -- using the 1,113 MCM conductor line type for the first several spans out of the substation. This change in design

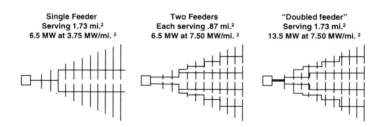

Figure 11.16 In cases of higher load density, feeders can be "doubled" as shown here, by combining their trunks with large conductor for a short distance out of the substation, but leaving configuration otherwise the same. This considerably eases feeder getaway problems, reduces overall feeder system cost, but worsens reliability.

Distribution System Layout

reduces substation cost by cutting in half the number of feeder breakers, getaways, etc., needed in the substation. It also increases feeder cost slightly by virtue of the larger line and its economics (the 1,113 MCM and its resultant segment loading of 13 MW lie outside the linear economical range at 12.47 kV, so that losses costs will be higher in spite of the much larger conductor). However, on balance, the overall cost will drop when reducing the number of feeders in this manner, for the savings at the substation more than pays for the additional feeder costs. While this savings can be noticeable, reliability will usually suffer -- a severe contingency will lock out 13 MW of peak load instead of only 6.5 MW, and switching during contingencies presents more of a challenge as well.

Optimal substation spacing shrinks as load density is increased

When combined into the total system cost, the impacts on the substation, sub-transmission, and feeder levels result in a changed total cost curve as shown in Figure 11.17. Generally, the cost per MW delivered drops over the range of all viable substation spacings. The two curves will eventually cross, at a point where the increasing variable costs at the feeder level equal not only the savings in feeder fixed cost, but the savings also seen at the substation and sub-transmission levels. In the example system, this occurs at a bit over ten miles.

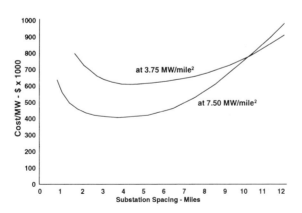

Figure 11.17 Cost per MW at the original load density of 3.25 MW/mile^2, and at 6.50 MW/mile^2, as a function of substation spacing.

The optimum substation spacing shrinks as load density is increased. The opposing economies of scale discussed earlier (the decreasing cost per MW at the sub-transmission & substation levels as spacing is increased, versus the decreasing cost/MW of feeders), which balanced at 4.56 miles when load density was 3.25 MW/mile^2, now reaches an optimum at 4.0 miles, as can be discerned in Figure 11.17. Among the discrete sizes that result when substations are built with 27 MVA increments of capacity, 3.72 miles ($407/MW) is best, with 4.16 miles (5 x 27 MVA at 6.5 MW/mile^2 -- $409/MW) being better than the original 4.56 mile spacing (which now costs $413/MW).

Optimum size for substations increases as load density is increased

Invariably, the optimum size of substations increases as load density grows and decreases as load density drops. At 6.50 MW/mile^2 substations in the example system are ideally composed of *four* 27 MVA transformers (108 MW of capacity) serving 78 MW (87 MVA) of peak load, whereas at 3.25 MW/mile^2 *three* 27 MVA transformers (81 MVA serving 58.5 MW peak) were best.

Economics dictate the design changes, not capacity limitations

It is vital to recognize that the reduction in optimum substation spacing is due to a change in the relative economies of scale among the various levels of the system caused by the shift in load density. A commonly held misconception among planners is that the move to smaller substation spacings as load density is increased is motivated by capacity limitations -- that smaller substation areas have to be used because substations and feeders large enough to serve the higher load density at a wide spacing are not available. That is not the case in this example: A viable T&D system with a substation spacing of 4.56 miles can be built for a density of 6.50 MW/mile^2; but one designed to a closer spacing (3.72 miles) is simply less costly overall. Spacing falls because the economies change in favor of smaller substation areas. This is a general result -- for most T&D system layouts, increasing the load density will result in a shift of economies so that the optimal design characteristic is closer spacing *and* larger substations. Basically, spacing and size "split the difference," each adjusting about halfway.

> Increasing the load density changes the relative economics of the various levels of the system, in favor of a slightly smaller substation spacing, with a slightly larger substation size.

Distribution System Layout

Changes in a T&D plan to adjust it to different load densities usually make little impact on reliability

Suppose that the system is re-designed as outlined above -- substations are now composed of 4 × 27 MVA transformers and feeders are increased in size to 13 MVA each so that there are still only nine, each serving two square miles. Then service reliability is basically unchanged. As observed earlier in this chapter, service reliability from a four-transformer substation is slightly better than from a three-transformer substation, but not significantly so -- reliability at the substation low-side buses improves very slightly. Service reliability at the feeder level is basically the same -- every feeder serves twice the number of customers and has twice the load as before, but it serves the same service area and has the same line exposure. Line segments are all composed of heavier conductor and hardware (for the sake of consideration here conductor size and hardware will be assumed to have no influence on reliability). Therefore, the revised plan serving twice the original load density can be expected to have very nearly the same overall service reliability.

However, if feeders are maintained at 6.5 MVA each, then there are twice as many (eighteen), each serving half the service area of the original plan (1.0 square mile) and with a correspondingly reduced exposure to outage. Service interruption as seen by the customers goes down, because there is less line length of exposure associated with each customer's service. On any given feeder, fewer outages occur and customers will enjoy an improved interruption rate.

Effect of Changing Primary Voltage

Figure 11.18 reviews the primary impact on basic feeder segment economics that are made by an increase in primary voltage (from Figure 6.13). As the primary voltage is changed:

The f_v fixed cost (Y-intercept) changes roughly in proportion to the square or cube root of the relative voltage level -- 25 kV costs about 35% more than 12.47 kV ($119,000 vs. $88,000), while 34.5 kV costs about 70% more, and so forth.

The width of the economical linear range grows in proportion to voltage (i.e., it doubles from (0-8.5 MW) for 12.47 kV, to (0-17 MW) for 25 kV).

The slope of the cost curve in the economical linear range, s_v is proportional to the reciprocal of the voltage change (linear cost is $45,250 MW for 12.47kV, but only one-half, $22,625/ MW, at 25 kV).

The net result is that cost rises for segments that are very lightly loaded, but falls for heavily loaded lines.

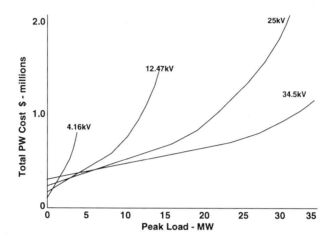

Figure 11.18 As primary voltage is changed, the shape of the power delivery cost curve changes, as shown. Doubling voltage, from 12.47 kV to 25 kV, increases f_v by about 35%, but reduces the slope by 50%.

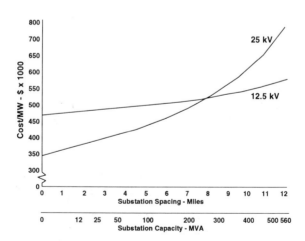

Figure 11.19 Cost of the feeder system if built with 25 kV primary distribution instead of 12.47 kV. The lower voltage is less costly in the example system application at all substation spacings of less than eight miles.

Distribution System Layout 475

The change in primary voltage will make no impact on sub-transmission-level costs, but it will alter the cost of the substation transformers and low-side equipment very slightly. These changes have a very small net impact, which will be ignored here.

Figure 11.19 shows the resulting impact on the feeder-system-cost-versus substation spacing diagram (at the base example system density of 3.25 MW/mile^2). The curve for a 25 kV feeder system is plotted along with that for 12.47 kV (from Figure 11.11). As covered earlier in this chapter, the vast majority of line segments in a feeder system are very lightly loaded. Therefore, at small substation spacings -- those dominated by fixed costs -- overall feeder system cost increases, because of the higher fixed cost of 25 kV feeder hardware. As can be can be seen in Figure 11.11, this occurs over most of the viable range of substation spacing -- the capability of 25 kV to carry load and reach greater distances at lower cost does not pay off until beyond eight miles spacing.

When combined with the cost curves at the substation and sub-transmission levels, the cost curve for the higher distribution voltage results in the composite system-versus-substation-spacing curve shown in Figure 11.20. Cost rises for all spacings less than eight miles -- at any spacing less than that, the higher fixed cost of the 25 kV distribution is simply not justified by the savings that it brings in actually moving power at the distribution level. Optimum spacing increases, to 6 miles (5.89 miles among the discrete spacings available with 27 MVA capacity increments). However, overall cost of the T&D system with 25 kV distribution (at its best spacing) is 6% greater than is was with 12.47 kV.

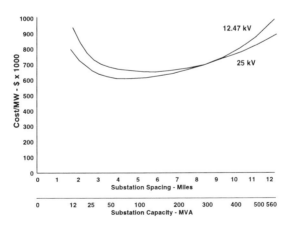

Figure 11.20 Cost/MW of the sub-transmission-substation-feeder system, as a function of substation size and spacing, with both 12.47 kV and 25 kV distribution systems. Optimum substation spacing has increased to six miles, and overall cost is 6% higher.

Cost interaction of substation size and spacing, primary voltage, and load density

If load density is increased sufficiently, then 25 kV becomes more economical than 12.47 kV. Figure 11.21 shows T&D system cost curves like those in Figure 11.20, but recomputed for a load density of 6.50 MW/mile^2 instead of 3.25 MW mile^2. At the higher density, 25 kV is clearly more economical, almost across the board. Now, the curves cross at just over two miles substation spacing, instead of at eight miles, as at 3.25 MW/ mile^2. For any substation spacing beyond two miles, the higher primary voltage provides superior economy. Optimum substation spacing is 5.26 miles (among the discrete sizes available with the 27 MVA units used in this example), and the optimum size substation is 216 MVA (8 × 27 MVA or 4 × 54 MVA) serving a peak demand of 157 MW (174 MVA).

The interaction between primary voltage and load density is quite interesting. At 3.25 MW/mile^2, the feeder fixed cost per MW was $347,000/MW, and the variable cost $85,000/MW-mile. Shifting to a higher voltage increases fixed cost by 35% and cuts variable costs by 50%. At 3.25 MW/mile^2, this increases fixed cost by $122,000/MW-mile and cuts variable costs by $42,500/MW-mile at a 4.56 mile spacing -- cost increase outweighs savings by nearly two to one. But at the higher load density, the fixed cost per MW of the 12.47 kV primary has

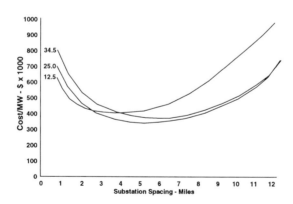

Figure 11.21 Total cost of T&D system utilizing 34.5 kV, 25 kV or 12.47 kV as primary distribution voltage, as a function of substation spacing, at 6.50 MW/mile load density. The 25 kV voltage is now more economical than 12.47 kV for all but very small substation spacings. However, 34.5 kV is not the most economical at any spacing.

Distribution System Layout

been cut by half, to only $174,000/MW, and the 35% cost penalty to move to 25 kV represents only $61,000. Meanwhile, variable costs have doubled because of the doubled load density, so that the savings wrought by using 25 kV also double, to $85,000 at a 4.56 miles spacing. Variable cost reductions now outweigh the increased fixed costs by a clear margin in favor of the higher voltage. Thus, overall T&D cost with 25 kV is lower than with 12.47 kV, a combination of the superior economy 25 kV permits at high load density, along with slight improvements in substation-level economy that come from the higher spacing allowed.

As shown, an even higher voltage -- 34.5 kV -- provides no further savings. The optimal substation spacing when using 34.5 kV, at 6.5 MW/mile^2 load density, is 6.45 miles (optimal size for substations is 324 MVA -- 6 × 54 MVA transformers). At that spacing, cost is about 8% higher than the best for 25 kV.

Other Variations In Design

Chapter 12 discusses a number of other variations in layout and design of the example system, from the context of planning the layout of the substations, which implies variation of sub-transmission, feeders, and other factors.

11.4 EXAMPLE OF SYSTEM DOMINATED BY VOLTAGE DROP, NOT CAPACITY

The example system described in section 11.2 was dominated by capacity and economic conductor usage concerns -- at the sub-transmission, substation, and feeder level, equipment was utilized at levels close to maximum capacity. Voltage drop was seldom a constraint. A majority of power distribution including most downtown, urban, and suburban systems fall into this category, where capacity, not voltage drop, ultimately constrains the equipment usage and system layout. However, electrical service in sparsely populated rural areas is usually dominated by voltage drop, not capacity constraints.

Figure 11.22 and Table 11.18 give an example rural power delivery system based upon a utility in the central plains area of the United States. The example area of 1,000 square miles contains 3,500 customers, each with a coincident peak load of 3.77 kW (4.2 kVA each at 90% power factor). Total load for the 1,000 mile region is 13.2 MW (14.67 MVA). Four substations each serve a quarter of the area, as shown. Each consists of four single-phase 1.75 MVA 69 kV to 22 kV transformers (three in service and one stand-by spare).

Distribution is 22 kV, overhead, Y-connected on wooden poles, built with 636, 336, 4/0 and #2 ACSR, with layout generally constrained to a grid of one mile spacing. Evaluated using the conductor economic sizing methods covered

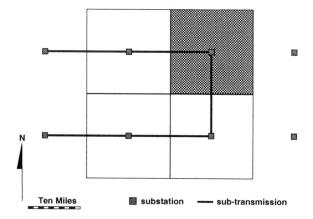

Figure 11.22 Slightly idealized concept of a rural delivery system. Sub-transmission at 115 kV delivers power to 5 MVA capacity substations for distribution on 22 kV feeders. Substations are 15.8 miles apart. Each substation service area (shaded area) is 250 square miles, with a peak load of 3.3 MW at 13.2 kW/mile^2.

Table 11.18 Characteristics of the Example T&D System

Peak demand density of 13.2 kW/ mile^2 (5.2 kW/sq. km)
   3.5 customers/mile^2 3.77 kW/customer average peak
90% power factor, for a peak load of 14.66 kVA/mile^2
Peak occurs annually as 87 periods of 4 hours each.

Annual value of	load factor	loss factor
sub-transmission	.67	.48
substation	.64	.41
feeder	.59	.36
service	.47	.27

Roads & property lines on a 5280 foot (1 mile) grid

69 kV sub-transmission
   277 MCM ACSR, 85 MVA maximum capacity

Substations rectangularly 15.8 miles apart (service areas
   of 250 sq. miles)
   peak load = 3.3 MW or 3.67 MVA
   3 × 1.67 MVA 115/22 kV transformers, plus one spare
   voltage regulators at substation with LDC
   high-side fused 1,000 MCM" bus" conductor/wood frame
   low-side fused/wood frame

22 kV primary distribution feeders, all OH
   conductor: 636 MCM, 336 MCM, 4/0, #2 ACSR

Distribution System Layout 479

in Chapter 7, 22 kV has a linear economic range from 0 to 14.5 MVA and an economical load reach of 5.8 miles (evaluated against a 7.5%, range A voltage drop requirement).

In this system, feeders need to achieve a maximum reach of 15.8 miles -- nearly three times their economical load reach. Voltage drop needs at this distance, in this example system, are met by four means. First, the utility uses a 9% voltage range as standard for normal operation, which effectively raises the economical load reach of the conductor set at 22 kV to 7 miles. Second, the conductors are de-rated from normal economical loadings by up to 50%, to obtain a ten-mile reach, which is sufficient for application in this system, as will be shown. Third, line drop compensation (voltage regulators) at the substation provide a constant voltage at a point six miles away from the substation. And fourth, feeder layout is done so that no customers are served for the first five miles of feeder from the substation, making it what is sometimes called an *express feeder* trunk. (This will be discussed later in this section.)

Sub-transmission and Substations

Both the sub-transmission and substation design in this system are far different than the urban-suburban system of section 11.2. Out of economic necessity, they are built to lower engineering and reliability standards than the prior example system -- all four substations in this example serve fewer customers and less load than just two feeders in the urban-suburban system. Cost heavily constrains design.

As a result, sub-transmission is 69 kV, post-insulator on wooden pole, with a single loop connecting all four substations to two feed points on the grid. The 69 kV line, whose total load is less than 1/4 its capacity, has the capability to feed all four substations from either end. Voltage drop during such contingencies is not a serious consideration for service quality, because it is compensated by the voltage regulation on the low-side of the substation.

The substation consists of wooden pole racks entirely, with fuses on both high and low side. Single-phase transformers are used -- three in service and a fourth in place with appropriate preparations made to switch it into service as a replacement within two hours of crews arriving on site. Voltage regulation, individually by phase, is arranged on the low-side.

Feeder System Layout

The feeder system is mostly "single-phase" feeders (see Chapter 9), laid out with a single large trunk design. Figure 11.23 shows a somewhat stylized and "theoretical" layout of the feeders, which nonetheless captures the essential elements of the system upon which this example is based. It is laid out in a large trunk design on a mile grid. The most interesting feeder-level design characteristic of this system is that no customers are served from a feeder for the first six miles of feeder run from the substation. For the first six miles out from the substation, feeders are "express" in that their sole function is to move the bulk power to the middle of "their" service territory. From there, laterals for service "backtrack," so that service to customers geographically close to the substation is done with feed from a point six miles out.

While this means that more than a minimum amount of conductor length is built, it also means that the line drop compensator at the substation can be set to maintain a constant voltage at that point six miles out on the express feeder. Voltage here is maintained at 126 volts (120 volt scale), which means it is overboosted at the substation, reaching 131 volts on a 120 volt scale at peak. In this configuration, no customer is more than ten miles from that controlled voltage point, and thus effective feeder reach must be only ten miles, not sixteen.

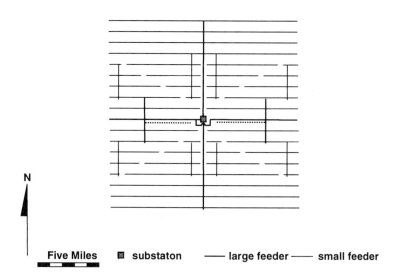

Figure 11.23 Circuit configuration for the 22 kV feeder system emanating from each substation, basically three "single-phase" feeders, arranged so that no customers are fed for the first six miles of "express" trunk out of the substation. Phase A serves the northern 1/3 of the service area, Phase C the southern third, and phase B portions to the east and west of the substation. Dotted lines show laterals that "backtrack" along trunks to provide service along the trunk route for the first six miles.

Distribution System Layout

Conductor Utilization and Achieving Load Reach

Each substation area consists of 292 miles of primary voltage line, with the conductor utilization by type shown in Figure 11.24, which shows little qualitative difference from that of the urban/suburban system -- over 75% is the smallest size conductor line type. Very little of this smallest conductor is loaded anywhere near its maximum capability, even as de-rated for a longer reach.

At a load density of 13.2 kW/mile, the longest laterals have a total load of less than 137 kW, well below the top of their economical loading range (single-phase #2 can deliver over half a MW at a seven-mile reach) and therefore convey a considerable reach advantage. As a result, worst voltage drop at peak load (without regulation) on these feeders would be only 14.2%, and less than four miles out of the 292 miles of distribution shown in Figure 11.23 has to be deliberately selected outside its normal economic range.

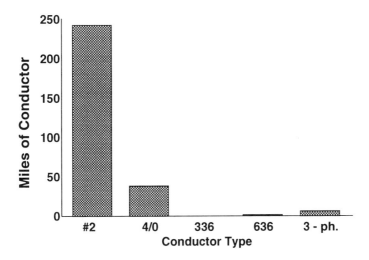

Figure 11.24 Miles of line by type for a rural substation service area. Only a small portion is not single-phase construction. Like the urban-suburban example system (Figure 11.4) the vast majority of lines are the smallest capacity type available.

Table 11.19 Equipment Costs in the Rural System - $ \times 1000$

Equipment	Initial	Annual
Sub-transmission, per mile	90	3
Substation site, including high-side	135	14
four transformers & low-side	115	6
Feeder, per mile of line	17 - 85	.5

Table 11.20 Cost of a Complete Substation of the Example System - $ \times 1000$

Equipment	Initial	O&M	PW	Losses	PW	Total
Sub-transmission, per sub. (20 miles)	1800	60	2316	41	353	2669
Substation	250	20	422	32	276	698
Feeder system	5500	143	6731	80	690	7421
Total =	7550	223	9469	153	1319	10788
Cost per kW peak, dollars	2287	68	2869	46	399	3269

Table 11.21 Cost of the Example Rural System by Level - Percent

Equipment	Initial	PW
Sub-transmission, per sub. (8.33 mi..)	24	25
Substation	5	6
Feeder system	71	69
Total =	100	100

Table 11.22 Annual Outage and Interruption Data for the Example Rural System. Frequency in Events/year and Duration in Minutes/year

Equipment	Frequency	Duration
No power, high-side	.09	.15
Insufficient transf. cap.	.03	.47
No power for low-side bus	.22	1.4
Feeder, at head of trunk	.25	1.4
Feeder, at end	1.9	1.9
Avg. lateral	2.3	3.1
Service transf.	2.4	4.6
Customer	2.6	5.5

Distribution System Layout

Cost and Reliability of the Rural System

Table 11.19 gives the cost of components of this system and Tables 11.20 and 11.21 describe overall cost of a substation and its associated sub-transmission and distribution components. Table 11.22 gives the estimated reliability of the system. A comparison of these tables to those in section 11.2 will reveal many differences from the urban system.

To begin with, the costs for line and substation construction are far different. Even allowing for differences in construction and engineering standards, the rural system costs are lower -- cost per mile of feeder built is less than half that of the urban system. Rural feeders are easier and quicker to build -- there are fewer permits and licenses to obtain, fewer nearby buildings and obstructions to slow down construction, and far less traffic passing by the construction site. The costs shown are representative -- usually construction in rural areas is about half of that in urban areas. Overall, as applied in this system (i.e., slightly de-rated for longer reach) the feeder conductor set has an f_v or $17,000/mile and a s_v of $85,000/MW-mile, giving an estimated PW cost of $7,180,000, 3.5% less than the actual feeder system cost of $7,421,000.[15]

But by far the most revealing comparison is total cost: this rural system has a cost per kW *five times* higher than the urban-suburban system (which delivered power to an area with 284 times the load density). There is an economy of scale in the load density of power distribution that no amount of innovation and brilliant design can completely overcome. Even though this rural system is built to lower reliability and design standards; even though its criteria specify a 9% rather than 7.5% voltage drop limit; and despite its cost per mile of feeder construction being much less, its cost per kW delivered is much greater than for a system delivering power at higher densities. Reliability of service is considerably less in this system than in that described in section 11.2. The reliability values shown actually do not reveal the complete magnitude of difference in design, because the rural system is in a plains region that has few trees to create either momentary or permanent outages (the urban system had a

---

[15] This s_v value is higher than might normally be expected both because the lines have been de-rated slightly for longer reach, but mostly because it has been calculated with only the single-phase line types. In a rectangular service area, average Lebesque distance is one half the maximum (which is 15.8 miles) -- the computation of cost estimate is an easy one. The difference in estimated and actual feeder cost -- $7.42 million (actual) minus the estimated least cost $7.18 million, or .24 million -- is a rough estimate of the cost added by building the feeder system in the non-optimal layout (with some laterals backtracking) which effectively avoided the use of line regulators. In actuality the difference is $220,000 PW, considerably less than the cost of adding four single phase line regulators and their lifetime O&M and losses costs to this system.

typical amount of trees). Were this rural system's terrain and foliage like the urban system, its total annual duration would be on the order of 6.5 to 7 hours.

Design Sensitivities of the Rural System

Both the fixed cost ($135,000) at the substation level and the incremental cost to add capacity (about $25,000 per MVA) are inconsequential compared to the cost of the feeder system. As a result, substation and sub-transmission costs play little role in cost sensitivity of this system to its substation spacing. Feeder cost is the major element of change. As substation spacing is increased, the feeder system has to move power farther, and its cost increases, particularly since more and more of it must built of line types de-rated (under-utilized from an economic loading perspective) in order to gain the required load reach. Figure 11.25 shows the cost per kW as a function of substation spacing for this system, and for systems built with 12.47 kV and 34.5 kV primary voltages also. Optimal spacing for this system, with its load density of 13.2 kW/mile^2, is 15 miles.

Figure 11.26 shows the effect of increasing the load density of the planned design. Even though load densities in a rural region may be orders of magnitude lower than in urban systems, the great distances power must be moved mean that

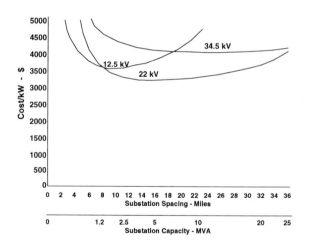

Figure 11.25 Cost per kW for the example rural system as a function of substation size and spacing, for three primary voltages.

Distribution System Layout

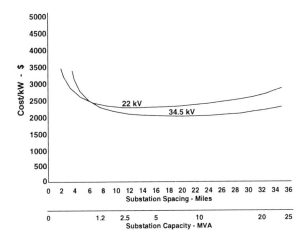

Figure 11.26 Doubling the load density on the rural system has the same qualitative effect as it did on suburban-urban systems. Overall cost/kW drops because fixed costs are spread over more load; optimal substation spacing decreases slightly, and higher distribution voltage proves more economical than before.

the same qualitative interaction between substation spacing, load density, and primary voltage level exists. Doubling load density to 26.4 kW/mile^2 justifies 34.5 kV -- the higher fixed cost of building it everywhere is more than justified by the savings it renders in moving power and in avoiding having to upgrade conductor above minimum size in order to buy load reach.

Overall Commentary of Rural Systems

Power distribution in regions with low load densities has an overall cost per kW that is much higher than that for regions of higher load density. Often, design standards and service criteria for rural systems must be lower than those for urban and suburban systems, because the marginal cost of quality is much higher than in urban systems -- in rural systems the cost of building to urban reliability levels is simply not justifiable.

Generally, the design of such systems is dominated by voltage drop considerations that include the expenses of installing voltage regulation and upgrading conductor size for lower voltage drop. Layout, as in the example given here (Figure 11.23), often includes unusual characteristics aimed at getting around voltage drop constraints.

Usually, the optimal primary voltage for rural application will be the lowest that can handle the load density without major amounts of conductor upgrade in order to "buy" lower voltage drop at peak. In the example system, 22 kV is just at the limit -- arranged very cleverly in layout as shown in Figure 11.23, it can handle the 15.8 mile maximum reach without any significant upgrade from only economical conductor sizing. It is aided in this task by the fact that even its smallest size line type is *very* lightly loaded compared to its most economic capacity, even at peak conditions, so much so that it results in much lower voltage drop (and a longer reach) than would normally be the case. However, its sensitivity to loading is emphasized by the fact that doubling the load -- to a mere 26.4 kW per square mile, results in unacceptable voltage drop without noticeable reinforcement to obtain longer reach. Thus, at higher load densities higher voltage is required (34.5 kV is more economical than 22 kV when density is doubled).

11.5 CONCLUSION AND SUMMARY

The numerical results given in this chapter are specific to the particular utility systems used as examples in this chapter. Many of the numerical results cannot be taken as generalizable recommendations -- for example, 4.56 mile substation spacing is often not the optimal substation spacing when using 12.47 kV distribution. However, the qualitative behavior identified throughout this chapter is generalizable, and the analytical methods used are recommended as planning procedures for the analysis of T&D systems in general. Specifically:

- A T&D system is composed of three major levels: sub-transmission, substation, and feeder. These three levels and their equipment, sizing, and location are heavily interrelated, so that changes at any one level heavily affect the others, and their economies interact in complex ways.

- A convenient, and recommended, yardstick to measure effectiveness of T&D systems is the cost per MW or kW served. This is what really matters -- what is the ultimate cost to deliver a kW of power?

- Substations and sub-transmission have a positive economy of scale -- the bigger, the better the cost/MW.

- Feeder systems have a negative economy of scale -- the bigger the area they have to serve, the higher the cost/MW.

Distribution System Layout

- The feeder system usually represents two thirds to three quarters of the total cost of the T&D system, and is the cause of between one half and two thirds of service interruptions.

- The vast majority of primary voltage lines in a feeder system are lightly loaded segments composed of the lowest capacity in the conductor set and (at least in "American" style system layouts) single phase.

- About one-third of the capital cost of building a T&D system consists of the cost of "running a wire" to every required location.

- While it is important to minimize losses, losses represent only a small portion of the cost of distributing power, even though they dominate conductor economic sizing evaluation.

- As substation spacing -- the average distance between substations -- is increased, substations must become larger, but fewer are needed. Substation size is related to the square of spacing -- double spacing and substations must be four times larger, on average.

- For any set of sub-transmission and primary voltages and substation equipment specifications, the resulting system T&D cost curve -- cost per MW served as a function of substation spacing/substation size -- will have an optimum spacing and substation size.

- As load density increases, the overall cost per MW served usually will decrease, the optimum spacing between substations will decrease, and the optimum size will increase.

- Selection of the primary voltage is a key element in achieving good economy of the entire T&D system, and is a balancing act between fixed and losses costs.

REFERENCES AND BIBLIOGRAPHY

J. K. Dillard, editor, *Electric Utility Distribution Systems Engineering Reference Book*, Westinghouse Electric Corporation, Pittsburgh, 1958 (slightly revised, 1965).

E. Lakervi and E. J. Holmes, *Electricity Distribution Network Design,* 2nd Edition, Peter Peregrinus, Ltd., London, 1995.

12
Substation Siting and System Expansion Planning

12.1 INTRODUCTION

The decisions about where to locate new substations and how much capacity to put at each one are the strategic moves in power delivery planning. For although substations themselves represent a minority of system cost, they define the overall character of the T&D system. They are the meeting place between transmission and distribution, a clue in itself to their importance. But more important, their locations define the end points of delivery need at the transmission system, as well as the starting points for the feeders at the distribution level. This last is particularly important, for the selection of a poor site from the feeder standpoint can increase the cost of the T&D system by a great deal, as will be demonstrated in this chapter.

This chapter examines substation siting and sizing -- the process of deciding *where* to built a substation, *what* capacity to install there, and *when* to build it. It also looks at the distribution planning process in general, much of which is driven by these decisions. This chapter begins in Section 12.2 with a look at substation siting and sizing and a number of basic concepts associated with them. In Section 12.3, the cost impacts of substation site and size selection on the T&D system as a whole are examined, including how siting, size, service

489

area definition, and other factors effect overall cost. Substation-level planning is both an art -- heuristic and non-numerical -- and a science -- quantitatively analytical. Sections 12.4 through 12.6 examine respectively the art, guidelines, and science of substation-level planning, giving examples and recommendations for how to plan the substation level as effectively as possible.

12.2 SUBSTATION LOCATION, CAPACITY, AND SERVICE AREA

Substations and Their Service Areas Tile the Utility Territory

The substation level, like other levels of a power delivery system, must cover the entire system. Power delivered to each customer must come through *some* substation -- if not one particular substation nearby, than from another somewhere farther away. Thus, the utility service territory is "tiled" with substation service areas, as shown in Figure 12.1. Each substation area consists of the part of the service territory that the substation serves, its area covered by the feeder system emanating from its low-side buses. Usually, the service area for a substation is contiguous (i.e., not broken into two or more separate parts), exclusive (substation areas do not partially overlap), roughly circular in shape, and centered upon the substation. However, there are sufficient exceptions to these characteristics that they should not be considered hard and fast rules.

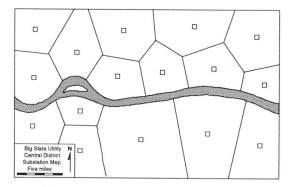

Figure 12.1 The utility service territory is divided into substation service territories (boundaries shown as solid lines), each served by a particular substation (small squares). Each substation must have sufficient capacity to meet the maximum demand of all customers in its area, and the feeder system emanating from its low-side buses must be able to deliver the required power to the customer sites.

Substation Siting and Distribution Expansion Planning

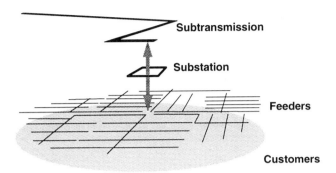

Figure 12.2 A "substation" as considered by the planner consists of four elements: its sub-transmission, the substation itself, the feeder system, and the customers. Electrical service cannot be provided without all levels functioning together. Substation planning is best done by considering the impact of any siting or sizing decision on all four levels.

A "Substation" Consists of Three "Levels" of Equipment

As stressed elsewhere throughout this book, effective planning requires that the planners keep in mind that all levels of the system are connected and that every part is only a small portion of the whole. This is particularly important with substations, whose importance derives not only from their cost, but indirectly from their influence on the cost and performance of other, more expensive portions of the delivery system. A "substation" consists of the four elements shown in Figure 12.2. Generally, the actual substation represents from five to twenty percent of the total cost for the combined "sub-transmission-substation-feeder" delivery system in its area, as shown in Table 12.1. (The data shown is from Chapter 11, Table 11.6, where its composition is fully explained).

Table 12.1 Cost of an Entire Substation, By Level - Percent

Equipment	Initial	PW
Sub-transmission, per substation	8	8
Substation, including site	16	22
Feeder system for substation area	76	70
Total =	100	100

There *is* an "Optimal" Substation Site, Size, and Service Area

For any existing or planned substation, there *is* a best location for it in the sense of economics, and a best location for it in the sense of electrical performance, reliability and service. Occasionally, these two locations are identical, but usually they are slightly different. Regardless, usually the economic aspect rules determination of location.

The optimal location for a substation from the cost standpoint is almost never the lowest cost site (i.e., cheapest land), but is instead the best overall compromise between all the cost elements involved in the substation -- land cost, site preparation cost, cost of getting transmission in and feeders out, and proximity to the loads the substation is intended to serve.

An optimal service area

Along with any particular substation site (be it optimal or not), there is an *optimal service area* -- the area (and load) around that site that is best served by it rather than any other substation, in order to keep *overall* system cost lowest.

Optimum "size" for the substation

There is an optimal "size," or capacity, for each substation. After taking into account everything -- the site, the service area, the constraints placed on it by the local subtransmission and feeder system, and the capacities of its neighbors, etc., this is the capacity to install at that substation site to achieve satisfactory service *and* maximum economy. The optimal service area and the optimal size for a substation are inter-related (Chapter 11).

Optimality is defined in terms of system needs

The need for a substation, and its optimum location, service area, and capacity are usually defined by what other substations *can't* do: A new substation is most needed when additional capacity is needed, and where there is no existing substation near the load at which to efficiently install the capacity. It is a substation's location *relative* to the its neighboring substations, and the customers, that most determines its overall economic value to the system.

The planners' challenge: find the optimum site, size, & service area

But regardless of why, for any particular situation, there *is* a best site, area, and capacity for a substation, the site which gets the job done, that meets all criteria and real world practicality, and whose impact on total system cost is a minimum. The challenge for distribution planners is to find this site, size, and area.

Substation Siting and Distribution Expansion Planning 493

The Perpendicular Bisector Rule

"All other things being equal," every customer in a utility system should be served from the nearest substation. Serving each customer from the nearest substation assures that the distribution delivery distance is as short as possible, which reduces feeder cost, electric losses costs, and service interruption exposure. Locating substations so they are as close as possible to the customers does likewise. There are a host of reasons why "all things are not equal" in most real-world situations, but as concepts, "serve every customer from the nearest substation," and "locate substations so they are as close as possible to as many customers as possible," are useful guidelines for optimizing site, size, and service area -- good concepts for the layout of a power delivery system.

The *perpendicular bisector rule* is a simple, graphical method of applying the concept "serve every customer from the nearest substation" to a map in order to determine "optimum" substation service areas and their peak loads. Applied in a somewhat tedious, iterative manner, it can also be used to determine where to locate a new substation to maximize its "closeness" to as many customer loads as possible. The method was widely used from the 1930s until the widespread application of computerized planning tools began in the early 1980s.

Although simple in concept and at best approximate, the perpendicular bisector rule is a useful qualitative concept every distribution planner should understand. Application of this rule to a service area map consists of several simple steps:

    a. draw a straight line between a proposed substation site and each of its neighbors

    b. perpendicularly bisect each of those lines (i.e., divide it in two with a line that intersects it at a ninety degree angle.

    c. the set of all the perpendicular bisectors around a substation defines its service territory

    d. the target load for this substation will be the sum of all loads in its service territory

This process is illustrated in Figure 12.3. Step (b) of this process determines a set of lines that are equidistant between the substation and each of its neighbors -- the set of all such lines around a proposed substation site encloses the area that is closer to it than any other substation. As a starting point in the planning process, this should be considered its preferred service area. The sum of all loads inside this set of lines defines the required peak demand to be served by the substation.

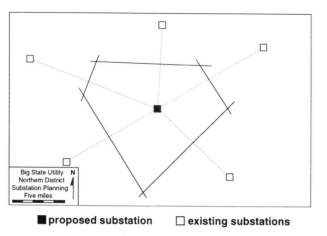

■ proposed substation   □ existing substations

Figure 12.3 The perpendicular bisector rule of identifying a substation service area identifies the set of all points equidistant between a proposed substation and its already existing neighbors. A line is drawn from the substation to each of its neighbors (dotted lines) and bisected with a perpendicular line (solid lines). Points inside this boundary are designated as served by the proposed substation. While simple, this method in one form or another was used to lay out a majority of substations currently in use worldwide. Modern substation siting computer programs use accurate analytical approaches that build upon the concept's guiding rule: serve load from the nearest substation.

The impact on the loadings of the nearby substations can be determined in a similar manner, by using the perpendicular bisector method to identify how their service area boundaries change, what area they "give up" to the new substation, and how much their load is reduced by the new substation taking a part of their service area away from them.

In traditional substation level planning (as practiced from the 1930s through the 1970s), this graphical approach or a similar method was used along with street and feeder maps to estimate substation siting and sizing needs. The substation planning process would generally begin by identifying one or more substations whose load was projected to exceed their capacity. A proposed substation site nearby would be identified, one in between as many overloaded substations as possible (Figure 12.3). It would have its service area identified in the process described above, and the load for that area estimated. This identified the preferred capacity requirement for the substation.

Substation Siting and Distribution Expansion Planning 495

The reduction in neighboring substation loads would be determined in a like manner. If the planner's goals were to reduce the load of a particular neighboring substation -- for example that directly to the east -- by 25 MVA, and this was not accomplished, then the proposed new site would be moved closer to that substation, which upon recomputation of the bisectors would tend to move more load toward the new substation and away from the overloaded one.

If the planner's goal were to find the "optimal" site for the new substation itself, that could be estimated by an iterative procedure based on the center of loads. Once a service area for the proposed substation was determined as described above, the center of load for that area was computed. This formed a revised substation site recommendation, from which bisectors and service area were recomputed, etc. This was done several times until the site and center of load converged to the same location.

In this manner, the initial planning of preferred sites and sizes for substations was done manually in traditional substation-level planning procedures. Many modern substation-siting computer programs, including those generally regarded as the most effective planning tools, apply this basic technique, but augmented by detailed, non-linear analysis that considers a host of exceptions and "things that aren't equal" in the determination of site and size. However, the basic concept used throughout those programs is as outlined here: each substation should serve those customers nearest it; its capacity should be determined so that it can fulfill this role and its location should be determined with relation to the other substations so that there is always a substation "close enough" to all load.

12.3 SUBSTATION SITING AND SIZING ECONOMICS

For planning purposes, the decision about whether to build a new substation, reinforce an existing one, or "get by" without any additions must include an assessment of all cost impacts at the sub-transmission, substation, and feeder levels. Some of these costs, including that of the substation itself, are not highly sensitive to changes in location -- given any reasonable site, just about the same equipment and labor will be required to build the same type of substation. Other costs vary a great deal -- feeder costs being pre-eminent in this regard, as discussed in the previous section.

All these many costs must be assessed before making a decision about whether to build a new substation or not. Chapters 10 and 11 focused in detail on the layout and costs of a substation and its associated parts. But in studies to determine the location of a possible new substation or aimed at identifying which of several existing substations will be reinforced, it is only spatially varying costs (those that change as location changes) that are important.

Table 12.2 Major Cost Sensitivities Involved in Substation Siting in Order of Typical Economic Importance to Overall Decision-Making

Type	Comment
Feeder impact (proximity to load)	Substations exist primarily to feed the feeders. The biggest impact in changing site and size is on the feeder system (as discussed in section 11.3). *This is almost always the dominating variable cost in substation siting studies.*
Subtrans. impact	Some sites are near available transmission lines or can be reached at low cost. Others require lengthy, or underground only access -- adding to cost.
Feeder getaway	Getting feeders out of a substation requires available routes with sufficient clearance. Confined or restricted sites mean higher costs in taking feeders underground or over non-optimal routes around nearby barriers to get power out of the substation.
Geographic	Nearby terrain or public facilities may constrain feeder routing raising costs. Close proximity to a large park or cemetery means feeders must be routed around them on the way to the load, which generally raises feeder costs.
Site preparation	The slope, drainage, underlying soil, and rock, determine the cost of preparing the site for a substation and of building the basic foundations, etc. The cost of transporting material to the site may also differ from one site to another by significant amounts. Esthetic requirements (fencing, landscaping) also vary.
Land cost	The cost of the land is a factor. Some sites cost much more than others.
Weather exposure	Sites on hilltops and in some other locations are more exposed to lightning and adverse weather than average, slightly increasing repair and O&M costs.

Substation Siting and Distribution Expansion Planning 497

This section looks at how the major costs involved in a power T&D system vary as a function of substation location, size, and service area. Table 12.2 lists the major cost *sensitivities* that must be addressed in considering substation site and size, in their usual order of importance. The first two -- feeders and sub-transmission -- and their interaction, will be discussed at length in the rest of this section. The others are discussed in Chapter 4, Planning Criteria, and elsewhere throughout this book.

Feeder Costs and How They Vary With Site and Size

Figure 12.4 shows one of the substations from the example urban-suburban system in Chapter 11. The system, and all of the costs and economic relationships developed in Section 11.2 will be used throughout this chapter to examine the cost interactions of substation siting and sizing with the entire distribution system. The substation shown serves an area of 15.6 square miles with a peak load of 58.5 MW (65 MVA) with an annual load factor of 61%. It is fed from the high side by 138 kV sub-transmission and uses nine 12.47 kV feeders on the low side to distribute power throughout its service territory. Overall cost is $623 per kW of peak load served. All substation boundaries in this example system satisfy the perpendicular bisector rule. Substations are hexagonally spaced at 4.56 miles, each substation serving 18 square miles with a load density of 3.25 MW/mile2, for a peak load of 58.5 MW.

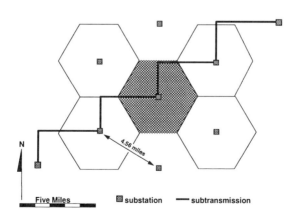

Figure 12.4 Example substation from Chapter 11, Section 11.2.

Moving a Substation Within Rigid Service Territory Boundaries

Suppose that the substation in Figure 12.4 is moved one mile to the east, while still being required to serve the hexagonal (shaded) area identified as its service territory in that figure (i.e., that area's boundaries and locations do not change). Moving this substation would have little impact on the sub-transmission costs (the new site is along the same transmission corridor) or the substation itself (the same exact substation would be built to serve the same load, assuming the new site costs no more or less than the old site).

The major cost impact effected by this change in location would be on the feeder system. The new site means the substation is closer to some loads (those to the east, and farther from others (those on the west side of its service territory). Figure 12.5 illustrates a way to estimate the impact on substation feeder system costs. The example uses square service areas because that makes the major interrelationship easier to picture. However, the rule of thumb developed is a useful guide in estimating impact in any shape of service areas. The rule of thumb developed in Figure 12.2 is

$$\Delta \text{ MW miles} = \text{load density} \times \text{length of edge} \times (\text{distance moved})^2 \quad (12.1)$$

Using the s_v value from Chapters 7, 8, and 9 to put a cost on this increase in MW-miles of power delivery burden gives,

$$\text{Cost impact} = s_v \times \text{load density} \times \text{length of edge} \times (\text{distance moved})^2 \quad (12.2)$$

Thus, moving the substation in Figure 12.4 one mile to the east, with a load density of 3.25 MW/mile, the "length of edge" as 4.56 miles, and the s_v as $45,250/mile, gives

Estimated MW mile $= 3.25 \times 4.56 \times 1^2$

$= 14.82$ MW-mile

Estimated PW cost impact $= \$45,250/\text{MW-mile} \times 14.82 \text{ MW-mile}$

$= \$671,000$

The estimated impact on feeder costs of this one-mile change in substation location is more than one-sixth the original cost of the substation ($3.98 million). In fact, equation 12.2 slightly underestimates the cost impact, because it applies the standard s_v, which applies only in the linear range and within the standard economical load reach of the feeder system. For this example system, with the substation moved a mile to the east, a good deal of the load along the western

Substation Siting and Distribution Expansion Planning 499

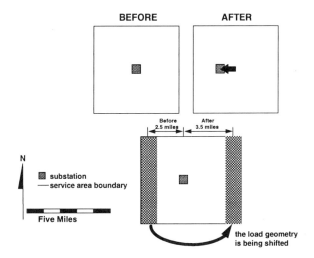

Figure 12.5 A simple method of estimating the impact of moving the substation within a fixed substation area, illustrated here with a square substation service area, five miles to a side. At the top, the situation before and after the planned substation is moved within its designated service area. Bottom: the change in substation location relative to the load is equivalent to "cutting" a swath from the eastern boundary, equal in width to the distance the substation is moved, and "pasting" it onto the western boundary. The net impact is that that amount of load is now farther from the substation by the distance the substation was moved. Thus, the increase in MW-miles of power delivery resulting is: (width of area in the direction perpendicular to the movement) times load density times (distance moved)^2.

edge of the substation's service area (about 15 MW) is now beyond the economic load reach of the 12.47 kV conductor set. Larger conductor and/or line voltage regulators will need to be used to correct the situation. Actual cost impact on the planned feeder system would be closer to $750,000.

The impact on reliability of service can also be estimated using a similar approach and data from Tables 11.9 and 11.10. The amount of load affected is 4.56 miles × 1 mile × 3.25 MW = 14.8 MW, or 14.8/58.5 = 25% of the substation load. Distance moved is one mile. Using the value of .48 interruptions per mile and .55 hour re-switching time for feeders given in Table 11.9, one could estimate an average increase of .48 events/mile × 1 mile × 25% of the load, or .12 events year average. Similarly, duration could be expected to increase by about .14. These are noticeable, if not large, increases (≈ 10%).

Impact of Moving a Substation Site When its Service Area Boundaries are Adjusted to Minimize The Cost Impact

Certainly, if distribution planners were forced to move a substation away from its theoretical optimum location, they should make whatever adjustments they could to the rest of the system layout in order to reduce the cost impact as much as possible. One option is to change the service area of the substation and its neighbors, rather than leave them fixed, as was the case considered above.

In Figure 12.6, the substation from the example system (Figure 12.4) has been moved one mile to the east as in the prior example, and the boundaries of its service territory and its neighbors' territories have been adjusted to minimize feeder impact cost (In this example this is tantamount to re-drawing them following the perpendicular bisector rule). The total area served by the substation changes little: the marginal amounts of loss on the west side and gain on the east are essentially the same, and the total area, and load, served by the substation remains the same. As in the previous example, the cost impact on the sub-transmission and substation itself would be small.

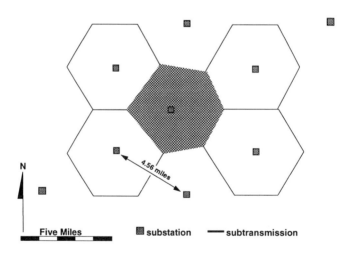

Figure 12.6 The substation in Figure 12.4 is moved one mile east, and its service area boundaries are "re-optimized" to minimize overall impact on the feeder system, slightly changing substation service area shape (but not its area, although the areas of its neighbors do change). All substation boundaries in that example system satisfy the perpendicular bisector rule, optimal in a situation where load density is uniform and there are no geographic constraints.

Substation Siting and Distribution Expansion Planning

Figure 12.6 illustrates a graphical "cut and paste" analogy leading to an estimation rule for feeder-level impact, when the substation is moved and the boundaries *are* adjusted. Thus, the impact of moving the substation in Figure 12.4 a mile to the east, when the boundaries of its service area are re-adjusted to stay half-way between it and each of its neighbors, is only half that when they cannot be re-adjusted. This ratio is a general result: readjusting substation boundaries when the site must be moved will usually halve the feeder-level cost impact of having to accept the non-optimal substation site. The rule of thumb developed from Figure 12.6 is therefore

$$\Delta \text{ MW miles} = \text{load density} \times \text{length of edge} \times (\text{distance moved})^2/2 \quad (12.3)$$

Using the s_v value from Chapters 7, 8, and 9 to put a cost on this increase in MW-miles of power delivery burden gives,

$$\text{Cost impact} = s_v \times \text{load density} \times \text{length of edge} \times (\text{distance moved})^2/2 \quad (12.4)$$

Thus, moving the substation in Figure 12.4 and 12.6 one mile to the east, and adjusting its boundaries, with a load density of 3.25 MW/mile, the "length of edge" as 4.56 miles, and the s_v as $45,250/mile, gives

Estimated MW mile $= 3.25 \times 4.56 \times 1^2 /2$
$= 7.41$ MW-mile

Estimated PW cost impact $= \$45,250/\text{MW-mile} \times 7.41$ MW-mile
$= \$335,000$

The net impact is roughly the cost of building two miles of feeder. Similarly, impact on reliability is also reduced by roughly half from the case where the boundaries are not moved.

However, an additional impact must be taken into account, one that cannot be estimated easily, that on rare occasions results in large cost penalties, but that in other cases can work to the planners' advantage: The loads of the neighboring substation in Figure 12.5 will change as a result of re-allocating substation areas with the movement of the boundaries. While the substation that is moved sees little if any change in its load, the substation to the east loses about 7.4 MW, while that to the west gains 7.4 MW. This means the utilization of the substation to the east changes from 80% (design target) to 70%, and that to the west increases to 90%. The amount of change in this particular example is probably tolerable if a bit worrisome from the contingency support standpoint in the western substation. In other cases, the shift in loads could lead to overloads or severe under utilization of equipment.

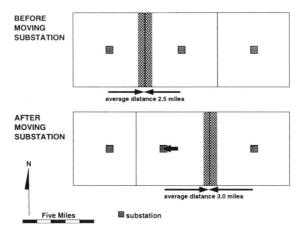

Figure 12.7 Top: three substation areas, each five miles square, with the substation centered in each. Bottom: the center substation has been moved one mile to the east, and its service area boundaries adjusted, each moving *one half mile*. The net effect is that a mile swath of load (shaded, top) served at an average 2.5 miles distance from the two substations along the eastern boundary, is cut from the eastern boundary and pasted onto the western boundary, at an average of 3 miles distance. The increase in MW-miles of power delivery is: (width of area in the direction perpendicular to the movement) × (distance moved)^2/2, or half the impact when boundaries are not moved.

Systems with higher voltage primary are less sensitive to deviations from optimal substation location

As primary voltage level is raised, the s_v cost per MW-mile drops. For example, doubling the primary voltage in the example system (Figure 12.4) to 25 kV cuts s_v by a factor of two, to $22,625. Of course, systems built with optimal layout at higher primary will have larger substation areas, so that the boundary width used in equations 12.1 through 12.4 will be greater. In Chapter 11, doubling the example system's primary voltage was shown to raise optimal substation spacing to 5.89 miles. Substitution of these revised figures in equations 12.1 and 12.2 shows that the estimated impact on cost for a 25 kV systems is nearly 1/3 less: Only $217,000 if the substation boundaries can be shifted when the planned substation site is moved by one mile, and $433,000 if they cannot be.

> *Doubling primary voltage cuts economic sensitivity to substation siting problems by about one-third.*

Substation Siting and Distribution Expansion Planning

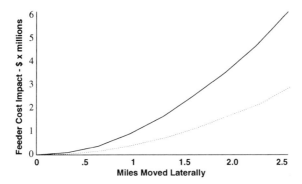

Figure 12.8 Feeder system impact as a function of the non-optimal location of a substation in the example system, for situations where the substation area boundary cannot be altered (solid lines) and can be (dotted line).

Systems with higher load density are more sensitive to deviations from optimal substation location

Conversely, as the load density of a planned system is increased, its costs become more sensitive to optimal substation location. Referring again to Chapter 11, doubling load density reduces optimal substation spacing to about 3.7 miles. Recomputing equations 12.1 through 12.4 for a 12.47 kV system using this value instead of 4.56 miles, and a load density of 6.5 MW/mile^2, gives $1.1 million if the boundaries cannot be moved, $.55 million if they can be.

Generalizations About Feeder Cost Impacts

Equations 12.1 through 12.4 indicate several aspects of substation siting that are generally applicable to any planning situation. First, whether the substation boundaries are re-adjusted or not, moving a substation away from its optimal location creates an impact on feeder system cost proportional to the square of the distance moved. This occurs in spite of the fact that these equations use a linear MW-mile cost factor s_v, (which is valid assuming planners can re-design the feeder system on both sides of the substation boundary). [1,2]

---

[1] The linearization gives fairly representative results up to about 1.5 miles movement in location, after which load reach problems increase cost considerably.

[2] In cases where planners cannot, impact is proportional to distance *cubed.*

Figure 12.8 shows the feeder cost impact, as a function of the distance the substation is moved, for both "immobile-boundary" and "moving boundary" scenarios. The values shown were computed using a detailed substation siting program, not approximations using equations 12.1 through 12.4. Small non-optimalities in location cause little cost impact, but large differences create very large cost penalties. Moved 2.6 miles (to the edge of the stationary service area), the cost impact in increased feeder costs is $5.5 million -- more than the cost of the substation itself ($3.98 million), and one million dollars more than an estimate based on the linearized approximation would predict -- indicating that load reach problems in that extreme state cost nearly $1 million to solve.

Equations 12.1 through 12.4 and the response of the examples here indicates the following guidelines with respect to trying to obtain the optimal substation site:

1. There is an "optimum" location for a substation from the standpoint of feeder system costs. Feeder cost to deliver the power to customers increases as the actual substation site is moved away from this location (Figure 12.9). This location is called the *center of feeder cost*.

2. Small deviations (less than 1/3 mile) in the actual substation site from the center feeder cost create insubstantial cost increases. (A 1/3 mile difference in the example carries a cost impact of $37,000, less than 1% of the substation's cost and less than .2% of the sub-transmission-substation-feeder cost).

3. Large deviations in actual site location from the center of feeder cost are very expensive -- moving the substation in the example 6,400 feet (1.2 miles) costs $1,000,000.

4. Whenever non-optimal sites must be accepted, planners should re-plan the associated service area boundaries and feeder network for the substation and its neighbors. This cuts the cost increase due to the non-optimal site in half.

5. Non-optimal sites near "hard" substation boundaries -- such as when a river or other geographic feature that cannot be moved forms one side of a substation's area boundary -- carry a much higher cost penalty than those near substation boundaries that can be re-adjusted. Planners should do everything possible to avoid having to accept non-optimal sites whenever the service areas are heavily constrained by geographic features.

Substation Siting and Distribution Expansion Planning

6. T&D systems with higher primary voltage are less sensitive to deviations in their substation location.

7. T&D systems with higher load density are more sensitive to deviations in their substation locations.

Impact of Changing the Area Size of an Otherwise Optimal Substation

If a substation has insufficient load to serve all the customers nearest to it, some of them must be served by substations farther away, raising feeder system costs. In cases where there are unusual or severe capacity or geographic constraints, this may be the best course for design, but regardless, it is important for planners to appreciate the impact that a lack of optimal substation sizing, as well as optimal siting, can make on the feeder system.

Suppose the substation in Figure 12.4 were reduced in capacity by 33%, from three to two transformers, necessitating a proportionate reduction in its loading -- from a peak load of 58.5 MW to 39 MW. This reduction in load could be

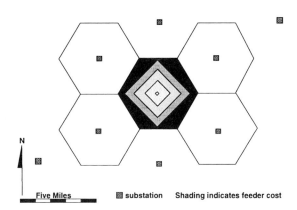

Figure 12.9 For every substation siting situation there is a "center of feeder" cost. Locating the substation at this point will minimize the feeder-level cost for delivering power (cost as discussed here includes any impacts due to interactions with feeder systems in neighboring substation areas). Degree of shading above shows cost impact of a site, and clearly illustrates the effect of grid routing constraints on the feeders.

accomplished by shrinking the substation service area as shown in the top of Figure 12.9, transferring 19.5 MW to surrounding substations. The average reduction in service area radius can be estimated as

$$\sqrt{\frac{58.5 \text{ MW}}{3.25 \text{ MW/mile} \times \pi}} - \sqrt{\frac{39 \text{ MW}}{3.25 \text{ MW/mile} \times \pi}} \qquad (12.5)$$

$$= .44 \text{ mile}$$

The computation above was based on the assumption that the area was a circle. Recall from Chapter 11 (see particularly footnote 12) that an adjustment must be made for the necessity of routing feeders through the grid. Therefore,

$$\text{actual distance change} = .55 \times \max \Delta X + \max \Delta Y$$

$$= 1.1 \times .44$$

$$= .55$$

The 19 MW transferred to the neighboring substations will have to be served over roughly .55 mile farther feeder distance. Assuming the feeder system can be re-planned for this additional requirement, the minimum PW cost impact will be approximately

$$\text{Cost impact} = \text{load} \times \text{distance} \times s_v$$

$$= 19 \text{ MW} \times .55 \text{ mile} \times \$45,250/\text{MW-mile}$$

$$= \$475,500$$

The load transfer scheme shown in Figure 12.9 (top), in which load is transferred equally to all neighbors, minimizes the feeder-level cost impact of the size reduction in this example problem. The $472,500 figure computed is the minimum cost impact that could be expected -- the lower bound of the possible impact. In reality, very often load transfers cannot be done equally to all neighbors, because all neighboring substations do not have sufficient capacity to pick up the addition load that entails   The worst case (upper bound on cost impact) is to assume that all the load must be transferred to one neighboring substation, as shown in Figure 12.10 (bottom). The maximum Lebesque-1 distances in the case shown are X = 1.91 miles (1.66 miles divided by cos 30°) and Y = 1.66 miles, thus the cost is

$$\text{Cost} = 19 \text{ MW} \times .55 \times (1.91 + 1.66) \text{mile} \times \$45,250/\text{MW-mile} \qquad (12.6)$$

$$= \$1.69 \text{ million}$$

Substation Siting and Distribution Expansion Planning 507

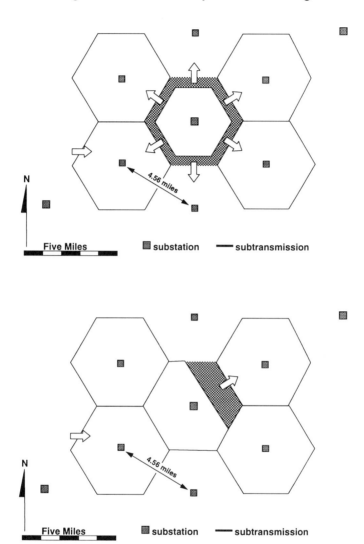

Figure 12.10 If the example substation has insufficient capacity, load must be transferred to neighboring substations, with a consequent increase in feeder cost. Top: an estimate of the lower bound of cost impact on the feeder system is to assume the load is transferred evenly to all neighbors. Bottom: worst case impact from a feeder cost standpoint is if all the load must be transferred to a single neighboring substation.

Thus, if a substation in the example system falls short of required capacity by one transformer, the cost impact on the feeder system is between $472,500 and $1.69 million, depending on if and how feeders and substation boundaries can be adjusted to try to mitigate the additional expenses.

There are many ways to compare this figure to other costs and cost impacts in the system. One of the most interesting is to contrast this cost penalty to the cost of the transformer capacity itself. In Chapter 11, Table 11.4, the cost for an additional 27 MVA transformer at a substation (i.e., capacity sufficient to serve 19 MW of peak load) is $775,000 -- this includes the transformer itself, all equipment required to support it, and their complete installation.

The $472,500 to $1.69 million cost penalty computed above does not occur because the transformer capacity has not been installed. This example assumed that the load *can* be served from the nearby substations, so that the required capacity is somewhere in the system.[3] It occurs because the capacity has been installed in the wrong place

As evaluated here, the importance of locating a substation transformer at the correct substation is somewhere between half and twice as valuable as the capacity itself. This result is generally applicable as a rule of thumb -- location and capacity have about the same value, and thus:

> *Nearly half of the value of substation capacity comes from having it in the right place.*

Sub-transmission-Substation Siting and Sizing Cost Interactions

The cost of building sub-transmission to a substation site can be considerable. In the example system (Figure 12.4), adding a mile of sub-transmission increases PW cost by $349,000. One way to reduce sub-transmission cost is to seek a site along an existing sub-transmission-level right-of-way, where there is sufficient capacity to serve the substation. For such sites, the "sub-transmission cost" is

---

[3] Somewhat unrealistically, this lower bound computation assumed that capacity was distributed among the neighboring substations. The upper bound assumed the capacity was all at one neighboring substation (i.e., the transformer was "mis-sited" one substation to the west of where it ideally should have been located. The important point is that the $775,000 has been spent putting the capacity into the system somewhere -- the $472,500 to $1.69 million cost penalty *is in addition to* the expense of the capacity, and *just* because that capacity has been put in the wrong place.

Substation Siting and Distribution Expansion Planning 509

essentially zero. However, in other situations several or even many miles of sub-transmission may have to be built to reach a site.

The length of line that may have to be built, can occasionally be much more than straight-line distance, or even the minimum grid-path from the grid feed point to the substation might indicate, due to esthetic, property, or other land-use restrictions. Unlike distribution, which can usually obtain overhead or underground easements along any road, transmission routing requires rights-of-way that have far heavier esthetic and land-use impacts, and thus it is far more difficult to route. As a result, some sites are approachable only via circuitous or restricted routes whose length is much greater than might first be supposed.

Generally, the cost to build sub-transmission is linear with respect to distance -- double the distance required and cost doubles, too. It is also generally non-linear with respect to capacity, but not in any general way -- every situation is different, with cost versus capacity function usually discontinuous.

Sub-transmission and Feeder System Cost Sensitivity Interaction

For sites not along an existing sub-transmission route, the cost of building sub-transmission to the site is roughly linear with respect to distance -- double the distance that must be covered and the cost doubles, too. On the other hand, the substation site-feeder system cost interactions discussed in section 12.3 are exponential -- move a substation site a short distance away from the center of feeder cost and the cost impact is very small; double that deviation and the cost quadruples, etc.

When combined, the two cost sensitivities indicate that the cost of building sub-transmission to reach all the way to the center of feeder cost is never justifiable. Figure 12.11 illustrates this concept. Shown are feeder-cost versus distance data from Figure 12.6, along with the cost of building sub-transmission at $349,000/mile PW (from Table 11. 4 and 11.5) as a function of distance. The break-even distance, assuming that planners can adjust substation boundaries to minimize feeder-system cost impact, is just slightly over one mile. If substation boundaries cannot be adjusted (for example, if they are defined by a highway or a river, etc.) then this break-even distance is half a mile.

Figure 12.12 shows an example application of this concept in a modified version of the example system. A "site cost map" including both feeder and sub-transmission costs has been developed from what was only feeder costs (Figure 12. 9) by adding the cost to build sub-transmission from the ROW directly north to every location. The optimal location is now one mile north of the center of feeder cost.

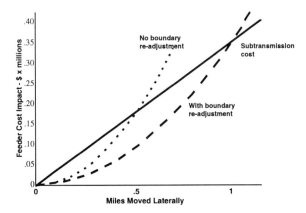

Figure 12.11 Cost of building sub-transmission compared to the cost penalty feeder-cost impact of non-optimality in location. Break-even distances are .5 and 1 mile.

The plot of sub-transmission and feeder costs in Figure 12.11, and the site cost map example in Figure 12.12, illustrate a general result: for distances less than the break-even distance, the sub-transmission cost needed to move a planned substation to the center of feeder costs exceeds any possible feeder-level savings that could be produced by reaching it. Thus,

> If the center of feeder cost is within .4 to .8 mile of an available substation site that has sub-transmission line access, then it is most economical to use that as the substation site.

> - If the center of feeder cost is more than .4 to .8 mile from a sub-transmission line, then the optimum location for the substation will be on a circle of radius .4 or .8 mile around the center of feeder cost, at the point nearest the sub-transmission line).

Substation Sites and Access Cost Interaction with Site Location

The substation site, itself, must have several characteristics in order to provide a suitable location for a substation and its equipment. Their costs can vary from one locale to another depending on the individual land parcel itself and the locale around it. First, there must be sufficient space for the substation equipment -- transformers, racks and buswork, breakers, control house, incoming transmission termination, etc. The space required will vary depending on the type, amount, and voltage levels of the equipment planned for the substation, and can range from several hundred square feet for a small, completely packaged-

Substation Siting and Distribution Expansion Planning

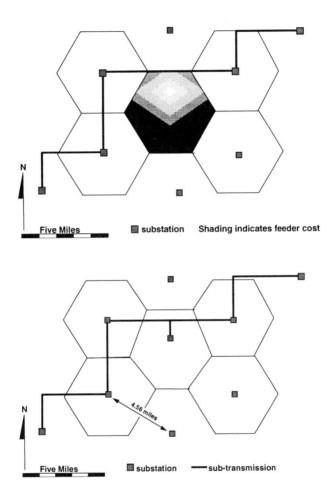

Figure 12.12 A substation planning situation slightly altered from Figure 12.4. Here, the best site to serve the hexagonal substation area in the center is being sought under circumstances where sub-transmission does not pass through the center of the area. Top, shading indicates cost -- to the feeder system costs from Figure 12.7 ("center of feeder cost") have been added the $349,000/mile cost of building sub-transmission from anywhere along the line skirting the northern edge of the proposed service area. The lowest cost point is north of the center of feeder cost by a break-even radius of one mile. Bottom: the final plan with the sub-transmission in place and the substation boundaries adjusted to minimize feeder-level impact. (The one mile break-even radius assumed they would be re-adjusted once the site was determined).

single-transformer substation to ten or more acres for a multi-transformer substation with very involved high and low-side buswork and switching.

In addition to having the room for the substation equipment, the site must be accessible with a transmission right-of-way, so that incoming transmission can reach the substation, and have sufficient room for feeder getaways -- routes out of the substation through which the manifold feeders can run. Any restrictions on routes in and out of the substation, or unusual clearance problems with either transmission or distribution, will increase cost if not rule out the location altogether, but given that it meets these requirements, it can then be used for the construction of a substation (see Chapter 10). On occasion, there are other considerations, including unusual esthetic and environmental requirements, which might impose additional cost at one site that is not realized at another.

And most important, the substation site must be obtainable at any price. Often, many of the best sites are not -- they have been purchased by other parties with other plans for them; community opposition may be quite strong against locating a substation at some sites; or they may be "off-limits" due to environmental or other public interest requirements. (See Planning Criteria, Chapter 4, for a more complete discussion of esthetic and environment attributes required of T&D facilities).

Given that a site is available, a matter of concern is its cost. Prices for land vary depending on location, attributes, and other factors. Generally, land that is close to load centers and accessible is more costly than land that is not. However, site cost is only one of many aspects of overall cost, and even a relatively expensive site represents just a small portion of overall substation cost. It is a mistake to let site cost drive the substation locating process (as it does in many cases); all too often the desire to save several hundred thousand dollars in real estate cost by moving the location of a new substation to a less expensive site results in impacts at the substation and feeder level that come to many times that amount.

Substation Site Cost Maps

Figure 12.13 shows maps of land parcel cost (the cost of the land itself) and complete site preparation cost (the cost of building access roads to the site, clearing the land, and preparing the site for a substation), and a total summation of land cost, site cost, feeder, and sub-transmission costs on a locational basis for the example problem (substation costs are not included because they do not vary depending on location in this particular planning scenario).

As can be seen in Figure 12.13, both land purchase and site preparation costs vary considerably from one location to another. These costs are usually not

Substation Siting and Distribution Expansion Planning

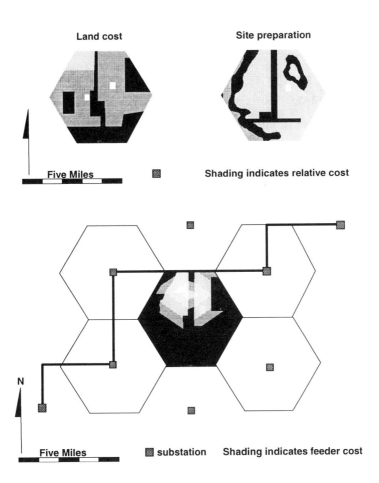

Figure 12.13 Maps of the cost of land, (top left), and site preparation cost (top right) show variation in cost depending on location. Shading indicates cost, with unavailable sites shaded completely. When the land and site preparation costs are added to the subtransmission and feeder costs of Figure 12.12, the resulting sum (bottom) shows the complete "substation cost variation as a function of location" map, which provides planners with identification of the optimal site according to their criteria and cost analysis. As a result of these considerations being added to the site evaluation, the optimal location moves very slightly to the north and west compared to Figure 12.12. If selected, this site would dictate very slight revisions in the (already revised) service area boundaries of the substation area shown in Figure 12.12.

computable for planning purposes from estimation formulae as are sub-transmission and feeder costs (transmission and feeder costs are functions of system parameters; land and site cost are not), and they have a very high spatial frequency of variation (i.e., cost can change by an order of magnitude in just a very short distance, unlike feeder and sub-transmission costs. In addition, land and site cost variations among acceptable and reasonable sites generally do not vary by such large amounts that they become the major factors in siting -- but their variations do have an effect, and should be taken into account.

Site Cost and Variation Contour Maps

A "site cost map," such as illustrated by the example problem in Figure 12.13 is a useful tool in identifying the best substation sites and developing an understanding of how costs and spatial relationships interact in any particular planning situation. Whether the data is developed and displayed explicitly as a shaded/colored map as shown in Figure 12.13, or used internally in some computerized siting analysis and optimization program, evaluation of total site cost variation and development of the *information* illustrated in Figure 12.13 is important in planning substation sites properly.

A useful variation in the display of the information is a cost variation contour map, as illustrated in Figure 12.14. This type of map is a very effective communication tool for planners. It shows in contours the "cost penalty" that a

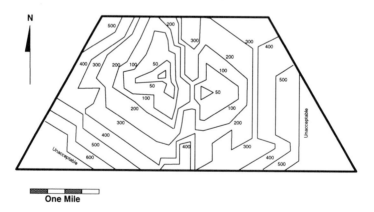

Figure 12.14 Cost contour map for the northern part of the target substation area, developed from Figure 12.13, shows the "cost penalty" that is paid as one moves away from the optimal substation site. Such a map helps the utility's land purchase agents decide which parcels of land, at different prices, are worth pursuing as substations.

Substation Siting and Distribution Expansion Planning 515

non-optimal site has for the company, and thus identifies when and if it is preferable to pay more for a site closer to the optimum. Provided to the land acquisition department, such a map identifies the relative merit of various sites that may or may not be available, allowing them to make judgments about which to pursue on a detailed basis with respect to the utility's total needs. Shown to management, such a map indicates where spending more for a particular site is justifiable when other parcels are available at less cost.

12.4 SUBSTATION-LEVEL PLANNING: THE ART

Despite the importance of substations, and although many aspects of their planning can be studied analytically, as in Section 12.3, substation planning remains as much an art as a science, in the sense that it cannot be reduced to *only* a set of rules and formulae, and that some individuals or groups of individuals are simply much better at it than others. This section attempts to present a coherent picture of the art involved in substation planning. This section describes a perspective and manner of strategic thinking which the author has found useful in substation planning, as well as several guidelines and helpful rules of thumb in the procedures for substation planning.

The "Game" of Substation Tile Planning

This analogy to a game has helped some planners understand substation planning and become better planners. Others think it is silly. It is offered for whatever it is worth. The process of planning the future of a T&D system can be likened to a game, in which the goal is to serve all the customers, the score used to determine quality of play is overall cost, and the moves consist of decisions about when, where, and how to add substation capacity. It is not a "zero-sum" game in the sense of mathematical game theory, but a positive-sum game where quality of play, attention to detail, and particularly innovation can yield big winnings and generate better than expected scores over the long run.[4]

Usually, the substation planning "game" is played as a series of iterative moves (annual T&D plans). It has its own unique rules (criteria, standards) and permits many unorthodox moves ("build nothing this year" or "split a substation and build two smaller ones rather than one large one" or even "install a mobile substation there for a year, then build a permanent one there").

---

[4] In a zero-sum game, the the sum of scores of all participants is a constant (i.e., zero) so one wins by taking away from another player or aspect of the game.

Placing a substation at a certain point in the service territory creates a source for the feeder system there, lowering the cost of running feeders to, and providing power for, the customers in the vicinity of the substation. Compared to plans that have no substation in that vicinity, feeder capital costs will be lower, losses will be lower, and service reliability will be higher. The price paid for these advantages is a higher cost at the sub-transmission level (required to get power to the substation) as well as the cost of the substation itself. Thus, the decision on whether to build a substation in an area of the system is quite simple: are the savings that it will provide in lower distribution costs worth its sub-transmission and substation-level price? Until they are, a substation should not be built. When they are, the site, size, and service area with the best overall economics should be built. The benefit that accrues from this can be determined using a variation on the siting analysis method developed earlier in this chapter.

Substations as "tiles" or "areas"

Earlier in this Chapter (Figure 12.1), it was observed that the substations must "tile" -- completely cover -- the utility service area. Substations can be likened to three-tiered elastic "tiles" (Figure 12.15), which the planner must fit together in the substation planning game, by twisting, stretching, or otherwise forcing them to fit, until as a group they somehow cover the entire service territory. Each "tile" consists of a sub-transmission route which must be attached to the system grid at appropriate places, the substation itself at the center, and the feeder system that covers the surrounding territory, all elasticly anchored to the actual substation site chosen by the planner. This concept of substation planning is based on three critical observations about T&D systems.

> As was illustrated by Figure 12.1, the substation service areas in a T&D system must completely cover, or "tile," the utility service territory, in the sense that their service area boundaries must meet and the union of all substation service areas must jointly cover all the utility service territory.

> As also discussed in section 12.2 (Figure 12.2), *a "substation" consists of three levels,* sub-transmission, substation, and feeders. To do its job it must have sufficient capacity in all three levels, in a compatible manner.

> The utility can *afford only a few substations,* so each one must be well-utilized. Additionally, while management, regulators, and the public (grudgingly) understand that electric utility facilities are necessary, asking for too many substations, particularly if planners cannot demonstrate they have used past sites well, is a recipe for a public relations disaster.

Substation Siting and Distribution Expansion Planning 517

"The optimal tile" and how it reacts to changes in design

For any system, there is an optimal target substation size (capacity), service area size, and substation spacing (as was discussed in Chapter 11) which defines a "optimal tile" -- *the target characteristics which the planner would like every substation to have.* Assuming feeders must be routed through a grid, the optimal shape for each substation "tile" is a diamond -- a square rotated 45° to the grid -- not a circle (circles do not fit together as tiles) or a hexagon (hexagon's fit together as tiles, but a hexagon, which is the optimum if feeders could be routed by Euclidean distance) is very slightly more expensive). The optimal site for the substation is at the diamond's center; the optimum service area is defined by the relationships explored in Chapter 11, but basically is set by the maximum economical load reach of the primary distribution voltage/conductor set (see Figure 12.15). The optimal size (capacity) is what is just enough to meet the needs to serve the load inside its territory.

All the elements of a tile -- shape, width, height, location of the substation, etc., are elastic, so the tile can be twisted, bent, or stretched into shape to fit among other substation tiles. However, increasing elastic tension indicates increasing cost. The sub-transmission route attached to the substation can be stretched, twisted around obstacles, and routed as necessary to bring power to the substation from anywhere, far or near --- increasing its length and/or using a circuitous route increases its cost.

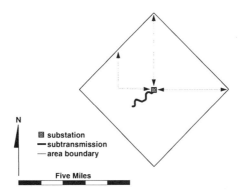

Figure 12.15 An optimum substation tile: a square (diamond-shaped) service territory, of Lebesque-1 radius equal to the economic load reach of the primary distribution system (dotted line, 3.3 miles for the 12.47 kV feeders from Chapter 7), a substation at the center, and a short sub-transmission line. All three dotted lines are the same length.

518 Chapter 12

Similarly, the feeder system can be stretched beyond its optimum (which increases cost) or cut short (which throws away capability, which generally increases cost somewhere else) so that the tile's shape fits any requirements. The planner's job is to design a set of substation tiles so they cover the service territory, leaving no gaps -- something like fitting the pieces of a picture puzzle together, as illustrated in Figure 12.16.

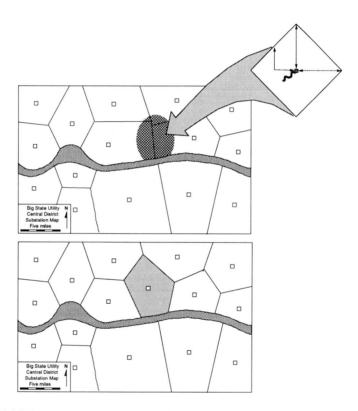

Figure 12.16 Top: a new substation is needed to serve an area where capacity shortfalls are expected (shaded area). Bottom: a new "tile" is inserted into the system, and its neighbors' service areas will depend on how well the distribution planners looked ahead and planned the entire set of substations to fit together. Here, the new substation is not optimal, but close -- its service area is vaguely diamond-like in shape. Shapes of other tiles are adjusted, too, as shown, in this case probably improving their economics as well (there is not enough information about subtransmission, site costs, and load locations displayed here to be certain), since the two substations on either side of the new substation were quite far from the center of their service territories.

Substation Siting and Distribution Expansion Planning

The feeder system, which can be likened to an elastic sheet, with the substation, anchored in the middle, and a natural (no tension) radius equal to the economic load reach, can be pushed to fit into a tight location, or stretched to meet a boundary farther away -- but either increases cost. Similarly, the substation belongs in the very center of the feeder sheet, but it can be pushed about inside its territory, again increasing elastic tension (cost).

Some set of substation tiles must be fitted together to cover the entire utility load pattern, with their boundaries butting together or against natural boundaries (Figure 12.16). The planner's job is to fit tiles together into an overall plan, spending the least on resources and causing the least amount on them and creating the least elastic tension (non-optimality cost).

12.5 GUIDELINES FOR SUBSTATION SITE AND SIZE APPLICATION TO ACHIEVE LOW COST

Whether one is referring to elastic tiles or real substations, there are a number of concepts behind good and bad application, as illustrated in Figure 12.17. The basic concept of achieving low cost is to utilize every substation "tile" as closely as possible to its optimum in size, shape, diameter, etc. -- anything else will increase cost. Figure 12.17 shows several guidelines for doing so. As with any complicated design endeavor, these various rules have to balanced carefully against one another, and worked around constraints of geography, etc.

Substation siting near service territory boundaries

Most utility service territories have limits set by the utility's franchise agreement, or geographic boundaries defined by natural features such as lakes, oceans, or other similar features. Regardless of whether natural or "man-made," these service territory boundaries constitute immovable substation area boundaries.

As discussed in section 12.3, when substation boundaries cannot be moved, costs due to non-optimal siting are doubled. Thus, from the standpoint of overall economics, *it is especially important to optimally locate and size substations near service territory boundaries and natural barriers.* Figure 12.18 shows two substations located on the boundary of the system. Fully half of their potential feeder system reach lies outside the utility service area, where it provides no value. A good deal of their fixed cost investment is therefore wasted. A better site for each is well inside the service territory, just within economical load reach of the boundary. As shown in Figure 12.19, this means there is a "band" of preferred substation sites, about $1/\sqrt{2}$ times the economical load reach of the distribution feeder system inside the service territory boundaries, and around any large geographic restrictions, as shown.

Figure 12.17 "Substation siting rules."

The perfect substation. Square service area oriented 45° to the road grid, "radius" equal to the economic reach of its feeders, substation in the center, with capacity sufficient to serve the load in the square service area, located on an existing sub-transmission ROW.

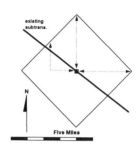

Lengthy sub-transmission construction is expensive, and should be avoided unless justified by the site's other savings. As mentioned in section 12.3, sub-transmission all the way to the optimal location from a feeder standpoint is never justifiable.

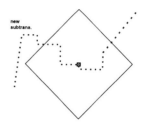

Poor aspect ratio. Serving the same-size area and load in a "longer-narrower" shape results in higher feeder costs.

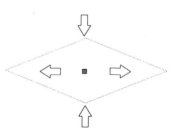

Substation not near center of load. As described in section 12.3, feeder system cost goes up if the substation is not in the center of the service area.

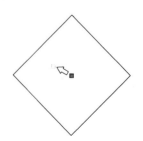

Substation Siting and Distribution Expansion Planning

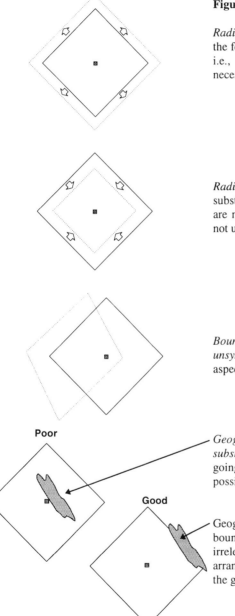

Figure 12.17 cont.

Radius greater than economic reach -- the feeder system will not be efficient -- i.e., cost/kW will be higher than necessary.

Radius less than economic reach -- the substation and sub-transmission system are not efficient -- their fixed costs are not utilized fully.

Boundaries moved drastically and unsymetrically. Compound problem: aspect ratio and substation not at center.

Geographic barrier/restriction near the substation -- increases feeder costs going around it to get to the load. If possible this should be avoided.

Geographic barrier on the substation boundary -- the impact of the barrier is irrelevant if the substations can be arranged so that the boundary overlaps the geographic restriction.

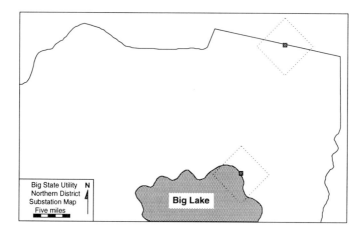

Figure 12.18 A substation located on the edge of the utility service territory, or right against a geographic barrier, is largely wasted. The two substations shown (shaded squares) cannot utilize half of their most economic load reach capability (dotted lines). Therefore, regardless of what capacity is installed, half of the investment made in their fixed cost (site, sub-transmission, etc.) is essentially wasted.

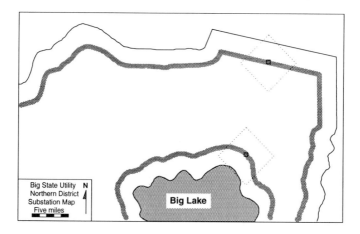

Figure 12.19 Optimal substation siting of the outer tier of substations in a utility is in a band $1\sqrt{2}$ times the economic load reach inside the substation service territory boundaries and around any large geographic restrictions.

Substation Siting and Distribution Expansion Planning 523

12.6 SUBSTATION-LEVEL PLANNING: THE SCIENCE

As pointed out earlier, substation planning is somewhat of an art, part of which can be likened to a game. A great deal of formal analysis can be applied to the identification, evaluation, and selection of alternative sites, sizes, and substation service areas. Usually, this analysis will have a limited context, or will involve optimization applied to only a limited number of factors. Hence, the art discussed earlier: the planners need to understand how all of these system factors fit together, and, in particular, where these analytical methods run into their limits. This requires intuition, judgment, and a type of artistic application to the portion of the problem not being analyzed.

But nonetheless, there are powerful analytical tools available for substation-level planning (as discussed in Chapter 16), as well as formal, analytical evaluation procedures for evaluation of sites and cost interactions, all providing useful evaluation and documentation of need. This section examines their overall application to the planning of substation siting, sizing and timing.

How Much is Adding a Substation at a Particular Site Worth?

If the analysis method of reducing capacity and transferring load to neighbors (section 12.3) is carried to its extreme, the capacity of a substation in the example system could be reduced to zero and all of its load transferred to the six neighboring substations, and the cost impact determined. The cost thus computed for "reducing capacity to zero" can then be compared to the savings that accrue from not building the substation, to identify the net value, or worth, of having the substation at that location

$$\text{Value of substation} = \text{cost without it} - \text{cost with it} \quad (12.7)$$

From equations 12.1 through 12.5, the cost impact with zero capacity would be calculated as

$$\sqrt{\frac{58.5 \text{ MW}}{3.25 \text{ MW/mile} \times \pi}} \quad (12.8)$$

$$= 2.4 \text{ miles}$$

The adjustment must be made for the necessity of routing feeders through the grid.

$$\text{actual distance change} = 1.1 \times 2.4$$
$$= 2.64 \text{ miles}$$

The substation's entire 58.5 MW must be transferred to the neighboring substations, over roughly a 2.64 mile greater average feeder distance. The example system's normal s_v of \$45,250/MW-mile for 12.47kV distribution is inappropriate in this case because of load reach limitations (see Chapter 7). Under normal circumstances (i.e., with the substation in place and serving 58.5 MW), the feeder system is called upon to deliver power over distances of up to 3.4 miles -- its maximum reach, with the average distance being about .75 of that (2.55 miles). The average distance required to serve this particular 58.5 MW, now that the substation is no longer there, is about 2.64 + 2.55 miles = 5.2 miles, or more than 50% beyond the distribution feeder system's economical load reach. The cost penalty for the reduction in loading to achieve greater reach is not as great as one might expect -- s_v increases by about 12%, to \$50,700. Thus,

$$\text{Cost impact} = \text{load} \times \text{distance} \times s_v \qquad (12.9)$$
$$= 58.5 \text{ MW} \times 2.64 \text{ mile} \times \$50,700/\text{MW-mile}$$
$$= \$7.83 \text{ million}$$

If the substation were not built at the site at all, the feeder system cost increase can be expected to be \$8,000.

How much would be saved? Although a complete substation was shown to cost over \$36 million PW (Table 11.5), the savings would be nothing like that. To begin with, that \$36 million PW includes about \$12 million PW for the laterals and small primary line segments of the distribution system, needed in *any* planning scenario where the customers are to be served. The substation capacity is not avoided, either, because again it is necessary in order to service the utility customers: the 81 MVA of substation capacity planned for the substation at this site is just located somewhere else -- somewhere not specified in this analysis but presumably split among the six neighbors.

Table 12.3 shows the costs that are avoided by "not building" the substation. These include \$2.9 million PW of sub-transmission cost, the substation site (\$1.5 million PW). Also shown is a differential transformer capacity cost: the first two transformers at a substation cost \$925,000 each, complete. However, subsequent transformers cost only \$775,000 each, complete. The first two transformers originally slated for this substation are located somewhere else in the example system -- i.e., at a substation that already had three transformers before the change in plan, and thus their capacity cost is \$775,000 each, for a total cost differential of \$300,000 less.

Substation Siting and Distribution Expansion Planning

Table 12.3 Cost Differential for "Not Building" a Three-Transformer Substation for 58.5 MW peak load - $ × 1000

Item	Cost	Savings
Sub-transmission, save 8.3 miles & losses		2909
Increased losses on other sub-transmission	935	
Substation site		1,475
Differential transformer cost		300
Additional feeder system cost	7,830	
	8,765	4,684

Value of substation , $8,765 - $4,684 = $4,081

The comparison in Table 12.3 assumes that the additional substation transformers would fit among the neighboring substations (not a certainty, and perhaps only at additional costs not included here), and that the load increase in picking up the missing substation's load could be tolerated by the neighboring substations' sub-transmission (an average of 16.6% each), and so forth. It also neglects a number of minor costs (costs from "overcrowding" at some substations, the savings that accrue from the larger economy of scale in transformer maintenance with more at any one site).

However, those and other factors neglected in the analysis are not germane to the central point of this example: the decision to build a substation at the site -- to pay the cost of putting capacity there rather than at other, already existing substation sites -- produces a net savings of about four million dollars. Variations in exactly where the substation goes can change the costs and savings, as discussed in section 12.3, affecting this net savings. But for the example system, the decision to locate capacity at a new substation rather than at other substations, has a benefit/cost ratio of about 2:1.[5]

---

[5] This analysis neglects entirely the impact on reliability of service. Interruption rate and duration for the customers in the planned (now not built) substation's target area, go up considerably, and so do those for the customers in the six surrounding substations (just how much depends on detail assumptions about how the feeder system is re-built). The cost impact of this from a value-based planning perspective can be anywhere from another $3.5 million to $8 million.

When Should a New Substation Be Added?

Re-computation of the values in Table 12.3 at a "break-even" load of 42.7 MW (rather than 58.5 MW) will show that for any peak load above 42.7 MW, the substation has a positive net value, while for peak loads below that the substation is not cost-justified. However, such analysis indicates little beyond a recommendation that the entire substation (all three transformers) be in place by the time the load reaches 73% (42.7/58.5) of its final load value. To begin with, the foregoing analysis assumes that if the substation is not built, its capacity is put somewhere else. A substation expansion plan cannot move capacity from one location to another easily and cheaply, putting it in one location and then moving it several years later.[6]

Substation timing is growth and capacity-driven

Most substation-level planning is capacity-driven: a growing load dictates that capacity be added to serve it. Given the load growth, there is no real question that the capacity must be added -- but the issue of whether it should be added at a new substation, or to an existing substation, depends on where the load is relative to the existing substations and the potential new site, *and* the rate of growth.

> *The decision to build a new substation centers on whether required capacity to meet new load should be added to existing substations or at a new substation. This depends on the location of the load and the load growth rate.*

In cases where existing substations have no remaining room for capacity additions, a new substation will have to be built. But such cases are very rare: there is nearly always some way to increase capacity at an existing substation. Even if relatively expensive, this cost is always less than the site, sub-transmission additions, and other fixed costs required for a new substation.

In addition, the load growth dictating the new capacity additions may be in a location not ideally reachable from any existing substation. For example, if the growth is occurring at point A in Figure 12.20, rather than at point B, its location provides an additional reason in support of building a new substation.

---

[6] Such "moving transformer" plans have been studied at several utilities and found to be un-economic except in very unusual circumstances.

Substation Siting and Distribution Expansion Planning

A new substation is a long-term investment: its placement in the system will reduce feeder capital construction costs in the near-term future (2 to 5 years ahead), and feeder losses, O&M, and service interruptions over the short and long-term (2 to 30+ years ahead). But its initial capital cost is immediate, and the investment will look justifiable only if sufficient future costs are evaluated to outweigh the immediate ones -- evaluation on the basis of present costs will always favor making the addition at an existing substation.

As an example, Table 12.4 compares the 30-year PW cost of building a one-transformer substation (27 MVA) to serve an 18.5 MW peak load, to the cost of adding that transformer elsewhere and serving the load over feeders from other substations. The substation site cost shows an adjustment ($300,000) from that shown in Table 12.3, and the sub-transmission costs a reduction of $250,000, due to building only enough facility to serve the one transformer. The substation is nowhere near justifiable, even over the long run; on the basis of 1/3 of the normal load it represents a PW loss of two thirds of a million dollars.

The decision on whether to build a new substation for the one transformer, or put the capacity at an existing substation depends greatly on the future load growth and its timing. Table 12.5 shows the economic comparison for a two-

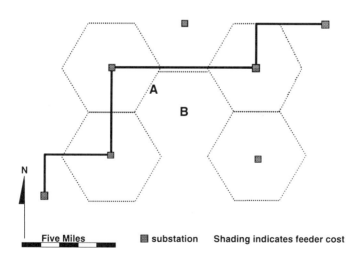

Figure 12.20 The economic incentive to build a new substation depends on where the load growth is relative to existing substations that can be expanded. Growth occurring at location A provides much less incentive for a new substation than growth at B.

Table 12.4 Cost Differential for "Not Building" a One-Transformer Substation for 18.5 MW peak load - $ × 1000

Item	Cost	Savings
Sub-transmission, save 8.3 miles & losses		2,054
Increased losses on other sub-transmission	110	
Substation site		1,175
Differential transformer cost		150
Additional feeder system cost	2,610	
	2,720	3,379
Value of substation, $2,720 - $3,379 = ($659)		

Table 12.5 Cost Differential for "Not Building" a Two-Transformer Substation for 39 MW peak load - $ ×1000

Item	Cost	Savings
Sub-transmission, save 8.3 miles & losses		2,650
Increased losses on other sub-transmission	454	
Substation site		1,475
Differential transformer cost		300
Additional feeder system cost	5,220	
	5,674	4,425
Value of substation, $5,674 - $4,425 = $1,249		

transformer substation serving 39 MW, rather than a one-transformer substation serving 18.5 MW. The economics are now solidly in favor of building the new substation, with a PW savings of nearly $1.25 million. The second increment of load provides a $1.908 million shift in net PW savings.

Thus, if the load were 39 MW, then the substation would be justifiable. Suppose the load were growing in two stages -- 18.5 MW now, 18.5 MW three years from now. A quick analysis using a PW factor of .90 produces an overall savings of

$$\text{Total benefit} = (\$659,000) + .73 \times \$1,908,000 \quad (12.10)$$

$$= \$732,000$$

Substation Siting and Distribution Expansion Planning 529

The substation, built now with one transformer and expanded with a second transformer in three years -- is less expensive than adding that capacity and serving the load from other substations, by almost three-quarters of a million dollars. In fact, if load growth is expected to continue at any rate such that the additional transformer capacity will be needed within ten years, then it makes economic sense to build this substation now with one transformer and add the second when needed.

Build/no build decisions: "When" should be based on comprehensive present worth analysis

While the foregoing analysis neglected several secondary and a number of tertiary cost considerations associated with the build/no-build decision, it illustrated the major concept in determining if and when a new substation should be built. When considering a specific substation siting decision, planners should perform a PW cost analysis of alternatives in detail, as illustrated here but taking into account all costs on both the "build" and "no-build" sides.

Substation Siting and Planning Interaction with Load Growth Behavior Dynamics

Most intuition, and many traditional rules-of-thumb about substation and feeder expansion are based upon the concept that the load density in a region or target service area increases at continuous, steady rate over a very long period of time.[7] This viewpoint is partially incorrect, particularly as applied to the substation and feeder level, and its application as the guiding principle in planning substation expansion leads to invalid conclusions about how to expand a distribution system at minimum cost.

When viewed on a wide, regional basis, such as over an entire metropolitan area, load growth trends do exhibit continuous and steady growth histories -- except for weather and economic recessions, annual peak load marches upward at a steady, never-ending pace year after year. But at the local level, at the size of substation and feeder service areas, load growth trends look quite different, both spatially and temporally.

---

[7] In particular, those relating to the handling of future load growth by "splitting" substation service areas in Chapter 3 of the Westinghouse "Green Book" (Electric Utility Distribution Systems Engineering Reference Book).

Blocky spatial growth behavior

First, an area the size of a substation service territory "fills up" in parcels, each developing from no load to complete development in a short period of time, until the that parcel is saturated -- then growth shifts to a nearby parcel. The result is a gradual filling in of a region at a block at a time, as shown in Figure 12.21.

Local "S" curve growth trend

Most of the parcels in an area the size of a substation service territory will fill in during a relatively few number of years (7-18, depending on growth rate and substation size) resulting in a temporal trend called an "S" curve, as shown in Figure 12.22. Growth before and after this period of relatively rapid growth is low. "S" curve load growth behavior is the typical and expected trend of load development at the substation and feeder level. It and other characteristics of load growth at the distribution level, are discussed in more detail in Chapter 15, particularly section 15.2. The reader is urged to become familiar with these load growth characteristics because they are typical -- matching them with appropriate, coordinated substation-feeder expansion is one key to low cost.

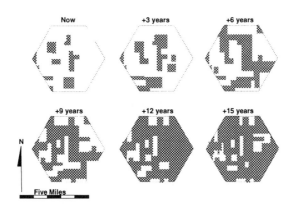

Figure 12.21 Load grows as developing parcels of land (small areas), not as a general, region-wide increase in density where all areas grow equally from zero load to full density. As a result, substation siting and feeder expansion must deal with delivering "full load density" to an increasing number of neighborhoods as shown here. This means that full feeder capability (maximum designed load and maximum designed distance) may be needed far sooner than predicted by the "gradually increasing density" concept.

Substation Siting and Distribution Expansion Planning

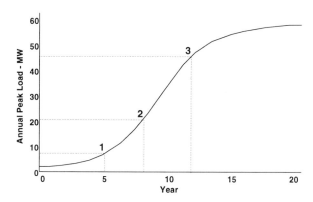

Figure 12.22 Local area growth dynamics usually result in the load in a substation's target service area growing in an "S" shape trend as shown here. This means that a substation's two increments of additional capacity are usually required at short intervals after initial construction: in the example system the substation is built in year 5, when peak load in the target service area is 7 MW, the first capacity upgrade is made three years later (peak 21 MW), and the third four years after that (peak 45 MW).

Capacity upgrades usually are needed within a short time after initial construction of a substation.

The spatial-temporal load growth behavior shown in Figures 12.21 and 12.22 has two implications for substation siting and the subsequent planning that flows from it. The first is that cases where building a substation is justifiable because a second transformer will be required after a few years, as described early in this section, is in fact typical. This is the situation for most new substations in metropolitan and suburban areas and not far different for rural expansion cases.

In addition to a typical "S" curve growth trend for a substation service area taken from the example system, Figure 12.22 shows the timing of the three transformer installations made at the substation. While the overall growth of the metropolitan region where this substation is located is only 1.6% annually, this local area is growing quite rapidly at the moment (see Chapter 15, Section 2). Most of its growth occurs in the decade between years 4 and 14, when over 80% of the load develops, starting from a peak load of only 5 MW annually, to 54 MW only ten years later (as with all the substation areas in the example system, load eventually reaches 58.5 MW).

The substation is built and the first transformer installed when load reaches 7 MW (the "1" indicated on the plot in Figure 12.22) -- prior to that the load is handled by splitting it among several neighboring substations that have traditionally served the new substation's target load area. The second capacity upgrade is required only three years later (the "2" indicated on the plot), when load has reached 21 MW (23.3 MVA), for an 86% utilization of the single 27 MVA transformer initially put at the substation. It is worth noting that the timing for this second transformer installation is the most critical of the three transformers to be installed at the substation. At this point in the "S" trend, load growth is 7 MW (7.8 MVA) per year -- a delay of even one year would result in severe overloads. The final increment of capacity (the "3" indicated on the plot) is required only four years later.

As a result of the "S" curve growth trend, the substation and its two transformer additions come within a seven year period, and a PW assessment of the build/no build decision for the substation comes down heavily in favor of building the substation. For although the initial investment to cover the fixed costs of site, sub-transmission, etc., is not justified by the load at that time (only 7 MW), the site's utilization will grow quickly to seven times that load level. The utilization of the substation looks better at that load level, and equally important to the analysis, the cost of trying to serve this rapidly growing, higher load level from nearby substations looks quite bad.

Purely from a substation standpoint, the PW economics look even better if the initial site and transformer installation are delayed as long as possible -- one or even two years beyond that shown in Figure 12.22 -- perhaps until the peak load in the region is 11 or even 14 MW. (At a PW factor of .9, a delay of one year in a one-transformer substation site's fixed costs will be worth about $320,000). However, the exact timing of the "build" decision is dictated by the second aspect of substation planning affected by the growth dynamics discussed above -- feeder reach requirements.

Feeder load reach may be a constraint forcing "early" construction of a new substation.

One step that planners want most to avoid is having to build facilities or install equipment that will be needed only for a short period of time. The blocky spatial load growth characteristic shown in Figure 12.21 presents planners of a new substation area with this dilemma early in the substation's lifetime -- in fact before it is built.

The classical (*and wrong!*) concept of load growth and feeder expansion was that the load everywhere in the substation area grows from low density to high at a steady but slow pace over many years. One result of such thinking about

Substation Siting and Distribution Expansion Planning

load growth is a concept that envisions feeders, when built, as having a load far less than their eventual (design) loading for the area they serve, so they have extra capacity, and extra reach, early in their lifetime. Thus, in combination with re-switching, they can be used to reach into the service area of a (yet unbuilt) substation to serve load there for many years until the substation has to be built. In this way, a substation's construction could be delayed for a decade or more as its target service area filled in, with the load supported from nearby substations.

But actually, the load in any target service area grows in parcels of near final load density, filling in by area by area, as was illustrated in Figure 12.23. Usually, some parts of the feeder system see the complete, full load they have been designed to serve within only a few years of their construction. The concept of using them to help defer substation construction just won't work because the reach is not there, and reinforcing them so they meet voltage-drop requirements at the great distances involved increases cost a great deal. For example, the area to the southeast of the substation site in Figure 12.23 has already developed fully by year six.

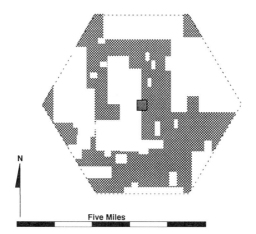

Figure 12.23 Load growth in the target substation area about six years into development. Some areas of load growth are very nearly completely filled in with full load density. In this case, those feeders heading to the southeast and south from the substation site are very close to their final load level in years 5-6. Any plan that tries to defer the substation by using "long distance" feeders from neighboring substations while staying within range A voltage drop criteria will have to "overbuild" feeders at additional cost, defeating their purpose of trying to produce a savings by deferring the substation.

Thus, if the feeders in the southeast part of this substation area are built to the levels of capacity and reach needed for the long run, then at this stage of growth (year 6), their voltage drop is fairly near the maximum permissible when delivering power from the substation site to the loads. There is little margin left for the voltage drop that would develop if they were split and fed from nearby substations.

The planners in such situations have two choices. First, they can reinforce the feeder system, both in the target substation area and in neighboring substation that will feed the target area, so that the feeders can do the job within permissible voltage drops limits. This means building parts of the feeder system in to a longer reach requirement, and reinforcing portions of the feeder trunks in nearby substation areas. This additional cost in year six can be estimated as: 9 MW (the estimated load) × 5.2 miles (the distance power must be moved from a nearby substation) times $7,500 (the difference in the $45,250 and $52,500 s_v values for 12.47 kV distribution at 3.3 and 5.2 miles reach), for a value of $339,000. This additional expense has little long-term value, and thus has to be judged against the gain of delaying the substation a year (which as shown above is about $320,000 PW).

On the other hand, the planners can build the substation, basically foregoing the savings of $320,000 that would accrue from delaying it another year. By providing a source close to the load and requiring no feeder reinforcements beyond those needed in the long run to serve load out of the substation, they avoid any feeder reinforcements.

The $339,000 and $320,000 values shown here are only rough estimates based on an approximation formula that considered only primary aspects of the various alternatives (but it turns out that the actual detailed assessment, like that recommended by these approximations, comes down slightly in favor of building the substation in year five, not six).

Approximate as this examination of the costs and alternatives was, it illustrated the *principle* of the substation timing decision -- the substation should be built when the costs of delaying its construction another year outweigh the savings of deferring in another year.

The actual decision of when to build the substation should be made with a comprehensive analysis of all factors involved, using the utility's time-value-of-money decision-making process, whatever that is and however it is applied. The load growth dynamics illustrated in Figures 12.21, 12.22, and 12.23 mean that very often this decision falls in favor of building the substation several years earlier than might be expected based purely on the basis of substation capacity economics and feeder considerations if the load growth were evenly distributed in both space and time.

Substation Siting and Distribution Expansion Planning

12.7 THE MOST IMPORTANT POINT ABOUT SUBSTATION-LEVEL PLANNING

Plan ahead.

The determination of the sites, sizes, and timing of substation additions and expansion in a power delivery system is *the* strategic planning for the power delivery system. It must be done while taking into account an amazing number of aspects of the T&D system, including the pattern of load and load growth; the interconnection and interrelation of all electrical levels with one another; the time value of money; natural barriers and limits to the system; existing facilities and capabilities, and a host of other important considerations. Detailed numerical analysis can provide useful guidelines and tools in substation planning, but many of the factors cannot be fully assessed by numerical methods, making substation planning partly an art, where experience, judgment, and natural inclination and talent play a big role in success.

There is no better example in all of utility planning for the adage "an ounce of prevention is worth a pound of cure" than at the substation level. Mistakes made in the planning of sites, sizes, and the service areas of substations create sweeping inefficiencies affecting wide areas of the system that are permanent (and they usually have a negative impact on reliability, too). Many such problems can be avoided easily, often at *no* cost, simply by looking ahead and organizing sites, resources, and schedules well.

If substation-level planning is done well and far enough ahead to provide direction and scope to the whole T&D plan, and if substation sites are well selected and if adequate space and access have been planned to allow sufficient capacity to be built when needed, then the T&D system will evolve in a relatively orderly, efficient, and economical manner, making the whole process look quite simple and even straightforward. But if such planning is not done, or is done poorly, then no amount of subsequent, short-term planning can make up for the lack of long-range structure and fit, and T&D planners will spend a good deal of their time on a series of constant capacity and voltage-drop crises -- what is often called "putting out fires."

REFERENCES

J. K. Dillard, editor, *Electric Utility Distribution Systems Engineering Reference Book*, Westinghouse Electric Corporation, Pittsburgh, 1958 (slightly revised, 1965).

M. V. Engel, et al, editors, *Tutorial on Distribution Planning*, IEEE Course Text EHO 361-6-PWR, IEEE, Hoes Lane, NJ, 1992.

13
Service Level Layout and Planning

13.1 INTRODUCTION

The service level, which consists of the service transformers, utilization voltage circuitry (sometimes called secondary), and service drops, is the final link in the power delivery chain. From the low side of the service transformers to the customer, the service level operates at utilization voltage. This voltage is too low to move power very far, or very efficiently, a fact that heavily constrains the design and layout of the service level. But despite the fact that the voltage is low, and most of the equipment is small "commodity" units, the service level is as essential as any other, and quite expensive and complicated in its own unique way.

This chapter looks at the service level and the planning of the service level. It begins with a look at the service level and its role in the power delivery process, in Section 13.2. Section 13.3 looks at the elements of the service level, and at several very different approaches that can be taken in its layout. Despite its low voltage and apparent simplicity, the service level is quite difficult to analyze and plan effectively, due to its load dynamics and the effects of coincidence, which are examined in Section 13.4. Section 13.5 focuses on the "table-based" engineering methods usually applied to design the service level, and at how these tables can be built to provide optimized designs.

13.2 THE SERVICE LEVEL

The service level of the power system feeds power directly to electric consumers. Power is obtained from the distribution feeders through service transformers, which reduce voltage from primary to utilization voltage (the voltage at which consumers use electricity -- see Chapter 4) and onto secondary circuits that directly feed the electrical consumers (Figure 13.1). The most common utilization voltages are 105 volts (in Japan), 120 volts (in the United States and nations that have adopted US electrical standards) and 230 or 250 volts (in Europe and many other nations).

Voltages in the range of 105 to 250 volts cannot move large amounts of power, and they cannot move even small amounts of power very efficiently over any great distance. For this reason, the service level consists of many small transformers, each providing power to only a handful of customers in its immediate vicinity, with secondary circuits operating at utilization-voltage to route power no more than a few hundred feet from each transformer to the property lines of each customer. Service drops are individually dedicated lines that take the power across the customer's property to his meter.

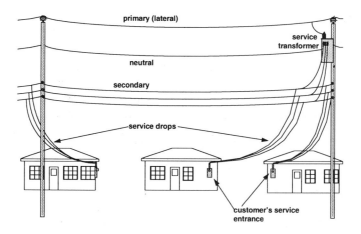

Figure 13.1 The service level consists of service transformers, secondary circuits, and service drops. Shown here is a typical overhead 120/240 volt (American) type of system fed from a single-phase lateral. (This particular transformer is shown without a lightning arrester or fused cutout, although it would probably have both in service.)

Service Level Layout and Planning

13.3 TYPES OF SERVICE LEVEL LAYOUT

"American" and "European" Layout

American and European distribution systems differ substantially in the layout of the secondary system, due to the different capabilities of their secondary voltage systems. Utilization voltage in European systems is 250 volts, in American systems, 120/240 volts. Engineering and analysis of American service level electrical performance is complicated by the dual nature of the voltage delivered. Most residential and light commercial service is provided as single-phase, three wire, 240/120 volt service, as shown in Figure 13.2. The 240 volt potential is obtained across two 120 volt line-to-neutral potentials that are 180° out of phase. Major appliances, such as water heaters, central air and heat, electric dryers, etc., are connected between the opposing 120-volt legs to provide 240-volt power. All other load in a household or business is connected across one or the other of the 120 volt legs. Normally, attempts are made when wiring a house or commercial building to split circuits (usually each room in a house has its own circuit) between the two legs of the incoming service so as to balance the load connected to each 120 volt leg (e.g., four rooms of an eight room house on each leg, etc.). If load on the two 120 volt legs is identical, there is no net flow onto the middle (neutral) conductor, and the power flow is essentially 240 volt.

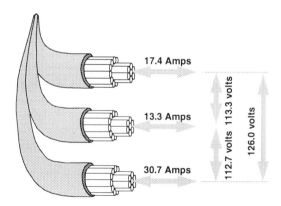

Figure 13.2 Standard practice in the United States for residential and light commercial applications is 120/240 volt (nominal) volt service. Three conductor service drops provide two 120 volt legs, with 240 volts between them. Shown here are typical current flows and voltages that result from residential service (see footnote, next page).

Table 13.1 Distance Power Can Be Moved at 2.5% Voltage Drop) for 4/0 Direct Buried Service Cable (600 V Alum.) as a Function of Loading - Feet

Load - kV	120/240 V 1-P	Typ. 120/240 mix	250 V 1-P	250 V 3-P
5	635	1230	2755	4800
10	315	630	1375	2375
25	130	260	550	950
50	-	120	275	475
100	-	-	-	235

However, in practice, actual customer loads are seldom completely balanced between opposing legs, and the service drops in most American systems, as well as the secondary circuits to some extent, see a mixture of 120 volt and 240 volt power flow from most residential and light commercial customer loads.[1] As a result, from the standpoint of electrical voltage drop and losses, America secondary systems behave somewhat like 120 volt systems, and somewhat like 240 volt circuits.

While the difference between American and European utilization voltages -- 130 volts -- may seem small, the ratio of European to American utilization voltage is more than two to one. This makes a very great difference in how far the secondary (low voltage) circuits in the two systems can move power before encountering voltage drop limitations. In most power distribution systems, the service level is allocated 2 or 3% voltage drop (the feeder system, by contrast, is allocated 7 to 10% (see Chapters 4 and 7 for a discussion of voltage drop standards and feeder-level voltage drop considerations).

The relatively small margin of voltage drop allocated to the service level, and the low voltages at which its circuits operate, means that the distance between the service transformer and the customer cannot be very great. As shown in Table 13.1, when operating at 120 volts, a typical secondary cable appropriate for moving 25 kW at 120 volts and 85% PF can move that amount of power only 130 feet before encountering a 2.5% voltage drop. Allowing for the typical mixture of 120/240 volt loads (second column in Table 13.1), an American type

---

[1] For example, at this moment the author's home, a typical suburban, two-story, five-bedroom house, is drawing 17.4 amps at 113.3 volts on one leg, and 30.7 amps at 112.7 volts on the other leg. This means a net flow of 13.3 amps on the central conductor of the UG tri-plex line leading to the author's home. Such imbalance is the rule, rather than the exception.

Service Level Layout and Planning 541

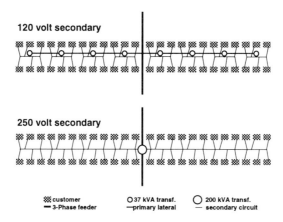

Figure 13.3 Restricted to a shorter secondary circuit load reach, American systems (top) must locate service transformers within 150-200 feet of customers, meaning many small transformers must be used. The higher European utilization voltage and 3-∅ circuits (bottom) mean secondary circuits can reach farther, and fewer but larger transformers are used instead. Shown here is service to 48 homes, each with a load of 6 kVA, as it could be arranged in either system. Note the European system requires no primary voltage lateral.

of service line can move power only 260 feet (including service drops).[2] But at 250 volts, this same cable can be depended upon to move power over 500 feet, almost two times as far, covering an area *nearly four times* as large. Many European urban and suburban systems use three-phase (416 volt phase-to-phase, 250 volt phase-to-ground) low voltage circuits, which can move power farther still, covering even larger areas, as shown in the last column of Table 13.1.

The distance that the secondary circuitry can move power is the major constraint on the overall layout of the service level, for transformers must be located within this distance from the customers they serve. As a result, American distribution systems tend to use very small service transformers -- typically 15 to 75 kVA -- each serving only a dozen or fewer households or business within a radius of 100 to 250 feet from their location. By contrast, a system utilizing 250

---

[2] This follows the author's rule of thumb that 120/240 volt service can generally reach twice as far as 120-volt service (normally, if voltage is doubled, reach at any load level would be *four* times as far, but imbalance among the 120 volt legs cuts this in half).

volt secondary can have each service transformer cover on average six times as many customers, and hence transformers in European systems are typically 100 to 500 kVA. Figure 13.3 illustrates the difference in layout of single-phase residential service for typical American and European systems. Many European systems utilize three-phase secondary circuits, which can reach even farther (see Table 13.1) and as a result, service transformers are often 1,000 kVA or more. The three-phase secondary circuits essentially fulfill the delivery function performed by single-phase laterals in American systems.

Network versus Radial Secondary

The vast majority of service-level systems are radial, as depicted in Figures 13.1 and 13.3. However, secondary networks are typically used in the very high-density downtown cores of large cities, and in other applications where extremely high reliability, at high cost, is preferred over more typical reliability and cost levels. The secondary network is fed from numerous service transformers, which are themselves fed from interlaced feeders (see Chapter 1, Figures 1.11 and 1.12 and accompanying discussion).

Secondary networks are expensive for a number of reasons. To begin with they require quite expensive protective devices -- network protectors -- to provide both fault isolation and prevention of backfeed in the event of equipment outages. Second, they are expensive to engineer, requiring network analysis for both normal and protective engineering studies. Third, in some networks, line segments are greatly oversized for a variety of reasons including contingency and load dynamics, which increase conductor cost by factors of two to four.

Distribution network engineering is well covered in several references, and will not be given any further space here -- less than 1/10 of one per cent of power in the United States is delivered over secondary networks. Most secondary is radial, although secondary may occasionally be operated as closed loops.

Overhead versus Underground

Like the transmission and distribution levels, the service level can be built either above or below ground. While the equipment used, and the detail engineering required differ, depending on overhead or underground application, the concepts, functions, and layout guidelines for service-level planning and operation are similar.

Often, in residential and suburban areas, where primary laterals and service lines can be directly buried, and pad-mounted transformers and switchgear can be used, underground service has a lower cost than overhead service would.

Service Level Layout and Planning

However, where duct banks must be used for all wires, and vault-contained transformers and switchgear are required, costs are much higher.

Usually, underground service is considered potentially more reliable, because underground lines and the pad-mounted or vault-contained transformers that go with them are less prone to weather, tree, and lightning damage. Regardless, operating results at most utilities indicate that overhead and underground systems differ substantially in their reliability. Generally, underground has far fewer outages, but any outages that occur last much longer.

Many Transformers and No Secondary versus Few Transformers and Lots of Secondary

Figure 13.4 illustrates two very different ways to lay out an American system's service-level -- what are called the "many transformers/no secondary" and the "few transformers/lots of secondary" approaches to service-level design. The "many transformers" approach is aimed at minimizing secondary level losses by eliminating as much of the utilization-level power flow distance as possible. Despite the short distances involved, and even when using very large conductor, losses at the 120-volt level are appreciable. The "many transformers" layout uses many small transformers, all located at or near the customers' property lines, to eliminate all utilization voltage circuitry except the service drops. Typically, a system laid out with this approach will have a customer to service transformer ratio of about two to one.

By contrast, the "few transformers" design typically has a customer to transformer ratio of somewhere around ten to one. This alternative approach uses far fewer, but much larger transformers. It achieves a considerable savings in transformers compared to the "many transformers" design, because there is a very great economy of scale among small service transformers (a 75 kVA transformer, installed, probably costs only twice as much as a 15 kVA, installed, despite having five times the capacity). However, what the "few transformers" approach saves in transformer costs, it gives away, at least in part, by requiring extensive service level secondary circuits (a noticeable but not outstanding capital cost) and having much higher secondary losses (they may nearly equal those on the entire feeder system).

Both types of approach are popular within the industry, and each design has many proponents -- utilities that have standardized on one or the other, and individual planners who will swear that one or the other is far better. It is important to realize that intermediate designs (for example, a layout utilizing a customer/transformer ratio of four to one) are possible. With properly selected and sized equipment, either approach, and all intermediate approaches to service layout, can provide good service.

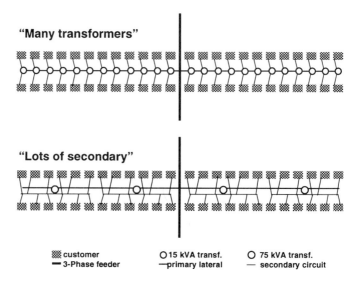

Figure 13.4 Two approaches to lay out of an American service level system. At the top, many small transformers are arranged so that no secondary circuitry is needed, only service drops, minimizing power flow (and losses) at utilization voltage. At the bottom, large conductor and lengthy secondary circuits permit the number of transformers to be reduced by a factor of six, with a consequent savings (the four 75 kVA transformers will be *much* less expensive than the 24 10 kVA transformers), but the lengthy secondary power flow will produce considerable losses.

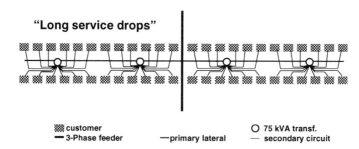

Figure 13.5 An alternative with the "many transformers" approach, particularly in underground applications, is to use no shared secondary circuits, but instead run dedicated service drops from the transformer to each customer.

Service Level Layout and Planning 545

Generally, one or the other of the "many transformers" or "few transformers" approaches *will* provide better economy in any particular application, but comparison of the electrical and economic performance of the two approaches, and selection between the two (or better yet, optimization of some intermediate design between these two extremes) for any specific application is surprisingly complex and difficult to accomplish. Most studies attempting to evaluate economy at the secondary level fail due to poor approximations of the load dynamics interaction with losses and capacity costs, as will be discussed in the next sections).

Long service drops

In underground distribution applications, it is quite common for secondary voltage circuitry to consist only of service drops, with no shared power flow among customers, as depicted in Figure 13.5. This is done mostly because it means construction requires no branching or "T" splices of the low-voltage cable, and also because it provides slightly superior voltage drop to customers at relatively long distances from the service transformer. However, it costs a bit more than shared circuitry. It also is thought to slightly reduce harmonics propagation from one customer to another in some cases (see Chapter 3).

13.4 LOAD DYNAMICS AND COINCIDENCE, AND THEIR INTERACTION WITH THE SERVICE LEVEL

The planning, engineering or the service-level is fairly straightforward (it will be discussed in more detail in section 13.5), except for one complication affecting nearly every aspect of analysis and evaluation. In particular, cost-reduction and electrical performance evaluation must take into account the dynamic nature of the electric load as seen by the secondary level.

Service-level equipment is very close to the consumer loads, operating in an electrical environment where there is no coincidence of electric load. As a result, the service drops, secondary circuits, and service transformers often see peak loads much higher than might be expected based on coincident load research data, and they can see significant shifts in their loading on a nearly instantaneous basis. The degree to which this load dynamicism affects economy and voltage behavior is not generally appreciated.

Figure 13.6 compares the non-coincident load curve for a typical all-electric household (left side, from Chapter 2, Figure 2.4), with the coincident load curve for the same customer sub-class. The coincident load curve has a peak load of 9.2 kW (this particular customer is an all-electric house of about 2200 square

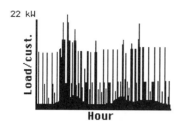 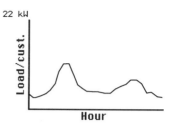

Figure 13.6 Left: winter peak day load curve of a typical 2500 square foot, all electric home in a suburb of a large metropolitan area in Texas (left) sampled on a five minute basis. This is the load that the service drops leading to this particular customer sees. Right: coincident peak day load curve for 2,500 square foot all-electric homes. This is the load the feeder serving this houses sees in combination with its other loads.

feet) and exhibits the smooth 24-hour load curve commonly used to represent coincident customer class load characteristics. As explained in Chapter 2, this daily load curve shape is what a feeder or substation, serving many customers, would see. For example, a feeder serving 707 of these homes would see a load curve of the shape shown on the right side of Figure 13.6, with a peak load of 707×9.2 kW = 6.5 MW, a daily energy demand of 112 kWhr, and daily load factor of 55%.

However, every one of the 707 service drops leading from the service transformer to each of those 707 customers would see a load whose behavior resembled that shown on the left side of Figure 13.6. Daily energy (area under the curve) would be 112 kWhr, but peak load is 22 kW, but instead of a smooth flow of energy peaking at 9.2 kW, the daily electrical usage would consist of erratic, apparently randomly timed "needle peaks" as high as 22 kW, as major appliances (heat pumps, electric heaters, cooking) turned on and off. Load factor is 23%. (the reader who does not understand the difference in load curve shapes, peaks, and load factors, and their cause, may want to refer to Chapter 2.)

Voltage Drop and Losses on Secondary Circuits Are Much Higher Than Estimates Based on Coincidence Load Curves

Figure 13.7 compares the service-level voltage drop that results from delivering the smooth 9.2 kW peak coincident load curve shown on the right side of Figure 13.6, versus the actual non-coincident load curve, and is based on 4/0 triplex UG

Service Level Layout and Planning

 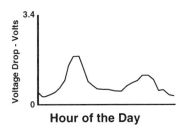

Figure 13.7 Right: voltage drop on the service drops serving the household load in Figure 13.6, calculated using a coincident load curve (right side of Figure 13.6). Peak voltage drop is only 2.4 volts (2% of a 120 volt range), the maximum permitted in the secondary system -- presumably service to this customer is within standards. Left: actual voltage drop on the service drops serving the household load in Figure 13.6, based on the non-coincident load curve (left side of Figure 13.6). Actual peak voltage drop is 3.4 volts, considerably beyond the 2.4 volt (2%) permitted. In addition, voltage drop at the service entrance varies by 32 volts. This violates this particular utility's flicker standards (3 volts).

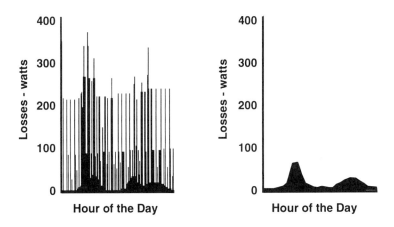

Figure 13.8 Left: summer peak day losses for service drops serving the house in Figure 13.6. Peak losses are 350 watts, total losses for the day are 1.45 kWhr (1.3%). Right: summer peak day losses for the service drops if computed using the coincident load curve in Figure 13.6. Peak losses are 62 watts, and losses for the day are .46 kWhr (.41%).

service drops. It results in a calculated voltage drop of 2.4 volts if estimated based on coincident load behavior (i.e., the smooth curve and the 9.2 kW peak). In fact, voltage drop reaches 3.4 volts, as shown, due to the non-coincident load behavior.

Thus, what is assumed to be service that is just within standards (2.4 volts is the maximum permitted at the service level for this particular utility) is, in fact, more than a volt beyond the permissible limit. In addition, Figure 13.7 shows that the voltage at the customer service entrance varies almost instantaneously by nearly 3 volts as various heavy appliances (heat pump, etc.) turn on and off, creating what is probably very noticeable lighting flicker in this household.

As would be expected due to their I^2R nature, losses display an even more dramatic difference between actual and "calculated with coincident load" results. Figure 13.8 plots the service-drop losses over the course of the peak day. Losses computed based on coincident load behavior (right side of Figure 13.8) have a peak of 62 W (.67%), with a daily total of .46 kWhr (.41%). Actual losses (left side of the diagram) peak at 354 watts and average 60 watts -- nearly what the coincident behavior analysis predicted for peak losses. Total daily losses are 1.44 kWhr for (1.3%), almost a four-fold increase.

The Most Common Pitfall of Distribution Engineering

Evaluating secondary and service drops performance and economy with coincident load curves, or with load data which does not fully represent actual non-coincident load behavior, is perhaps the most prevalent mistake made in the design of power distribution systems throughout the United States and much of the world. Although most distribution planners are aware that service-level load behavior is not coincident, that individual household and small business peak loads are higher than would be predicted by coincident load curves, and that load and losses factors at the secondary level are much worse than coincident behavior would suggest, they are not aware of the *degree* to which actual load behavior varies from the smooth behavior predicted by coincident load curves, nor do they generally appreciate the impact these differences have on their system performance and economy.

Very often, this error is compounded by load research and load curve analysis -- done largely on a class basis by marketing and rate departments interested only in coincident load behavior -- that institutionalizes use of only coincident load curve data corporate-wide. It is further exacerbated by the fact that coincident load curve behavior is difficult to measure. Many load curve measuring methods cause a "filtering" of the raw data that resembles the effects of coincidence, so that even if individual household load curves are measured, the result might be what could be termed "semi non-coincident" load curve data.

Service Level Layout and Planning

The load curves so recorded may underestimate true peak load and load and losses factors on the secondary system by up to 16%.[3]

Load Coincidence Effects on the System

Figures 13.7 and 13.8 and the foregoing discussion looked at the impact of load curve coincidence on one set of service drops. Figure 13.9 reprises the load curve plots shown in Chapter 2, Figure 2.8, but labeled with the type of service level equipment that typically sees each load curve shape. As explained in that chapter, diversity of load and load curve shape occurs over large groups of customers, but small groups of customers produce load curves that are intermediate, with load shifts, peak loads, and losses factors somewhere between those of the coincident and non-coincident extremes shown in Figure 13.6.

Most of the equipment and circuitry at the service level, particularly in American types of systems, serves between one and ten customers. Therefore, it has loadings that are more non-coincident than coincident -- with many high, short, square needle peaks and considerably lower load and losses factors than coincident load. As a result, voltage drops and losses throughout the service level are higher than might be expected if computed based on coincident loads.

Table 13.2 gives typical differences in voltage drop and losses that occur in various equipment in the service level, caused by the load behavior. As a result, in actual service, the low voltage and high losses illustrated for the customer in Figure 13.7 and 13.8 might actually be even more below standard than indicated earlier in this chapter, because voltage drop and losses in the service transformer feeding the service drops, and in the lateral feeding the service transformer, might be higher than calculated if based upon coincident load data.

15-minute load data is often not "fast" enough

Many utilities use load research data based on individual customer data recorded on a 15-minute basis. The shorter sampling periods capture more of the "needle peak" and non-coincident load behavior at the customer level as compared to sampling on an hourly basis (see Chapter 2, Figure 2.12). However, 15-minute sampling of load curve data is not sufficient to guarantee

---

[3] See Chapter 2, "Measuring and Modeling Load Curves," for a summary, and Willis, 1996, Chapter 2, Section 2.3 for more detail. Further, very detailed investigation of these effects is available in two Master's degree theses done at New Mexico State University (Hale, 1993; Howell, 1995).

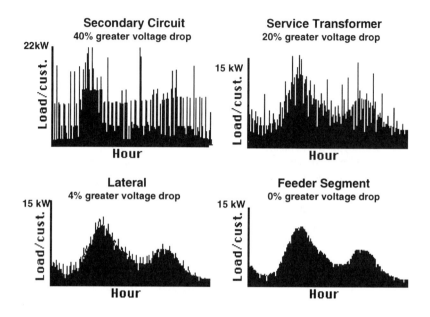

Figure 13.9 Non-coincident load behavior does not affect only service drops. As explained in Chapter 2, load curves for small groups of customers also exhibit aspects of non-coincident behavior. As a result, a secondary circuit segment serving two customers (upper left) may see voltage drops as much as 40% higher than expected if calculated using coincident load curve data, while service transformers can have 20% difference.

Table 13.2 Typical Percentage Error between Actual Voltages and Losses and Values Computed with Coincident and "15-minute" Load Curve Data

Equipment Type	Using coincident curves voltage drop	losses	Using "15-min." curves Voltage drop	losses
Service Drops	50 %	200 %	16 %	40 %
Secondary Circuits	40 %	140 %	10 %	25 %
Service Transformers	15 %	33 %	10 %	8 %
Laterals	3 %	5 %	none	none

Service Level Layout and Planning

that non-coincident effects are fully considered in evaluation of the service level. To begin with, many common statistical analysis methods remove or "filter" some of the non-coincident load behavior from the recorded data.[4] As a result, load curve data sampled at 15-minute intervals may not be showing true 15-minute load curve behavior.

In addition, 15-minute sampling of load curve data is simply not fast enough to "see" the complete effects of non-coincident load behavior on a customer and very-small-groups-of-customers-basis. Household and small-business appliance duty cycles are often shorter than 15-minutes, and the accompanying needle peaks are much shorter. To capture completely load curve behavior and peak magnitude on a household or small commercial customer basis requires *one or two minute sampling* (sampling on a five-minute interval basis comes very close but misses peak heights by up to 20%).

Coincidence effects are the first suspect for service or economy problems

Whenever service quality or evaluated economy (i.e., actual voltage drops and losses costs) fall short of expectations based on engineering evaluation, load coincidence effects of the type shown in Figures 13.7 and 13.8 should be among the first suspected causes. Recommended policy is to always regard peak load and coincident load curve data, even if measured with detail on an individual customer basis, as approximate. Usually, actual load behavior is more dynamic, with higher and shorter needle peaks, than modeled, even when extreme detail and resolution has been used to measure the load correctly (Willis, 1996).

An increasingly prevalent problem

Traditionally, the mismatch between predicted and actual voltage drop and losses on the service level has not been critical to, or even noticed by, most electric utilities, because equipment at the service level was over-specified with respect to coincident load curve values. Such over-specification was done partly because of the traditional engineering conservatism in the power industry, but also because the type of coincidence effects discussed here were suspected, although data did not exist to analyze them quantitatively.

---

[4] See Willis, Vismor, and Powell, 1985; or Willis, 1996, Section 2.3, for a discussion of how averaging, addition, and other common statitical load curve analysis processes can mimic the effect of coincidence, making recorded load curve data smoother, and measured peak values lower, than they actually are. The fact that both characteristics of the analysis and actual coincidence casuse similar effects creates considerable problems.

As electric usage gradually grows, capacity and impedance margins of equipment installed long ago, which were once adequate to cover the problems caused by non-coincident load behavior, are no longer sufficient. Service voltage problems, and losses costs at the service level can also escalate faster than increasing peak load and kWhr sales values might indicate.

> *Non-coincident load behavior at the service level means that voltage drop and losses' costs there are very likely worse than predicted. Measuring non-coincident load curve data accurately, and applying it to engineering studies, requires high resolution data sampling and analytical detail often not available to distribution planners.*

13.6 SERVICE-LEVEL PLANNING AND LAYOUT

The Service Level Is a Significant Expense

Despite its low voltages, short distances, and the "commodity" nature of its equipment (service transformers and conductor are generally bought in bulk, on the basis of lowest bid), the service level is a significant one in the power system. To begin with, over 90% of a typical utility's sales pass through the service level -- it is the final link to the majority of revenue. In addition, this level shapes, more than any other except perhaps the primary feeder level, the quality of service experienced by the customer.

And finally, the service level represents a major expense, for while its individual elements are without exception among the least expensive equipment purchased by the utility, they more than make up for this in volume -- there are about 250 times as many service transformers in a typical power system as there are feeders. And losses at the low voltages associated with the service level can be quite high, to the extent that in a few systems they exceed those of the entire primary distribution system.

Table 13.3 shows the cost data for a typical 37 kVA, pad-mounted transformer, of the type that might be selected on the basis of capacity and losses costs as appropriate to serve four residential customers with a peak (coincident) load of 6.19 kW each (7.74 kVA each -- the customers in the example system from Chapter 11, with an uncorrected PF of 80%) through direct-buried long service drops. PW cost per kW served works out to $201. Table 13.4 gives similar values for a 500 kVA, three-phase, European service transformer serving sixty of those same residential customers through a 250 volt secondary system.

Service Level Layout and Planning 553

Table 13.3 Typical Costs Associated with a Typical 37 kVA Pad-Mounted 12.47 kV/120/240 Volt Transformer and Associated Lines and Equipment to Serve Four 6.2 kW (Coincident Peak) Loads - Dollars

Item	Initial cost	Annual cost	PW
37 kVA pad mount transformer	430		305
Pad & other materials	27		27
570 ft. 4/0 jriplex, junc. box, materials	700		700
Installation labor	890		890
O&M&T		35	302
Secondary losses		62	538
Transformer losses		102	878
Total	1,922	199	3,640
Cost per coincident kW peak	78	8	147

Table 13.5 Typical Costs Associated with a Typical 500 kVA Pad-Mounted 11 kV/250 Volt Transformer and Associated Lines and Equipment to Serve Sixty 6.19 kW (Coincident Peak) Loads - Dollars

Item	Initial cost	Annual cost	PW
500 kVA vault type transformer	3,700		3,700
Vault, entry ducts & other materials	915		915
1200 ft. 500 quadraplex, 6000 ft. 2/0 duplex and associated materials	9,000		9,000
Installation labor	22,300		22,300
O&M&T		810	6,975
Secondary losses		2,900	24,969
Transformer losses		2,050	17,650
Total	35,915	5,670	85,509
Cost per coincident kW peak	97	15	230

Table 13.6 Cost of Chapter 11's Example T&D System by Level, Including the Cost of the Service Level - Percent

Equipment	Initial	PW
Sub-transmission, per sub. (8.33 mi..)	7	7
Substation - 3 × 27 MVA transf.	13	18
Feeder System - 230 mile primary	62	56
Service Level - 9450 customers	18	19
Total =	100	100

554   Chapter 13

Overall, European service configuration is about 50% more expensive due to higher capital and higher losses costs as compared to the American system, but it avoids a portion of the laterals built in American systems, a savings not included in Table 13.4.

The important point borne out by Tables 13.3 and 13.4 is that, American or European style, the service level is expensive. At $147/kW peak load served, the entire service level for one substation area in Chapter 11's example suburban T&D system (58.5 MW peak load) would cost $8.6 million PW, making it nearly as expensive as the substation and sub-transmission levels combined, and roughly one third the cost of the primary feeder system.

Table 13.5 gives the breakdown of total power delivery cost for the example system, including the service level (this is Table 11.6 with the service level added). The service level represents slightly more than 18% of the capital cost of power delivery, and roughly 19% of the PW cost. Its higher proportion as PW than capital costs indicates that it has higher than average losses as compared to the other levels of the system -- not surprising considering its very low voltage nature.

"Table-Based" Planning

The vast majority of service-level equipment is specified and laid out by technicians and engineering clerks based on rules and tables developed in "generic" case studies of typical situations, in what the author terms "table-based engineering." Very little of the service level as installed is actually "engineered" in the sense that detailed power flow analysis and equipment selection studies are done on a case-by-case basis, as is typical at the feeder, substation, and transmission levels. The reason is that a utility cannot afford the labor, data collection, and analytical cost to do so -- a large and growing electric utility may have to engineer tens of thousands of new service connections annually, as compared to only a few dozen new feeders and perhaps one or two substations. Beyond this, the benefits of detailed engineering on a case-by-case basis, over what can be done with good table-based engineering, are small, and generally an effort to improve on them is not justifiable.

Figure 13.10 shows a service transformer selection table (top) and a service drop selection table (bottom) for underground residential service, taken from a utility in the southwest United States which uses the "few transformers" and "long service drops" service-level layout philosophy. This utility has standardized on a set of five sizes of a single-phase pad-mount transformer -- 25, 37, 50, 75, and 100 kVA, and three sizes of UG service drops -- 2-7, 4/0-19, and 500-37 triplex 600 volt rated cable. These tables are listed here as examples only -- they are not necessarily recommended selection criteria.

Service Level Layout and Planning

```
UG DB SERVICE TRANSFORMER
R2A: Residential -- homes w.o. electric heat
```

Average (sq.ft)	Number of homes served														
	1	2	3	4	5	6	7	8	9	10	11	12	13	14	15
< 1500	25	25	25	25	25	37	37	50	50	50	75	75	75	75	75
to 2000	25	25	25	25	37	37	37	50	50	75	75	75	75	100	100
to 2500	25	25	25	25	37	37	50	50	75	75	75	75	100	100	100
to 3000	25	25	25	37	37	50	50	75	75	75	75	100	100	100	NP
to 3500	25	25	25	37	50	50	50	75	75	75	100	100	NP	NP	NP
to 4000	25	25	25	37	50	50	75	75	75	100	NP	NP	NP	NP	NP

UG-RN2   Revised 8-16-76

```
UG DB SERVICE DROP
R2A: Residential -- homes w.o. electric heat
```

House size (sq. ft)	Distance from Trans. (including drops)						
	0-25	50	75	100	125	150	200
< 1500	2-7	2-7	2-7	2-7	2-7	4-0	4-0
to 2000	2-7	2-7	2-7	4-0	4-0	4-0	4-0
to 2500	2-7	2-7	4-0	4-0	4-0	4-0	4-0
to 3000	2-7	4-0	4-0	4-0	4-0	500	500
to 3500	2-7	4-0	4-0	4-0	500	500	NP
to 4000	4-0	4-0	500	500	500	NP	NP

UG-RN3   Revised 6-23-69

Figure 13.10 A transformer selection table (top) and a service drop sizing table (bottom) for selecting equipment to serve residential (non-all-electric) homes. The real planning of the service level goes into building the tables and the rules that go with them, not in their application. These tables and their values are shown only as examples of the types of tables used in "table-based engineering," and are not necessarily recommended equipment or selection criteria. Note that at the time of this writing (early '97), neither table had been revised in over 20 years.

556   **Chapter 13**

When laying out or revising the service level plan for a neighborhood, designers consult equipment selection tables like those shown in Figure 13.9 to determine which of several sizes of transformer should be used in each application. The layout of an UG service level plan (Figure 13.11) would proceed in the following steps

1. The designer is given a plat of the new neighborhood by the developer, showing the lot lines, the structure footprints, and giving the square footage for each house.

2. An open lateral loop is laid out following the back property lines, using standard cable size for the lateral, or perhaps selected from a table based on the number and size of homes served, as shown.

3. Groups of adjacent homes on the plat are circled, forming groups of homes. The designer's goal is to group as many homes as is

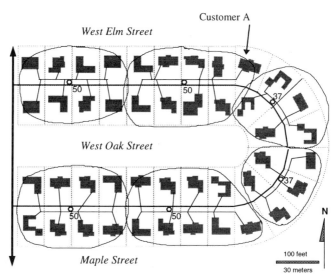

Figure 13.11 Plat of a new housing development showing the initial layout of the feeder system, with a loop lateral, houses grouped into six groups each served by one transformer, and routes for the service drops. Homes average 2500 square feet. Dotted lines represent property lines. Grouping of homes, done by hand as indicated, leads to identification of the sizes of the transformers (circles, with kVA capacity indicated) from the top table in Figure 13.9.

Service Level Layout and Planning

possible within an area whose maximum radius does not exceed the voltage drop reach limits of the service drops, so that a single transformer can be located in the center of each group.

This is a purely judgment-based step, and a source of occasional mistakes. For example, in the plat shown in Figure 13.10, the designer has circled groups of homes (clockwise, starting from the upper left), of 8, 9, 5, 5, 9, and 8 homes. Slightly better economy would result from a grouping of 8, 8, 6, 6, 8, and 8. This would shorten the secondary runs (particularly to the customer indicated by "A" by about 50 feet) and also make a slight improvement in efficiency by taking better advantage of the economy of scale of coincidence (which will be discussed later in this chapter).

4. The service transformer for each group of homes is selected based on average size and number of homes, using the service transformer selection table (top of Figure 13.10). The particular table shown gives recommended transformer size based on number and size of houses, but some utilities' tables are based upon estimated total load or other factors -- practices vary.

5. The transformer's location is determined by the designer. The goal is to locate it on the lateral loop, as close to the load-weighted center of the group of homes as is possible. Usually, this is also a completely manual, judgment-based step, without prescribed tables or procedure. Good judgment, based on study and experience, generally provide for quite satisfactory siting in this step.

6. Service drop routing distance from the transformer location to each home is determined (this utility uses a "long service drops" layout -- see Section 13.4). Based on this length and the house size, the conductor size is selected from the table at the bottom of Figure 13.10.

Given good tables and rules, the resulting layout can be excellent with respect to both electrical and economic performance, but the quality of the equipment selection and system layout is only as good as the tables and the rules developed for their application. Considering the amount of money spent on the service level, a utility's distribution planners should put effort into careful evaluation and design of the tables, and also the rules and procedures used to apply them.

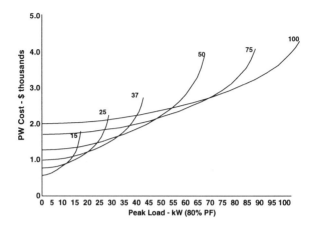

Figure 13.12 Economic evaluation of PW cost as a function of coincident peak load when serving residential customers, as evaluated for a utility in the southwestern United States for various size service transformers.

Transformer and Service-Level Conductor Inventory

As with primary feeder systems, the first order of business in planning the service level is to assure that a good set of basic building blocks is available and that their specification and utilization will be evaluated correctly. An appropriate set of service transformers, and an appropriate set of low voltage line types, along with their recommended loading limits, can be determined in a manner *qualitatively* similar to that explained in Chapter 7 for primary feeder conductor set selection. However, several differences in evaluation procedure must be applied for the results to be applicable to the service level, and the results need to be interpreted in tables for the table-based engineering. Figure 13.12 shows evaluation of the PW economics of the service transformer set used in one utility system.

Simple, Accurate Table Design

Equipment selection tables and table-based layout procedures should be simple and foolproof, not because the people applying then aren't capable (many distribution designers are very good at what they do), but because the whole point of table-based design is to simplify and reduce design effort, and to assure uniformity of application of the best equipment selection and application. In

Service Level Layout and Planning 559

addition, unambiguous and easy-to-follow tables and procedures make it easier for the inevitable new employees to learn the system and apply it correctly. This lowers cost and improves quality.

Tables and procedures should always be based upon data that is available to the designers (e.g., number of houses, square footage, distance in feet), as opposed to using values that require interpretation, translation, or calculations based on data values (e.g., coincidence factors, load factors, peak responsibility factors). They should be unambiguous in their interpretation by the user.

The span of all tables should cover all possible situations -- even if improbable. As an example, the service transformer selection table in Figure 13.10 has appropriate entry spaces for up to 15 houses, of up to 4,000 square feet. It is doubtful if any one service transformer would ever serve 15 hours of such size (it is quite unlikely that so many would fit within the reach distances of service drops from one transformer). However, the table should cover any eventuality that might arise, because mistakes or inconsistencies may develop when interpretation or interpolation outside the bounds of the tables is required.

Tables should clearly indicate when the situation or application is not permitted. Note that both tables in Figure 13.10 have "NP" (Not Permitted) in some instances. This indicates that, for one or more reasons, the application is not permitted and service must be rearranged or design changed to avoid equipment applications that fall into those categories.

The economic evaluation method used to determine the members of the transformer and conductor sets can also be applied to ascertain the values in the engineering tables. In general, to provide an accurate set of tables, it must be applied several times, varying the customer type load shapes used. for example, all-electric and non-all electric loads and load characteristics differ enough in most systems that service-level equipment needs are far different. A set of tables should be developed for laying out all-electric homes, and another for residential service in areas where gas distribution is available or gas appliances are prevalent. More generally, load characteristics differ sufficiently from one customer class to another, that the most satisfactory table-based service and cost minimization results are achieved when tables are developed for each customer class or major customer subclass (e.g., residential all-electric).

Computer-based tables and service-level guides

While tables and procedures as described here can be applied efficiently by hand, they lend themselves easily to implementation by computers. This may involve nothing more than listing them as tables, plots, and graphs available on the computer, for "manual" application, or perhaps adding some spreadsheet and table-look-up capability to ease their application. On the other hand, such tables and procedures work well as the core of a "rule-based" expert-system (artificial intelligence) program, which can be developed to shorten the design process, apply selection criteria more precisely, and identify and change its rules to a wider range of exceptions and unusual situations. Such computerized systems can improve design quality and reduce labor requirements. However, computerization should not disguise the fact that the final quality of service and economy is only as good as the analysis and evaluation methods that went into selecting the table values.

Interaction of Sizing Tables with Coincidence of Load

The chief difficulty in developing accurate and effective tables to specify equipment in table-based engineering is assessment of the coincidence load effects, of the type discussed in Section 13.4. Unlike the evaluation of primary conductor done in Chapter 7, *at the service level load factor and curve shape change significantly as a function of the amount of load served.* These changes are sufficiently dramatic that accurate evaluation of equipment application and layout can be achieved only by assessing these changes.

The major reason to assess load coincidence is to improve cost-minimization, not because there are doubts about whether service-level equipment may fail in the presence of the high, non-coincident peak load behavior. Generally, equipment at the service-level could be sized based only on coincident peak load, and provide adequate reliability and durability in service. However, optimal reduction of cost can only be done through assessment of non-coincident load behavior.

For example, the single household load curve on the left side of Figure 13.8 has needle peaks that reach 22 kV, but a typical 15 kVA service transformer could serve this load quite easily. When doing so, its internal temperature build up would correspond almost exactly to what would occur if it were serving a completely coincident, smooth load curve (right side of Figure 13.8) which peaks at only 9.2 kW (11.5 kVA at 80% PF). Temperature rise inside the transformer takes time, even when severely overloaded, and the needle peaks never last long enough to result in really stressful temperatures. Similarly, the ability of low-voltage line segments to handle the needle peaks is not a dramatic concern either if they are sized to handle the coincident load. In general, if sized to handle the coincident peak expected as a result of a customer or group of customers, a line or transformer can "get by" without damage when serving the actual, non-coincident load curve.

But its losses may be much more than expected (as was illustrated in Figure 13.8), so much higher due to non-coincidence of load (as opposed to what they

Service Level Layout and Planning 561

would be if serving the coincident load curve), as to dictate a shift to larger equipment solely to reduce their cost.

For example, suppose a 100 foot segment of 750-61 triplex is serving ten of the customers whose peak day load curve is diagrammed in Figure 13.6. Although each customer's individual load curve is choppy and has a needle peak of 22 kW and a losses' factor of only about .10, due to diversity of load this line segment would see a fairly smooth load curve, quite similar to that shown in Figure 13.9 for "lateral." It would have a peak load of only about 115 kW, and an annual losses' factor of about 33%. Annual losses would be about 8200 kWhr (2%), with an annual losses' cost (at 3.25 cents per kWhr) of $266.

Based on that result, a planner might expect that a 100 foot segment of 2/0-19 triplex (which has roughly five times the impedance of the 750-61 triplex), serving two of these customers would develop about 1640 kWhr of losses per year for an annual losses cost of about $53. But instead, due to the non-coincident load shape, losses are 3000 kWhr, with an annual cost of $103 -- nearly twice as much per customer.

> *Due to load coincidence effects, there is equivalently more incentive for distribution planners to specify a larger conductor in order to reduce losses when serving only a few customers, than when serving many.*

In this particular case, the best choice for the utility is to "double up" the conductor for the ten-customer service. Since 750-61 is the largest standard triplex available, loading is reduced by splitting the load among two 750-61 triplex conductor sets, each serving five customers. This cuts losses costs by 50% at a capital cost increase of about 33%. However, the best "upgrade" for the line serving a single customer is to double it too, and increase capacity by a factor of two, from 2/0 triplex serving two customers, to two sets of 4/0-triplex, reducing losses by nearly 85% at a capital cost increase of 40%.

Qualitatively, this is a very general result, applicable to nearly all utility systems, but the numerical results vary from one utility to another. The important point is, that when building tables such as shown in Figure 13.10, the evaluation should use peak and losses' factors appropriate for the groups of customers involved: a completely non-coincident curve for single customer evaluation; a smooth, completely coincident one for groups of 25 or more; and values representing "intermediate" load coincidence for customer counts in between (Figure 13.9 shows several intermediate curves). This applies both to conductor selection and service transformer selection.

Furthermore, as was illustrated in Figure 13.7, voltage drop is also affected by load coincidence in a similar manner. The voltage drop on the 2/0-19 line serving two customers in the example above will be much worse than it will be on the 750-61 line serving ten customers, even though the ratio of customers matches the ratio of impedances (5:1). Thus, tables such as shown in the bottom of Figure 13.10, which specify line selection on the basis of load served and distance, need to be based on accurate coincident behavior analysis, too.

A slightly beneficial economy of scale

Often, one result of load coincidence is a noticeably increasing economy of scale in service transformer application. Note in Figure 13.13, that the "linear" range of transformer application from Figure 13.12 is actually a curve with a slightly *decreasing* slope as peak load is increased. The curve shown is for service transformer application to residential customers (not quantitatively the same as those load characteristics shown in Figure 13.8, but with qualitatively similar load behavior nonetheless).

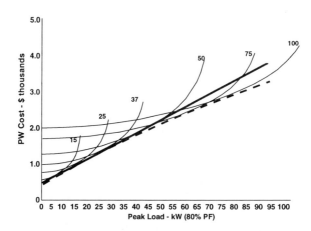

Figure 13.13 The PW cost-versus-peak load performance of the service transformer set (dotted line) is better than linear behavior (solid line) due to the improved losses' characteristics of the coincident loads of large groups of customers. This means there is more incentive than might be expected to use a "few large transformers" layout. Not all systems or customer classes exhibit the degree of increasing economy of scale shown here (amounting to nearly 14% -- the most extreme case the author has seen).

Service Level Layout and Planning

In addition to any economies of scale due to non-linearity of costs (i.e., installed cost for a 100 kVA transformer may be only two times that of a 25 kVA transformer), grouping customers in large blocks improves the losses factor of the resulting load curve, lowering losses' costs. Thus, there is a noticeable economic bias -- typically 5 to 10% beyond what might be expected based on coincident load behavior, in favor of the "few large transformers" layout over the "many transformers" layout, and against the "long service drops" approach and in favor of using shared (but much larger) secondary conductor where feasible.

Complicated layout decisions

It is impossible to generalize recommendations about service-level layout and equipment selection. The wide range of possible variations of load characteristics *and coincidence behavior* from utility to utility and from class to class, means that it is impossible to determine whether a few- or many-transformers approach is best, or to determine any of the quantitative details of equipment selection and loading without very comprehensive analysis.

The examples given above indicate only some of the complexity involved. The service level sends mixed signals with respect to economy -- there is a definite increasing economy of scale in utilization of equipment (Figure 13.13) and yet the best specific practice in the example of secondary-line sizing given above was to split routes into smaller groups of customers, for both large (10 customers) and small (2) groups. This is typical -- determination of the layout of the service level requires detailed evaluation and compromise between conflicting economies, in addition to careful assessment of voltage drop constraints. Selection between basic layout styles, evaluation of equipment suitability and loading standards, and development of design tables must be done almost on a case by case basis, for every customer class, and for every micro-climate (different climatic regions of the service territory). All of this detail and effort is worthless unless load coincidence is properly and accurately assessed.

13.6 CONCLUSION

The service level of most power delivery systems is composed of tens of thousands of individually simple and inexpensive equipments, and is usually laid out in relatively simple configurations according to standardized rules and tables of equipment selection and utilization. Voltage levels are very low, so that the losses are relatively quite high and mistakes or non-optimalities in design carry a very heavy economic penalty. Most important, load behavior is quite complicated. The important points about the service level are:

- Power cannot be moved far at utilization voltages. Voltage drop and high losses that result from low-voltage power flow heavily constrain design decisions.

- Service-level design is normally accomplished through table-based engineering and layout procedures.

- Quality service and cost minimization depend on thorough and accurate design of the various tables and procedures used in the table-based engineering.

- Adjacent units of equipment can see widely differing load curve characteristics even though they are connected electrically.

- Equipment evaluation, layout and loading guidelines are heavily influenced by load coincidence effects, which must be taken into account to optimize design properly.

REFERENCES

J. K. Dillard, editor, *Electric Utility Distribution Systems Engineering Reference Book,* Westinghouse Electric Corporation, Pittsburgh, 1958 (slightly revised, 1965).

R. E. Hale, *Development of a Radial Distribution Coincidence Power flow Algorithm,* Masters Thesis, Electric Utility Management Program, New Mexico State University, December 1993.

V. A. Howell, *A Study of the Effects of Coincidence of Load on Secondary Distribution Systems,* Masters Thesis, Electric Utility Management Program, New Mexico State University, December 1995.

H. L. Willis, *Spatial Electric Load Forecasting,* Marcel Dekker, New York, 1996.

H. L. Willis, T. D. Vismor, and R. W. Powell, "Some Aspects of Sampling Load Curves on Distribution Systems," *IEEE Transactions on Power Apparatus and Systems,* November 1985, p. 3221.

14
Planning and the T&D Planning Process

14.1 INTRODUCTION

The objective of T&D planning is to provide an orderly and economic expansion of equipment and facilities to meet the electric utility's future electrical demand with an acceptable level of reliability. T&D planning involves determining the future needs for the T&D system, including the correct sizes, locations, interconnections, and schedules for future delivery system additions and changes. In addition, implicit in almost all planning is that the cost should be minimized. Such planning is a difficult task, compounded by continuing pressures to shrink design margins, reduce labor and operating costs, and delay capital expenses as long as possible.

Regardless of what or why something is being planned, there are certain common aspects to all planning and certain functions inherent in all planning. Section 14.2 looks at the planning process itself, beginning with a discussion of its goals and proceeding through an examination of its basic steps. Following this, the differing goals and functions of short- and long-range planning are presented in section 14.3. An important aspect of effective planning is analysis of uncertainty in future events and planning for their possibility -- the multi-scenario approach to address such uncertainty is presented in section 14.4. The T&D planning process, and each of its steps, is examined in section 14.5. Section 14.6 concludes the chapter with a summary of useful strategic guidelines for T&D planning.

14.2 PLANNING: FINDING THE BEST ALTERNATIVE

A Five-Step Process

Planning is a decision-making process that seeks to identify the available options and determine which is the best. Applied to electric utility planning, that process seeks to identify the best schedule of future resources and actions to achieve the utility's goals. Usually, among the major concerns of planning are financial considerations -- minimize cost, maximize profit, or some similar goal. In addition, service quality and reliability are almost always considerations, too. And often, other criteria are important, including environmental impact, public image, and a host of factors that the utility must cope with on a daily basis. But regardless, planning is the process of identifying alternatives and selected the best from among them.

The planning process can be segmented into the five steps shown in Figure 14.1. Each step is an important part of the process of accomplishing the goals of any type of planning, particularly utility T&D planning. Any of the five steps, poorly performed, will lead to poor decisions, a poor plan, and ultimately failure to attain those goals.

Figure 14.1 Planning involves five steps, as shown.

Planning and the T&D Planning Process

Step 1: Identify the Problem

Perhaps the most notorious way to fail a college mathematics or science exam is to "solve the wrong problem." Few students make it from matriculation to graduation in college without at least once being admonished by an indignant professor who exclaims, "Nice job, but unfortunately you didn't read the question and you solved the wrong problem -- no credit." Similarly, planners can do a brilliant job in their evaluation and selection of a plan, but fall short of the mark because they, too, solved the wrong problem.

Before embarking on data collection, study, and analysis, every group of planners should identify explicitly the nature of the "planning problem." Does the present project concern only the siting and sizing of a single new substation on the north edge of town, or is it really concerned with how the company will serve growing electric needs in the northern half of the city? Is the study of a problem feeder concerned only with fixing voltage flicker and fluctuation problems on that feeder, or does it concern identifying equipment problems and determining a fix that will subsequently be applied to the entire system? Is the problem to apply DSM and T&D to minimize expansion cost in a growing area of the system, or to set a standard and precedent for IR T&D planning company-wide?

Time devoted to explicitly framing the planning situation at hand will be time well spent. It is recommended that the planning problem and scope be formally written, as "this project involves finding the lowest-cost way to provide substation-level capacity, fully compliant with company criteria, to meet the load growth in the northside operating region, specifically with regard to the expected development in the West Oaks area."

Step 2: Identify the Goals

"You never get there unless you know where you're going" -- a clear identification of the *goals* makes it more likely that they will be met. A common mistake made by many planners is to begin their activity without explicitly identifying the goals for the particular planning situation at hand, or identifying how these goals might have changed from prior planning situations.

Corporate mission and goals

A company's "Mission Statement" succinctly defines the overall corporate objectives, while its economic goals explain its financial priorities. Usually, mission and financial goals have been incorporated in the economic guidelines and design procedures that the planner applies in his work. Regardless, while

Missions Statements are often taken lightly, a review of the company's mission and financial goals provides the planner with strategic insight into its planning guidelines and economic evaluation formula. Table 14.1 lists mission statements from four electric utilities. Planning goals for T&D, for DSM, for IRP, *will* be different among these utilities, because they have different priorities for what their executive management wants to accomplish. Mission statements are qualitative assertions of the overall philosophies and goals that lead to the quantitative engineering and economic criteria used in distribution planning. By understanding *what* their company is trying to do, the T&D planners have a better understanding of *how* to achieve those goals.

Table 14.1 Mission Statements of Four Electric Utilities

"Our company will be the premium regional provider of electric power."

> Recognizing that its current financial situation prohibits competing on the basis of price, this utility has decided to make quality and service its hallmark. Achieving lowest *possible* cost is not the goal; achieving low cost while meeting high service standards is.

"Provide economical electric power for the prosperity of the region."

> This municipal utility has a long-standing tradition of low rates, a way of attracting new industry (i.e., growth) to the region. Plans that invest a good deal to improve quality are simply "not with the program." Marginal quality improvements in a new plan are permissible, only if they lead to lower cost.

". . . maximize return on equity."

> This criterion means the utility wants to earn the most on what it currently owns. Utilization of *existing* T&D facilities is the key. This utility is more reluctant to invest by borrowing than a utility that wants to maximize return on investment. Plans that call for massive capital spending may be unacceptable.

". . . maximize return on investment."

> This mission makes no direct distinction between equity and debt -- apparently the utility is willing to borrow, if it increases what it can earn on that investment. A plan calling for considerable capital outlay is permissible, if that investment yields a good return.

Planning and the T&D Planning Process 569

Special goals

Occasionally, there are additional goals unique to a particular plan or situation. For example, it might be important to accommodate the esthetic and land-use preferences of local community leaders, or to use equipment or rights-of-way and sites already in inventory. Usually, such special goals are unique to one particular case, caused by some unusual situation or set of circumstances. Table 14.2 lists several special goals from T&D planning cases the author has done.

Table 14.2 Examples of Special Planning Goals or Criteria

New facilities will be located inside the county boundaries.

> Here, the utility wants visible proof that it is investing in and increasing the tax base of its particular community. Where possible, new substations and facilities must be built within the county, even though there may be reasons to locate switching stations and other equipment outside the county.

New plan will include nothing over 46 kV "transmission."

> The utility's management knows that 46 kV and below are classified as distribution voltages and require a shorter, easier, and less costly approval process. The planners' challenge is to develop a workable, efficient plan that uses what is, in effect, a distribution voltage for transmission.

Substation sites must be "developed" immediately upon purchase.

> Substation sites will be cleared, fenced, and have equipment -- spare transformers or breakers -- put on the site. Management's purpose in this policy is "disclosure" -- it is very clear from that moment on that the site is a utility substation, as anyone buying and developing land around it can see. This initial cost must be factored into plans.

Expansion plan must include "recourse."

> In this case, the utility's executive management was concerned that a new, untried DSM program might fail to meet its projected load reduction goals. They wanted a guarantee that there was a "T&D fallback" plan to reinforce delivery capability quickly and economically if DSM fails, completely or partially, to meet its goals.

Expansion plan must include "targeted DSM."

> Here, for political reasons, the utility chose to demonstrate its willingness to apply targeted DSM to reduce peak demand in the downtown area. Such DSM programs and their projected load reductions had to be factored into the plan.

Step 3: Identify the Alternatives

No amount of detailed evaluation and selection in subsequent steps can choose an option if it is never considered in the first place. In many ways, this third step is the most critical part of planning: identifying what *could* be done -- the alternatives open to the planner. What options are available? What variations on these options would be possible? This function encompasses determining the *range* of possibilities for solving the problem.

This is where the majority of great "planning disasters" occurs -- the mistakes that lead, weeks later, to statements like, "Why didn't we think of that?" For, while apparently the least challenging part of planning, a good deal of skill, and breadth of thinking, and time, is required to identify the range of possibilities. Time and resources are always limited, and the temptation is strong to assume that one can see all the options just by "a quick look at the problem." But this is seldom the case: a brilliant solution to an unusual problem most often comes from recognizing that there are also unusual options available to solve it.

A common failing: too short a planning horizon

Failing to look far enough ahead tends to limit options by default, contributing to poor planning As a simple example, consider the generation options shown in Table 14.3. These options have lead times that range from three to 12 years. The fact that generation can be added with lead times as short as three years means that a utility *can get by* while never looking more than three years ahead in its generation planning. But that limits it to future additions that include only small diesel, gas turbine, combined cycle, and photovoltaic units, which may not make for the most economic mix of generation. If it adopts a policy of planning further ahead, it does not limit its options, but can perhaps plan a better system.

Short planning periods often limit a utility's options with regard to power delivery planning and lead to much higher costs, generally due to an inability to obtain good substation sites. A common lament among T&D planners is, "We could have done a much better job, but the good sites and land were already taken." What is usually left out of this complaint is that the planning period used for distribution expansion evaluation was only three to four years ahead. The utility is losing out to other people interested in acquiring sites for other purposes, people who are looking ahead more than three or four years.

By looking only a few years ahead, the utility reduces the number of options available to it. Yet despite the best sites and routes being committed to other purposes, the distribution planners *will* find a way to do the job, but the resulting plan will usually have *both* higher costs and lower reliability than would have been the case had better sites been obtained.

Planning and the T&D Planning Process 571

Table 14.3 Lead Times for Generation Units

Type	Lead time
Nuclear	12+
Coal > 600 MVA	9
Hydro > 50 MVA	9
Tidal/Ocean current	8
Fluidized bed coal	8
Coal to 600 MVA	7
Windfarm > 3 MVA	6
Hydro < 50 MVA	6
Coal < 400 MVA	5
Combined cycle > 100 MVA	4
Photovoltaic > 3 MVA	3
Combined cycle < 100 MVA	3
Diesel < 15 MVA	2
Gas turbine < 20 MVA	2

Within reason, the recommended approach to any type of planning is to lengthen the planning period sufficiently to include all possible alternatives. In the case of the generation planning shown in Table 14.3, this would mean 12 years, or if nuclear were not considered to be viable, at least nine years ahead.

Having a sufficiently long planning period does not in and of itself assure good planning, but it is a necessary step. For distribution planning, a *minimum* planning period of five to seven years is normally recommended. Within this, the utility can identify sites if and why and where it needs to obtain them early, before they are unavailable. Savings accrue both because the utility obtained the site early at a lower cost, but most important because the site at the correct location *is* obtained. Often, if the utility waits, a site at the best possible location is not available at any price. By using a lead time long enough to give it these options, the utility is performing the first function of the planning process correctly -- assuring itself that it considers all possible alternatives.

The downfall of planners: failing to consider all the alternatives.

Step three in the planning process -- identifying all the alternatives -- is where a surprising number of both manual and computerized planning procedures fail. When looking at or reviewing their planning procedure, T&D planners should study if and how it assures them that it does consider all possible options.

Planners need to constantly remind themselves to be open-minded, and not to have pre-conceived notions about what and where the solution lies. In addition, the procedures and computer programs used need to be reviewed to ascertain that they do not limit the options considered.

Example of a computer program that fails to look at all options

A good example of how a planning process can fail because it does not look at all options is a method of computerizing "capacitor optimization" which was popular in the 1980s. This will serve as an example of how many planning processes fail to generate sufficiently wide "alternatives scope."

Optimization procedures generate, evaluate, and search through a large set of alternative solutions to a problem, selecting the best solution based on their mathematical criteria. However, like manual methods (and just plain lazy planners) a computerized method can fail to do a good job if it does not consider all the alternatives.

In this example, the distribution planner wants to determine how many, what sizes, and where capacitor banks should be located on a particular feeder in order to minimize losses and meet voltage and power factor criteria. The "nodal scan" method approaches the problem of locating and sizing capacitor banks on a feeder such as shown in Figure 14.2 with the following approach:

1) Using some form of starting rule, estimate how many kVAR might be needed (what this formula is and how it determines kVAR is not important to this discussion).

2) By trial and error evaluation, try a capacitor bank of that size at every node on the feeder, and compute voltage drop and economic losses. Select the node that gives the best result. Put the capacitor bank there.

3) Recompute the starting rule, acknowledging the existence of this new capacitor bank, to determine if additional kVAR is needed. If so, repeat the process of trial and error, trying capacitor bank number two at all nodes. Pick the best node for its location, and compare this solution to the "single capacitor" solution obtained in step 2. If capacitor bank number two improved the situation, leave it there, and repeat this step for capacitor bank number three. If this solution is worse, forget this capacitor, and stop the analysis.

In the feeder shown in Figure 14.2, this method decides that bank number one belongs at point A, and bank number two at B.

Planning and the T&D Planning Process

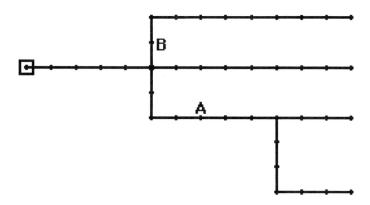

Figure 14.2 A 40-node feeder, a candidate for one or more capacitor banks. Eventually, a computerized planning procedure decides to put one bank at A, and a second at B.

At the time of this writing (1997) there are a dozen computer programs throughout the power industry that used some variation on this heuristic search theme. They vary greatly in the type of load flow and the details of their search rules, but all use the basic search rule given above: they check all possible nodes to see where capacitor bank one should be, then, with that first bank in place, they check all possible nodes to see where the second capacitor bank belongs. What they fail to consider is situations for capacitor bank two where capacitor bank one is anywhere *but* at the single location that was determined for it in step 2. Thus, they do not generate *all the alternatives* that are possible, as shown in Figure 14.3.

The "solution space" for the 40-node feeder in Figure 14.2 has 1,681 possible combinations of two capacitor bank locations. Yet the "optimization" procedure examined only 82, or less than 2% of the possible combinations. As a result, this "optimization" method is unreliable. Occasionally, it gives very good results, when the optimal solution is for only a single bank, or when it just happens to stumble on the correct answer. But most often it gives poor results, because it has not examined every possible option.

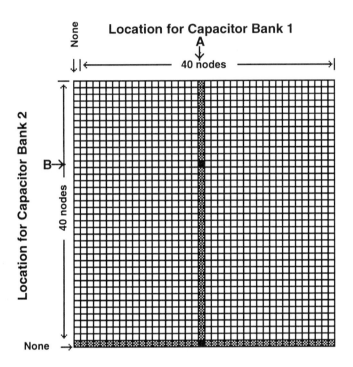

Figure 14.3 The 1,681 combinations (41^2) of possible locations for two capacitor banks on a 40-node feeder ("none" in the axis represents the case "no capacitor") are represented here as squares in a 41 by 41 grid. The computerized capacitor-siting procedure described in the text examined only those alternatives that are shaded. The shaded bottom row indicates the first iteration -- the procedure scanned all possible locations for capacitor bank one when capacitor bank two wasn't there (none), settling on node A as its selection. It then examined all possible locations for capacitor bank B, with capacitor bank one already at location A (shaded vertical column). It never identified the many combinations of siting that are unshaded. By failing to identify them, it most likely failed to find the best solution.

Planning and the T&D Planning Process

"Do nothing" should always be one of the alternatives considered

One of the most serious mistakes made by planners is to assume that the best alternative involves doing something. Occasionally, it is both less expensive and better from a service standpoint to delay any commitment or changes in the T&D system -- to do nothing. Even when this is not the case, "nothing" should always be included in a planning report as one of the reviewed options. This permits the planners to explicitly justify why something must be done, by showing that "doing nothing" would violates criteria or requirements, or lead to very poor economy.[1] Evaluation of "nothing" is a key element of justifying a plan's conclusions. A planner should *always* include "nothing" in the explicit evaluation of options, documenting why it fails to meet requirements or economy.

Using judgment to delete alternatives prior to evaluation

Given prudence and reasonable care, a planner is not making a mistake to use his or her judgment to delete options as soon as they are identified. All planners do this to some extent, and experienced planners can reduce the work required in subsequent steps by rejecting possible solutions as soon as they are identified. Applying judgment where one is confident it is accurate enough for evaluation purposes seldom leads to planning mistakes. Judgment lets a planner down if it is applied to *assume quickly* that all options have been identified, without time having been spent to review and check this conclusion.

Step 4: Evaluating the Alternatives

All alternatives should be evaluated against a common and comprehensive evaluation standard, one that considers every aspect pertinent to any option, and one that addresses all the goals. For power delivery planning this means evaluating alternatives against criteria and attributes that represent the utility's requirements, standards and constraints. For DSM programs this means matching customer needs with marketing standards and guidelines.

Very often, the actual planning process will combine the evaluation and selection functions (this will be discussed later in this chapter) in a process designed to reject alternatives quickly and with minimum effort. Regardless, evaluation is a key function of these methods and should be examined carefully

---

[1] A common argument is that evaluation of "nothing" is so obvious that it does not need to be included. But if that is the case, why not include and document its evaluation?

to make certain that it too, is done completely. Alternatives should be evaluated completely, with criteria that apply to them, and so that nothing is overlooked or biased.

Alternatives must be evaluated against both criteria and attributes. *Criteria* are requirements and constraints the plan *must* meet, including: voltage, flicker, and other service quality standards, contingency margin rules, summer and winter loading limits, safety and protection standards, operating guidelines, service and maintainability rules, and all other design standards and guidelines. A true criterion must only be satisfied: if the utility's voltage standards range from 1.05 to .95 per unit, then a particular design alternative is acceptable if it has voltages as low as .95 per unit, but no lower.

On the other hand, if the planners treat a feeder whose lowest voltage is .96 as better than one that reaches .95, to the extent that it would be selected as "best" even if it cost a bit more, then voltage drop is not a criterion, it is an attribute. An *attribute* is a quality that is to be *minimized* (or maximized) while still meeting all criteria. Criteria and attributes are discussed in detail in Chapter 4 The point here with respect to planning is that there are many criteria against which the plan needs to be assessed, but usually only one attribute.

That attribute is cost. One goal of nearly any T&D planning is to minimize cost. And while voltage drop, protection coordination, and other requirements are quantities that the planners would like to see exceed standards by comfortable margins, they are usually treated as criteria -- targets that only have to be met.

Table 14.4 Some Typical Criteria and Costs Evaluated in T&D Planning (see Chapter 4)

Voltage standards	Equipment loading standards
Protection standards	Contingency margin requirements
Clearance and safety standards	Construction method requirements
Esthetic impact and standards	Pollution/toxicity of materials
Maintenance restrictions	Flexibility of future design
Site cost (including legal costs)	Site preparation
Design cost	Equipment costs
Labor cost (including permitting)	Maintenance and operations cost
Taxes and insurance costs	No-load losses costs
Load-related losses costs	Contingency loss of life costs
Salvage cost/value at end of lifetime	Liability/risk costs

Planning and the T&D Planning Process

Traditional power delivery planning is *single attribute planning,* in that only one attribute (cost) is to be minimized. The evaluation step consists of determining if each alternative meets all the criteria (if it doesn't, it's unacceptable no matter how low its cost). Among those that are acceptable on all counts, cost is determined.

Cost is a multi-dimensional attribute, *all* aspects of which must be included in the evaluation: equipment, site costs, taxes, operations and maintenance, and losses. Table 14.4 lists some of the more common criteria and cost factors that need to be assessed in distribution evaluation. Chapter 6 discusses cost evaluation in more detail.

Where mistakes occur in the evaluation step

Mistakes that take place in the evaluation function generally occur because the planners do not check to see if their methodology meets three requirements:

Does the evaluation consider all criteria and factors that are important to the goals? For example, if a planning goal is to have recourse on the T&D side in case a DSM program fails to meet its reduction targets, then the evaluation used must be able to measure "recourse" -- not necessarily an easy task. Otherwise, there is no way to assure this goal is achieved, and certainly no way to evaluation options based on it.

Does the planning method evaluate all criteria fully with respect to all resources? For example, some planning methods evaluate reliability only with respect to T&D outages and connectivity. When evaluating DG and DS, the reliability method must analyze and accommodate lack of fuel availability and failure of the units themselves.

Does the planning method treat all options equitably? An evaluation method that considers all factors affecting some options but not all factors affecting others provides a biased, unrepresentative comparison of options. For example, at a municipal utility, a planning method that neglects the cost of future taxes might be considered suitable (the utility doesn't pay taxes to itself, so tax is a moot issue for facilities located inside its territory). However, many municipal utilities serve areas outside their municipal boundaries, or occasionally find themselves having to locate facilities outside their boundaries. When considering such situations, the evaluation must include the value of any taxes that would have to be paid to other taxing authorities. Only in this way can the advantage/cost of sites inside the territory be assessed properly against the advantages and costs of those outside.

Step 5: Selecting the Best Alternative

This step involves selecting the best alternative from among those that are evaluated. In many planning procedures, this function is combined with the evaluation function, and it is difficult to identify where evaluation ends and selection begins. In fact, efficient planning methods that minimize the time and effort required to complete the planning process tend to combine steps three through five into one process that gradually converges on the best alternative.

But regardless of if and how the evaluation and selection functions are combined and performed, the planner should explicitly study the planning process, and determine *how* it selects the best alternative, and *if* this is done in a valid manner. The most important points are to assure that *the definition of "best"* truly matches the goals and value system being used, and that the evaluation/selection method is *capable of distinguishing between alternatives* in a valid manner.

Definition of best

What is the best alternative? This depends on the goals for the particular situation. Going back to an example discussed earlier, if one goal in a particular planning situation is to have recourse to reinforcing the system economically at a later date, then the evaluation process must measure this quality in each alternative, and the selection process must give weight to the resulting measures to assure this criterion is met. Alternately, if one of the goals is to meet the esthetic requirements of the community, these must be defined, alternatives must be evaluated against that criterion, and those determinations used in selecting the best one.

A surprising number of planning mistakes, or inefficiencies, is made in the selection process, because the planner does not assess alternatives on the basis of what is important to the planning goals of the particular planning situation. The planning method's selection function should be examined to determine:

> *Does the definition of "best" match the planning goals?* The fact that alternatives were evaluated on the basis of a particular attribute in function 2 does not mean that it is weighted properly in the selection phase. For example, a common failure in T&D planning is to assess the value of electric losses in a distribution plan, but then fail to acknowledge their value in selecting the best design.
>
> *Can the planning method accurately distinguish between alternatives?* Merely because it evaluates alternatives on the basis of a particular criterion or attribute does not mean that a planning method does so

Planning and the T&D Planning Process

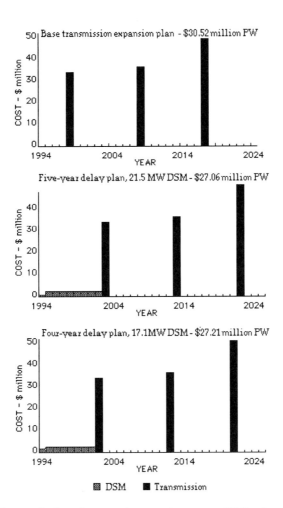

Figure 14.4 Screen displays from an integrated resource T&D planning program, showing expected annual expenditures for a base T&D expansion plan and two integrated DSM and transmission plans. Height of bar shows expenditures by year (in constant dollars). The first alternative plans buys 21.5 MW of load control peak reduction to obtain a five-year deferral of the transmission projects (tall budget spikes) for a projected savings of $3.46 million PW. Alternative two buys a four-year deferral, for a projected savings of $3.32 million. While the planning method used was accurate to within 2% (enough to determine that an IRP alternative is preferable, since the savings shown are about ten per cent in either case), it is questionable whether it is accurate to the point that the evaluated difference *between* the two IRP alternatives -- $146,000 out of $30,000,000, (a difference of .5%) is meaningful.

with the accuracy required to select among the alternatives. If the "best" alternative scores 2% better in evaluation than any other (e.g., costs 2% less), but the evaluation process is only accurate within 3%, then selection of that "best" option is questionable. In many planning cases the margins in cost, environmental impact, or other attributes among alternatives can be small. Error range in the evaluation and selection among close cases must be small enough to assure that the determination of "which is best" is accurate.[2]

Figure 14.4 illustrates this problem with a practical example. Two integrated resource plans are being evaluated to determine if either would be preferable to a base, all-T&D plan, and which plan is the best. As evaluated, the two integrated resource alternatives have a difference in savings of $146,000 PW out of more than $27 million PW. Is the planning method accurate enough to make this determination on a reliable basis?

This example brings up a common problem with many computerized planning procedures. Like any analytical method, computer programs produce only so many "significant digits" of accuracy in their computations and contain approximations and other limitations to their accuracy. Planners often fail to consider this and interpret as significant minor differences in computed evaluations, even when those differences are well within the accuracy limitations of the algorithms.

Many widely used planning programs have approximation ranges of five to ten percent. The substation siting and selection method developed in EPRI project RP-570, and widely used in the 1980s, assessed site and substation costs by estimating feeder and losses costs with a linear evaluation of "between substation transfers." Error in feeder-level costs analysis was often greater than ten per cent (Willis and Northcote-Green, Masud, EPRI EL-1198). Another approach of computerized substation planning, the service area optimization, methods evaluates feeders and losses with more detail, using a piece-wise linear "synthetic feeder analysis." Usually, its results are accurate to within plus or minus three percent (Crawford and Holt, Willis, 1988). That accuracy advantage makes service area optimization much better at *reliably* selecting the best site, one reason it gradually became a preferred planning method, but it still has an "approximation band" of which the planners should be aware.

---

[2] Accuracy in evaluating factors and costs that *are the same* among alternatives is *not* necessary for selection. For example, an accurate estimate of right-of-way cost is not needed to decide between two alternative transmission designs that would use the same route and right-of-way width.

Planning and the T&D Planning Process

14.3 THE FUNCTIONS OF SHORT- AND LONG-RANGE PLANNING

There are many sounds reasons to plan, but lead time -- the fact that it can take one or more years to accomplish a planned activity -- is the reason that planning has to be done. The time it takes to prepare, build, and install facilities sets the minimum acceptable planning period for any activity. If it takes up to five years to order materials, obtain permits, survey, build, test, and bring on line a new substation, then a utility has to look at least five years ahead in order to make certain it will have new substation facilities by the time they are needed. There may be reasons to plan farther into the future, but if the lead time for a particular system element is five years, then the utility must plan that element at least five years in advance.

Thus, the minimum possible planning horizon is a function of the lead time, which varies depending on the level of the power system -- the bigger the equipment the longer it seems to take to get new facilities in place, as shown in Table 14.5, which shows typical lead time periods for various levels of the power system. As discussed earlier, short lead times limit the alternatives available to the planner and reduce the effectiveness of the planning process. *Effective, minimum-cost planning means looking far enough ahead to allow time for all reasonable alternatives, not just the one with the shortest lead time.* The lead times shown in Table 14.5 are the recommended periods that encompass this concept, and are somewhat longer than those is needed just to get by.

Table 14.5 Typical Minimum Lead Times at Various Levels of the System

Level	Years Ahead
Generation	13
EHV Transmission	9
Transmission	8
Sub-transmission	7
Substation	6
Feeder	3
Lateral	.5
Service level	.1

Short-Range Planning: Decisions and Commitments

The purpose of short-range planning is to make certain that the system can continue to serve customer load while meeting all standards and criteria. It is driven by the lead time -- it must identify all commitments and purchases that need to be made now in order to provide for service-within-standards in the future. The product of the short-range process is a series of decisions -- the selection of various alternatives over other ones -- the expansion of the system.

Commitment should never be made until the lead time has been reached -- if it takes four years to permit, order equipment, construct, test and bring on-line a substation, then there is no reason to start that process five years in advance of need. Conditions or needs could change in the intervening year. But if the lead time is four years, then at four years ahead the decision must be made. Good or bad, the utility has to commit either to building or waiting, otherwise waiting any longer means that it selects the "do nothing" option by default.

Thus, the product of the short-range planning process is a set of Planning Authorizations, basically specifying exactly what facilities and equipment are to be bought, where they will be put, how they will be fit into the system as well as what marketing commitments and plans will be enacted. Short-range planning initiates the resource allocation process, which fills in the detail and produces the land acquisition requests, drawings, permit requests, materials lists and construction authorizations, hiring, re-organization, marketing, advertising, customer survey, and other efforts necessary to make it happen.

> *One purpose of short-range planning is to assure that lead-time requirements are met. The output of the short-range planning process is a set of decisions and specifications for future changes to the system.*

The short-range plan

The "short-range plan" is a schedule of additions, enhancements, reinforcements and changes that are to be made to the system and that have been authorized and committed. Although there is always recourse in any plan, the short-range plan is composed of decisions which have, in effect, been made and "locked in." As such, it tends to be very "project oriented" -- the short-range planning process leads, ultimately, to a series of separately identifiable project authorizations, each committing to a particular addition(s) to the system, with its own scope of work, schedule, budget and justification (Figure 14.5). These project authorizations are the hand-off from the planning process to the engineering and construction process.

Planning and the T&D Planning Process 583

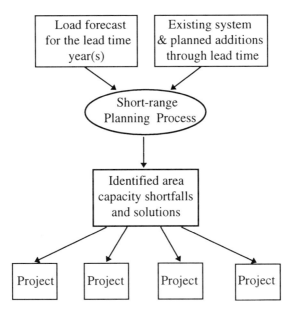

Figure 14.5 The short-range planning process consists of an initial planning phase in which the capability of existing facilities is compared to short-term (lead time) needs. Where capacity falls short of need, projects are initiated to determine how to best correct the situation. The load forecast is primarily an input to the short-range planning process -- i.e., it is used to compare with existing capacity to determine if and where projects need to be initiated.

Long-Range Planning: Focus on Reducing Cost

Why look beyond the lead time? Among several important reasons, the most important is to determine if the short-range decisions provide lasting value. Unlike short-range planning, long-range planning -- planning that looks beyond the lead time of the equipment being studied -- is not concerned with making certain that the system can "continue to serve customer load while meeting all standards" but rather in making certain that those decisions that are made have a low present worth cost and fit the long-term needs.

Consider the planning situation indicated in Figure 14.6, which shows a coastal region with five small cities. Customer load growth in this region requires that transmission capability be expanded within five years. Since this

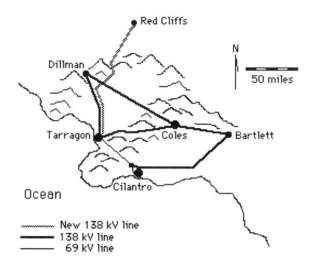

Figure 14.6 Within five years, *something* must be done to meet this region's growing power needs. Among the options is the new Tarragon to Red Cliffs 138 kV circuit shown above. Studies confirm the new line will solve the immediate need for new capacity, but will it prove a good investment and a useful system element over its lifetime? To answer these questions T&D planners must look at the line over a period beginning when it goes into service and extended into the future through a good portion of its lifetime -- they must do long-range planning.

happens to be the lead time for transmission, a decision must be made soon. The short-range planning goal is to identify additions or changes to the system that will satisfy the immediate need, which is for greater power import capability. Among other alternatives, these short-range studies confirm that the Tarragon-Red Cliffs line shown in Figure 14.6 would solve that immediate problem.

How can the utility's planners know if this new line is a good investment, with a low lifetime cost and a high utilization over its lifetime? That lifetime begins when the new line is completed and put in place -- in five years, at the end of the lead time. Thus the short-range studies are of no use toward this end.

To assess the overall present worth of a proposed addition, and determine if it is a viable long-term asset, it must be studied over a significant portion of its lifetime, beginning five years from now and extending a decade or more into the

Planning and the T&D Planning Process

future. This is long-range planning. Its focus is on the equipment once it is put in service. Its purpose is to determine if the equipment will have a long-term utilization within the system and it provides a good present worth. Given typical economic factors, and considering the uncertainties in predicting conditions over the very long-term, a ten-year period generally is considered the minimum for such economic evaluations.[3] For major projects like the example in Figure 14.6, an assessment of the facility's role in the system during the first 15 to 20 years of its lifetime is recommended. This long-range planning period *begins* at a time, already five years into the future, when the new line is first put into service.

> *The purpose of the long-range plan is to assure all short-range decisions have lasting value and contribute to a minimum-cost system. The output of the long-range planning process is evaluations of the economy and viability of the short-range recommendations.*

The long-range plan

To study a new transmission line, substation, or other planned expansion item over a future period of from five to 25 years ahead, the utility's planners need to have a good idea of the future conditions throughout the time the new line will be in service. This includes a forecast of the future load levels it will be called upon to serve, as well as a description of the system within which it will operate. Almost certainly, this new line is not the only addition being considered for the next 20 years. In this particular case, new generating plants and additions to existing ones could increase the need for transmission, while other planned additions to the transmission grid could influence the performance of the Tarragon-Red Cliffs line, and change its role and justification in the system. In the more general case, the value of a new system addition depends greatly on both the load it will serve and the other equipment that will be added to the system along with or after it is in place. Evaluation should take into account the larger impacts of these future events.

The utility planners need a long-range plan to provide such background information. This long-range plan must lay out the economics (value of losses,

---

[3] When using typical discount rates (e.g. 11%) analysis of a ten-year period of service captures 66% of the present worth of a level investment with a 40-year lifetime, while a 15-year period captures 80% of its present worth. Thus 10 and 15-year periods that begin at the lead time see a majority of the present worth aspects of an investment, but not all.

etc.) and operating criteria (usually assumed to be the same as today's) that will be applied by the utility in its future operations. It must specify the load that will be served (long-range load forecast), and it must specify *the other T&D additions* that will be made during the period, so that their interaction and influence on the performance and economy of any proposed short-range addition or change can be fully assessed over a good portion of its lifetime.

The long-range plan does not need great detail

A long-range plan does not need full and complete detail in equipment, location, and design, as does a short-range plan. It needs only enough detail to permit evaluation of the economics and "fit" of short-range decisions regarding long-range system requirements.

As an example of detail that is *not* needed, identification of the exact routes, turns, tower-locations, and spans of other future transmission lines is not needed in order to evaluate how they will interact with the Tarragon-Red Cliffs line. Their interaction with the Tarragon-Red Cliffs line can be assessed using approximate data on their location, routes, capacities, costs, etc. A long-range plan needs only a certain amount of detail, sufficient to its purpose. Effort beyond this is wasted (Willis and Northcote-Green, 1983).

The functions of the long-range plan

Thus, the long-range plan provides *a backdrop against which the short-range decisions can be evaluated.* But in addition, it provides several other useful functions, as listed below. The first is the long-term economic evaluation function discussed above, the reason why a long-range plan *must* exist if good economy of investment is to be assured. But there are other valuable contributions that the long-range plan makes to the planning effort, and to the utility as a whole.

> *Forecast of long-range distribution budget.* T&D, DSM, and IRP planners are not the only planners in a utility. Corporate finance requires careful planning, too, in order to arrange debt, cash flow, and purchases so that the company maintains a competitive position, stays solvent, pays dividends, and attracts investors. One input to financial planning is a projection of the future T&D and DSM expenses required in order for the utility to do its job and stay in business. The long-range IRP identifies distribution items whose capital costs, operations and maintenance costs, and losses costs can be estimated for this purpose.

Planning and the T&D Planning Process

Identification of long-term direction and strategy. A long-range plan provides a clear direction and long-term strategy for the T&D system's future. It may not be the best direction or strategy, but good or bad, it shows engineer, manager, operator, marketer and executive alike, the current vision of the T&D system's future. It provides a point of comparison and critique, so that flaws are exposed and improvements identified. As a result, the plan tends to evolve and improve over time. If there is no long-range plan, this cannot happen.

A basis for evaluation of new ideas or changes in procedure. An important aspect of the long-range plan's role as an identification of the utility system's future, is its role as a sounding board for new ideas. Proposals for changes in standards or system design are difficult to evaluate unless they can be compared on a fair basis against present standards and future needs. The long-range plan provides a basis for doing so, a "base case" to which new proposals can be compared, whether they suggest adopting a new primary voltage (23.9 kV versus the current 12.47 kV), using larger conductors (795 MCM versus 550 MCM), or making some other change in design standards or operating rules. This is especially true in integrated studies. How can one show if and how DSM could save T&D expenses unless one has a base T&D plan for comparison?

Coordination of planning among levels of the power system. As will be discussed in more detail in section 14.5, the plans for transmission, substation, and distribution levels of the system must fit together well -- they must be coordinated. This is particularly true if long-term present worth of the entire system is to be optimized. This goal is best achieved with a long-range plan that shows how these various levels will fit together and that exposes any mismatches or inconsistencies among their various individual long-range and short-range plans.

14.4 UNCERTAINTY AND MULTI-SCENARIO PLANNING

Unlike short-range planning, long-range planning's product is not a decision *per se*, but the long-range plan, which in many ways is an end in and of itself. An important point for planners to bear in mind is that long-range elements in their plan do not have to be built, or committed to, or even decided upon, for years to come. This means that the utility can change its mind, or beyond that, be of two or more minds at the same time. *There can be more than one long-range plan.*

Uncertainty in predicting major events is a major concern in long-range planning. Will a possible new factory (employment center) develop as rumored, causing a large and as yet unforecasted increment of growth? Will the bond election approve the bonds for a port facility (which would boost growth and increase load growth)? Situations such as these confront nearly every planner. Those of most concern to distribution planners are those factors that could change the location(s) of growth.

In the presence of uncertainty about the future, utility planners face a dilemma. They want to make no commitment of resources and facilities for load growth that *may* develop, but neither can they ignore the fact that there are lead times required to put facilities in place, and that the events *could happen*. Given the reality of lead times, planners must sometimes commit without certainty that the events they are planning for will, in fact, come to pass.

Ideally, plans can be developed to confront any, or at least the most likely, eventualities, as illustrated in Figure 14.7. *Multiple long-range plans,* all stemming from the same short-range decisions (decisions that must be made because of lead times) cover the various possible events. This type of planning, called multiple-scenario planning, involves explicit enumeration of plans to cover the various likely outcomes of future events. It is the only completely valid way to handle uncertainty in T&D forecasting and requirements planning.

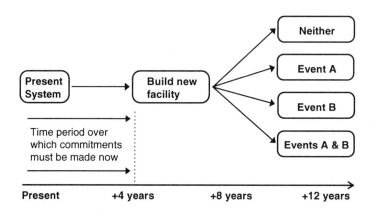

Figure 14.7 Multi-scenario planning uses several long-range plans to evaluate the short-range commitments, assuring that the single short-range decision, fits the various long-range situations that might develop. This shows the scenario variations for the case involving two *possible* future events.

Planning and the T&D Planning Process

Uncertainty in T&D growth forecasts cannot be addressed by planning for the expectation of load growth

One of the worst mistakes that can be made in T&D planning, integrated or otherwise, is to try to circumvent the need for multi-scenario planning by using "average" or "probabilistic forecasts" to develop a single "middle of the road" plan. This approach invariably leads to plans that combine poor performance (as evaluated with respect to both electrical and customer service) with high cost. As an example, the forecast in Figure 14.8 shows spatial forecasts for a large city. At the top are maps of load in 1994 and as projected for 20 years later. The difference in these top two forecast maps represents the task ahead of the T&D planners -- they must design a system to deliver an additional 1,132 MW, in the geographic pattern shown. This is a formidable task, even though they have a total of 20 years to implement whatever plan they decide upon.

At the bottom, in Figure 14.8, are two alternate forecasts that include a possible new "theme park and national historical center." Given voter approval (unpredictable) and certain approvals and incentives from state officials (likely, but not certain), and continuing commitment by the entertainment-industry mega-company that wants to build the park (who knows?) the theme park/resort area would be built at either of two locations. It would not only generate 12,000 new jobs but bring other new entertainment and tourism industries to the area, causing tremendous secondary and tertiary growth, and leading to an average annual growth rate of about 1% more than in the base forecast during the decade following its opening.

Thus, the theme park means an additional *260 MW* of growth on top of the already healthy expansion of this city. The maps in Figure 14.8 show where and how much growth is expected under each of three scenarios -- no park (base forecast, considered 50% likely), site A (25% likely) or site B (25% likely). As can be seen, the locations of the additional scenario growth will be very much a function of where the new park locates. Hence, the where aspect, as well as the how much, of the utility's T&D plan will change, too. T&D equipment will have to be re-located and re-sized to fit the scenario that develops.

The planners need to study the T&D needs of each of the three growth scenarios and develop a long-range plan to serve that amount and geographic pattern of load. Ideally, they will be able to develop a short-range plan that "branches" after the lead time, to three different plans tailored to the three different patterns of load growth. This may not be possible, but the planners will never be able to do it unless they study the situation and try.

One thing the planners do *not* want to do is form a single forecast map based on a probability-weighted sum of the various possible forecasts, an "expectation

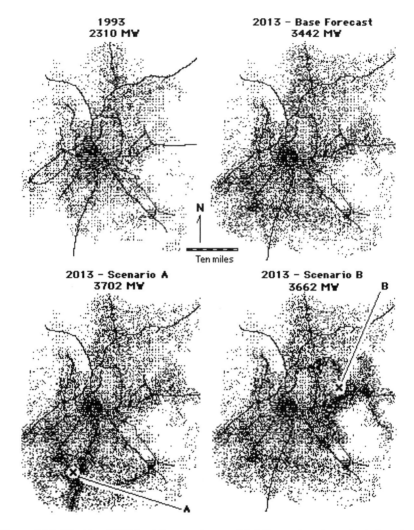

Figure 14.8 The base T&D load forecast (upper right) for an American city is based on recent historical trends. However, given voter approval of a $145 million bond issue, as well as approval by state and federal authorities, a "major theme park" will be built within five years at one of the two locations shown. The economic growth generated by the park could mean about 260 MW of additional load growth by 2013, in either spatial distribution shown, depending on the park's location. Utility planners should explicitly consider each of the forecasts, as described in the text, and develop plans accordingly (Forecasts A and B differ as to total amount of load because a good portion of the park's growth falls outside the utility's service area, to the east, in scenario B).

Planning and the T&D Planning Process 591

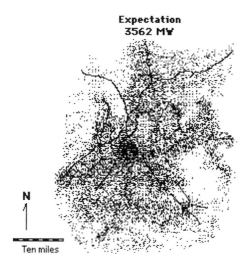

Figure 14.9 The probability-weighted sum of the three forecasts in Figure 14.8. This is perhaps the *worst* forecast the T&D planners could use to develop their plans.

of load growth" map. Such a map is easy to produce. Figure 14.9 is the probability-weighted sum of the three forecast maps shown in Figure 14.8.

While mathematically valid (from a certain perspective, at least) this forecast map will contribute nothing useful to the planning effort -- it represents a future with 1/4 of a theme park at location A, 1/4 at location B and the other half somewhere else. That is one scenario that will *never* happen. Using it as the basis for T&D planning will lead to plans that spread resources too thinly over too much territory. In this case, it would lead the utility to plan additional T&D capacity to handle about half of the load difference represented by the theme park -- capacity to serve about 140 MW of load -- split between the south part of the system (site A) and the east (site B). No matter which scenario eventually develops, the system plan based on contain several major and potentially expensive flaws:

1) If the base case actually develops, then the planners have "wasted" facilities and resources capable of serving about 140 MW of capacity, installing them when and where they are not needed.

2) If Scenario A develops, the planners have wasted the capacity additions they made in the east part of the system (capable of serving about 70 MW). In addition, they have put in place around site A facilities that fall short of the capacity needed there by 75%.

3) If Scenario B develops, the planners similarly have wasted the capacity additions made in the south (equal to about 70 MW). In addition, they have put in place around site B facilities that fall short of capacity requirements by 75%.

Planning with the forecast shown in Figure 14.9 guarantees only one thing: the T&D plan will be inappropriate no matter what happens. In particular, if either of the alternate scenarios occurs, the plan is very non-optimal. Recall from Chapter 1 that rebuilding T&D facilities to a higher capacity costs more than building them in the first place. In this case, with too little capacity installed at either site A or B, if the theme park develops, the utility will find that its costs in either case are *more* than if they had built a plan for the base case (no park) in the first place.

Some announcement about the theme park will be forthcoming within two to three years. Thus, the uncertainty will be resolved long before most of the facilities required are put in place. But the poor planning would still have a significant impact. In two years, *all* of the evaluations based on the long-range T&D plan -- all of the assessments of how new facilities will match long-term needs, all of the decisions among alternatives based on future economics, and all of the estimates of future budgets -- will be wrong.

The impact of this poor planning will be significant and long-lasting. An assessment of the uncertainty and risk using a method presented elsewhere (Willis and Powell; Tram *et al.*) indicated that the expectation-of-probabilities approach would actually "mis-spend" something on the order of $1.5 million PW in this case. That is actually worse than if the planners simply ignored the possibility and planned using only the base forecast (about $1 million PW predicted impact on actual expenses). Multi-scenario planning reduces the risk of dollars mis-spent to less than one quarter million.

Regardless of how they plan, whether they use the expectation of load growth or just the base forecast, the T&D planners in this case will have a chance to change direction once the uncertainty is resolved -- they will patch up their plan and somehow "muddle through." But a much better way to handle uncertainty is to recognize that there *will* be a need to change direction and to plan for *that*. This is the essence of the multi-scenario approach and the concept that was illustrated in Figure 14.7.

Planning and the T&D Planning Process

14.5 THE T&D PLANNING PROCESS

Figure 14.10 shows the T&D planning process as it is often represented, consisting of five stages, including a load forecast (customer needs), followed by coordinated steps of transmission, substation, distribution, and customer-level planning. The exact organization and emphasis of these individual planning steps will vary from one utility to another. A particularly effective organization of these steps will be presented later in this chapter. What is important is that all are accomplished in a coordinated manner. Other chapters in this book will focus on the data, procedures, and analytical techniques needed in each step. Here, the emphasis is on what each of these steps is and how it fits into the overall planning process.

Of the five stages, only the load forecast involves a process over which the utility has no control. The other four concern resources over which the utility has considerable (although often not total) freedom of action and investment. Each of the five stages consists of both short- and long-range planning portions, of length shown in Table 14.6. For the load forecast, "short-range" planning consists of producing a single "base" forecast -- a non multi-scenario "most accurate" forecast for use in all short-range T&D planning. This need cover no more than a decade ahead, that being the longest short-range planning period for any of the levels shown in Table 14.6. For the other four stages, short-range planning is aimed at achieving recognition of problems and definition of solutions within their lead time requirements.

Long-range planning is focused on evaluating the utilization and investment economics of all short-range projects, using a present-worth perspective or something similar. Short-range projects and decisions are judged against this sounding board of long-range performance. The load forecast can and often will be multi-scenario. Working for that, the long-range planning steps of all four subsequent stages are merged into a single, *long-range T&D plan,* with appropriate and coordinated representation of all system levels.

The short-range planning periods shown in Table 14.6 are recommended for normal utility circumstances. The long-range planning at each level begins at that short-range lead time and extends through a period at least twice and as far as four times farther into the future. *Coordination of focus, plan, and criteria among these five steps is the single most important quality-related issue in T&D planning.* Each of the major levels of the power system -- transmission, substation, distribution, customer, is connected to at least one other level. Its economy, performance, and flexibility of design interact with that other level, to the extent that no decision at one level is without noticeable, and often major, influence on the others. Coordination of method and criteria among the levels of the system is vital to successful, minimum-cost planning.

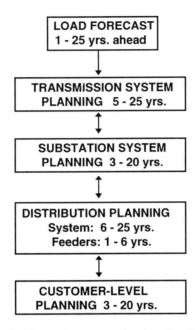

Figure 14.10 The basic T&D planning process involves five discrete stages.

Table 14.6 Typical Short and Long-Range Planning Periods for T&D System Planning

System level	Planning period - years ahead	
	Short-range	Long-range
Large generation* (> 150 MVA)	10	30
Small generation* (< 50 MVA)	7	20
EHV transmission (500 kV and up)	10	30
Transmission (1115 kV to 345 kV)	8	25
Sub-transmission (34 kV to 99 kV)	6	20
Distribution substations	6	20
"Feeder System" (service area) planning	6	20
Primary three-phase feeders	4	12
Laterals and small feeder segments	1	4
Service transformers and secondary	.5	2
Customer level (DSM, RTP)	5	20

* Generation time periods are for "traditional" planning in a vertically-integrated utility planning within a least-cost monoploy-franchise regulatory structure. Generation in a competitive environment might be planned within a profit-payback structure with shorter time periods.

Planning and the T&D Planning Process

Spatial Load Forecast

The planning process is driven by a recognition that future customer demands for power, reliability, and services may differ from those at the present. The load forecast projects future customer demand and reliability requirements, which define the minimum capability targets that the future system must meet in order to satisfy the customer demands. Forecasting for T&D planning requires a spatial forecast, as illustrated in Figure 14.11. The forecast of location of customer demand is as important in T&D planning as the forecasting of the amount of load growth: both are required in order to determine where facilities need to be located and what capacity they must have.

Timing. Usually, spatial forecasts project annual figures (peak, etc.) over a period into the future or from three to five times the lead time of the equipment being planned -- a spatial forecast for substation planning (five year lead time) would be carried out through 15 to 25 years into the future.

Spatial resolution Spatial forecasts are accomplished using the small area technique (Figure 14.12), dividing the utility service area into hundreds or perhaps tens of thousands of small areas and forecasting the load in each.

As a general rule of thumb, the small areas must be no larger that about 1/10th the size of a typical equipment service area at that level, in order for the forecast to have sufficient "where" information for planning. Thus, spatial resolution needs are less at higher voltage levels -- transmission can be planned with "small" areas that are perhaps 25 squares miles in size, whereas substation planning must be done on a mile basis and distribution on a 40 acre (1/16 square mile) basis.

Quantities forecast. A T&D forecast projects the peak demand (usually with respect to very specific weather called "design conditions") at every locality, since these local peaks are what the T&D system must satisfy. In addition, simulation methods (see Chapter 15) forecast customer type (residential, commercial, industrial) and annual energy requirements (kWhr).

Some spatial forecasts also include an end-use based small area analysis, in which usage of electric energy by end-use (heating, water heating, lighting, etc.) is used to provide estimates for marketing and demand-side planning. Advanced methods may use end-use analysis to estimate the customer value of reliability of service. Cumulatively, these forecast characteristics of load location, amount, customer type and number, reliability and service requirements, set goals that the future system must meet. If the existing system cannot meet them, then additions or changes must be made to upgrade its capability.

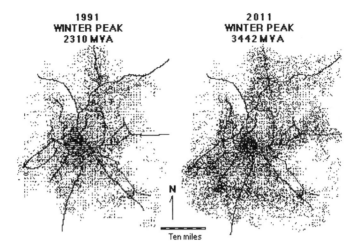

Figure 14.11 A spatial forecast begins the planning process. It maps the location and type of present (left) and future (right) load, identifying the change in capability targets for the future system.

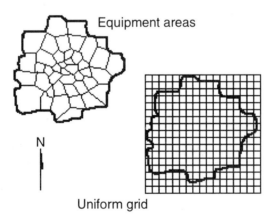

Figure 14.12 A load forecast for T&D planning (such as that shown in Figure 14.11 above) is accomplished by dividing the service territory into a number of small areas -- either equipment service areas or the cells of a square grid -- and forecasting customer demand in each.

Planning and the T&D Planning Process 597

Table 14.7 The Spatial Load Forecast Step

Purpose: to provide forecast of power delivery requirements for T&D planning

Timing: selected years to cover the period from 1 to 25 years ahead, usually 1, 2, 3, 5, 7, 10, 15, and 20 years ahead.

Products: small area forecast by year of peak demand, annual kWhr, customer types, and other factors (e.g., end-use, reliability)

Coordinated with: Corporate forecast, customer-level planning (DSM impacts forecast). Marketing plan. Critical input to distribution plan.

Tools used: Spatial load forecast method (simulation or trending). End-use analysis of load/customer value. Corporate forecast database.

Coordination with corporate planning. By far the most important aspect of the spatial load forecast is that it be consistent with the corporate, or "Rate Department" forecast, which projects sales and revenues for the utility as a whole. The T&D forecast must be based upon the same assumptions and forecast base as that.

There are a number of different analytical approaches and computer algorithms that are applied to spatial forecasts for T&D planning. What is best for any given situations depends on planning needs, data, and other details specific to a utility. Chapter 15 presents more details on spatial forecasting requirements and application, and reviews a number of popular spatial forecasting techniques. Its important elements are summarized in Table 14.7.

Transmission Planning

The high-voltage transmission grid of a power system is usually planned in conjunction with generation planning and usually in cooperation with, or perhaps as part of, the regional power pool. Only part of the criteria and goals at this level relate to delivering power. Certainly one goal of transmission planning is to be able to deliver the required customer demand, but, in addition, the transmission system is planned to provide stability and freedom of generation dispatch -- designers generally have three goals: a) serve the load, b) provide sufficient electric tie strength between generation plants so that the loss of any plant does not unduly disturb system stability, c) assure that transmission constraints do not limit choice of the pattern of generation that can be run. These last two goals mean that a good deal of the transmission system is planned from the perspective of generation: permitting generation to be controlled and dispatched as if on a single infinite bus, and providing stability against major plant disturbances. These aspects of transmission planning are not directly concerned with delivery, but the generation-grid level of the system.

Planning of the "generation-grid" combination, which includes most of the high voltage transmission, is an activity quite apart from what might be termed "sub-transmission" planning -- the planning of transfer capability to distribution

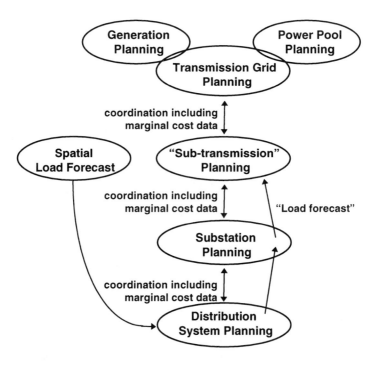

Figure 14.13 Transmission planning consists of two distinct levels or types, grid planning which is related to generation and power pool performance, and sub-transmission, or delivery planning. The "load forecast" for sub-transmission planning comes from the substation-distribution planning process.

Planning and the T&D Planning Process

substations, which is the only type of planning relevant to power delivery planning. In general, sub-transmission planning reacts quite little with grid planning -- the grid is planned and presented as a "given" to the sub-transmission planning process. This is the most limited "coordination" among any of the levels in a power system.

Figure 14.13 shows the recommended relation of sub-transmission planning to the other stages of utility system planning. Key aspects of this relationship include:

> *Transmission encompasses two planning functions:* "generation-grid" planning which focuses on establishing a strong grid with security and dispatch freedom and meeting all power pool requirements, and "sub-transmission" planning, which includes all planning activities related to moving power to substations so that it can be distributed to customers. Only "sub-transmission planning" is overtly a power delivery planning function.

> *Forecast of future load for transmission planning* comes through the distribution planning process. Fully half of the statistical variation in future loads due to "load growth" is a function of distribution planning (Willis and Powell, 1985). To the transmission-sub-transmission planning process, the "loads" to be served by the system being planned are the high-side loads of the distribution substations. Exactly what and how much load is collected to a particular substation depends greatly on how the distribution planning is done -- whether load growth ends up at one substation, or another, or is split between them, which depends on the distribution planning.
>
> This "high side substation load forecast" is best produced by starting with the spatial forecast, assessing the preferred patterns of distribution substation service areas in the distribution planning/substation planning stages, and then passing the target substation loads to the transmission planning process. If particular loads prove difficult to serve, transmission planning can provide feedback to distribution planning, asking if alternate possibilities can be considered.

> *Coordination with substation planning.* The transmission system must reach every substation. Therefore, very clearly the sub-transmission planning stage needs to be coordinated with the siting, capacity planning, and scheduling of substations (which of course has to be coordinated with distribution system planning, as shown in Figure 14.13). This coordination involves planning and scheduling of the

Table 14.8 Sub-Transmission Planning

Purpose: short and long-range planning of facilities to deliver power to the distribution substations.

Timing: short-range -- 3 to 7 years; long-range -- 5 to 20 years.

Products: short-range - sub-transmission project schedule coordinated with substation and transmission grid plans. Long-range - a base sub-transmission plan.

Coordination with: "transmission system" planning (EHV, grid, generation), substation-level planning (must deliver power to substations).

Tools used: transmission network load flow and short circuit, route optimization software, combined sub-transmission-substation optimization programs.

sub-transmission and substation levels in a very coordinated manner in which cost impacts among the levels are "traded" back and forth. In many utilities, these functions are combined into one "sub-transmission level" planning process. Given that this is coordinated well with distribution planning, this is a good approach.

Table 14.8 summarizes the important elements of this part of the T&D planning process.

Substation Planning

In many ways, the substation level is the most important level of the system, for two reasons. First, this is where the transmission and the distribution levels meet. To a great extent this level fashions the relationship between T and D -- how and where cost, contingency, or capacity at one can be traded or played against the other, and how expansion of both can be coordinated.

Second, substation sites set very definite constraints on the design and routing of the transmission and distribution systems -- transmission lines must "end" at the substations and feeders must begin there. Substation location heavily impacts the cost and performance, and particularly the available design freedom, at both T and D levels. Substation capacity limits also heavily influence both transmission (how much incoming capacity is needed) and distribution (what amount of power can be distributed out of a substation).

Planning and the T&D Planning Process

As a result, regardless of whether substation planning is done as a separate planning function (about 20% of utilities), or as part of transmission planning (60%), or as part of distribution planning (20%), it should be interactively coordinated with both T and D planning. This means that communication should be explicit and frequent enough that glaring differences in marginal costs of expansion at the T, S, or D levels will come to the attention of the planning establishment in the utility and be dealt with by appropriate changes to the plan.

For example, in Figure 14.14, distribution planners have made plans to expand Edgewood substation from 55 to 90 MVA capacity in order to serve a peak load that is expected to grow from 49 to 80 MVA due to load growth in the area indicated. Both Kerr and Edgewood substations are close enough to the load growth to pick up this new load, and either could be expanded with roughly the same plan and cost -- addition of another 27 MVA transformer and associated equipment, at a cost of $750,000 PW.

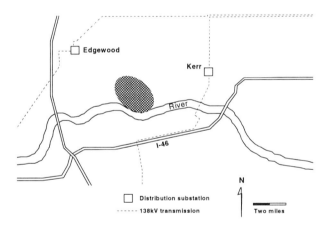

Figure 14.14 Development of new residential and commercial customers in the area shaded is expected to add 31 MVA in load over the next decade. Either Edgewood or Kerr substation could be expanded at about the same cost to serve this growth by adding a new 27 MVA transformer and associated buswork and control equipment. However, expanding the feeder system out of Kerr would be expensive -- present feeder getaways are saturated, the system is underground and many duct banks are congested, and several geographic barriers make feeder runs quite lengthy. By contrast Edgewood's all-overhead feeder system would be straightforward to expand, costing $1,900,000 less. Despite this, the overall optimal alternative is to expand Kerr, because transmission limitations at Edgewood would cost more to overcome than their savings at the distribution level.

However, from the perspective of the distribution system, expanding the feeder system out of Kerr substation instead of Edgewood would cost an extra $1,900,000 PW. Thus, Distribution Planning selected Edgewood for addition of new feeders and equipment to serve the load growth. But suppose that expanding transmission delivery capability at Edgewood by the required 31 MVA will require extensive reinforcement of towers and conductor, at a cost of $2,600,000 PW, whereas the changes needed at Kerr to accommodate a similar increase in peak load are minor and would run only $325,000 PW. Then it is best for the utility to decide to serve the new load out of Kerr, since even though Edgewood is the least expensive distribution option, when overall T-S-D cost is considered, it comes to $600,000 PW less to serve the load growth from Kerr.

The planning process at many utilities can not accommodate this very simple, common-sense approach to cost minimization, because the procedural "hand-offs" between transmission, substation, and distribution planning groups does not convey enough information, nor allow for sufficient back-and-forth cooperative communication, so that cost differences like this are noted and the planning process can recognize if and how to adjust to them. At many utilities, the distribution planners will pass no details other than the projected total load for a substation -- "We need 80 MVA delivered to Edgewood." At others the reverse happens, with transmission planners setting limits on substation capacity due to transmission constraints, without information ofnwhether these could be eased if economically justifiable based on distribution costs.

Substation planning is important chiefly because of this requirement, that as the meeting place of transmission and distribution it defines their cost interaction. Planning procedures at this level ought to reflect this relationship and foster the required communication. There are at least three possible avenues to do so.

Planning meetings between transmission and distribution planning can be scheduled frequently and set up so that a free exchange of planning details, not just results, occurs.[4]

---

[4] The author is well aware of the institutional barriers that exist within many large utilities to this type of communication. In the case of one large investor-owned utility, the managers of the transmission and distribution groups, both of whom had been in their positions for over four years, had never met and had exchanged only a handful of formal "form reports" during their respective tenures. Under de-regulation, there could well be formal procedures defining the limits of and requiring such cooperation. Regardless, the results of such communication are significant savings, and worth considerable cultural adjustment. Situations like the Kerr-Edgewood example occur more often than is recognized.

Planning and the T&D Planning Process

Table 14.9 Substation Planning

Purpose: short and long-range planning of facilities to control, route, and transform power from T to D.

Timing: short-range -- 3 to 7 years; long-range -- 5 to 20 years.

Products: short-range: substation project schedule coordinated with sub-transmission and distribution plans; long-range - long-range plan.

Coordination with: sub-transmission planning, distribution planning.

Tools used: AM/FM-GIS systems, substation selection optimization applications, combined sub-transmission-substation-feeder system optimization application.

Use of software for the sub-transmission and distribution planning functions that accepts marginal or incremental cost data for the other level, or for the substation level that accepts such data for both. For example, some service-area based optimization algorithms applied to substation-distribution system planning can accept transmission system cost curves for incoming capacity, cost which is included in the optimization (see Chapters 12 and 16).

A system of common planning procedures and nomenclature that sets typical target values for every level of the system and reports the incremental cost of all projects at any level, to all planners at other levels in routine, periodic reports. This would distribute to distribution, substation, and sub-transmission planners information for their use on the costs at other levels of the system. This provides the information, but not the incentive, for cooperation.

Frankly, application of all three of the above measures is best so as to lead to the most coordinated planning, which will minimize the potential for planning errors. Table 14.9 summarizes substation planning.

Feeder Planning

There are two very different aspects of "feeder planning." The first is feeder system planning, which involves the overall planning of the distribution feeder system in conjunction with the substations. The second, feeder planning, involves the detailed layout of feeders and their specification to a detailed engineering level of itemization and precision.

Feeder system planning

What might be called "strategic" feeder planning involves determining the overall character of the feeder system associated with each substation. This includes determining the number of feeders from the substation, their voltage level, the overall service area served by the substation as a whole and the general assigned routes and service areas and peak loads for each. Additionally, operating problems (voltage, reliability, contingency switching) and overall costs may be estimated (budgetary).

Generally, feeder system planning is done in the substation long-range planning period, usually from six to 20 years ahead, and with node resolution of perhaps only 20-30 nodes feeder -- well beyond any lead time for feeders but with considerably less detail than needed for short-range feeder planning. Feeder system planning is *only* a long-range planning activity. Its purpose is to provide "distribution system" feedback to planning decisions or alternatives being considered at the substation and sub-transmission level.

As was mentioned earlier, the interplay of sub-transmission, substation, and feeder costs can be significant. In particular, the value of the primary distribution system associated with any particular substation probably outweighs the value of the substation itself. The 30-year PW of feeder losses on a heavily loaded system may exceed the value of both. Substation planning, particularly siting and sizing, can often be the economic equivalent of the tail wagging the dog -- small changes in location or planned capability can affect the cost of the feeder system dramatically -- in some cases a shift of one mile in the location of a suburban substation can have a PW impact of over $1,000,000.

Table 14.10 Feeder System Planning

Purpose: long-range planning of the feeder system, mainly so that feeder impact of substation-level decisions is assessed for substation planning.

Timing: 5 to 20 years.

Products: evaluation of feeder-level cost and performance impact of all substation-level decisions; long-range feeder level cost estimates for budgeting.

Coordination with: substation planning, customer-level planning.

Tools used: multi-feeder (multi-substation) optimization programs, combined substation-feeder system optimization application.

Planning and the T&D Planning Process

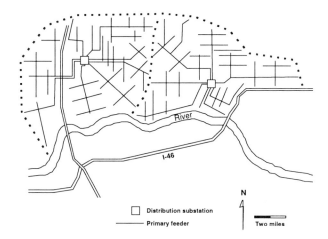

Figure 14.15 Multi-feeder system planning evaluates the performance of the entire feeder system associated with a set of substations. Its goal is to capture completely the interaction of substation level and feeder level constraints and costs to determine the overall substation-feeder optimum. A big part of this boils down to assigning "optimal service areas" to all substations while acknowledging capital and operating costs and electrical and geographic constraints at both the substation and feeder levels simultaneously, while determining the boundaries (dotted lines) between the substations.

Feeder system planning is often called multi-feeder planning (Chapter 9), to distinguish it from feeder planning (Chapter 8), which includes functions that can be done on a feeder by feeder basis. To be of maximum impact, multi-feeder planning should cover multiple substations, assessing how the feeder system interacts with substation capacities and costs -- optimizing substation service areas while considering both substation and feeder costs (Figure 14.14). Chapters 11 and 12 discuss techniques for simultaneous substation-feeder system planning in greater detail.

Short-range feeder planning

Short-range feeder planning involves determining in detail the routing of each feeder and identifying its equipment to near engineering precision. In many utilities, feeder planning and feeder engineering are merged into one function -- "Distribution Planning" produces completed feeder plans replete with detailed equipment specifications and route maps, pole locations, and all construction details. It also provides the final authorization and scheduling for construction.

Table 14.11 Feeder Planning

Purpose: short-range planning of the feeder system, to produce project definitions and authorization for feeder additions and enhancements.

Timing: 1 to 5 years.

Products: feeder system project specifications, schedule, and budget.

Coordination with: feeder system planning, substation planning, customer planning, and construction.

Tools used: "feeder design" CAD systems, feeder optimization programs (either single or multi-feeder), AM/FM and GIS systems.

Generally, feeder planning can be done on "a feeder at a time" basis, often using only load-flow and short-circuit analysis tools augmented by engineering judgment to determine alternative designs in a "trial and error" approach to find the best feeder layout. However, many modern CAD systems for feeder planning have load flow and short circuit applications that can be run on a number of feeders at once, and augment these with optimization utilities to help refine conductor size specifications, capacitor/regulator plans, and other aspects of each feeder's layout. Feeder planning is covered in detail in Chapters 8-11. Table 14.11 summarizes its essential elements.

Customer-Level Planning

Traditional supply-side-only planning included very little if any "customer-level" planning functions. Regulated utility integrated resource planning of the type widely practiced in the late 1980s and early 1990s includes a great deal -- various DSM options optimized on a TRC, RIM, or other resource-value basis (see Chapter 6). While the role of IRP and DSM, in particular, in a de-regulated power industry is unclear, the fact that some energy efficiency, advanced appliance, and load control/TOU technologies can prove cost-effective is not in doubt. For this reason, some sort of customer-level planning can be expected in the future. Under the right circumstances, DSM, DG, or DS may be of value to a regulated or unregulated electric supplier/distributor. Therefore this level of planning will be a part of many utilities, and will, at least obliquely, impact T&D planning.

The interaction may be much more than just oblique, because DSM and DG programs often have a very substantial impact on T&D, with more than half their potential savings being in avoided costs among the sub-transmission,

Planning and the T&D Planning Process 607

substation, and distribution levels. Usually, optimization of the benefits of a DSM or DG program includes some assessment of T&D impacts and location.

Customer-level planning often has a very long planning period, much longer than distribution planning, and often as long as that of generation. There are four reasons why both its short- and long-range planning periods are longer than almost all other levels of the system:

Lead time. While DSM programs can often be approved and implemented nearly immediately, it can take years of slow progress to obtain meaningful levels of participation -- rates of 50% may take a decade or more to obtain. This "ramp up time" has the same effect as a long lead time.

Customer commitments. In starting a DSM program, a utility is beginning very visible customer-interaction. Such programs are not entered into lightly and are often studied over the very long-term to make certain the utility can feel confident it will commit to them for the long-term -- starting such a program and then canceling it several years later is not only costly, but leads to customer confusion and lack of confidence in the utility.

Proxy for generation. DSM is often traded against generation in integrated resource planning. Comparison of the two resources needs to be done over the time span used for planning of generation, which means a present-worth or similar economic comparison over a 20 to 30 year period.

Customer-level planning generally begins by adding more detailed, end-use and customer value attributes to the load forecast -- forecasts by end-use and appliance category and of reliability value functions. It then tries to optimize

Table 14.12 Customer Level Planning

Purpose: short- and long-range planning of customer-level resources including DSM, DG, DS, RTP and TOU rates.

Timing: 1 to 25 years.

Products: schedule of customer-side resource projects/programs; long-range plan of DSM/DG/DS targets.

Coordination with: all levels of planning, corporate forecast/rate plan, spatial load forecast. Note: this planning is often done by corporate or rate department planners.

Tools used: end-use load models. Integrated resource (T&D or otherwise) optimization models. Customer response models. Value-based planning analysis. Econometric/demographic models. Spatial customer/load forecast models.

PW cost of providing end-uses to the customers by juggling various DSM and T&D options. Generally, optimization methods function by balancing marginal cost of the various customer-side options against marginal cost curves obtained for the supply side. DSM and DG evaluation may also be part of value-based planning in which the utility and customer value functions are jointly optimized (see Chapters 5 and 6).

The Coordinated, Multi-Level, Long-Range Plan

As was discussed earlier in this chapter, to study the "system fit" and economic viability of short-range alternatives over their lifetime, the utility's planners must have a good idea of the conditions under which those alternatives would be operating for a majority of their lifetime. They need a long-range plan to provide a *background* for evaluation of equipment in the short-range plan.

This long-range plan must lay out the economics (value of losses, etc.) and operating criteria (usually assumed to be the same as today's) that will be applied by the utility in its future operations. It must specify the load (long-range load forecast) the system will be expected to meet, and it must specify how the T&D planners expect to handle that load growth -- what new facilities will be installed in the long run to accommodate the future pattern of load growth. Most important, it should specify how the various long-range plans at the sub-transmission, substation, feeder, and customer levels will fit together. The economics of interaction of all four levels, leading to a least-cost system, is assured by evaluation of long-term PW value in a coordinated manner.

The long-range plan is its own product

Long-range T&D planning is one of the few legitimate instances where a process's major goal is itself. The long-range plan, along with its accompanying budget estimates and indicators of economic factors for the future system, *is* the major goal of the long-range planning process. More important, it is crucial that this plan's development be ongoing, that it be maintained and continuously updated in order to provide long-range evaluation, guidelines, and direction to T&D planning and investment decisions.

Detail is not needed in a long-range plan

The long-range plan needs only enough detail to permit evaluation of the economics and "fit" of short-range decisions against long-range needs. Effort to provide detail beyond this is wasted (Tram et al.). The coordinated planning process is shown in Figure 14.16. Its is summarized in Tables 14.13 and 14.14.

Planning and the T&D Planning Process

Table 14.13 The Coordinated Multi-Level Plan

Purpose: to assure coordination and optimization among all system levels.

What: A coordinated multi-level long-range plan which includes essential elements from the long-range plans of the sub-transmission, substation, distribution and customer levels.

Coordinated with: everything.

Tools used: attention to detail and good communication.

Table 14.14 Purposes of the Long-Range Plan

Definition of "future" for evaluation of short-term planning decisions.

Identification of mismatches between levels -- coordination of plans.

Forecast of long-range budget needs (for financial planning).

Identification of long-term direction and strategy.

A basis for evaluation of new ideas or changes in procedure.

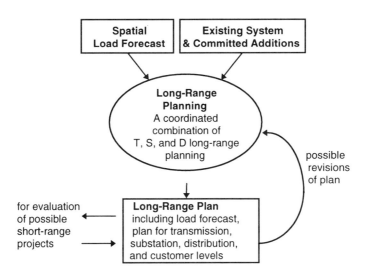

Figure 14.16 The coordinated multi-level long-range plan, where the major goal is the maintenance of a long-range plan that identifies needs and development of the system, so that short-range decisions can be judged against long-range goals. Here, the load forecast is a part of the plan, and there is no "project oriented" part of the activities.

14.6 SUMMARY AND RECOMMENDED T&D PLANNING PROCESS

Planning is the process of identifying, evaluating, and selecting alternatives. It is driven by two inputs: a recognition that future conditions or needs may differ from those of the present (forecast) and the fact that time is required to put corrective measures in place to meet those new conditions and needs (lead time).

Short and long-range planning have very different functions. Short-range planning is focused on making sound decisions by the lead time, so that planned additions and changes are made in time and do provide the desired results. It leads to projects -- discrete systems additions each with a scope, budget, justification, and identity of its own. Long-range planning is motivated by a desire to see that short-range planning decisions have lasting value, and that the planning of diverse elements of a business (e.g., transmission, distribution, substations) is coordinated to the sum in a least-cost system. The primary product of and the basis for long-range planning is the long-range plan, an estimation of the elements expected to be added over the long-term -- in T&D cases this means over the next five to 20 years.

A long-range plan extends beyond the lead time and thus, most of its elements do not have to be committed to any time soon. Therefore, there can be more than one long-range plan, or put another way, the long-range plan can have multiple-scenario variations to explain how it will handle uncertainty through adjustment of what is added to the system. Multiple scenario planning is the recommended way to address uncertainty in planning that deals with siting and routing issues, because planning methods based on probability-weighted forecasts or plans invariably lead to inappropriate commitment of resources.

Flow of the T&D Planning Process

T&D planning consists of five stages -- load forecast, transmission, substation, distribution, and customer level planning. Each has its long and short-range elements. Most important is the coordination of the planning at every level with the other levels --the electric system is interconnected and changes and additions at one level affect the economy and performance of the system at other levels.

The recommended interaction of the five T&D planning stages with one another and other planning functions within a vertically integrated utility is shown in the flowchart in Figure 14.17. Ovals indicate planning functions or processes, rectangles indicate products of the planning process, and arrows indicate information flow. This T&D planning process has several vital characteristics:

- *The T&D planning process has three major inputs:* a) the corporate "Rate and Revenue" forecast for future load growth for the system, which forms the

Planning and the T&D Planning Process

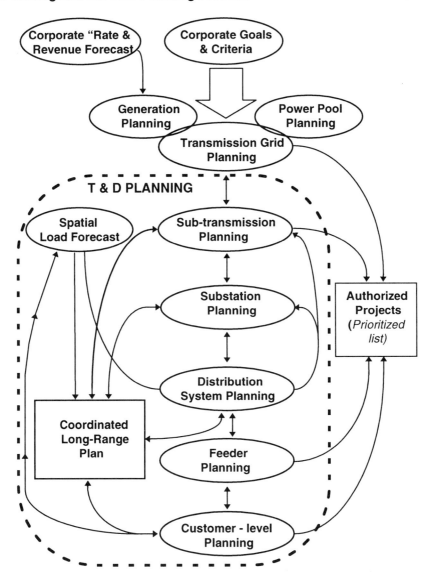

Figure 14.17 The overall T&D planning process includes load forecast, and planning functions for transmission, substations, distribution, and customer levels, as well as maintenance of a coordinated long-range plan encompassing all the levels. Its inputs are the corporate forecast & goals & critiera, its outputs are the short-range projects.

basis for the spatial (T&D) load forecast, b) corporate goals and criteria, shown as a broad arrow in Figure 14.17 to indicate this information flows to all planning functions directly, c) the generation-grid plan, which is handed to the T&D planning process more or less as a "given."

- *T&D planning's only "output" is the list of authorized projects* -- which cumulatively form what might be termed the "short-range plan." The T&D planning process is focused on producing short-range project decisions of high quality, timeliness, and low cost. Long-range planning and all other parts of the long-range planning process exist for and should focus on contributing to the value of these decisions.

- *The long-range plan is internal to the T&D planning process.* To a great extent, the long-range plan exists only so that the short-range planning process can be done well. Certainly, information contained in the long-range plan -- future ROW needs, long-term budget estimates, plans for future equipment purchases -- is passed on to other departments from time to time and used in other functions. However, the long-range plan exists primarily so that the T&D planners can do long-range planning, and it is internal to their process. The long-range plan needs only detail and comprehensiveness sufficient for assessing short-term projects with respect to future viability and economy -- a common mistake is to put too much time and effort into refining the long-range plan's fine points to a level similar to the short-range. For example, a long-range plan does not need to show the precise location, give detailed equipment specifications, and design data for a substation that is still ten years beyond the lead time. It is sufficient for the plan to say, "We expect to built a distribution substation with two 25 MVA transformers and associated switching and control equipment at a location near here."

- *Transmission grid and sub-transmission planning are different processes.* Given that the utility has any significant amount of both, these two very different processes should be split into separate functions, as this promotes focus on each of the two types and their unique goals and characteristics.

- *There is no "short-range" feeder* system *planning.* Multi-substation/multi-feeder distribution planning is the long-range planning aspect of feeder planning. If done correctly, it assures that each planned feeder in the short-range plan contributes to long-term system needs. That is its sole goal.

- *There is no long-range "feeder planning."* Feeder-at-a-time planning exists only as a short-range, engineering-type function. Distribution feeders should not be planned individually for the long-range -- their role as part of the switched primary feeder "network" precludes that. But once long-range

Planning and the T&D Planning Process

analysis has determined that they are viable parts of the overall plan, they can be planned on an individual basis, for only the short-run.

- *Customer-level planning and the load forecast interact.* Figure 14.17 shows customer-level planning as part of the T&D planning process. At many utilities this is not the case -- DSM, TOU, RTP, and DG planning are done in conjunction with Rate and Revenue planning. However, a majority of savings from customer-side programs are realized at the T&D levels, and spatially targeted resources are often the key to economy and effective reliability enhancement from DSM and other similar resources. Focus on such opportunities is best done through the T&D planning process, which is the only part of the utility planning process that has experience with and focuses on locational-based (spatial) resources and constraints.

REFERENCES

Electric Power Research Institute, "Research into Load Forecasting and Distribution Planning," report EL-1198, Electric Power Research Institute, Palo Alto, CA, 1979.

M. V. Engel et al., editors, *Tutorial on Distribution Planning,* Institute of Electrical and Electronics Engineers, New York, 1992.

H. L. Willis, *Spatial Electric Load Forecasting,* Marcel Dekker, New York, 1996.

H. N. Tram et al., "Load Forecasting Data and Database Development for Distribution Planning," *IEEE Trans. on Power Apparatus and Systems,* November 1983, p. 3660.

H. L. Willis and R. W. Powell, "Load Forecasting for Transmission Planning," *IEEE Transactions on Power Apparatus and Systems,* August 1985, p. 2550.

15
Forecasting T&D Load

15.1 SPATIAL LOAD FORECASTING

In order to plan an electric power delivery system, the T&D planner must know *how much* power it will be expected to serve, and *where* and *when* that power must be delivered. Such information is provided by a *spatial load forecast*, a prediction of future electric demand that includes location (where) as one of its chief elements, in addition to magnitude (how much) and temporal (when) characteristics. The spatial forecast depicted in Figure 15.1 shows expected growth of a large city over a 20-year period.

Growth is expected in many areas where load already exists -- facilities in those areas may need enhancement to higher capacity. Growth is also expected in many areas where no electric demand currently exists. There, the utility will need to install new equipment and facilities. Equally important, growth will not occur in many other areas -- facilities built there would be wasted. And in a minority of areas, peak electric demand may actually *decrease* over time, due to numerous causes, particularly deliberate actions taken by the utility or its customers to reduce energy consumption (Demand Side Management, DSM).

Effective planning of the T&D system requires that such information be taken into account, both to determine the least-cost plan to meet future needs, and in order to assure that future demand can be met by the system as planned.

615

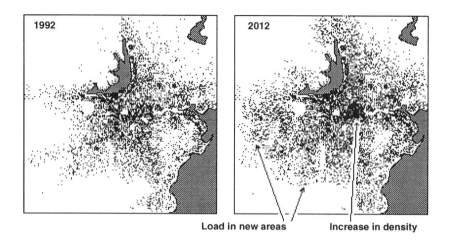

Figure 15.1 Maps of peak annual demand for electricity in a major American city, showing the expected growth in demand during a 20-year period. The area shown is roughly 30 miles on a side. Growth in some parts of the urban core increases considerably, but in addition, electric load spreads into currently vacant areas as new suburbs are built to accommodate an expanded population.

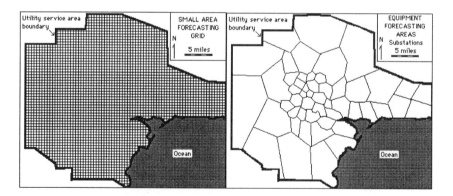

Figure 15.2 Spatial load forecasts are accomplished by dividing the service territory into small areas, either rectangular or square elements of a uniform grid (left), or irregularly shaped areas, perhaps associated with equipment service areas such as substations or feeders.

Forecasting T&D Load

Small Area Forecasting

Geographic location of load growth is accomplished by dividing the utility service territory into *small areas*, as shown in Figure 15.2. These might be irregularly shaped areas of varying size (the service areas of substations or feeders in the system) or they might be square areas defined by a grid. The forecasts shown in Figure 15.1 were accomplished by dividing the municipal utility service territory into 60,000 square areas, each 1/8 mile wide (10 acres).

A Series Including Interim Years

Usually T&D planning is done on an annual peak basis, the planners forecasting and planning for the annual peak demand in each year. This usually consists of detailed, year-by-year expansion studies during the short-range period, and plans for selected years in the long-range period (with results interpolated for years in between). A typical set of planning periods might be 1, 2, 3, 5, 7, 10, 15, 20, and 25 years ahead. This set of forecast years accommodates the short-range planning needs, which must distinguish growth from year to year during the lead time period) and long-range planning needs, which require less timing and more long-term vision on eventual development. To accommodate this need, a T&D forecast generally produces a series of spatial peak load forecasts of selected future years, as shown in Figure 15.3.

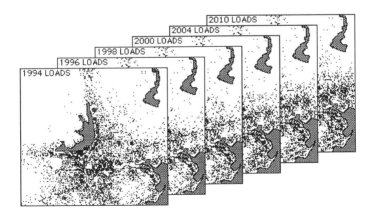

Figure 15.3 Generally, a spatial load forecast produces a series of "load maps" representing annual peak load on a small area basis, for a select set of future years.

15.2 LOAD GROWTH BEHAVIOR

Two Causes of Load Growth

Peak demand and energy usage within an electric utility system grow for only two reasons:

1. *New customer additions.* Load will increase if more customers are buying the utility's product. New construction and a net population in-migration to the area will add new customers and increase peak load. With more people buying electricity, the peak load and annual energy sales will most likely increase.

2. *New uses of electricity.* Existing customers may add new appliances (perhaps replacing gas heaters with electric) or replace existing equipment with improved devices that require more power. With every customer buying more electricity, the peak load and annual energy sales will most likely increase.

There are no other causes of load growth. Similarly, any decrease in electric demand is due to reductions in either or both of these two factors. Regardless of what happens to the load, or how one looks at load growth or decline, change in one or both of these two factors is what causes any increase or decrease in peak and energy usage.

The bulk of load growth on most power systems is due to changes in the number of customers. North American electric utilities that have seen high annual load growth (5% or more) have experienced large population increases. Houston in the 1970s, Austin in the 1980s, and the Branson, Missouri area in the 1990s, all experienced annual increases in peak load of 5% or more, due almost exclusively to new customers moving into the service territories.

Load growth caused by new customers who are locating in previously vacant areas is usually the focus of distribution planning, because this growth occurs where the planner has little if any distribution facilities. Such growth leads to new construction, and hence draws the planner's attention.

But changes in usage among existing customers are also important. Generally, increase in *per capita* consumption is spread widely over areas with existing facilities already in place, and the growth rate is slow. Often this is the most difficult type of growth to accommodate, because the planner has facilities in place that must be rearranged, reinforced, and upgraded. This presents a very difficult planning problem.

Forecasting T&D Load

Load Growth at the Small Area Level

When viewed at the small area basis, electric load growth in a power system appears different than when examined at the system level. This phenomenon is fundamental to distribution load studies, affecting all types of forecasting methods, whether grid-based or equipment-oriented, and regardless of algorithm.

Consider the annual peak electric load of a utility serving a city of perhaps two million population. For simplicity's sake, assume that there have been no irregularities in historical load trends due to factors such as weather (i.e., unusually warm summers or cold winters), the economy (local recessions and booms), or utility boundary changes (the utility purchases a neighboring system and thus adds a great deal of load in one year). This leaves a smooth growth curve, a straight line that shows continuing annual load growth over a long period of time, as shown in Figure 15.4.

When divided into quadrants, to give a slight idea of where the load is located, the city will still exhibit this smooth, continuous trend of growth in each quadrant. The total load and the exact load history of each quadrant will be slightly different from the others, but overall, each will be a smooth, continuous trend, as shown in Figure 15.5.

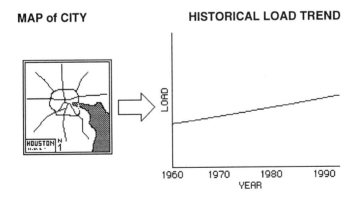

Figure 15.4 Annual peak load of the large, example city discussed here, over a 50-year period is relatively smooth and continuous when irregularities due to weather and economic cycles are removed.

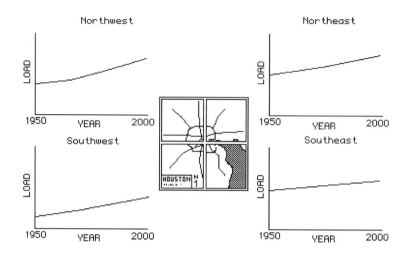

Figure 15.5 Dividing the city into four quadrants and plotting the annual peak load in each results in a set of four growth curves, all fairly smooth and showing steady growth.

Subdivide again, dividing each quadrant into sub-quadrants, and examine the "load history" of each small area once more, looking for some typical pattern of behavior. Again, the resulting behavior is pretty much as before. The typical sub-quadrant has a long-term load history that shows continuing growth over many years.

If this subdivision is continued further, dividing each sub-quadrant into sub-sub-quadrants, sub-sub-subquadrants, and so forth, until the city is divided into several thousand small areas of only a square mile, something unusual happens to the smooth, continuous trend of growth. Each small area has a load history that looks something like that shown in Figure 15.6, an S curve, rather than a smooth, long-term steady growth pattern. The S curve represents a history in which a brief period of rapid growth accounts for the majority of load.

When analyzed on a small area basis, there will be tens of thousands of small areas, and every one will have a unique load growth history. Although every small area will vary somewhat, the typical, or average, growth pattern will follow what is known as an S curve -- a long dormant period, followed by rapid growth that quickly reaches a saturation point, after which growth is minimal. The S curve, also called the Gompertz curve, is typical of small area, distribution-level load growth (EPRI, 1979). Each small area will differ

Forecasting T&D Load

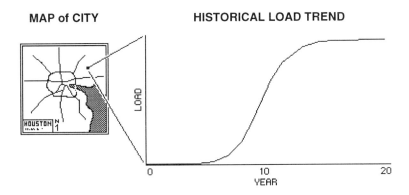

Figure 15.6 Example of the typical growth behavior of a mile-square small area within the city. Once divided into "small enough" areas, growth in any region will display this characteristic. The 640 acre area experiences almost no growth for many years, then a period of rapid growth that lasts only a decade or slightly more will "fill it in."

slightly from the idealized, average behavior, but overall, the S curve shown in Figure 15.6 represents load growth behavior at the distribution level very well.

The S curve has three distinct phases, periods during the small area's history when fundamentally different growth dynamics are at work.

Dormant period. The time 'before growth", when no load growth occurs. The small area has no load and experiences no growth.

Growth ramp. During this period growth occurs at a relatively rapid rate, because of new construction in the small area.

Saturated period. The small area is "filled up" -- fully developed. Load growth may continue, but at a very low level compared to that experienced during the growth ramp.

What varies *most* among the thousands of small areas in a large utility service territory is the *timing* of their growth ramps. The smooth overall growth curve for the whole (Figure 15.4), occurs because there are always a few, but only a few, small areas in this rapid state of growth at any one time. Seen in aggregate -- summed over several thousand small areas -- the "S" curve behavior averages out from year to year, and the overall system load curve looks smooth and continuous because there are always roughly the same number of small areas in their rapid period of growth.

This explanation should not surprise anyone who stops to think about how a typical city grows. It began as a small town, and grew outward as well as upward: most of its growth occurred on the periphery -- the suburbs. The average small area's "life history" is one of being nothing more than a vacant field until the city's periphery of development reaches it. Then, over a period of several years, urban expansion covers the small area with new homes, stores, industry, and offices, and in a few years the available land is filled -- there was only so much room. Then, growth moves to other areas, leaving the small area in its saturated period. Growth patterns in rural areas are similar, merely occurring at lower density and sometimes over slightly longer S periods.

Of course, the actual characteristics of growth are not quite this simple. Often growth leapfrogs some areas, only to backtrack later and fill in regions left dormant. Sometimes a second S growth ramp occurs many years later, as for example, when an area that has been covered with single family homes for several decades is suddenly redeveloped into a high rise office park.

However, the S curve behavior is sufficient to identify the overall dynamics of what happens at the distribution level. Examining this in detail leads to an understanding of three important characteristics of growth.

1. *The typical S curve behavior becomes sharper as one subdivides the service territory into smaller and smaller areas.* The average four square mile area in a city such as Denver or Houston will exhibit, or will have exhibited in the past, a definite but rather mild S curve behavior of load growth, with the growth ramp taking years to fill in. The average growth behavior at one square mile resolution will be sharper -- a shorter growth ramp period. The average, typical load growth behavior will be sharper still at 160 acres (small areas 1/2 mile wide), and so forth, as shown in Figure 15.7. The smaller the small areas become (the higher the spatial resolution) the more definite and sharper the S curve behavior, as shown in the figure.

   Carrying the subdivision to the extreme, one could imagine dividing a city into such small areas that each small area contained only one building. At this level of spatial resolution, growth would be characterized by the ultimate S curve, a step function. Although the timing would vary from one small area to the next, the basic life history of a small area of such size could be described very easily. For many years the area had no load. Then, usually within less than a year, construction started and finished (for example's sake, imagine that a

Forecasting T&D Load

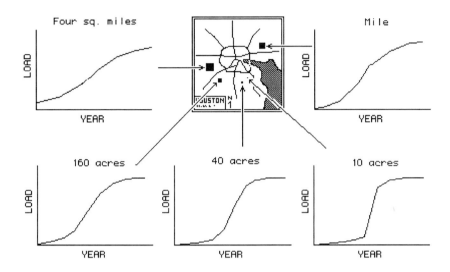

Figure 15.7 As small area size for the load growth analysis is decreased, the average small area load growth behavior becomes more and more a sharp S curve. Quantitative behavior of this phenomenon depends on growth rate, demographics, and other factors unique to a region, and varies from one utility to another. Qualitatively, all utility systems exhibit this overall behavior: load trend becomes sharper as small area size is reduced. Also see the discussion in Chapter 12 with regard to load growth behavior and its impact on substation planning (Figure 12.21).

house is built in the small area), and a significant load was established. For many years thereafter, the annual load peak of the small area varies only slightly -- the house is there and no further construction occurs.

2. *As the utility service territory is subdivided into smaller and smaller areas, the proportion of small areas that have no load and never will have any load increases.* A city such as Phoenix or Atlanta will have no quadrants or subquadrants that are completely devoid of electric load. When viewed on a square mile basis (640 acre resolution) there will likely be very few "completely" vacant areas -- a large park or two, etc. But when chopped up into acre parcels, a significant portion

of the total number will be "vacant" as far as load is concerned, and will stay that way. Some of these vacant areas will be inside city, state, or federal parks, others will be wilderness areas, cemeteries or golf courses, and many will be merely "useless land" -- areas on the sides of steep mountains and inside flood-prone areas -- where construction is unlikely. Other vacant areas, such as highway and utility rights-of-way, airport runways, and industrial storage areas, are regions where load is light and no "significant" load will ever develop.

3. *The amount of load growth that occurs within previously vacant areas increases as small area size is decreased.* When viewed at low spatial resolution (i.e., on a relatively large-area basis such as on a four by four mile basis) most of the growth in a city over a year or a decade will appear to occur in areas that already have some load in them. But if a city or rural region is viewed at higher spatial resolution, for example on a ten acre basis, then a majority of new customer growth over a decade occurs in areas that were vacant at the beginning of the period

Thus, as spatial resolution is changed, the character of the observed load growth changes, purely due to the change in resolution. At low resolution (i.e., when using "large" small areas) load growth appears to occur mostly in areas that already have some load in them, as steady, long-term trends, and few if any areas are or will remain vacant. Such steady, omnipresent growth is relatively easy to extrapolate.

However, if this same situation is viewed at high spatial resolution, most of the growth appears as sharp bursts, lasting only a few years. Many small areas have no load, and the majority of load growth will occur in such vacant areas, yet not all vacant areas will develop load -- some will stay vacant. This type of behavior is relatively difficult to trend because it occurs as brief, intense growth, often in areas with no prior data history from which to project future load.

The reader must understand that the three phenomena discussed above occur *only* because the utility service territory is being divided into smaller areas. By seeking more spatial information (the "where" of the distribution planning need) by using smaller areas, the planner changes the perspective of the analysis, so that the very appearance of load growth behavior appears to changes. High spatial resolution (very small area size) makes forecasting load a challenge, calling for forecasting methods unlike those used for "big area" forecasting at the system level.

15.3 IMPORTANT ELEMENTS OF A SPATIAL FORECAST

Forecasting Accuracy

Accuracy is naturally a concern in any forecasting situation particularly in the short-range planning time-frame. To determine how well a particular load forecast satisfies T&D needs, it is necessary to determine how well it answers the "*where?*" "*how much?*" and "*when?*" requirements of planning. A forecast or plan can be entirely correct as to amount and timing of growth, and yet lead to major inadequacies and unsound economies purely because it expends its resources in the wrong places -- it forecasts the "where?" wrong.

Since "where?" is such an important aspect of T&D planning, it might be useful to evaluate mistakes in meeting planning requirements by measuring error in terms of distance (location) rather than magnitude (kVA). Suppose a particular forecast method has an average *locational error* of 1/2 mile when forecasting load growth five years ahead -- on average it location of new growth by 1/2 mile from where it eventually develops. Such an error statistic would tell a planner something useful about the forecast method: An error of one-half mile would not dramatically impact substation planning (substation service areas are generally about 20 square miles), it would be useless for detailed feeder segment planning, where the average service area is roughly 1/2 mile^2.

The concept of locational error is quite important. It can provide powerful intuitive insight into developing quality distribution plans. Its quantitative application can be useful in evaluating planning methods, risk, and multi-scenario sensitivity.

Yet location error is only one aspect of error, equally important as magnitudinal error (mistakes in forecasting amount). Ideally, T&D forecast accuracy should be evaluated with an error measure that simultaneously evaluates the impact of locational and magnitudenal mismatches as they relate to T&D planning needs. Such error measures, called the U_x measures, involve considerable mathematical manipulation, but are simple in concept (Willis, 1983). They compute on the basis of a series of tests of a forecasting or planning method, a single-valued error metric which is proportional to the percentage of the T&D planning budget those errors are likely to "mis-allocate." Most forecasting methods fall between 7-45% in forecasting ten years ahead, so there is an incentive to evaluate them carefully.[1]

---

[1] The original technical paper (Willis, 1983) provides a good discussion, but perhaps the best discussion of theory and application is in chapter five of *Spatial Electric Load Forecasting*, H. L. Willis, Marcel Dekker, 1996.

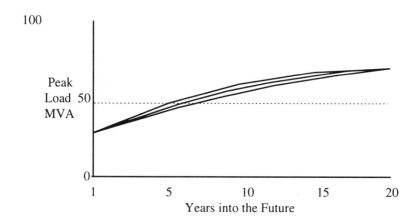

Figure 15.8 Existing T&D facilities in a 27 square mile area will be intolerably overloaded when the peak load reaches 47 MVA (dotted line), and a new substation must be added, costing $6,500,000. A year's delay in this expense means a present worth savings of $500,000, but delaying a year too long means poor reliability and customer dissatisfaction. Therefore, accuracy in predicting the *timing* of growth -- in knowing which of the three trends shown above is most likely to be correct -- is critical to good planning. All three projections agree on the eventual "build out" peak load of 75 MVA, but differ in their rate of growth.

Short-Range and Long-Range Forecasting

Chapter 14 discussed the short- and long-range time periods, and the far different planning needs within each time frame, something that will be examined at length later in this book. Not surprisingly, the short- and long-range periods have drastically different forecasting needs.

Recall that short-range planning is motivated by a need to reach a decision, to commit to a particular installation or type of construction, and to do so at the lead time. Letting the lead time for additions slip past without making appropriate plans is perhaps the worst type of planning mistake. Therefore, what is needed most in short-range planning is a reliable "alarm" about *when* present facilities will become insufficient, as shown in Figure 15.8.

Long-range planing needs are somewhat different. No commitment needs to be made to the elements in a long-range plan, so timing is less important than in short-range planning. Since the long-range plan evaluates whether and how

Forecasting T&D Load

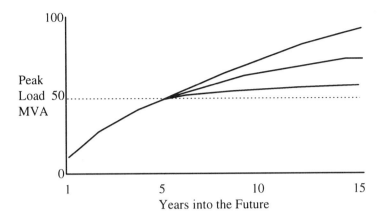

Figure 15.9 Long-range planning requirements are less concerned with "when" and more sensitive to "how much will eventually develop." The three forecasts shown here, all for the same substation area, lead to the conclusion that the new substation is needed in five years, but would lead to far different conclusions about how short-range commitments can match long-range needs.

short-range commitments fit well into long-range needs, capacity and location are more important than timing. Unlike short-range planning, long-range T&D planning requirements for a spatial load forecast are oriented less toward "when" and more toward "how much," as shown in Figure 15.9. For long-range planning, knowing *what* will eventually be needed is more important than knowing exactly *when*.

The chief reason for the multi-forecast approach is that it is often impossible to predict some of the events that will shape the location of long-range distribution load growth. In such cases, the planner is advised to admit that these factors cannot be forecasted, and to run several "what if" studies, analyzing the implications of each event as it bears on the distribution plan. This is called *multiple scenario forecasting*.

Multiple Scenario Forecasting

The locational aspect of electric load growth is often very sensitive to issues that simply cannot be predicted with any dependability. As an example, consider a new bridge that might be planned to span the lake shown to the northwest of the city in Figure 15.10, and the difference it would make in the developmental patterns of the surrounding region. That lake limits growth of the city toward the northwest. People have to drive around it to get to the other side, which is a considerable commute, making land on the other side unattractive as compared to other areas of the city. As shown in Figure 4.13, if the bridge is completed it will change the shape of the city's load growth, opening up new areas to development and drawing load growth away from other areas. The reader should make note of *both* of the bridge's impacts: it *increases* growth to the northeast, and it *decreases* growth in the eastern/southern portions of the system.

But the construction of this new bridge might be the subject of a great deal of political controversy, opposed by both environmentalists concerned about the damage it could cause to wilderness areas, and by community groups who believe it will raise taxes. Modern simulation forecasting methods can accurately simulate the impact such a new bridge will make on load growth, forecasting both of the "with and without" patterns shown in Figure 4.13. But these and other forecasting methods cannot forecast whether the bridge will be built or not.

No bridge   Bridge across lake

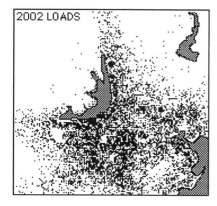 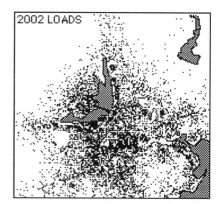

Figure 15.10 Left, the central area of the forecast shown in Figure 15.1, with the lake to the northwest of the city standing as a barrier to growth in that direction. At the right, a new bridge spans that lake, bringing areas on the other side into close proximity to downtown and drawing growth away from the southeast and east sides of the system.

Forecasting T&D Load

A future event that causes a change in the amount or locations of growth is called a *causal event*. Very often, critical causal events such as the bridge discussed above simply cannot be predicted with any reliability (Willis and Tram, 1993). A decision to lower taxes in the inner core of a large city, in order to foster growth there and avoid "commercial flight" to the suburbs, is a similar case. Other situations might be a major new factory or large government facility, announced and planned, but still uncertain as to exact location and timing within the utility service territory.

In such situations the planner may be better off admitting that he cannot forecast accurately the causal event(s), and do "what if" planning to determine the possible outcomes of the event. By doing multiple-scenario forecasts that cover the variety of possible outcomes, the planner can alert himself to the consequences of such unpredictable events. At the very least, he can then watch these events with interest, against the day a decision must be made, knowing what impact they will have and having an idea how he will respond with changes in his plan. Hopefully though, he can rearrange those elements of his plan that are sensitive to the event, to minimize his risk, perhaps pushing particularly scenario-sensitive elements into the future and bringing forward other, less sensitive items.

In the example shown in Figure 15.10, the net impact of the bridge is to shift growth from the east and south areas of the city to the area west of the lake. Either way, three new substations are needed by the year 2002. However, their locations differ depending on the outcome. If the bridge is completed, load growth will move outward past the lake. One substation will be needed there and only two, not three, will be needed further in toward the city.

The challenge facing the planner is to develop a short-range plan that doesn't risk too much in the event that one eventuality or the other occurs. He must develop a short-range plan that can economically "branch" toward either eventuality -- a plan with *recourse*. Such planning is difficult, but not impossible. The role of multi-scenario forecasting is not to predict which outcome will develop (that is often impossible) but to show the planner what *might* happen and provide a foundation for analysis of risk among different scenarios. The multi-scenario approach, leading to the development of a plan with the required amount of recourse built into it, is the heart of modern distribution planning, and is discussed in Chapter 14.

Spatial versus just Small Area Forecasting

Any forecast done for T&D planning must apply some form of the small area approach, even if it is merely an implicit "geographic" assumption that all areas will grow at the same average rate as the system forecast. Spatial forecasting methods are a subset of small area methods, in which the forecast of each area is coordinated with -- made consistent with all assumptions and method -- that for all other areas. This improves both forecast accuracy and applicability. While all spatial forecast methods are small area methods, some approaches that utilize the small area approach are not spatial methods.

Representational Accuracy

A load forecast -- spatial or otherwise -- is *not* an attempt to forecast future load most accurately. Forecasts designed to support good planning often contain biases meant to improve the planning process, and thus they are not attempting to forecast future load with minimum error.

As an example, consider weather normalization of forecasts and data. If a forecast of future load were an attempt to project peak load as accurately as possible, weather normalization would adjust the forecast to average (most likely) annual peak day conditions. Yet typically, forecasts for electric system planning are normalized to "worst weather expected in ten years" or similar weather-related criteria used in planning (i.e., hottest summer, coldest winter).

This is done because the power system is not designed to "just get by" under average conditions, but instead to "just get by" under the harshest conditions that are reasonably expected every ten years. The forecast is used to predict load under these weather conditions, which are not the most likely, but which will occur with enough frequency to be important. On average, such weather will develop only one year in ten, so that on average, the forecast will be high nine years out of ten. Numerous methods to do such weather normalization exist, and are covered elsewhere (Willis, 1996).

In addition, the ability to show how the forecast may vary as assumptions about future conditions are changed -- a multi-scenario capability -- is useful in all long-range planning, and essential in some situations. The key is to be able to represent the conditions under which growth and load occur.

The author refers to the types of accuracy needed for planning described here as *representational accuracy* -- a forecast's ability to represent load growth under the conditions needed for planning criteria and goals. A forecast method must be judged against such planning forecast needs, not actual load development. Statistics and error measures that evaluate a forecast method's accuracy directly against actual load are useful, but do not measures its ultimate value as a planning tool. What is most important is if it can *represent accurately* load under conditions that are specified as part of the T&D planning scenario and criteria.

Forecasting T&D Load

Corporate Forecast

Almost all utilities maintain a division or group within the rates department whose sole function is to forecast future sales and revenues, usually by application of comprehensive econometric models to customer and historical sales data. Regardless of how it is performed, this activity produces the forecast of most interest to the company's executives -- the forecast of future revenues on which budgeting and financial planning are based. All forecasts used for electric system planning must be based upon or corrected to agree with this forecast. There are several reasons for this recommendation.

First, based on the spatial forecast and subsequent steps in the planning process, the T&D planners will ask their executives to commit future revenues to T&D expenses. The executives have a right to expect that the forecast and plan that is telling them they must spend money (T&D planning forecast) is based upon the same assumptions and growth as the forecast that tells them they will have that money to spend (revenue forecast). Second, most rate and revenue forecasts of customer count, energy sales, and peak loads for the system as a whole are as accurate as anything the T&D planners could produce themselves. Why re-do all that work?

However, it is important to bear in mind that the rate and revenue forecast was developed under a different set of representational goals and within a different context. Adjustment of its values to account for coincidence of area peaks, differences in weather correction requirements, and other similar factors may be necessary. Such adjustments are valid if they are fully documented and can be justified.

Forecast of Magnitude in Space and Time

The amount of future peak load is the primary forecast quantity for T&D planning. Locations of this demand are important in order to determine the sites and routes for facilities; timing is important in order to determine the required schedule of additions. T&D equipment must meet peak demands on a local basis, not just satisfy average demand. Therefore, a spatial forecast must project expected peak demands in each small area *throughout* the system.

Spatial Analysis Requirements

The amount of "where" information, or spatial resolution in a forecast must match the T&D planning needs: Locations of future load growth must be described with sufficient geographic precision to permit valid siting of future T&D equipment. Regardless what type of small areas are used (Figure 15.2), they must provide the required spatial resolution. Reliable location of load to within ten square mile areas might be sufficient for a particular planning purpose. Then again, another might require location of future load to within 1/2 square mile (160 acre) areas. There are analytical methods that can determine the required spatial resolution based on T&D equipment characteristics and load densities (Willis, 1983 and 1996).

Coverage of the Entire Region

While T&D planning applications might require forecasts only for high growth areas where augmentation or new construction will be required, a small area forecast of the entire service area is recommended. First, this permits the multitude of small area forecasts to be summed and compared to statistics on the total system -- a reasonable way to check against the Rate and Revenue forecast. Second, a spatial forecast that covers the entire service territory is a step toward assuring that nothing is missed in the utility's T&D planning: growth doesn't always occur where expected, and a system-wide small area forecast often catches trends before they become apparent. In addition, some of the best (most accurate, most flexible) spatial forecasting methods work only when projecting the growth of all small areas over a large region: they will cover the entire service area even if only a portion is of interest.

Temporal Detail

Load forecasts for T&D planning must project peak annual demand. In addition, many T&D planning situations will require forecasting of seasonal peak loads. Spatial analysis often reveals that while the system as a whole has a peak in winter, certain areas peak in the summer. For this reason, projecting both summer and winter peak loads is common in spatial forecasting.

Forecast of hourly loads for the peak day -- 24 hourly loads for the peak day may be desirable in some cases; it permits analysis of the timing of coincidence of peaks among different areas of the system, as well as analysis of the length of time demand expected to be at or near peak -- important where capacity margin is slim.

Annual energy (the area under the entire 8760 hour annual load curve) is an important additional element, because it is useful in determining the economic sizing of equipment -- a part of minimizing the system cost of losses. Knowledge of the general characteristics of the annual load curve shape -- in particular its annual load and loss factors -- is useful in computing these economic capacity targets and for estimating the cost of losses accurately.

Forecasting T&D Load

Table 15.1 Required Characteristics of Spatial Load Forecasting Methods Used in T&D Planning

Characteristic		Percent of applications
Forecast annual peak		100%
Forecast off-season		60%
Forecast total annual energy		66%
Forecast some aspects of load curve shape (e.g., load factor)		50%
Forecast peak day(s) hourly load curves		35%
Forecast hourly loads for more than peak days		25%
Forecast 8760 hour load curves		<1%
Power factor (KW/kVAR) forecast		20%
Base forecasts updated	- at least every three years	100%
	- at least every year	85%
	- at least twice a year	20%
Forecast covers period	- at least three years ahead	100%
	- at least five years ahead	95%
	- at least ten years ahead	66%
	- at least twenty years ahead	33%
	- beyond twenty years ahead	20%
Spatial forecast "controlled by" or adjusted so it will sum to the corporate forecast of system peak (coincidence adjustment may be made)		85%
Forecasts normalized to standard weather, economy		75%
Customer classes forecast in some manner		35%
End-use usage forecast on small area basis		20%
DSM impacts forecast on a small area basis		15%
Small area price elasticity of usage forecast		5%
Customer "value base" (need for reliability) forecast		1%
Multiple-scenario studies done in some manner		66%

Forecasts done for DSM planning invariably need hourly data for the peak day and some off-peak days (or longer) periods, both to identify when the peak occurs as well as to help assess how long and how much load must be reduced to make a noticeable impact on expected T&D expansion costs. On very rare occasions, as when planning distributed generation for local peaking purposes, a forecast for all 8760 hours in the year may be required.

Customer Class Identification

The type of load (customer class) is often an important factor in T&D planning, particularly in any study that involves DSM assessment. Traditionally, basic distinctions of customer class (residential, commercial, industrial) have been used by distribution planners because they broadly identify the expected load density (kW/acre), load factor, equipment types, and power quality issues on an area basis. Spatial forecast methods based on forecasting customer type and density have been used since the 1930s and computerized since the middle of the 1960s (see Engel, et al., 1992).

Today many spatial forecasts distinguish among sub-classes within residential (apartments, small homes, large homes), commercial (retail, offices by low-rise and hi-rise, institutional), and industrial (various classes and purposes), typically using between nine and 20 customer classes.

Table 15.1 lists a number of categories that are important in T&D forecasting, along with the percentage of utilities where it is an aspect of planning. These values are based on the author's experience and opinion and, while approximate, are indicative of general requirements and application.

15.4 TRENDING METHODS

Trending methods extrapolate past load growth patterns into the future. The most common trending method, and the method most often thought of as representative of trending in general, is multiple regression used to fit a polynomial function to historical peak load data and extrapolate that function into the future. This approach has a number of failings when applied to spatial forecasting, and a wide variety of improved methods have been applied to extrapolate load for T&D forecasting, some involving modifications to the polynomial-regression approach, others using completely different approaches.

Trending Using Polynomial Curve-fitting

Trending encompasses a number of different forecasting methods that apply the same basic concept -- predict future peak load based on extrapolation of past

Forecasting T&D Load

historical loads. Many mathematical procedures have been applied to perform this projection, but all share a fundamental concept; they base their forecast on historical load data alone, in contrast to simulation methods, which include a much broader spectrum of data.

Most utility planners and forecasters are familiar with the concept of curve-fitting -- using multiple regression to fit a polynomial function to a series of data points, so that the equation can be used to project further values of the data series. Not surprisingly, this technique has been widely used as a distribution load-forecasting method. Generally, it is applied on an equipment-area basis (see Figure 15.2) such as substations or feeders.

In general, the curve-fit is applied to extrapolate annual peak loads. There are two reasons for this. First, annual peak load is the value most important to planning, since peak load most strongly impacts capacity requirements. Second, annual peak load data for facilities such as substations and feeders is usually fairly easy to obtain -- most electric utilities maintain readings on maximum feeder, substation bank and major customer loads on an annual basis.

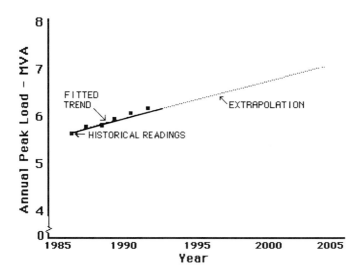

Figure 15.11 Trending methods project future load values by trending -- extrapolating the long-term trend of past load values into the future.

Consider a planner who has annual peak load data on each of his company's substations, going back for the past ten years. He wants to forecast future loads by trending -- finding a polynomial equation that fits each substation area's historical load data and then extrapolating that equation to project future load growth into the future. There are a wide number of polynomials that he could use for the curve-fit, but among the most suitable for small area load forecasting is the four-term cubic equation,

$$L_n(t) = a_n t^3 + b_n t^2 + c_n t + d_n \qquad (15.1)$$

$$= \text{annual peak load estimate for substation n for year t}$$

where n indexes the substation areas, n = 1 to N areas.

t indexes the year, beginning with t =1 for the first year of load history.

a_n, b_n, c_n, d_n are the coefficients of the particular polynomial for substation n.

The coefficients that fit this equation to a particular substation's load history can be determined using multiple regression applied to the substation's load history. For each substation, n, the technique can find a unique set of coefficients, a_n, b_n, c_n, d_n.

All the substations could share the same basic equation, but the coefficients will vary from one feeder to another, tailoring the equation to each particular substation's load history. Alternatively, a forecaster could apply a different equation to each substation: a cubic polynomial to one, a second order to another, a forth order to a third.

In the case discussed here, multiple regression curve-fitting would begin with a parameter matrix ten elements high (for the ten years of data) by four elements wide (for the four coefficients to be determined). If the same polynomial equation is being fitted to each substation area, then this matrix is constant for all, even though one would determine different coefficients for each substation.

Each column in this matrix is filled with the values of its particular parameter. For example, the first column is filled with the values 1 through 10

Forecasting T&D Load

(because there are ten years, t = 1 to 10) cubed, because the first parameter of the polynomial is a cubic term, t^3. The second column is filled with one through ten values, squared, since the second parameter is a squared term, and so forth.

$$P = \begin{vmatrix} 1 & 1 & 1 & 1 \\ 8 & 4 & 2 & 1 \\ 27 & 9 & 3 & 1 \\ 64 & 16 & 4 & 1 \\ 125 & 25 & 5 & 1 \\ 216 & 36 & 6 & 1 \\ 343 & 49 & 7 & 1 \\ 512 & 64 & 8 & 1 \\ 729 & 81 & 9 & 1 \\ 1000 & 100 & 10 & 1 \end{vmatrix}$$

Substation n's annual peak loads for the past ten years are placed into a matrix ten elements high by one column wide:

$$L_n = \begin{vmatrix} l_n(1) \\ l_n(2) \\ l_n(3) \\ l_n(4) \\ l_n(5) \\ l_n(6) \\ l_n(7) \\ l_n(8) \\ l_n(9) \\ l_n(10) \end{vmatrix}$$

The coefficients a_n, b_n, c_n, d_n that best fit the polynomial to the load history are determined by the matrix equation

$$C_n = [P^T P]^{-1} \cdot P^T L_n \tag{15.2}$$

Once the actual values of the coefficients, are determined, they can be used to project future load, merely by placing them in equation 11.1, and solving it for values of t greater than ten. Solving with a value of t = 11 gives the projected value for the year following the last historical data point, t = 12 gives the value for two years beyond, and so forth, producing a projection of the future loads year by year.

Curve-fitting does not necessarily need to use consecutive years of data, either. Going back to the original cubic equation, 11.1, suppose that of the ten years of past data, the third and fourth year for a particular substation are missing. Those can simply be left out of that substation's curve-fit analysis. In this case, the L and P matrices must both be changed to have eight instead of ten rows. L becomes

$$L_n = \begin{vmatrix} l_n(1) \\ l_n(2) \\ l_n(5) \\ l_n(6) \\ l_n(7) \\ l_n(8) \\ l_n(9) \\ l_n(10) \end{vmatrix}$$

and P becomes an 8 by 4 matrix, missing the two rows for years 3 and 4:

$$P = \begin{vmatrix} 1 & 1 & 1 & 1 \\ 8 & 4 & 2 & 1 \\ 125 & 25 & 5 & 1 \\ 216 & 36 & 6 & 1 \\ 343 & 49 & 7 & 1 \\ 512 & 64 & 8 & 1 \\ 729 & 81 & 9 & 1 \\ 1000 & 100 & 10 & 1 \end{vmatrix}$$

The other steps in the calculation remain precisely the same. In order to apply the curve-fit to all N substations in a particular utility's service territory, it is necessary to perform only this calculation on each substation. A computer program to perform this procedure can be quite simple.

Forecasting T&D Load

What polynomial is best?

In the example given above, a four-coefficient cubic equation was used, but multiple regression will work with any polynomial equation, as long as the number of data points (years of load history) exceeds the number of coefficients.

For example, instead of the equation in the earlier example, the equation shown below could be used:

$$L_n(t) = a_n t^3 + b_n t^2 + c_n t + d_n + e_n t^{-1} \qquad (15.3)$$

in which case five coefficients must be determined in the curve-fitting, not four. The P matrix could change, becoming ten by five elements wide, and the coefficient matrix, C, will have five elements, instead of four, but otherwise the matrix equation for the solution, and any computerized procedure, will be as outlined earlier.

Any polynomial, with any number of coefficients, can be fit to the historical data, as long as the number of data points exceeds the number of coefficients. In all cases, matrix P has a number of rows corresponding to the number of years of historical data, and a number of columns corresponding to the number of coefficients.

Regardless of the type or order of the polynomial, the multiple regression method determines the set of coefficients that minimizes the fitted equation's RMS error in fitting to the historical data points. The RMS error is the sum of squares of the errors between the equation and the data points, as shown in Figure 15.12. RMS tends to penalize large residuals (difference between fitted point and actual point) more than small ones, and thus forces any minimization method to find coefficients that may give many small errors but few large ones.

For example, if the coefficients found for equation 11.3 are 3, 2, 1.5, 4, and 6, then

$$L(t) = 3t^3 + 2t^2 + 1.5t + 4 + 6/t \qquad (15.4)$$

will pass through the data points with minimum RMS error.

It is important for any user of polynomial curve-fit to realize just what "minimum fitting error" means in this context. First, the error is minimized only within the context of the particular polynomial being curve-fit -- no other set of coefficients will do as well with *this particular polynomial*, but another polynomial, perhaps one with the term t^4 substituted for t^3, might give less fitting error.

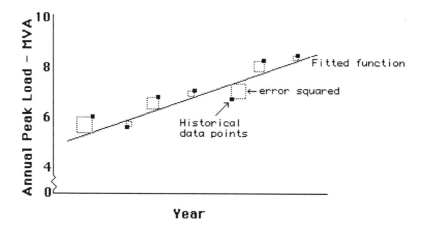

Figure 15.12 Multiple regression curve-fitting determines a function that passes through the data points, minimizing the sum of the squares of the error between the function and the data points.

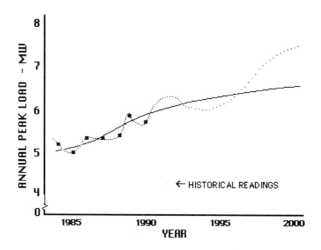

Figure 15.13 A fitted function's ability to pass close to historical data points is no guarantee that it will be more accurate than another curve that has a higher fitting error. A high order equation (dotted line) fits the historical data shown much better than a third order equation (solid line), but clearly provides a poorer forecast.

Forecasting T&D Load

In fact, fitting error can almost always be reduced by using a higher order equation. But minimizing fitting error in this manner means nothing in terms of actual forecasting ability, as shown in Figure 15.13. Forecast accuracy is related to the equation's ability to predict *future* values closely, not pass through past ones. In actual forecasting situations, equations with more than four terms generally perform worse than third-order equations such as equation 11.1 (see Electric Power Research Institute, 1979; Meinke, 1979).

Beyond this, the reader should note that numerical methods that minimize RMS fitting error are popular because they are easy to apply, not because RMS is inherently the best error measure to minimize. The use of methods that minimize RMS is so widespread that many people are unaware that other fitting measures exist or can be applied. There *are* curve-fitting methods that minimize the sum of the errors (the mean error, not the root mean square error) and other fitting error measures.[2] However, none of these is nearly as easy to compute as the standard regression method which minimizes RMS, and some are much more involved. Thus, they are seldom used.

Research into trending applications has shown that no equation works best in all situations. However, the cubic log equation is slightly superior to others in general applications.

How many years of data?

A considerable body of research has shown that when working with typical distribution data and cubic or cubic log polynomials and fitting to only historical data, *the most recent six years of data* give slightly better results than any other historical sample, including seven, eight, or *more* years of data.

Improvements to Multiple Regression Curve-fitting

Trending is poor at high resolution spatial forecasting

The spatial forecast error associated with a multiple regression polynomial curve-fit -- or almost any trending method for that matter -- increases very rapidly as small area size is decreased. Trending methods generally do not do well when applied to small area forecasting, largely because they cannot handle the S curve growth behavior well.

---

[2] The author's favorite for many applications is an "R^0" curve-fit method that minimizes only the *maximum* residual -- it solves for the polynomial coefficients that give the smallest possible value for the largest value of the residuals.

Trending has a basic incompatibility regarding small area analysis. In any form in which it can be applied, trending implicitly assumes that the character of the system being extrapolated remains the same. If the boundaries of a small area remain the same, if the load growth doesn't pass through one or more of the S curves transitions as it grows, then nothing changes in the character being extrapolated, and the trending method is valid, and gives generally satisfactory results.

But when the service areas, growth characteristics, or controlling factors change, as they often do at the small area level, trending encounters one or more of three problems:

1) "S" curve behavior is difficult to trend. As described earlier, small area growth behavior undergoes two transitions in behavior, from "dormant" to "growing," and from "growing" to "saturated." Both are associated with a major change in the slope of the expected growth curve.

2) When applying trending on a uniform grid basis, typically a good deal of growth occurs in "vacant" areas (those without any load history). Without no extrapolatable data from which to work, trending is inaccurate in forecasting future load growth in such areas.

3) Service areas may change. If applied on an equipment-area basis, all areas will have some history to trend, but historical data may include changes in service area, due to load transfers. For example, when a new substation is built, its service area is "cut out" of existing substation service areas by transferring load to it, away from its neighbors. This disturbs the historical trends, making forecasting difficult.

These three problems can be partially rectified by using modifications to the basic regression based curve-fit approach outlined here.

Horizon year load data

Figure 15.14 shows that a polynomial curve-fit to historical data can yield quite different results, depending on exactly where in the S curve pattern of load growth the load history happens to lie. In many cases, the resulting forecasts

Forecasting T&D Load

obtained by applying curve-fitting at the feeder level or below can be truly ridiculous, with negative loads.

One way to reduce this type of overextrapolation is to use what are called *horizon year loads* -- estimated future load values put into the data set to be curve-fit (i.e., treated like the historical data). Very simply, these are the *planner's guess* at the eventual load level -- the value that load growth will eventually reach, in another 15 or 20 years. Nothing else in the multiple regression method changes. In the computation, the horizon year loads are treated exactly like historical data, and the coefficients are determined and the fitted function applied precisely as outlined earlier.

Horizon year estimates improve short-range forecast accuracy considerably, even if the estimates are not highly accurate. Long-range accuracy is not improved much by the horizon year loads, because long-range values depend mostly on the horizon-year estimates, which are random, or judgmental. However, since trending is seldom used for long range planning, this is not a serious limitation. *Horizon year load estimates are recommended in all trending applications.*

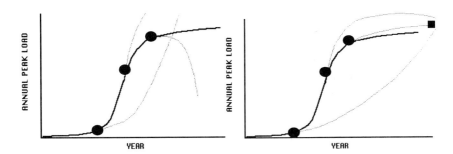

Figure 15.14 The "S curve" growth behavior seen at the small area level means that trending results will be very dependent on how much load history is available. Among the many ways this deficiency can be addressed, use of horizon year loads is the simplest effective measure. Here, far different forecasts result depending on which of three points (round dots) in the S curve are used as the most recent history of load growth. At the right, applying a horizon- year load (square dot) improves forecasting -- while the three trends shown are not highly accurate, they are much better than those at the left.

Projecting load growth with hierarchical vacant area inference curve-fitting

One of the major problems with trending is that it cannot establish any trend in a "vacant area" -- one where no load history exists. When faced with this "historical trend" extrapolation methods yield only one result, a flat projection whose value remains zero. Yet very often development will start in a vacant area and a significant load will develop rapidly. (All small grid areas with load in them were vacant at some time in the past).

This particular problem is encountered in grid-based approaches, but seldom in equipment-oriented forecasts. The reason is that in an equipment-oriented forecast there are usually no vacant areas -- all parts of the service territory are assigned to some substation or feeder, even if that area is vacant. Using non-zero horizon load estimates provides a clue to trending, indicating that the load in a vacant area is expected to grow eventually, but lead to very poor forecasts. The forecast is completely a function of the horizon year load, and very sensitive to its value.

Vacant area inference (VAI), which is based on the simple idea illustrated in Figure 15.15, can improve trending results in these situations a good deal. In that diagram, one of four neighboring small areas has no load history, and therefore cannot be forecast directly. The other three small areas have a load history, and can be trended. The steps in applying the VAI in this case are:

1. Apply trending, with a horizon year load, to forecast each of the three small areas that have a load history.

2. Add the load histories of *all four* small areas together. Since the vacant area has no load history, this is the same as adding the load histories of only those three that do. Note that this sum of their load histories is the load history of the entire block (all four).

3. Apply multiple regression to extrapolate the load history sum, (this is the sum of the load histories of the entire block), using as the horizon year load value the sum of *all four* small area's horizon year loads. The resulting trend is the forecast load for the entire block.

4. Subtract from this "block forecast," the projections for each of the three small areas that were forecast in step 1. What remains is the projection for the cell with no load history.

Forecasting T&D Load

The VAI method results in a reasonable improvement in forecast accuracy over the basic trending method, when applied to forecasts where a significant amount of growth might occur in vacant areas. Equally important, the VAI concept can quite easily be worked into a computer program that automatically performs the above steps on all vacant cells in a small area grid. The method is a hierarchical top-down procedure, which begins with a forecast of the entire service territory, obtained by adding all the small area load histories and extrapolating their sum.

The VAI method then breaks the small area data set into four quadrants, forecasting each by adding together the load histories in only that quadrant, and extrapolating that summed trend. It then subdivides each quadrant by four and repeats the process, and so forth. If at any time the method encounters a quadrant or small area with an all-zero load history, it applies the four steps above to infer its load history. The method continues subdividing by quadrants until it reaches the small area level, at which point it stops. For more details see Willis and Northcote-Green (1982).

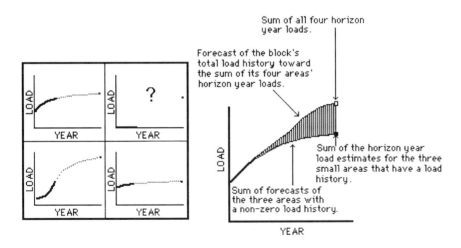

Figure 15.15 Vacant area inference is most easily accomplished as shown here. The vacant area is trended in combination with a block of its neighbors that have load histories. The vacant area's forecast, corresponding to the shaded area above, is the difference between the two trends obtained by: (1) trending the block's load history toward the horizon year load estimates of the three areas that have a load history, and (2) trending it toward the sum of all four horizon year loads. The vacant area's forecast accuracy is thus quite sensitive to the quality of the horizon year estimates, but in all cases it is an improvement over not taking this approach.

Load transfer coupling (LTC)

Most often, trending is applied to equipment-oriented areas, such as feeders or substations. In such cases, it is plagued by errors due to load transfers hidden within the historical data, as illustrated in Figure 15.16. The load histories of many feeders and substations in most utility systems contain such transfers. This is particularly true in growing areas of the system -- those areas where accuracy is paramount. Any new feeder or substation has a significant load from day one; that load did not simply spring into being on the day the substation was put into operation, but instead, was transferred to the new facility by re-switching loads from neighboring substations or feeders. As a result, both the new substation and the neighboring ones have significant switching shifts in their load histories.

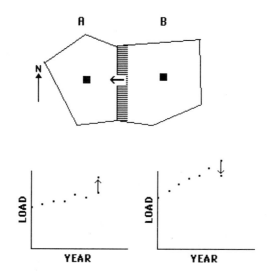

Figure 15.16 A transfer of load from substation B to its neighbor (map at top of figure) is accomplished by switching the service in the area shown from feeders emanating out of substation B to feeders served by A, and is done in order to keep peak load at substation B within its capacity limit. This transfer impacts the load histories, increasing the load at A (left) and decreasing the load for B (right), changes that are intermingled with any load growth. Trending cannot distinguish the cause of any change in load -- it responds to the changes wrought by the load transfer, treating it as if that change should be extrapolated.

Forecasting T&D Load 647

To compound this problem, there may be more than one transfer in the load history of any particular substation or feeder. These cause significant forecasting errors, because changes in trends due to transfers are treated by extrapolation in exactly the same manner as changes due to growth. Historical data for any T&D system is usually rife with load transfers. It is often difficult, and sometimes impossible, for the planner to obtain accurate historical data on load transfers, and more difficult still to remove them from historical data, to the point this is usually not practical.[3]

It is possible to solve the multiple regression extrapolation for two or more areas simultaneously, in a way that identifies the areas as subject to a possible transfer so that the extrapolation removes most of the forecast degradation caused by the transfer (Willis et al, 1984). What is particularly useful with this method, called Load Transfer Coupling (LTC) regression, is that it does not need to be given the amount of load that was transferred, or even its direction (which area was from and which was to). LTC can be applied as an "automatic procedure" that detects and removes load transfers impact from a trending forecast. Details on its implementation are given elsewhere (Willis, 1996).

Trending Methods That Do Not Use Regression

Multiple regression is not the only method of curve-fitting that can be applied to small area load forecasting, although it is by far the most widely used (due to its ease of implementation, not necessarily any superiority in results). Many other mathematical approaches have been applied. Many produce results no better than multiple regression curve-fitting. Generally, those that do so gain an improvement through one of two approaches, either by forcing all extrapolated curves to have something close to an S curve shape, or by using some other trend beyond just load history to help the extrapolation.

Template matching pattern recognition

"Template matching" method is a unique S curve extrapolation method that functions without curve-fitting or regression. Rather than extrapolate load

---

[3] In theory, records are maintained on every load switched, but in practice assembling the data requires a great deal of time. Very often the exact load transferred is not recorded, or if available those data are of little value. For example, a load switch may have been made in the spring, months before summer peak. Operations personnel may have recorded that they transferred 800 kVA between two feeders, based on a reading taken at the time. How much load does this represent at peak? 900 kVA? 1000 kVA? Any estimate is only that, an estimate.

histories, template matching tries to forecast a small area's load by comparing its recent load history to "long-past" histories of other small areas. Figure 15.17 illustrates the concept -- small area A's recent load history (last six years) is similar to the six years of load history that occurred 16 years ago in small area B. Therefore, small area B's past load growth trend is used as the forecast trend (template) for the future of small area A's load growth.

Template matching gives no better forecast error than regression-based curve-fitting, but has two advantages that are sometimes important: it is very insensitive to missing or erroneous data values, and it can be implemented in only 16-bit integer arithmetic with few multiply or divide functions. Its only disadvantage is that it requires much longer load histories -- which is seldom a problem Its numerical simplicity made it quite popular in the mid 1980s, as it was ideally suited to the limited capabilities of the first generation of personal computers, such as the original Apple. While it continues to be widely used on older PCs by many utilities worldwide, the capability of modern PCs of even modest cost to handle large floating point matrix calculations makes its computational advantages negligible (Willis and Northcote-Green, 1984).

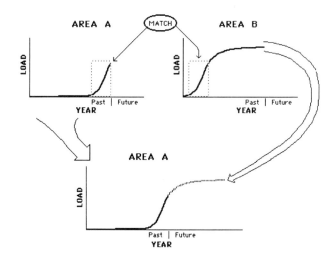

Figure 15.17 The template matching concept. Small area A's recent load history is found to match small area B's load history of many years ago. Small area B's load history is then used as the "template" for area A's forecast.

Forecasting T&D Load

Multivariate Trending

In some forms mutlivariate trending does use regression -- lots of it -- but its major difference from other trending methods is its non-regression-based steps. Many trending methods attempt to improve the forecasting of electric demand by extrapolating it in company with some other trends, such as customer count, gas usage, or economic growth. The concept behind *multivariate trending* is:

a. Establish a meaningful relation between the variables being trended (e.g., electric load is related to number of customers).

b. Trend both variates subject to a mathematical constraint that links them using the relationship established. In this way, the trend in each variate affects the trend in the other, and vice versa.

The larger base of information (one has two historical trends and an identified relationship between the two variables, so that the base of information is more than doubled) leads to a better forecast of each variate. Many electric forecasting methods have been developed with this approach, and multivariate methods forecast with noticeably lower error levels than any type of polynomial curve-fitting to historical peak load data. Perhaps the most ambitious and well-documented multivariate method applied to spatial forecast was a computer program called *"Multivariate"* developed as part of EPRI project RP-570 (see Electric Power Research Institute, 1979).

Despite the accuracy advantage, multivariate trending never enjoyed widespread use. First, it never matched the accuracy possible from simulation methods. Second, it does not have the numerical simplicity and use ofeconomy of other trending methods, being as expensive to apply as simulation methods.

Geometric and cluster-based curve-fit methods

Beginning in the mid to late 1980s, research on trending methods turned away from multivariate methods, and concentrated on developing improved methods that preserved the traditional simplicity and economy of trending, while improving forecast accuracy, particularly in the short-range T&D planning period (1 to 5 years ahead), where multi-scenario capability is not a big priority. This led to a number of trending methods with substantially better forecast accuracy than regression-based curve-fit. In fact, the best of these methods can match the forecasting accuracy of multivariate approaches, while retaining most of the simplicity and economy of operation sought in trending methods.

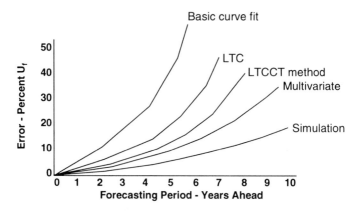

Figure 15.18 Relative accuracy of various trending methods in forecasting feeder-level capacity needs, as a function of years into the future from a utility in the southern United States, 1985 to 1995. All methods suffer from exponentially increasing levels of forecast error as it is extended further into the future, but the rate of error increase with time varies considerably.

Typical of these newer trending methods is the LTCCT forecast method developed by the author and two colleagues (see Willis, Rackliffe, and Tram, 1992). LTCCT (Load Transfer Coupled Classified Trending) combines geometric analysis, LTC regression, template matching, clustering, horizon year loads, and regression-based curve-fitting in one algorithm. Several other trending methods have been developed with similar "combinations" of regression, clustering, and template matching, and give similar performance improvements. All give roughly equivalent results. Figure 15.18 compares this method's forecast accuracy with several others

15.5 SIMULATION METHODS FOR SPATIAL LOAD FORECASTING

Simulation-based methods attempt to reproduce, or model, the process of load growth itself in order to forecast where, when, and how load will develop, as well as to identify some of the causes driving that growth. Unlike trending, simulation is ideally suited to high spatial resolution, long-range forecasting, and to the needs of multi-scenario planning. Most important, when applied properly, it can be much more accurate than the very best trending techniques.

Forecasting T&D Load

However, simulation methods require more data and more user involvement than trending methods. While data collection is seldom a burden given the capabilities of modern computing systems to gather and manipulate land-use data bases, geo-coded demographic data, satellite imagery, and a host of other sources, simulation methods do require three to ten times the effort of trending methods. In many cases, this additional cost is more than justifiable, due to their greateHr accuracy and representational precision.

De-coupled Analysis of the Two Causes of Load Growth

Simulation addresses the two causes of load growth by trying to model, or duplicate, their processes directly but *separately*. As was described in section 11.2, electric load will grow (or decrease) for only two reasons:

>Change in the number of customers buying electric power

>Change in *per capita* consumption among customers

Simulation addresses possible changes in customers and possible changes in *per capita* consumption using separate but coordinated models of each. This is quite a contrast to the trending techniques discussed in section 15.4, which treated all changes in load as of the same cause (just a change in load, without explanation beyond the fact that it happened).

	Customer Growth	Per capita Growth
Spatial Analysis	Handled by spatial land-use model	not done
Temporal Analysis	not done	Handled by temporal end-use model

Figure 15.19 Simulation applies separate models to the analysis of customer count as a function of location and customer *per capita* consumption of electricity as a function of time. It thus "de-couples" load growth analysis in two ways, by cause (customers, *per capita*), and by dimension (spatial, temporal).

There is a further distinction of scope of analysis in almost all simulation methods. One part of the model handles the spatial analysis, the other all the temporal (hourly, seasonal) analysis. Customer modeling is done on a *spatial* basis, tracking where and what types and how many customers are located in each small area. The *per capita* analysis does not consider location, only *temporal* variation in usage as a function of customer type, end-use, and time of day, week, and year. Thus, simulation de-couples not only the causes of load growth, but also the dimensions of the forecast, as shown in Figure 15.19.

Land-Use Customer Classes

Simulation methods distinguish customers by *class*. Both the modeling of customers and the modeling of *per capita* usage are done on a customer class basis with various residential, commercial, and industrial classes and subclasses (Table 15.2). The customer and per-capita models are coordinated by using the same customer classes in each. Classification of customers by type allows both the customer and *per capita* models to distinguish customers by different types of behavior. Three basic important types of behavior can be distinguished:

> *Rate classes.* Land use classes are closely related to the rate classes used in customer billing systems, making it easy to interface them with utility customer, rate, and load research data.

Table 15.2 Typical Land Use Types Used in Simulation (Tram, 1983)

Class	Definition
Residential 1	homes on large lots, farmhouses
Residential 2	single family homes (subdivisions)
Apartments/townhouses	apartments, duplexes, row houses
Retail commercial	stores, shopping malls
Offices	professional buildings
High-rise	tall buildings
Industry	small shops, fabricating plants, etc.
Warehouses	warehouses
Heavy industry	primary or transmission customers
Municipal	city hall, schools, police, churches

Forecasting T&D Load 653

Per capita consumption classes. Residential, commercial, and industrial customers differ in how much and why they buy electric power, in daily and seasonal load shapes, and in growth characteristics and patterns.

Spatial location classes. Residential, commercial, and industrial customers seek different types of locations within a city, town, or rural area, in patterns based on distinct differences in needs, values, and behavior, patterns that are predictable. The term "land-use" is often applied to simulation methods because variability of land usage is the most visible and striking of the three types of distinctions made by class definitions used in forecasting models.

The spatial patterns of load use development are the basis for the spatial forecasting ability of simulation methods. Within any city or region, land is devoted to residential use (homes, apartments), commercial use (retail strip centers, shopping malls, professional buildings, office buildings and skyscrapers), industrial use (factories, warehouses, refineries, port and rail facilities), and municipal use (city buildings, waste treatment, utilities, parks, roads). Each of these classes has *predictable* patterns of what type of land they need in order to accomplish their function and explainable values about why they locate in certain places and not in others. Their locational needs and preferences can be utilized to forecast where new customers will most likely locate in the future.

For example, industrial development is very likely to occur alongside existing railroads (93% of all industrial development in the United States occurs within 1/4 mile of an existing railroad right-of-way). In addition, new industrial development usually occurs in close proximity to other existing industrial land-use. These two criteria -- a need to be near a railroad and near other industrial development -- identify a very small portion of all land within a utility region as much more likely than average to see industrial growth. Addition of a number of other, sometimes subtler factors can further refine the analysis to where it is usually possible to forecast the locations of future industrial development with reasonable accuracy.

By contrast, residential development has a tendency to stay away from railroads (noise, pollution), and industrial areas (poor esthetics, pollution). But residential development has its own list of preferences, too -- close to good schools, convenient but not too close to highways, near other residential areas -- that can be used to further identify areas where residential development is most likely. This list is different from industrial's preference list, but just as distinct and just as applicable to forecasting the locations and patterns of growth.

In fact, every land-use class has identifiable, and different, locational requirements. No American would be terribly surprised to find tall buildings in the core of a large city, or a shopping mall at the suburban intersection of a highway and a major road. On the other hand, most people would be surprised to find low-density housing, large executive homes on one to two acre lots, being built at either location. It wouldn't make sense.

Modern simulation methods project future customer locations by utilizing quantitative, spatial models of such locational preference patterns on a class basis to forecast where different land uses will develop (Willis and Tram, 1992). They use a separate forecast of *per capita* electric usage, done on the same class definitions, to convert that projected geography of future land use to electric load on a small area basis. In so doing, they employ algorithms that are at times quite complex and even exotic in terms of their mathematics. But the overall concept is simple, as shown in Figure 15.20: forecast where future customers will be, by class, based on land-use patterns; forecast how much electricity customers will use and when by class; then combine both forecasts to obtain a forecast of the where, what, and when of future electric load.

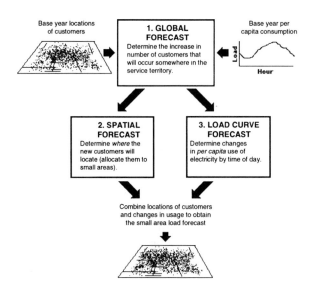

Figure 15.20 Simulation methods project customer locations and *per capita* consumption separately, then combine the two to produce the final small area forecast.

Forecasting T&D Load

Overall Framework of a Simulation Method

The land-use based simulation concept can be applied on a grid or a polygon (irregularly shaped and sized small areas) basis. Usually, it is applied on a grid basis, which the author generally recommends. A grid basis assures uniform resolution of analysis everywhere. In addition, some of the high-speed algorithms used to reduce the computation time of the spatial analysis work only if applied to a uniform grid of small square areas.

Simulation methods are generally iterative, in that they extend the forecast in a series of computational passes, transforming maps of customers and models of *per capita* consumption from one year to another as shown in Figure 15.21. Forecasts covering many future years are done by repeating the process several times. An iteration may analyze growth over only a single year (i.e., 1997 to 1998) or may project the forecast ahead several years (1998 to 2001).

Simulated land-use transitions mimic the "S" curve shape

Simulation methods have proven quite good at forecasting vacant area growth. In fact that is among their greatest strengths compared to other spatial forecasting methods. However, they are not particularly good at forecasting gradual, incremental growth on a small area basis -- as in representing that load

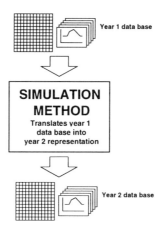

Figure 15.21 Simulation is applied, as shown, transforming its data base from representation of year t to a projection of customers and consumption in year t+p.

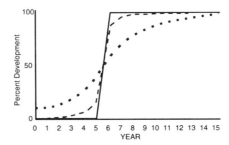

Figure 15.22 Most simulation methods forecast growth as step functions (solid line). This matches small area growth characteristics as seen at high spatial resolutions.

in a particular area will go from vacant to only 20% in year 1, then reach 50% developed by year 2, 80% the year after, but then growth will slow, so that it will only achieve 100% a number of years later. Instead, simulation methods forecast a single year transition from vacant to fully developed, as shown in Figure 15.22 (Brooks and Northcote-Green, 1978). Thus, they are not very suitable for low resolution spatial forecasting, where the typical small area S-curve may have a growth period (rise time) of more than ten years. Instead, they work better at high spatial resolutions (small areas of 10 acres or smaller), where typical growth rise times are short, and thus compatible with single year step functions, as shown in Figure 15.22.

Three steps are common to all simulation methods

More than 50 different computerized methods have been developed to apply the simulation method. These vary from simple and approximate to quite involved numerical approaches. However, there are some common steps that must be accomplished one way or another in the forecasting of spatial customer location:

Global customer counts. A spatial model must account for the overall total of each customer class within the utility service territory as a whole. Generally, these "global totals" are an input to the simulation method. Most utilities have a forecasting group separate from T&D planning -- usually called Rate and Revenue Forecasting or Corporate Forecasting -- that studies and projects total system-wide customer counts. The author strongly recommends that all electric forecasts used for electric system

Forecasting T&D Load

planning be based upon, driven by, or corrected to agree with this forecast in an appropriate manner, as described in section 1.3 (see Lazzari, 1965 and Willis and Gregg, 1979).

Interaction of classes. The developments of the various land-use classes within a region are interrelated with respect to magnitude and location. The amount of housing in a region matches the amount of industrial activity -- if there are more jobs there are more workers and hence more homes. Likewise, the amount and type of retail commercial development will match the amount and type of residential population, and so forth.

These interrelationships have a limited amount of spatial interaction, too. If industrial employment on the east side of a large city is growing, residential growth will tend to be biased heavily toward the east side, too. In general, these locational aspects of land use impact development only on a broad scale -- urban model concepts help locate growth to within three to five miles where it occurs, but no closer (see Lowry, 1964, and Willis, 1996).

There are myriad methods to model the economic, regional, and demographic interactions of land-use classes. All are lumped into a category called "urban models." Many are appropriate only for analyzing non-utility or non-spatial aspects of land-use development, but some work very well in the context of spatial load forecasting.

One such method is the Lowry model, which represents all land-use development as a direct consequence of growth of what is termed "basic industry" -- industry that markets outside the region being studied (Lowry, 1964). In the Lowry model concept, growth starts through the development of "basic industry," such as refineries, factories, manufacturing, and tourism. These create jobs, which create a need for housing (residential development), which creates a local market demand for the services and goods sold by retail commercial, and so forth.

Locational land-use patterns. Ultimately the spatial forecast must assign land-use development to specific small areas. While overall system-wide totals and urban model interactions establish a good basis for the forecast, it remains for the spatial details to be done based on the distinction of land-use class locational preference as described earlier in this chapter.

The determination of exactly where each class is forecast to develop has been handled with a wide variety of approaches, from simple methods utilizing a few rules interpreted by the planner's judgment, to highly specialized, computerized pattern recognition techniques that forecast land-use location automatically based upon pattern recognition and artificial intelligence methods But one way or another, a spatial forecast must ultimately assign customer class growth to small areas and whether judgmental or automatic, acknowledge and try to reproduce the types of locational requirements and priorities, and model how they vary from one class to another, as discussed earlier in this chapter.

Most simulation methods accomplish the three tasks described above with an identifiable step to accommodate each, with the steps usually arranged in a top down manner. A top down structure can be interpreted as starting with the overall total (whether input directly or forecast) and gradually adding spatial detail to the forecast until it is allocated to small areas. The forecast is done first as overall total(s) by class, then each total sum of growth for the service area is assigned broadly on a "large area basis" using urban model concepts, and finally the growth on a broad basis is assigned more specifically to discrete small areas using some sort of land-use analysis on a small area basis.

Alternatively, a bottom up approach can be used in which the growth of each small area is analyzed and the forecast works upward to an overall total, usually in a manner where it can adjust its calculations so that it can reach a previously input global total (the input Rate and Research forecast). The Bibliography has several papers that describe such algorithms in great detail.

Per capita consumption is usually forecast with some form of customer-class daily load curve model, often on an end-use basis (as discussed in Chapter 2). Regardless of approach, the *per capita* consumption analysis must accommodate two aspects of load development on a customer class basis:

Forecast of demand per customer and its coincidence with other classes. Both the actual peak per customer as well as the contribution to peak (the two may occur at different times) need to be forecast.

Future changes in the market share of various appliances, appliance efficiency, and usage patterns from year to year need to be included in the forecast. This can be done by "inputting" the class curves as previously output from a separate end-use analysis effort, or by end-use analysis within the simulation method itself (Canadian Electric Association, 1982).

Thus, the structure of most simulation methods is something like that shown in Figure 15.23.

Forecasting T&D Load

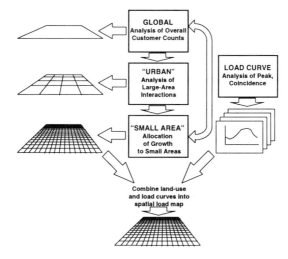

Figure 15.23 Overall structure of a simulation method. In some manner every simulation method accomplishes the steps shown above, most using a process framework very much like this one.

Land-Use Growth: Cause and Effect

This section summarizes the central tenet of the Lowry model with a hypothetical example of industrially driven growth. A city, town, or rural region can be viewed as a "socio-economic machine," built by man to provide places for housing, places to work, places to shop for food, clothing, and services, and facilities that allow movement from one place to another. Cities and towns may differ dramatically in appearance and structure, but all provide for their population's residential, commercial, industrial, transportation, and cultural needs in functional proportion to one another. Urban models are representations of how these different functions interact. There are literally dozens of different approaches to modeling how a city, town, or agricultural region functions. Some are simple enough that they do not need to be computerized to be useful, whereas others involve quite complicated numerical analysis.

To illustrate the concepts of urban growth and urban modeling, consider what would happen if a major automobile company were to decide to build a pickup truck factory in an isolated location, for example in the middle of western Kansas (a sparsely populated region in the middle of the United States), one hundred miles from any large city.

Having decided for whatever strategic reasons to locate the factory in rural Kansas, the automaker would probably start with a search for a specific site on which to build the factory. The site must have the attributes necessary for the factory: located on a road and near a major interstate highway so that it is accessible, adjacent to a railroad so that raw materials can be shipped in and finished trucks can be shipped out, near a river if possible, so that cooling and process water is accessible and also permitting barge shipment of materials and finished trucks. Figure 15.24 shows the 350 acre site selected by the automaker, which has all these attributes.

In order to function, the pickup truck factory will need workers. Table 15.3 shows the number by category as a function of time, beginning with the first years of plant construction and start up, through to full operation. Once this factory is up to full production, it will employ a total of 4,470 employees. Since there are no nearby cities or towns, the workers will have to come from other regions, presumably attracted by the prospect of solid employment at this new factory. Assume for the sake of this example that the automaker arranges to advertise the jobs and to help workers re-locate.

These workers will need housing near the factory -- remember the nearest city or town is a considerable commute away. Using averages based on nationwide statistics for manufacturing industries, the people in each of the employment categories are likely to want different types of housing in roughly the proportions shown in Table 15.4.

Using that data, one predicts that the 4,470 workers listed in Table 15.4 will want a total of 39 executive homes, 2,656 medium-sized homes, and 1,775 apartments. Using typical densities of housing type this equates to 57 acres of executive homes, 830 acres of normal single-family homes, and 291 acres of apartments, for a total of 1,178 acres (nearly two square miles) of housing.

Assume for the sake of this example that a road network is built around the site, and that the 1,178 acres of homes are scattered about the factory at locations that match both "residential land-use needs and preferences" and are in reasonable proximity to the factory, as shown in Figure 15.25.

Based on nationwide statistics for this type of industry, there will be 2.8 people in each worker's family (there are spouses and children, and the occasional parent or sibling living with the worker). This yields a total of 12,516 people living in these 1.173 acres of housing. These people need to eat, buy clothing, get their shoes repaired, and obtain the other necessities of life, as well as have access to entertainment and a few luxuries. They will not want to drive several hours or more to do so. They create a market demand, for which a

Forecasting T&D Load

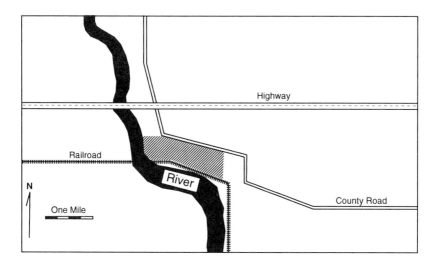

Figure 15.24 The hypothetical factory site at an isolated location (shaded area).

Table 15.3 Expected Pickup Truck Factory Employment by Year

Employment category	Year					
	2	4	8	12	20	25
Management	8	20	30	30	30	30
Professional	40	180	240	240	240	240
Skilled labor	250	1400	3100	3100	3100	3100
Unskilled Labor	300	900	1100	1100	1100	1100
TOTALS	598	2500	4470	4470	4470	4470

Table 15.4 Percent of Housing by Employment Category

Employment category	Large houses	Med-Sm. houses	M.F.H. (apts, twnhs)
Management	50	45	5
Professional	10	82	8
Skilled labor	2	70	30
Unskilled Labor	0	25	75

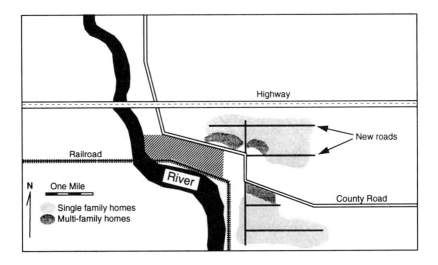

Figure 15.25 Map showing the location of the pickup truck factory, to which has been added 1,173 acres of housing, representing the residential land-use associated with the homes that will be needed to house the 4,470 workers for the factory and their families.

cross-section of stores, movie theaters, banks, restaurants, bars, TV repair shops, doctor's offices, and other facilities will develop to serve them, as shown in Figure 15.26. There will also need to be schools to educate the children, and police and fire departments to protect all those homes, and a city hall to provide the necessary infrastructure to run these facilities. And of course, there will have to be an electric utility, too, to provide the electricity for the factory, as well as for all those homes and stores.

These stores, shops, entertainment places, schools, and municipal facilities create more jobs, but there is a ready supply of additional workers. Based on typical statistics for factories in North American, one can assume that of the 4,470 factory workers, 3,665 (82%) will be married, and of those 3,665 spouses, about 2,443 (two-thirds) will seek employment. However, the stores, shops, schools, and other municipal requirements just outlined create a total of nearly 5,000 jobs (more than one for every one of the original factory jobs!) for a net surplus of 1,400 jobs unfilled.

Forecasting T&D Load

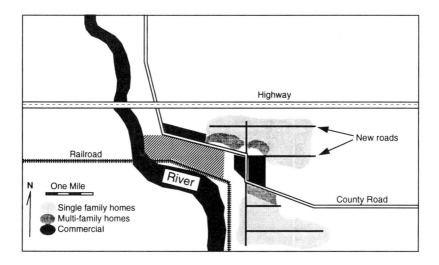

Figure 15.26 Map showing the factory, the houses for the factory workers, and the locations of the commercial areas that develop in response to the market demand for goods and services created by the factory workers.

This rapidly growing little community will need even more workers to fill these jobs. Again, using typical values:

> At 1.55 employees per family (Remember, both husband and wife work in some families), these 1,400 additional jobs mean a further 1,400/1.55 = 903 families to house, for an additional population of 2.8 × 903 families = 2,529 *more* people.

> Assuming these 903 families have a cross-section of housing demands like the original 4,470 factory workers,[4] one can calculate a need for a further 8 executive homes (12 acres), 537 normal homes (168 acres), and 358 apartments (59 acres).

---

[4] Actually, this isn't a really good assumption. The cross-section of incomes, and hence housing demands, for employees in these categories is likely to be far different than for the factory workers. However, to simplify things here, it is assumed the same net percentage of housing per person as with the factory.

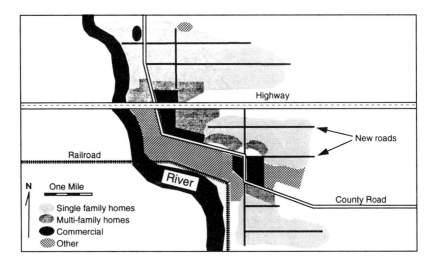

Figure 15.27 The ultimate result of the factory, a small "factory town" with housing, commercial retail, commercial offices, schools, utilities, and everything else needed to support a population approaching 20,000. This community "earns its living" building pickup trucks -- the factory is the driving force behind the entire community's economy.

These additional 2,529 people create a *further* need for shops and entertainment, police, fire, and municipal services, generating yet another 868 jobs, requiring housing for yet another 560 families (1,569 people), requiring a further 7 acres of executive homes, 104 of normal homes, and 37 acres of apartments. And those people require even more stores and services, creating jobs for yet another 347 families.

This cycle eventually converges to a total population (counting the original factory workers and their families) of slightly more than 19,000 people, requiring a total of 87 acres of executive homes, 1,270 acres of medium-size homes, and 547 acres of apartments, with 396 acres of commercial and industrial services sector-- a small community of very nearly four square miles, not counting any parks, or restricted land.

Essentially, the pickup truck factory, if built in the middle of Kansas, would bring forth a small community around it, a "factory town" which would eventually look something like the community mapped in Figure 15.27.

Forecasting T&D Load

Table 15.5 Total Land Use Generated by the Pickup Truck Factory in Acres

Land use class	Year					
	2	4	8	12	20	25
Residential 1	12	49	87	87	87	87
Residential 2	170	710	1,270	1,270	1,270	1,270
Apartments/twnhses.	73	305	547	2547	547	547
Retail commercial	22	94	168	168	168	168
Offices	8	34	61	61	61	61
Hi-rise	3	12	22	22	22	22
Industry	12	50	89	89	89	89
Warehouses	6	27	48	48	48	48
Municipal	1	5	8	8	8	8
Factory	350	350	350	350	350	350
Peak community MVA	16	34	73	77	77	77

In a manner similar to the analysis carried out here, the intermediate years of factory start up, before all 4,470 workers were employed, could be analyzed to determine if and how fast the community would grow. The employment numbers from Table 15.3 on a year by year basis can be analyzed to yield the year by year land use totals shown in Table 15.5. Working with projections of electric demand provided by the factory (doubtless, the automaker's architects and plant engineers can estimate the electric demand required by the factory quite accurately), and using typical *per capita* electric demand data from mid-North America, this information leads to the forecast of total load given in the last line of the table.

A common shortcut used in electric load forecasting is to convert a Lowry type model as discussed above to a geographic area basis, in which all factors are measured in acres (or hectares, or square kilometers) instead of in terms of population count. The factory of 350 acres caused a demand for:

> 57 acres of executive homes, or .163 acres per factory acre
> 830 acres of single-family homes, or 2.37 acres per factory acre
> 291 acres of multi-family housing, or .83 acres per factory acre

This means that on a per acre basis, the factory causes:

> .163 acres of executive homes per factory acre
> 2.37 acres of single-family housing per factory acre
> .83 acres of multi-family housing per factory acre

Similar ratios can be established by studying the relationship between other land use classes, as for example, the ratios of the land-use development listed in Table 15.5. The Lowry model can then be applied on an area basis, without application of direct population statistics. Little loss of accuracy or generality occurs with this simplification. Techniques to apply this concept are covered in several references (Lowry and Willis 1996).

A workable method for customer forecasting

The example given above showed the interaction of industrial, residential, and commercial development in a community, and illustrated how their relationships can be related quantitatively and in a cause-and-effect manner. This example was realistic, but slightly simplified in the interests of keeping it short while still making the major points. In an actual forecasting situation, whether done on a population count or a geographic area basis, the ratios used (such as 2.8 persons per household, 1.55 workers per household, etc.) could be based on local, industry-specific data (usually available from state, county, or municipal planning departments), rather than generic data as used here.

In addition, the actual analysis should be slightly more detailed. In the preceding example, it was assumed that the employees in the retail commercial and industrial areas had the same *per capita* housing and market demands as the new factory workers. In fact, this is unlikely. Salaries at the factory are probably much higher than for the average retail or service job, and the skilled workers for the factory will tend to have a different distribution of ages, family types, and spending habits than workers in retail and service industries. Thus, the additional housing and market demand created by the retail and services sector of the local community will most likely have different *per capita* housing and market impacts than did the factory. To account for this local variation, retail- and services-specific data could be obtained to refine such an analysis, or generic data based on nationwide statistics could be used in the absence of anything better. However, the basic method presented above is sound; only the data and level of detail used would need to be refined to make it work well in actual forecasting situations.

Locational forecasting

While not highlighted during the example discussed above, it is worth noting that Figures 15.24 through 15.27 showed *where* the growth of residential, commercial, and industrial development was expected. The locational aspects of land-use fall naturally out of such an analysis. Land-use locational preference patterns of the type discussed earlier can be used to derive realistic patterns of

Forecasting T&D Load

development. Application of such "pattern recognition" is a key factor in most simulation methods.

Quantitative Models of Customer Class/Land-Use Interaction

To a great extent, the factory's impact on development around it would be independent of where it was built. Dropped into a vacant, isolated region such as the one used in the example above, its effects are easy to identify. However, if that factory were added to Atlanta, Syracuse, Denver, Calgary, San Juan, or any other city, the *net result* would be similar -- 4,470 new jobs would be added to the local economy. This increase in the local economy would ripple through a chain of cause and effect similar to that outlined above, leading to a net growth of about 19,000 population, along with all the demands for new housing, commercial services, and other employment that go along with it.

The development caused by the factory would be more difficult to discern against the backdrop of a city of several million people, a city with other factories and industries, and with other changes in employment occurring simultaneously. But it would still be there. A complicating factor would be that the new housing and commercial development caused by the factory would not necessarily happen in areas immediately adjacent to it, but would have to seek sites where appropriate vacant space was available, which might be farther away than in the example given.

A further complication would be that if local unemployment were high, or if other mitigating circumstances intervened, some of the 4,470 jobs created by the factory might be absorbed by local unemployment. Determination of the net development impact would be more complicated, requiring a comprehensive analysis of those factors. However, in principle the train of development given in section 7.3 is a valid perspective on how and why land-use development occurs and one that works well for forecasting.

The small community of 19,000 pictured above earned its living building pickup trucks. Only that segment of the local economy brought money into the community. The jobs in grocery stores, gasoline stations, shoe repair shops, bars, doctors offices, and other businesses that served the local community did nothing to add to that. Thus, the fortunes of this small town can be charted by tracking how it fares at its "occupation." If pickup trucks sell well and the factory expands, then the city will do well and expand. If the opposite occurs, then its fortunes will diminish.

So it is with nearly any town or city. Its local economy is fueled by only a portion of the actual employment in the region, and a key factor in forecasting its future development is to understand what might happen to this basis of its economy. Large cities generally are a composite of many different basic industries -- a city like Edmonton, or Boston, or St. Louis, or Buenos Aires "earns its living" from a host of local basic industries. By definition, these basic industries are any activity that produces items *marketed outside* the region -- in other words an activity that brings money in from outside the region. In the example given earlier, only the truck factory is basic industry. Double the factory's employment, the Lowry concept states, and the entire town will eventually double in size. Double the number of grocery stores, shops and movie theaters (none of which "market" outside the local economy) and nothing much would happen, except that eventually a few grocery stores, shops and theaters would go out of business -- the town "earns its living" from the factory and its local population will only support so many stores.

Moreover, whether a city is growing, shrinking, or just changing, its total structure will remain in proportion to the local "basic industrial" economy, in a manner qualitatively similar to that outlined in section 7.3, whatever the causes of that change. *If a city is going to grow, all of its parts will grow in an interrelated, and ultimately predictable, manner related to the local basic industry.*

Urban models

The concepts of urban modeling covered here can be applied to predict how a city will grow, and in what proportions its various classes of land use will respond to such stimuli as; the construction of a new pickup truck factory or a large government research facility, or for that matter, how the loss of a steel mill or the closing of a military base will reduce the local community. This particular concept is useful, and has wide application, but there are situations where it needs modification or interpretation.

In addition, an "urban model" *per se* does not involve the detailed pattern recognition needed to apply the "preference list" location analysis of siting discussed above, which is the key to accurate identification of growth locations. And of course, neither provides any electric consumption analysis -- that must be done with some sort of (usually) end-use based load curve analysis.

Computerization of Simulation

While it is possible to apply simple simulation concepts through a very labor-intensive manual method, application to the extent necessary to exploit fully simulation's accuracy advantage requires computerization on an extensive and analytically involved scale. While a widely diverse range of computerized simulation approaches have been developed for T&D forecasting, most share a similar layout, consisting of spatial and load curve modules, as shown in Figure

Forecasting T&D Load

15.28. The spatial module usually operates in successive stages of increasing spatial resolution: in the case of Figure 15.28, global (no spatial resolution), large area (urban model) and small area (preference pattern recognition). The load curve module consists of a calibration submodule, which adjusts load curves for a closest fit to observed feeder/small area loads, and a load curve model -- usually an appliance-based end-use model -- for analysis and prediction of future customer class load curve shape.

More than 50 different simulation methods are described in various technical literature, and the structural details of the more popular methods are given in *Spatial Electric Load Forecasting* (Willis, 1996). While they vary in structure, data needs, and complexity of analysis, in general their accuracy exceeds anything that can be dependably expected from trending methods. Figure 15.29 shows the relative accuracy in forecasting future T&D loads as a function of time, for several popular load forecast programs.

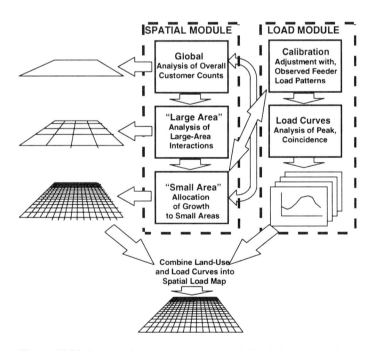

Figure 15.28 Layout of a typical computerized simulation approach.

670   **Chapter 15**

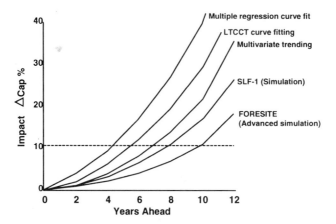

Figure 15.29 Forecast error -- based upon the U_x metric, the percent of total T&D capital budget "at risk" due to forecast errors -- as a function of forecast period for five popular forecast methods. Advanced simulation methods can forecast ten years ahead with accuracy comparable to what curve-fitting can provide four or five years ahead.

15.6 SELECTING A FORECAST METHOD

Although a wide variety of spatial forecasting approaches and computer programs to implement them are available, the selection of a T&D forecasting method begins with a choice between the two basic types of method: trending or simulation. While simulation provides better accuracy in nearly all cases (Figure 15.29) as well as an advantage in representational multi-scenario capability, it is much more expensive and time consuming than trending. And while accuracy is always a desirable trait in a forecast method, the effort to achieve it is not always justifiable based on the planning needs. On the other hand, often this additional effort *is* justifiable based on the T&D capital and operating savings that accrue from the improved forecasting and its resultant more accurate planning.

Table 15.6 lists seventeen forecast methods along with their popular names and technical references. Figure 15.30 compares their cost and forecast accuracy as determined in a series of tests. As can be seen, higher performance costs more.

Forecasting T&D Load

Table 15.6 Seventeen Spatial Forecasting Methods

Method		Type	References
1	Mult. regr. polynomial curve-fit	Trend	Meinke
2	Ratio shares polynomial curve-fit	Trend	Cody
3	Template matching "S" curve-fit	Trend	Willis and Tram
4	Mult. regr. polynomial fit (VAI)	Trend	Willis & Northcote-Green
5	Template matching "S" fit (VAI)	Trend	Willis and Tram
6	3-D extrapolation/urban Center	Trend	Lazarri
7	Urban center extrapolation/cust. class	Trend	Lazarri, Brooks
8	Multivariate extrapolation & clustering	Trend	Wilreker, EPRI 1979
9	Load transfer coupled & geometric	Trend	Willis et al, 1992
10	Manual land-use (manual simulation)	Manual	Willis, 1996, Chapter 8
11	Computerized form of "manual" simulation	Simul.	Scott
12	Land inventory/cust. class load curve	Simul.	Brooks & Northcote-Green
13	Pattern recognition/cust. class load curve	Simul.	Gregg
14	Multivariate cust. class load curve	Mixed	Willis, 1996, Chapter 10
15	Urban model/patt. recog./ end-use model	Simul.	Willis, Finley, Buri
16	"Road link" version of (15) for rural study	Simul.	Willis, Rackliffe, Tram
17	Very high-resol. (2 acre) version of (15)	Simul.	Willis, 1996, Chapter 10

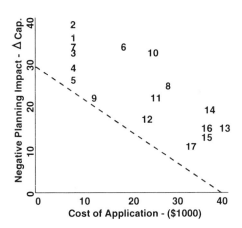

Figure 15.30 Data from the evaluation in section 10.3. SL&P planners pick methods 5, 9, 12, and 17 because these methods lie closest to the dashed line and thus represent the best cost-performance combinations. To this they add method 11 because they believe it may be better than rated here for their particular situation.

Table 15.7 Factors Influencing T&D Forecasting Method Selection

Accuracy and Applicability -- Forecasts that lead to the correct planning decisions are the whole purpose of forecasting.

Planning Period -- How far into the future can the method reliably project loads, and how far are forecasts needed?

Resolution Required -- What detail in location is needed to meet the "where" needs of the T&D planning process?

Corporate Forecast Compatibility -- Consistency with the corporate forecast is essential for internal credibility.

Compatibility with Other Planning -- In particular marketing, DSM, and integrated resource planning.

Changes in Factors and Scenarios -- Changes in the causes of growth, and changes in future conditions may be uncertain.

Data Requirements -- What data are available? What will it cost to get, maintain, and update more data as required?

Robustness -- The ability to tolerate reasonable "noise" in the input data is a practical consideration.

Cost and Resources -- What labor, information technology, and other resources are required?

Documentability and Credibility -- The method should be fully documented and have a credible track record.

Communicability of Results -- Methods vary in how easy they are to understand and explain to those not involved.

Forecasting T&D Load

Criteria Influencing Selection

Before deciding upon the type of computer program and data base to be applied to a utility's distribution load forecasting needs, its planners should assess their requirements in order to assure that they select a method that matches their resources, data, and needs. Growth rates, lead times, planning difficulty, requirement for detail, externalities, budget, and other criteria vary from utility to utility. Table 15.7 lists factors that are important in the selection process.

Accuracy and applicability

Clearly, accurate forecasting of future electric T&D demand is a primary consideration in the selection of a forecast method. The value of forecast accuracy should be assessed in terms of expected gain (in dollars) from the improved forecast. The present worth of this gain can then be matched against the expected cost of the method.

Planning period

If T&D planning is to be done over a ten to 20-year period, then both long-range and short-range accuracy are necessary, as is a multi-scenario forecast ability, because uncertainty in future conditions is the only certainty in the long term. On the other hand, if five years is the extent of planning applications, then the long-range forecast accuracy and multi-scenario forecast capability of more advanced methods are hardly worth the additional cost.

Spatial resolution requirements

How much "where" information is required to support the T&D planning? Higher resolution generally costs more money. There is a reason high resolution forecast techniques exist: their results are necessary and cost-justified in many cases. However, a planner should determine the minimum acceptable spatial resolution required to support the company's T&D planning goals, and focus on achieving good accuracy at that small area size. Forecast methods that work with a smaller size area will only be useful if the smaller area size improves accuracy versus the required resolution, or if it makes data collection, verification, or program use easier.[5]

---

[5] Which can be the case. Many basis simulation methods are easier to calibrate and their data base is easier and quicker to obtain and verify if done at high resolution (Tram et al., 1983).

Compatibility with the corporate forecast

In general, a spatial forecast should base its overall, global trends on the utility's "corporate" forecast, used as a "driving input" (overall totals) for simulation or as a "total adjustment" target for trending. The reasons for this *strong recommendation*, along with several *caveats* and a discussion of adjustments that can legitimately be made to the corporate forecast to increase its accuracy with regard to spatial forecasting, were discussed earlier in this Chapter.

Very often, the format of a corporate forecast -- what classes of customer it identifies, how its data relate to peak demands; energy, seasonal, and temporal load variations; if, how, and how precisely it can be geo-coded and related to small areas -- can have an impact in shaping the selection of a spatial forecast method. Ideally, the spatial load forecasting method should use the same classes, format, and coordinate mapping system used in the corporate forecast. Practically, some differences will no doubt exist, but they should be studied beforehand and the cost, time, and accuracy implications (if any) taken into account. The most important point is that the two forecast methods be compatible and and their results consistent.

Compatibility with other planning functions

Increasingly, T&D planners are being asked to coordinate their planning with other activities in their company. This is particularly true if the utility practices integrated resource planning (IRP), in which T&D and demand-side management (DSM) expenses are compared and balanced for effectiveness. The load forecast serves as the common ground upon which a coordinated plan of all the resources and concerns is studied. To perform this expanded function, the T&D planning forecast must be compatible and consistent with both T&D planning needs and those of other functions such as DSM, service quality planning, distributed generation planning, and market/opportunity planning.

Usually, this means that some form of end-use load analysis is necessary, as most of the other planning mentioned in the paragraph above includes customer-class usage based evaluation. Among other issues, this implies an expansion of the number of types of customer classes required in the T&D forecast. Regardless, the best time to "build in" compatibility with other planning and forecasting functions is before the method, classes, spatial resolution, time periods, level of detail, and other factors for the spatial load forecast have been established.

Forecasting T&D Load

Changes in growth conditions and multi-scenarios

These two items are closely related, but in fact they stem from slightly different planning concerns. To begin with, it is often recognized that the future conditions shaping growth will be different than they were in the past. For example, a large military air base in the region may be scheduled to close. Even in the presence of continued growth in other segments of the regional economy, a base closing will have a negative impact in several ways, but it could have positive impacts in others, particularly if the base site is sold and made available for development.

Regardless, it is clear that since conditions will be *different* than in the past, trending may be inappropriate, for it implies that system conditions will be similar to what they were in the past. A simulation method, at least a basic type that can react to changes in conditions such as an Air Base's closing, would be a better choice.

A somewhat different situation develops when planners are uncertain about if, whether, and how conditions may change in the future. The need to model variations in future conditions also precludes trending methods and means a simulation method should be used. However, not all simulation methods are capable of multi-scenario modeling -- they vary greatly in ease of use and range of conditions they can embrace. A simulation method that can model base case conditions with the local air base deleted may not be suitable for multi-scenario analysis involving shifts in population demographics, highway construction schedules, and other changes. Table 15.8 lists the types of changes in growth factors that can invalidate the basic assumption buried inside all trending methods, that the *process* of growth and change observed in the historical data will remain constant in the future.

Table 15.8 Factors Indicating Extrapolation May Be Unreliable

Seasonal peak changing from summer to winter or vice versa.
Major DSM or load control program sponsored by utility.
Addition of or change in major employer(s) in region.
Addition or major new highway, bridge, or mass transit.
Relocation of a port, airport, or other major infrastructure.
Significant change in relative property tax rates among sub-regions.
Changes in restrictions to growth (environmental, zoning).
Region "uses up" available land and must change growth pattern.

Data requirements

Available data and the cost of data gathering and verification are key factors in any selection process. Data collection is the major component of cost with some forecast methods, particularly simulation. If the data required for a particular method are simply not available, then that precludes that method's use. But very rarely is data absolutely unavailable -- usually almost anything needed for spatial forecasting can be gathered with sufficient effort. Therefore, the cost of various data sources must be weighed against the benefits that accrue from their enabling better forecast methods to be used.

Sources, quality, and timeliness of available data vary tremendously among electric utilities. Some have timely and accurate data on customers, land uses, end uses, appliance market shares, demographics, employment and labor, system load curves and equipment readings, and everything else required by the most advanced simulation methods, already on hand and available company-wide from a central information system. In such cases the incremental cost of preparing the spatial and temporal data bases required for application may be very low. However, a few utilities are at the other end of the data spectrum, with customer data limited only to billing data, and with little more in the way of load data than annual peak readings from "tattletale" current meters on major substations and transmission lines -- readings that are often imprecise because of infrequent checking and calibration, and which never provide the time of peak, but only the amount. Here, the cost of "upgrading" to even one of the advanced forms of trending might be considerable. Therefore, the cost of data collection for spatial forecasting must be judged with regard to utility-specific availability, conditions, resources, and support costs, and weighed against the need and benefits of its usage. In searching for the lowest cost data/forecast solution, planners should not forget an often attractive alternative to developing data in house -- buying it from third party suppliers. In almost all cases, the land-use, customer, demographic, appliance, and load curve data necessary to support any of the simulation or multivariate methods is available for purchase from commercial suppliers.

The most common error in evaluating data needs, however, is failing to look for economical sources of the data needed within the utility. T&D planners are most familiar with the historical peak feeder and substation load readings available at nearly every utility, and unfamiliar with customer, load research, and marketing data. In many cases, the land-use, zoning, demographic, and other spatial information needed for simulation is available at low cost within the utility, or from local governments or other utilities.

But there are cases where essential data may not be available, and the higher data needs for simulation *always* represent a higher handling and verification

Forecasting T&D Load

cost. In addition, availability of data developed for another planning purpose does not guarantee that it will completely meet T&D planning needs. Issues of timeliness (how recent is the most recent data?), periodicity (how often is the data source updated?), definitions (does "peak demand" here have the same meaning as when used in T&D planning?), and verification quality (exactly what does "verified accurate" mean to the people who gathered the data?) need to be examined, and the cost to modify or bring the data up to T&D requirements *on a continuing basis* determined.

Robustness

Robustness -- insensitivity to noise in the data or the occasionally missing data point -- is a desirable trait in any practical software system, for data are never perfect and occasionally far from it. Insensitivity to data error depends on both the inherent robustness of the mathematical procedures themselves (some types of calculations are more sensitive to error than others) as well as how those procedures are implemented as a computer program.

Spatial load forecasting algorithms vary widely with regard to how well they can tolerate noisy data and missing data points. Among trending methods, polynomial curve-fitting methods, particularly those with high order (cubic or higher) terms, tend to be relatively more sensitive to noisy data (and particularly missing data points), while the clustering and pattern recognition methods are more robust. With curve-fitting methods, a useful tactic in the presence of noisy data or missing data elements is to reduce the order of the polynomial being fitted -- for example, substituting a second order polynomial for the more often used cubic order function.

Generally, spatial simulation methods are *very* robust with regard to noisy small area data, mildly incorrect geographic data of any type, and missing data points in the load curves. In this regard, they are among the most tolerant of planning methods available. A point generally not appreciated is that *spatial frequency domain implementation* of the small area preference map calculations not only decreases computation time, but also cuts round-off error and sensitivity to data noise substantially (Willis and Parks, 1983, and Willis, 1996). Simulation methods that use distributed urban pole computations are also much more robust than those that do not.

However, the data sensitivity of the various urban models and the end-use load curve models employed in simulation methods varies widely. Some urban models developed for other industries and forecasting purposes are not robust with respect to electric utility applications (Wilreker, 1977; Carrington, 1988).

Single-stage Lowry models are generally more robust than multi-stage, but multi-stage models may still be preferable.[6]

Some commercial spatial load forecasting and T&D planning software systems include filtering routines, which at the option of the user can be applied during both the frequency domain computations and during temporal load curve analysis, to detect and "repair" certain types of data noise and omissions. Benchmarks indicate these work well, but their exact nature are considered proprietary by their manufacturers. To the author's knowledge no published reference on these very interesting additions to simulation exists.

Cost

As shown earlier in this chapter, cost of application varies as much as accuracy among available spatial forecasting methods. A forecast method must fit within the budget and should be selected with an eye on overall forecasting resource limits. *Absolutely nothing* contributes to poor forecasting accuracy more than attempting to use a method that exceeds the available budget -- the shortcuts forced on data collection, calibration, and usage when insufficient time and resources are available destroy accuracy and usefulness.

If the available budget does not permit quality collection and verification of the data, proper training, and "learning curve" time for the staff, while still leaving sufficient resources for application and interpretation of results, then it is best to accept a less demanding method even if it is inherently less accurate or applicable.[7]

Documentability, credibility, and communicability

Forecasts are the "initiator" of the planning process -- they identify a need for future facilities and are often viewed as the "cause" of the capital expenses put into the future budget. Very deservedly, they receive a good deal of attention

---

[6] This must be kept in perspective. A robust technique shows relatively little change in computed results when the input data are contaminated by errors. Therefore, a method that produces poor results with good data, and the *same* poor results with error-filled data, will be considered robust. For example, in the presence of increasing amounts of data error, a multi-stage Lowry model's accuracy degrades more quickly than a single stage model's. But while the single-stage is considered very robust, the multi-stage is still generally more accurate, regardless (see Willis, 1996).

[7] Actually, the recommended procedure is to determine which method is best overall and then convince management to devote the resources required. Often, however, budget and resources are constrained and planners must do the best they can with what is available.

Forecasting T&D Load

from upper management, who seek confidence that the needs identified in the forecasts are real and that the expenses are necessary. The resulting plans often call for rights-of-way and sites that will be opposed by local community groups and interveners on the basis of cost, inconvenience, and esthetics. Lastly, regulatory agencies pay close attention to both expenses and public concern, so that the forecast that initiates the planning process will be examined carefully and its results questioned heavily.

For these reasons, a very important aspect of forecast method selection deals with documentation, credibility, and communicability. Nothing is better with regard to these criteria than a forecast method with a long history of proven success in similar forecast situations, and with a format and output that are easy to communicate and that appeal to intuition and familiar insight.

Documentability. At a minimum, every power delivery planner should make certain that all methodology is fully documented. This means having an identified methodology and procedure for its application, including:

1. A spatial forecasting method using a fully disclosed procedure, hopefully one well-proven and published by other parties, not just the planners' own company

2. Documented sources of data and standards for its completeness, timeliness, and verification

3. A procedure for tracking and documenting "What we did" in the course of preparing a forecast

Ad hoc forecasting and seat-of-the-pants methods do not meet these requirements. Neither do non-analytic methods, computerized or not. A computer method that is well-documented in technical and scientific literature is a good start toward meeting the goals listed above, but the utility must establish procedures for items 2 and 3, and follow through consistently.

Credibility. Documentation does not make a utility immune to criticism or prevent the occasional "defeat" in the public arena. However, solid documentation and record-keeping of the forecast method, the procedure for its application, and the quality control that surrounds both go a long way toward bolstering credibility.

Nothing contributes quite as much to the credibility of a forecast as the methodology's proven track record of success at other utilities. Publication of the method and its evaluation in major reviewed technical journals such as *IEEE Transactions* is also meaningful -- the peer-review process used in such journals maintains a quality control on the legitimacy of technical claims and the validity of testing procedures used.

Communicability. Usually, opposition to a utility plan will focus on the forecast, not the plan itself. Neighborhood leaders opposed to a transmission line through their backyards are more likely to argue that the forecast is wrong (i.e., the line isn't needed) than to quibble with the engineering calculations that identified the line as the best solution to the forecasted demand. In such situations, a forecast procedure that is difficult to explain, or whose mathematics are impenetrable without special study, is a severe liability.

Simulation methods have a very great advantage with regard to communicability. Unlike trending, which relies on equations and numerical methodology that at its simplest is vastly complicated to the layman, simulation can be explained with the aid of colored land-use maps, charts of usage that list appliances that most people use everyday, graphs showing population growth, and similar exhibits that have a broad appeal and are easily understood. Even the pattern recognition, urban pole, and urban models within a simulation algorithm can be summarized in a way that appeals to the experience and understanding of most people, at least sufficiently to convey a strong sense of what the forecast method did and why the predicted growth and change was projected to occur.

Not only does such communicability help convince others of the validity of a forecast, it is often a great aid to the planners themselves. Nothing helps identify set-up errors, transposed data, or mistakes in application faster than a strong intuitive appeal in the format of the forecasted data. A forecast that communicates well, both in complete output and in appeal to intuition and experience, is valuable for this reason alone.

REFERENCES AND BIBLIOGRAPHY

C. L. Brooks, "The Sensitivity of Small Area Load Forecasting to Customer Class Variant Input," in *Proceedings of the 10th Annual Pittsburgh Modeling and Simulation Conference,* University of Pittsburgh, Apr. 1979.

C. L. Brooks and J. E. D. Northcote-Green, "A Stochastic Preference Technique for Allocation of Customer Growth Using Small Area Modeling," in *Proceedings of the American Power Conference,* Chicago, University of Illinois, 1978.

Canadian Electric Association, "Urban Distribution Load Forecasting," final report on project 079D186, Canadian Electric Association, 1982.

J. L. Carrington, "A Tri-level Hierarchical Simulation Program for Geographic and Area Utility Forecasting," in *Proceedings of the African Electric Congress,* Rabat, April 1988.

E. P. Cody, "Load Forecasting Method Cuts Time, Cost," *Electric World,* p. 87, Nov. 1987.

Electric Power Research Institute, "Research into Load Forecasting and Distribution Planning," report EL-1198, Electric Power Research Institute, Palo Alto, CA, 1979.

Electric Power Research Institute, "DSM: Transmission and Distribution Impacts," Volumes 1 and 2, Report CU-6924, Electric Power Research Institute, Palo Alto, CA, August 1990.

M. V. Engel et al., editors, *Tutorial on Distribution Planning,* Institute of Electrical and Electronics Engineers, New York, 1992.

J. L. Garreau, *Edge City,* Doubleday, New York, 1991.

J. Gregg, et al., "Spatial Load Forecasting for System Planning," in *Proceedings of the American Power Conference,* Chicago, University of Illinois, 1978.

A. Lazzari, "Computer Speeds Accurate Load Forecast at APS," *Electric Light and Power,* Feb. 1965, pp. 31-40.

I. S. Lowry, *A Model of Metropolis,* The Rand Corp., Santa Monica, CA, 1964.

J. R. Meinke, "Sensitivity Analysis of Small Area Load Forecasting Models," in *Proceedings of the 10th Annual Pittsburgh Modeling and Simulation Conference,* University of Pittsburgh, April 1979.

E. E. Menge et al., "Electrical Loads Can Be Forecasted for Distribution Planning," in *Proceedings of the American Power Conference,* Chicago, University of Illinois, 1977.

R. W. Powell, "Advances in Distribution Planning Techniques," in *Proceedings of the Congress on Electric Power Systems International,* Bangkok, 1983.

C. Ramasamy, "Simulation of Distribution Area Power Demand for the Large Metropolitan Area Including Bombay," in *Proceedings of the African Electric Congress,* Rabat, April 1988.

B. M. Sander, "Forecasting Residential Energy Demand: A Key to Distribution Planning," IEEE Paper A77642-2, IEEE PES Summer Meeting, 1977.

A. E. Schauer et al., "A New Load Forecasting Method for Distribution Planning," in *Proceedings of the 13th Annual Pittsburgh Modeling and Simulation Conference,* University of Pittsburgh, April 1982.

W. G. Scott, "Computer Model Offers More Improved Load Forecasting," *Energy International,* September 1974, p. 18.

H. N. Tram et al., "Load Forecasting Data and Database Development for Distribution Planning," *IEEE Trans. on Power Apparatus and Systems,* November 1983, p. 3660.

H. L. Willis, "Load Forecasting for Distribution Planning, Error and Impact on Design," *IEEE Transactions on Power Apparatus and Systems,* March 1983, p. 675.

H. L. Willis, *Spatial Electric Load Forecasting,* Marcel Dekker, New York, 1996.

H. L. Willis et al., "Load Transfer Coupling Regression Curve-fitting for Distribution Load Forecasting," *IEEE Transactions on Power Apparatus and Systems,* May 1984, p. 1070.

H. L. Willis and J. Aanstoos, "Some Unique Signal Processing Applications in Power Systems Analysis," *IEEE Transactions on Acoustics, Speech, and Signal Processing,* December 1979, p. 685.

H. L. Willis and J. Gregg, "Computerized Spatial Load Forecasting," *Transmission and Distribution,* May 1979, p. 48.

H. L. Willis and J. E. D. Northcote-Green, "Spatial Load Forecasting -- A Tutorial Review," *Proceedings of the IEEE,* February 1983, p. 232.

H. L. Willis and J. E. D. Northcote-Green, "Comparison of Fourteen Distribution Load Forecasting Methods," *IEEE Transactions on Power Apparatus and Systems,* June 1984, p. 1190.

H. L. Willis and T. W. Parks, "Fast Algorithms for Small Area Load Forecasting," *IEEE Transactions on Power Apparatus and Systems,* October 1983, p. 2712.

H. L. Willis and H. N. Tram, "A Cluster-based V.A.I. Method for Distribution Load Forecasting, *IEEE Transactions on Power Apparatus and Systems,* September 1983, p. 818.

H. L. Willis and H. N. Tram, "Distribution Load Forecasting," Chapter 2 in *IEEE Tutorial on Distribution Planning,* Institute of Electrical and Electronics Engineers, Hoes Lane, NJ, February 1992.

H. L. Willis and T. D. Vismor, "Spatial Urban and Land-Use Analysis of the Ancient Cities of the Indus Valley," in *Proceedings of the Fifteenth Annual Pittsburgh Modeling and Simulation Conference,* University of Pittsburgh, 1984.

Forecasting T&D Load

H. L. Willis, M. V. Engel, and M. J. Buri, "Spatial Load Forecasting," *IEEE Computer Applications in Power*, April 1995.

H. L. Willis, L. A. Finley, and M. J. Buri, "Forecasting Electric Demand of Distribution System Planning in Rural and Sparsely Populated Regions," *IEEE Transactions on Power Systems,* November 1995, p. 2008.

H. L. Willis, J. Gregg, and Y. Chambers, "Small Area Electric Load Forecasting by Dual Spatial Frequency Modeling," *in Proceedings of the IEEE Joint Automatic Control Conference,* San Francisco, 1977.

H. L. Willis, G. B. Rackliffe, and H. N. Tram, "Forecasting Electric Demands of Distribution System Planning in Rural and Sparsely Populated Regions," *IEEE Transactions on Power Systems,* November, 1995.

H. L. Willis, G. B. Rackliffe, and H. N. Tram, "Short Range Load Forecasting for Distribution System Planning--An Improved Method for Extrapolating Feeder Load Growth," *IEEE Transactions on Power Delivery,* August 1992.

H. L. Willis, H. N. Tram, M. V. Engel, and L. Finley, "Selecting Algorithms for Distribution Optimization," *IEEE Computer Applications in Power,* January 1996.

V. F. Wilreker et al., "Spatially Regressive Small Area Electric Load Forecasting," in *Proceedings of the IEEE Joint Automatic Control Conference,* San Francisco, CA, 1977.

T. S. Yau, R. G. Huff, H. L. Willis, W. M. Smith, and L. J. Vogt," Demand-Side Management Impact on the Transmission and Distribution System," *IEEE Transactions on Power Systems,* May 1990, p. 506.

16
Distribution Feeder Analysis

16.1 INTRODUCTION

This chapter is the first of two that examine the application of analytical tools for distribution planning. An amazingly diverse range of analytical tools are available for use by distribution planners. Often embodied in computer programs, numerical and non-numerical analysis methods are to distribution planning what "power tools" are to carpenters and construction workers. Used properly, they reduce the time required to complete basic tasks involved in the work, and they produce neater, cleaner, and more precise results than could be accomplished manually. As a result, they permit a skilled and knowledgeable worker to produce more work faster, and to undertake difficult projects, or to apply a particularly intricate technique, that would otherwise be impractical.

Standardization, Consistency, and Documentability

In addition to the quicker turn-around and improved precision, the use of computerized tools *standardizes* the planning process. Standardization assures consistent application of method and leads to equitable evaluation of all alternatives. Often, the ability to document impartiality in choosing among alternatives is of paramount importance to a utility operating in a tight regulatory framework. Computerization provides demonstrable impartiality of method and permits documentation on all evaluations to be provided. Of course, standardization also means that any flaws or approximations in technique are applied universally. Thus, it is important to make certain that standardized techniques are appropriate, and particularly to be aware of any limitations or exceptions in their use.

Finally, it is essential to bear in mind that power tools are no substitute for knowledge and skill -- in fact they increase the level of expertise required, for the planner must understand the tool and its application as well as distribution planning.

Performance Simulators and Decision Support Methodology

Tools for distribution planning can be grouped into two distinct categories. The first are performance simulators, covered in this chapter. These are procedures, very often embodied in elaborate computer programs, that predict the behavior or response of equipment or distribution systems to a particular set of conditions. A distribution load flow program, which predicts the expected voltage, current, and equipment loadings throughout a power system, when given as input the expected loads and initial conditions, is the prototypical performance simulator. Chapter 17 covers a second type of analytical tool -- decision support methods -- which assist the planner in evaluating and selecting from the many possible alternatives to a planning situation.

Chapters 16 and 17 are *not* treatises on numerical methods, algorithms, or computer programs. Instead, the emphasis throughout is on the issues pertinent to the selection and application of numerical analysis methods; how and why each is needed; what is important in its application; and how to use each type of tool to best advantage.

Performance Simulators

A good deal of the effort in distribution planning, particularly in short-range planning, involves determining if a particular part of the distribution system can perform within requirements in some manner. For example, a planner might want to know whether a particular feeder can serve the projected peak load three years ahead while staying within the utility's equipment loading and service standards, and while requiring no reinforcement and no switching of load to nearby feeders. There are two ways to find out. The planner can wait three years to observe the feeder during that peak period, in which case it will be too late to correct the situation if the answer is "no, the feeder can't meet the load," or a simulation can be done to evaluate the future situation. If the simulation indicates that the answer is "no," the simulator could be used further, to evaluate various possible

Distribution Feeder Analysis

solutions to the problem, in order to establish that the proposed "fix" will in fact do the job.

The simulator used in such a case would be a load flow, the most ubiquitous of power system analysis tools, which computes the expected voltages, current flows, and loadings at points throughout the network. Properly applied, and assuming that model, data, and analytical method selected are appropriate to the planning requirements, the load flow will tell the planner if, where and why voltages or loadings will be out of alignment with standards in the future network.

Many other simulators are needed at various phases of the distribution planning and engineering process, some of which are shown in Table 16.1. While the load flow predicts behavior under normal circumstances, short circuit (fault current) analysis methods simulate performance under abnormal conditions, and are needed for evaluation and design of protection schemes. Motor start simulators compute how voltages and loadings vary during the transient phase of starting (and sometimes stopping) large motors. Other simulators are used in planning, and a great deal more, not shown in the Table 16.1, in T&D engineering.

Table 16.1 Simulators Typically Used in Distribution Planning and Engineering

Type of Simulator	Usual Purpose	Used in Planning	Used in Engineering
Load flow	Predict voltage, loading for candidate designs, conditions	Yes	Yes
Short circuit	Predict fault current levels for protective coordination	Rarely	Yes
Predictive reliability	Predict the expected rate and duration of service interruptions	Yes	Rarely
Motor start	Predict voltage drop upon starting large loads	Yes	Yes
DSM simulator	Predict reduction in load due to DSM program	Yes	Rarely
Harmonic load flow	Predict propagation of non-standard frequency power	Rarely	Yes
Sag and tension	Evaluate clearances and tension required for lines and structures	Rarely	Yes
Dynamic loading	Compute maximum peak that can be safely sustained under load curve	Yes	Yes

Load flow, reliability -- and other performance simulators used in distribution planning -- work with models of various elements of the distribution system, as well as with several "models" of electrical or mechanical behavior. For example, a load flow requires a circuit model for the system -- a description of the lines, equipment, and loads. It needs additional models to explain the behavior of line segments, transformers, regulators, and so forth. A model of electric flow is necessary, too. Depending on the level of detail needed, that could be anything from a simplistic "DC" representation of power flow, to a very detailed representation of poly-phase AC power flow.

16.2 MODELS, ALGORITHMS, AND COMPUTER PROGRAMS

Models are descriptions of the behavior, response or structure of some element of the distribution system, used as a proxy for that element during analysis and evaluation of planning options. Performance simulators usually involve the combination of several models of various elements and behaviors of the distribution system.

Often, the models used involve equation(s), whose solution is best computed by some numerical method -- an *algorithm* which is usually implemented by a *computer program* on a digital computer for quick, accurate computation. As a result, the terms "model," "algorithm," and "program" tend to become blurred, and many people come to regard the three terms as interchangeable. Yet this is not the case.

For example, a utility's planners may decide to model electrical flow in their system as occurring over single lines (one-line equivalent) rather than three actual phase conductors, and with $I_{ij} = V_i/(Z_{ij}+Z_l)$, where the current I_{ij} flowing from point i to j is modeled as a function of a voltage, V_i, at point i, divided by the sum of the line and load impedances, Z_{ij} and Z_l. This basic one-line AC-model of power flow becomes the planners' model of electrical behavior.

In order to apply this model to their network, the planners must apply the equation above for the various segments and loads in the system. This means that, one way or another, they must solve a set of simultaneous equations to compute voltages and currents. The particular way selected is the algorithm. Many approaches to this numerical problem are applicable, including Gauss-Seidel, Newton-Raphson, or Stott de-coupled algorithms, to name just three. Any of these, properly applied, can provide more than an adequate solution for the analysis.

Once a particular algorithm is selected -- for example a Newton-Raphson load flow -- many different types of computer programs can be written to apply it, these programs differing in ease-of use, data base structure, error checking, compatibility with other software, and a host of other important issues. Yet, while these programs may differ greatly in such important features, all would apply the Newton-Raphson algorithm to solution of a one-line AC-model, providing, in essence, the same level of results.

While many aspects of algorithm selection and computer program development are important, from the standpoint of distribution planning the *model is more important than the solution algorithm.* In the example given above, the one-line model means planners will not see the effects that load imbalance has on performance and economy, and that their understanding of the performance and economy issues related to two- and one-phase elements of their system will be approximate at best. The differences in analytical results that come from choosing a one-line model instead of a full three-phase model, or even those that come from choosing a partial instead of a full three-phase model, are much more dramatic than any differences affected by a decision to use one algorithm or another, or any decision about program design.

Planners Should Focus on Models, Not Algorithms or Programs

The foregoing discussion is not meant to imply that programs and algorithms are not important. However, it is the selection and application of the model that defines what will and will not be represented in the planning; that defines the details that can and cannot be considered during analysis; and that fixes the limits of accuracy and applicability. A dazzling computer program, implementing a clever algorithm, will produce nothing of value if built upon a model that is inadequate when measured against the planning requirements.

The selection of the models used, and their details, is a pure planning function: there are technical support experts and information systems professionals to handle the details of algorithm and computer implementation, but the specification of *how* the system will be represented during its planning, and *what* will be considered, and *when*, is at the very foundation of distribution planning. It is easy for the planners' attention to be diverted from such considerations by issues related to the algorithms and computer issues. These are often more intellectually stimulating, technically glamorous, or just plain more fun than the work involved in setting out precisely the definition and details of the engineering and economic models to be used in planning. Frankly, it is not uncommon to see exotic algorithms and excellent programming applied in support of models that only marginally meet the utility's planning needs, because the planners forgot their basic focus -- *get the job done as economically as possible* -- *and simply wanted to play with the technology.*[1]

---

[1] This is often called "sandboxing" in the software industry.

> *It is the planners' first responsibility to make certain that the models used in planning are appropriate to their task, and to understand their limitations, shortcomings, and range of applicability.*

16.3 CIRCUIT MODELS

The most basic model for distribution performance analysis *is the representation of the distribution system itself.* In computerized analysis systems, this is essentially the distribution analysis database, sometimes called the applications database. However, planners should note that paper circuit maps also constitute a "model" of their circuits -- a printed representation of what is located, where, and how it is connected to other equipment. Database and paper map are only approximations of reality, showing certain features and neglecting others. Considerable attention should be devoted to the specification of the circuit model's content and detail for performance simulation, for the planners' evaluation of future electrical and economic performance can have no more accuracy or detail than that provided by the model itself

In modern distribution planning software systems, the same basic circuit model provides the basis for all performance simulators (i.e., load flow, short circuit, reliability, and economic analysis) of the distribution system. It contains representations of lines, loads, and equipment, along with *connectivity* -- information to link them together so that electrical flow can be traced. Connectivity is necessary for most planning applications, including electrical (load flow and fault current), reliability analysis, pricing studies, and certain types of customer service and distributed resource planning applications. However, there are a few applications (i.e., maintenance and property cost analysis) that do not require it.

Phases and Phasing

The most basic decision in building a circuit model for distribution planning relates to how much detail is used in representing these differences in phasing. Distribution systems consist of three-, two-, and one-phase equipment and loads. Circuit models can represent all aspects of phasing, or only some, or may neglect entirely the existence of phases and phasing.

Distribution Feeder Analysis

One-line models of electric circuits neglect phasing entirely, representing all distribution circuit elements and loads as "phase-less" single-line elements as shown in Figure 16.1. In a system composed only of three-phase equipment, with balanced loads, impedances, and connections, this is tantamount to analyzing one of three identical phases. As a result, one-line models are occasionally termed "single-phase models." However, for distribution applications one-line models are best thought of as "phase-less models," because they not only neglect phase imbalance where it occurs, but they also mix phases and phase structures, so that the resulting model represents no single phase but is instead an average of whatever is present, in both structure and loading. For example, when a single-phase lateral branch is connected to a three-phase feeder trunk, as shown in Figure 16.1, the implied result is that the single phase's current is split equally among all three phases, as represented in Figure 16.4. This is not a highly accurate depiction of reality.

Three-phase circuit models represent circuit elements explicitly as three-, two-, or single-phase equipment and loads. Additional details often included are a distinction of whether the equipment is Y- or delta-connected. Y-connected models are sometimes referred to as four-wire models, because they have provisions for four wires (three phase conductors and a neutral). Models limited to three wires can represent only delta-connected circuits completely. They are often used to represent Y-connected circuits by assuming the neutral current is zero, a tolerable approximation in many cases (a three-wire model is still much more accurate for distribution purposes than any one-line equivalent model).

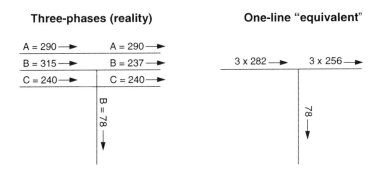

Figure 16.1 Three-phase representation (left) and one-line representation (right) of part of a three-phase feeder trunk connected to a single-phase lateral branch. Values shown are amps, arrows indicate direction of power flow. The one-line representation not only assumes balanced loads and flows, but it also "splits" single-phase current equally among three phase elements and mixes three- and single-phase circuitry, as depicted at the right.

Aggregated phase models, whether three- or four-wire, represent each line, load, or equipment unit as a single entity, with the number of phases of the equipment being an attribute of the element, and connectivity explicitly identified by phase. In an aggregated phase model, a lateral line segment consisting of two phases, connected to the three-phase feeder trunk at phases B, and C, would be represented as a single entity (type: line segment), with two phases present, connected at the trunk to phases B and C.

Disaggregated phase models, by contrast, would represent this two-phase lateral as two separate lines in two separate models. A disaggregated phase circuit model is basically three separate single-phase models, representing phases A, B, and C respectively, with no explicit continuity between them. Typically, such models are used only where a utility is applying a modification of a one-line equivalent circuit analysis method to three-phase distribution. Separate models (databases) are developed for phases A, B, and C (often with some sort of software system in place to coordinate and assure consistency) with each phase analyzed one at a time. This places limitations on, or barriers to, the electrical detail that can be modeled (i.e., mutual coupling of phases cannot be modeled easily, etc.), and is somewhat cumbersome to use, although clever programming and innovation can make the system quite tolerable to use.

Node Resolution

The detail in representing the distribution system can be measured as node resolution -- the ratio of circuit database nodes, or entities, to customer meters. Node resolution is a useful measure for comparison of the representation accuracy of different circuit models. Figure 16.2 shows two models of the same feeder. One has 20 nodes, the other an order of magnitude better: more than 200 nodes. However, both are significant reductions in detail from reality: the feeder itself has 4603 poles, one line regulator, two capacitor banks, and ten line switches, and delivers power to 240 service transformers serving 1255 customers. Roughly 8,000 nodes would be required to represent explicitly every customer meter, every unit of equipment, and every pole-to-pole line segment as a separate entity. More than 2,700 nodes are required to model completely every electrical detail of the feeder.[2]

---

[2] The number of nodes required for complete *electrical* modeling is less than for complete modeling because often several spans have identical conductor, or have no branches, equipment or load: they can be lumped into one segment with *no* loss of electrical detail. However, detail for other applications may be lost. For example, reliability analysis is potentially more accurate if an actual pole/span count is used, because failures are somewhat related to actual pole count, not just line length, etc.

Distribution Feeder Analysis

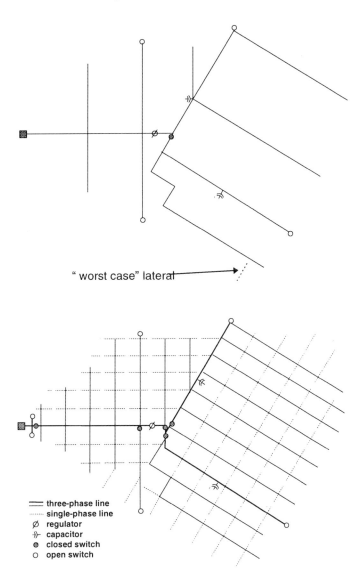

Figure 16.2 Two approximate depictions of a feeder show differences in modeling detail with respect to the number of nodes: at the top, a twenty-node model, at the bottom, a circuit model with ten times the "node resolution." Actual nodes (not shown), are the end points of segments.

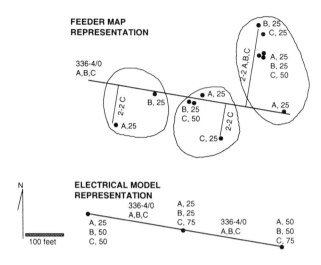

Figure 16.3 Two models of a portion of a feeder, used by utility planners. Top, the representation used in maps (paper or electronic) shows the feeder trunk and gives its phasing (A,B,C) and conductor sizes (e.g., 336 MCM phase conductor, 4/0 neutral), and shows laterals and twelve transformers (dots) by phase and capacity. Total transformer capacity is 375 kVA. Bottom, the electrical representation deletes the laterals and uses only two segments and three nodes. The transformers have been "collected" into three groups (circled areas at top), assigned to the nodes, but the total of 375 kVA is retained.

A greater number of nodes in a circuit model does not assure greater accuracy, but it provides greater potential for accuracy, for two reasons. First, a large node count provides more detail in representation of circuit topology and equipment. Second, most algorithms (e.g., load flow, reliability prediction), compute results only at nodes -- more nodes mean more locations where information is available about performance.

Omission of Laterals and Use of Load Collector Points

Typically, representation of every detail pertinent to both the primary and service levels requires a node resolution of 2 to 8 nodes per customer meter. Very few circuit models have this level of detail and very few utilities find the effort to build such detailed models economically justifiable. Models used in planning analysis usually reduce node resolution by an order of magnitude.

Distribution Feeder Analysis

Most often, reduction is made through an explicitly identified node reduction procedure. This may be computerized, as for example when circuit models are formed by extraction of information from AM/FM or GIS systems. However, worldwide, most circuit models are developed by digitization, and node-reduction procedures are applied manually as the circuit data is entered.

Chief among those entities not represented in most circuit models is the service level: it is common to represent only the primary-voltage level. In addition, laterals and short branch segments are often deleted, as are other small load points, with small loads lumped to *load collector nodes,* as illustrated in Figure 16.3.

When applying significant node reduction (on the order of a factor of ten, or more) to circuits for electrical or reliability simulation, a recommended practice is to include one "worst case lateral" in each feeders' representation, as shown in the 20-node example (top) in Figure 16.2. This leaves data from which the estimated worst voltage drop and outage rates expected on that primary feeder can be computed.

Whether computerized or manual, it is best to develop formal rules for what and how circuits are to be reduced and approximated for modeling, and to provide all planners with copies of these "reduction rules." Formal rules are necessary for any computerization of the node-reduction process, but their written distribution to the planners is not. However, providing that information helps them understand exactly what level of approximation they are using in their analysis. Formal rules should be developed, distributed, and applied in manual reduction also, in order to assure consistency of results among all planners.

Increasing Detail in Circuit Models

Distribution circuit models used for planning have shown a gradual but long-term trend of increasing representation detail in phasing and for node resolution, as shown in Figure 16.4. The information shown is based on surveys of customers in the United States and Canada obtained by the author from 1975 through 1997. The phasing and node resolution shown represents actual usage, not necessarily the maximum capability of the data and analysis software systems being used. Detail used in distribution performance simulation models is gradually increasing, due both to the increasing performance/price capabilities of computing and database systems, as well as the greater availability of efficient data-gathering systems for distribution. This trend is driven by the needs for greater detail that come about in a de-regulated, competitive electric industry.

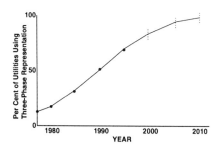 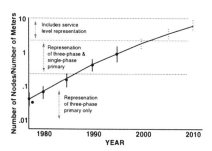

Figure 16.4 Detail of circuit models used for short-range distribution planning in North American utilities, 1977 through 2010. Left, percent of utilities using three-phase rather than one-line equivalent models. Right, node resolution (average ratio of nodes/customer meters on a feeder) in circuit representations.

Line Segment Impedance Models

Every element of an electric circuit simulation has a model of its electrical behavior that describes the interaction of current and voltage -- most often a model of current flow through it as a function of voltage applied. The simplest, useful electrical model for a line segment is to represent it as a single-phase element with ends at nodes i and j, and an impedance Z_{ij}. As shown, it is serving a load at node j with impedance Z_l and is supplied with voltage Vi at node i. The current I_{ij} flows through the segment impedance Z_{ij}; the voltage drop and current on the segment are:

$$I_{ij} = I = V_i/(Z_{ij}+Z_l) \qquad (16.1)$$

and the voltage drop along the line segment is

$$V_{ij} = I_{ij} * Zij \qquad (16.2)$$

The power conveyed to the load is

$$P_l = |V_j| |I_{ij}| \cos\theta = V_{max}I_{max} \cos \qquad (16.3)$$

where |Vj| indicates the rms value of the phasor Vj and θ is the phase angle shift between current and voltage. Such *a one-line equivalent model*, sometimes

Distribution Feeder Analysis

called a *single-phase model,* is compatible only with a one-line circuit model and computational method.

As discussed earlier in this chapter, distribution systems consist of unbalanced three-phase and two- and one-phase elements, to the extent that one-line equivalent models are often not highly accurate representations. A generalized line segment model for three-phase unbalanced analysis can include up to 25 impedances, representing the self- and mutual-impedances among up to five conductors -- the three phase conductors, the neutral, and the ground. Generally, the ground flow is left out of all but very special applications, resulting in reduction to nine (delta) and sixteen (Y-connected) values required for full modeling.

In a three-phase line model, the voltages and currents are represented by vectors (four-tuples in the case of wye-connected circuits, three-tuples in delta circuits) and the line impedances by a four-by-four (wye) or three-by-three (delta) matrix.[3,4] For a wye-connected circuit, the impedance matrix becomes

$$Z = \begin{vmatrix} Z_{aa} & Z_{ab} & Z_{ac} & Z_{an} \\ Z_{ba} & Z_{bb} & Z_{bc} & Z_{bn} \\ Z_{ca} & Z_{cb} & Z_{cc} & Z_{cn} \\ Z_{na} & Z_{nb} & Z_{na} & Z_{nn} \end{vmatrix} \quad (16.4)$$

Diagonal terms are the basic phase impedances, off-diagonal terms are the mutual impedances -- the mere existence of phase B influences current flow in phase A, and vice versa, and current flow in phase A influences voltage in phase B, etc.

Some models used in distribution planning and engineering neglect mutual coupling and represent all off-diagonal terms as zero. This greatly reduces complexity of data and computation, but introduces noticeable inaccuracies, particularly in representing rural distribution.

In company with vectors of phasors representing voltages and currents, the same basic equations 16.1 through 16.3 discussed for a single line model apply in this case. For example, equation 16.3 is now a matrix equation involving considerably more computation.

---

[3] Often, wye-connected models neglect representation of the neutral and use only a three-by-three representation of impedances.

[4] At only a very small cost in analytical accuracy it is possible to reduce the four-by-four matrix used for wye-connected circuits to a three-by-three representation for delta models, using the Kron reduction.

$$\begin{vmatrix} V_{aij} \\ V_{bij} \\ V_{cij} \\ V_{nij} \end{vmatrix} = \begin{vmatrix} Z_{aa} & Z_{ab} & Z_{ac} & Z_{an} \\ Z_{ba} & Z_{bb} & Z_{bc} & Z_{bn} \\ Z_{ca} & Z_{cb} & Z_{cc} & Z_{cn} \\ Z_{na} & Z_{nb} & Z_{nc} & Z_{nn} \end{vmatrix} \times \begin{vmatrix} I_{aij} \\ I_{bij} \\ I_{bij} \\ I_{nij} \end{vmatrix} \quad (16.5)$$

The most accurate way of computing the various elements of the impedance matrix is to derive them directly from Carson's equations (Carson, 1936), based on the actual conductor spacing. A very comprehensive discussion of impedance matrix computation and application, is included in Kersting (see References).

In most cases, distribution planning applications, including short-range planning, can be minimally met with computations and line-section models that neglect impedances imbalance caused by deviations of line spacing from perfect delta configuration. However, engineering applications generally cannot, and the best results are always obtained by using a full representation (Kersting).

Equipment Models

The transformers, regulators, capacitors, line drop compensators, and other elements of the distribution system can be modeled in varying levels of detail, from simplistic to greatly detailed. A thorough discussion of the possible variations for such models is far beyond the scope of this chapter -- it would consume a book unto itself.

Generally, distribution planning, even short-range planning, requires less detail in the modeling of equipment behavior and control than distribution engineering. Key aspects of equipment are: phase-by-phase representation (because of unbalanced flows), capacity, and basic electric behavior. Thus, the recommended minimum models for short range planning are as described below.

Transformers

Transformers best represented on a phase-by phase basis, because while three-phase transformers usually have identical impedance and capacity per phase, many three-phase transformer banks are composed of three single-phase transformers of (potentially) different sizes and impedance's. Core losses are best represented as a shunt impedance parallel to the winding losses in series with load.

Representation of hysterisis, electrical losses due to pump and fan loads for FOFA transformers, as well as dynamic models of changes in impedance or capacity with operating temperature, hence load -- are generally unnecessary for planning.

Distribution Feeder Analysis

Voltage regulators

Line regulators and line drop compensators are also best represented on a phase-by-phase basis, in the same manner as the transformers. For planning purposes, many of the details needed for engineering, such as explicit representation of steps in a regulator, and the slight changes in capacity that occur as upstream voltage level is changed, are not necessary. (They *are* for engineering, however).

Capacitors

Capacitors are best represented on a phase-by phase basis, too. They should be modeled as an impedance, not a constant injection. A 600 kVAR capacitor bank is an impedance that injects 600 kVAR, given than the voltage is 1.0 per unit. It is *not* a device that injects 200 kVAR, regardless of voltage. As voltage drops, so does VAR injection. It is important to model this distinction in planning, and vital in any engineering studies.

Equipment controls

Line drop compensators and load tap changers and switching equipment are best represented on a phase-by-phase basis, for even if they are balanced as to capacity and type, their loading can be quite unbalanced, affecting performance.

Wound, tap-changing devices can be approximated for planning purposes as "step-less" devices which maintain voltage at a control bus at a constant level, regardless of upstream voltage level. Controls can be modeled simply, as just showing the device keeping its control bus voltage at the specified target as long as upstream voltage does not stray outside a specified range.

Capacitor switching can be modeled as on-off as voltage or power factor requires -- switching control does not have to be modeled.

Switches

For most distribution planning studies, switches can be represented on a phase-by phase or gang-operated basis, with representation of current-carrying capacity as their only attribute. For automated distribution and reliability studies, switching times, probability of switching, and other attributes may be required.

Too Much Equipment Detail is Futile

Planning requires less detail in precise equipment representation than engineering, and the recommendations given above are sufficient to meet legitimate planning purposes, but not some precision engineering functions. Many utilities are combining the short-range planning and engineering functions, in order to streamline planning and engineering and reduce labor requirements. In such cases, the level of detail recommended above is minimally sufficient for planning and basic system engineering purposes, if equipment details are then studied on a unit by unit basis prior to specifying their details for installation and control system settings, etc. This approach is recommended: that combined planning-engineering functions be performed by doing the planning with the level of detail recommended above, followed by engineering on an equipment-specific basis, for one reason -- usually, planners do not know the details of the equipment they are specifying during their planning.

For example, their studies may call for voltage regulator at a particular point. During the early planning phases, planners may not know the exact capacity they need. Later on, they will identify this and other aspects of the regulator's requirements. However, they will not know the exact manufacturer and model type because that will not be known until the unit is purchased. Thus, details in representation of the regulator in their planning models, to a level of detail that would distinguish impedance, control response characteristics, and other among different manufacturers or models, would be speculative in the planning process, and effort to enter and use such data not more than a waste of effort.

16.4 MODELS OF ELECTRIC LOAD

Load models for distribution planning fall into two classes. The first, used in performance simulators during short-range planning and engineering applications, represent the load under specified conditions (e.g., annual peak period) and model the load and its behavior with respect to those conditions. Into this category fall the models of load used during load flow, motor starting, harmonics generation and sensitivity, and other similar applications -- models whose purpose is to *represent the electrical behavior* of the loads.

The second class of load models includes those used for long-range planning, their purpose to model the load with respect to capacity planning and economic considerations. Most of these distribution planning models treat the load as a quantity independent of the planning process or influence of the utility, to be forecast or analyzed (by a spatial load forecast or other means -- see Chapter 15). These will be discussed later, in Chapter 17.

The model of electrical load used in a load flow performance simulation has as much, or more, to do with the accuracy and applicability of the computed

Distribution Feeder Analysis

results than any other model except perhaps the circuit representation itself. The best load model represents load as either delta or Y connected, as distributed among the three phases in any proportion, and as being composed of any type of any mixture of constant power, constant current, or constant impedance loads. This results in up to nine "loads" present at any one node in the circuit, as shown in Figure 16.5. Each of these elements can have its own power factor.

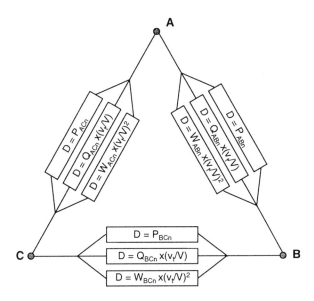

Figure 16.5 General model recognizes that there are nine separate load models at a three-phase node. Shown here, the demand at a delta-connected node, n, consists of three types of loads connected between each pair of phases -- constant power, constant current, and constant impedance, designated as P, Q, and W respectively. Conceivably each could have a distinctly different power factor. So whether entered as load and power factor, real and imaginary load, or magnitude and angle, this load model requires *eighteen* data input values. From a practical standpoint, without loss of real accuracy, the model can be simplified and data requirements reduced to twelve values, by deleting the constant current representations. Building a computer program to accommodate this level of detail is simple; populating its database with accurate data is extremely difficult.

Constant Power, Constant Current, and Constant Impedance Loads

Many electric appliances and devices have an electric load that varies as the supply voltage is varied. Generally, loads are grouped into three categories depending on how their demand varies as a function of voltage: constant power (demand is constant regardless of voltage), as a constant current (demand is proportional to voltage), or as a constant impedance (power is proportional to voltage squared). The load at a particular point might be a mixture of some proportion of all three.

It is quite important in both planning and engineering to model the voltage sensitivity of load correctly. For example, incandescent lighting, resistive water heaters and cooking loads, and many other loads are constant impedance loads. On a feeder with a 7.5% voltage drop from substation to feeder end (the maximum permitted under typical "range A" voltage drop criteria -- see Chapters 4, 8, and 9) a constant impedance load will vary by up to 14.5% depending on where it is located on the feeder. The same set of incandescent lights that create 1 kW of load at the feeder head would produce only .855 kW at the feeder end.

On the other hand, 1 kW constant power is 1 kW at 1.0 per unit voltage, at .97 PU, and at .925 PU (i.e., at the maximum voltage drop permitted under typical range A voltage drop limits of 7.5%). Induction motors, controlled power supplies, as well as loads downstream of a tap-changing transformer appear to the power system as relatively constant power loads and are generally modeled in this manner.[5]

Generally, loads are represented with a type and load at nominal (1.0 PU) voltage. Thus, a 1 kW constant power load is referred to as "1 kW, constant power" and a constant impedance that produces 1 kW at 1.0 PU is referred to as a "1 kW, constant impedance load." From any practical standpoint, it is not necessary to model constant current loads, however, the distinction between constant power and constant impedance loads is important.[6] Since voltage drop on a feeder depends very much on the load, the manner of representing load is critical to computing voltage drop accurately. For example, a three-phase 12.47 kV line built with 336 MCM phase conductors (of the type discussed in Chapter 7) can serve a 10.72 MW constant impedance load at a distance of two miles, while staying just within "range A" 7.5% voltage drop limits. The "10.72 MW constant impedance load" will create a demand of only 9.36 MW due to the voltage drop (9.36 MW = $(1 - .075)^2 \times 10.72$ MW). Thus, the feeder stays within

---

[5] Only steady-state voltage sensitivity is important in distribution planning, and is considered here. Planning generally does not consider transient voltage sensitivities.

[6] A 50%/50% mix of constant voltage and constant power loads represents a constant current load within 4% over the range .9 to 1.1 PU.

Distribution Feeder Analysis

voltage criteria if serving a "10.72 MW constant impedance load." However, voltage drop rises to 8.59 volts, more than a volt beyond what the criteria permits, if the load is a 10.72 MW constant power load.

The cost of writing and putting into place software to represent loads as constant power, constant impedance, or any mixture thereof is minor compared to the cost of obtaining the data for such a model. Usually, the mixture is estimated in one of the two ways:

> All loads, except special ones, are designated as a default mixture of power and impedance loads. In the absence of measured results to the contrary, the recommended general rule of thumb is to use a 60%/40% constant power/constant impedance mixture for summer peak loads, and a 40%/60% split for winter peak loads.

> The proportion of constant power and constant impedance loads at each location in the power system is estimated based on a spatial end-use appliance model (see Chapter 15, or Willis, 1996).

Phase Assignments of Load Models

Clearly, a one-line equivalent algorithm or computer program can represent loads only as phase-less, so that planners with access to only such tools are restricted to using a phase-less model only. However, computer programs and analytical programs that can represent imbalance in loads, assigning loads in any amount to any of the three individual phases present at a node, are widely available. Despite this, many utilities do not have accurate data about the individual phase assignments of single-phase service transformers or loads -- a single-phase service transformer located on a three-phase feeder trunk could be served by any of the three phases, etc., and records are not kept about which phase actually serves the transformer.

There is merit and advantage in using an imbalanced, three-phase analytical method or program, even when phase-assignment data is not available. In American distribution systems, since over 75% of circuit miles are single-phase (see Chapters 9 and 11) most loads can be assigned implicitly, they are obviously served by the phase of their single-phase lateral. It is only necessary to know the phase-assignment of each lateral -- something that can be determined in two or three hours of field observation for overhead lines -- to gather the data to assign all single-phase laterals correctly, and about 75% to 80% of loads correctly. Loads at all three- and two- phase nodes can be split evenly among the phases present at a node. While approximate, error in analytical results using this approach is typically reduced to only about 25% to 33% of that of one-line equivalent models.

Power Factor

Generally, constant impedance loads have high power factors, on the order of 90%, while constant power loads (most typically induction motors or controlled power supplies) can have power factors as low as 60%. Recommended values in the presence of no information to the contrary are 95% for impedance loads, and 75% for constant power loads. (See Chapter 9, Section 9.6).

Usually, power factor data is available only on a feeder basis, measured at the substation, and for large loads. However, power factors for small customers and generic models of typical nodes must be based on load survey and end-use analysis.

Assigning Loads to Individual Nodes

With very few exceptions, actual load values represented at nodes during short-range planning and distribution engineering are estimates. Only a small portion of loads -- usually large and critical loads -- are ever measured for peak demand and power factor -- most are measured only in terms of total kWhrs sold over a lengthy (monthly) billing period. Even for those loads that are measured for demand and power factor, information on coincidence of their peak values is unavailable -- it is very unlikely that they peaked at the same time.

Thus, almost all loads in the distribution circuit model are estimates based on the data that is available, usually kWhr sales (billing) data, peak demand for major loads, and measured peak data on a feeder or major branch basis, obtained from substation meters or SCADA. As a result, most nodes are assigned loads based on an allocation, where a large total, usually that measured on a substation or feeder basis, is allocated among the nodes using one of several procedures.

Allocation based on connected transformer capacity (TkVA)

Suppose a feeder has 15,300 kVA of service transformers connected to it, and a measured peak load of 7,650 kVA. Every node could be assigned a load equal to half its connected TkVA -- a node with 150 kVA of transformers is given 75 kVA, one with 25 kVA of transformers is given 12.5 kVA, and so forth. Such assignment of loads can be done on a phase-by-phase basis if load measurements and connected transformer capacity is available on a phase-by-phase basis.

Such allocation is only approximate, and purely pragmatic. Transformer capacity has no significant influence on load, but experience and judgment indicate that usually transformers are sized based on the expected peak load. Thus, TkVA is a known quantity somewhat proportional to expected peak load.

Distribution Feeder Analysis

More important, for most utilities, no other quantity known at every node has anywhere near as good a correlation to load as TkVA.

Usually, the allocation procedure is applied only to small loads. Large, or all "spot" loads, or loads known in value -- are determined prior to allocation and directly input to any computer program. For instance, perhaps in addition to the 7,650 kVA measured at the substation in the example feeder discussed above, two large loads are known, having been measured at maximums of 475 kVA and 800 kVA, and have 750 and 1000 kVA of transformers, respectively. These two nodes are assigned their measured loads. The remaining nodes on the feeder are then assigned a load equal to (7650-1275)/(15300-1750), or .435 times their connected transformer kVA.

Losses should be accounted for in any such allocation. They were not in the allocation described above. The 7,650 kVA metered at the substation presumably included losses, which could be 6% or more of the total. What is desired from the allocation is a set of loads assigned to the nodes that results in a load flow computation showing 7,650 kVA of load at the substation. Many distribution feeder analysis computer programs have features to perform such load allocation, using their load flow engine in an iterative procedure to adjust loads proportional to TkVA (or some other nodal attribute) until the computed total load equals the metered amount.

Approximately two-thirds of distribution load models use some sort of TkVA-based allocation as described here, often slightly modified or with special features to take advantage of other data available or to accommodate other factors the planners believe are important.

Estimates based on customer billing records

Rather than estimate load based on transformer capacity, load estimates can be based upon actual billing records. As a simple example, suppose that on the feeder where load was metered at 7,650 kVA, sales for the peak billing period were 2,900,900 kWhr. A particular node may serve ten customers, whose consumption during that period totaled 15,000 kWhrs. Then this node will be allocated 15,000/2,900,900 × 7,650 kVA, or 39.567 kVA as the modeled peak load, and other nodes similarly allocated "their share" of the 7,650 metered peak based upon their portion of the total kWhr sales.

This allocation method generally gives superior results to allocation based on TkVA alone, but it requires an additional item of data not available to many utilities -- data that ties customer records to locations (service transformers or nodes) in the circuit model.

Again, as was the case with allocation based on TkVA, this can be done on a phase-by-phase basis; typically large or other measured loads are excluded and assigned directly; and losses should be accounted for with an allocation that computes them based on load-flow representation.

The allocation is usually based on a more complicated formula than merely metered kWhr as used above. To begin with, different classes of customer may have different contributions despite having the same kWhr, values which are determined based on load research data. For example, load research data may have determined that a small commercial establishment with 1200 kWhrs is most likely to have a load during peak of 3.3 kVA, whereas a home with this level of sales has a peak of only 3 kVA.

In addition, some load allocation models base the estimate of the peak load at a node on the number of customers, in addition to their total metered kWhr and their customer-type data. What is being addressed in such models is coincidence effects as discussed in Chapters 2 and 13. As described in Chapter 2, due to coincidence effects, the measured load for a group of ten residential customers, each with 1200 kWhrs of sales, will be less than the measured peak for any subgroup of four of them, which will be less than the average peak load measured on an individual basis. Therefore, many billing-data driven load allocation models use the number of customers at a node as one factor in the estimation of load, so that the estimate of peak load at the node is adjusted for coincidence based on group size.

Such estimates-based-on-group-size may be more accurate than those computed without group size data -- generally they do result in load values that are closer to actual measured peaks, *but they are inappropriate for performance simulators* unless a "coincidence load flow algorithm" is being used (these will be discussed later in this section). Such algorithms are very rare.

Adjustment of loads to correspond to the level of coincident peak measured at each point is not the goal of load allocation -- adjustment of the loads to reflect contribution to feeder peak load is the goal. Loads should be estimated for the same level of coincidence, regardless of the number of customers at the node -- generally this means they are adjusted to the feeder-peak level. "Coincidence load flows" and coincidence interaction with simulation are discussed later in Section 16.6.

> *Load allocation based on customer count and non-coincident load data on a nodal basis does produce more accurate estimates of local peak loads, but such loads are inappropriate for feeder performance simulations.*

Distribution Feeder Analysis

Load estimates based upon end-use or detailed customer models

Long-range planning tools such as the load models described in Chapter 15 and elsewhere (see Willis, 1996, and Willis and Rackliffe, 1994) often use very detailed representations of customer appliances and usage patterns. Generally, if such models are detailed and accurate, and if customer type and count is known by node, these models can provide the best approach for determining the apportionment of load between constant power and constant impedance load types, and the power factor, on a node-by node basis.

In some cases, loads computed directly from such models are assigned to circuit model nodes for performance simulation. It is recommended, however, that they are used only as allocation factors, which are then used in company with billing kWhr to allocate the feeder peak. The reason is that such models were not originally developed for node load allocation and they are not robust in this application.

Spatial customer and end-use appliance models often are implemented in very impressive software packages with comprehensive numerical engines and extensive graphics and user-interface features. They use extensive customer, demographic, and geographic databases, and when set up in their most comprehensive form can give the impression that the planner "sees all and knows all." In their designed application, forecasting and analysis for planning decision making, such models are amazingly robust. The best spatial models have little if any model error -- they are among the very best examples of tools whose error sensitivities and robustness have been engineered for their applications.

However, such robustness does not extend to using them directly for allocation of load on a node basis. In such applications they can produce large errors. Their use to determine and set allocation factors for one of the two types of allocation methods given above is recommended, and generally gives the best results possible.

16.5 MODELS OF ELECTRICAL BEHAVIOR

Computation of the voltage drop, power factor, and current flow on the distribution system is a fundamental part of power distribution system planning, resulting in "load flow" results, as shown in Figure 16.6. Analysis of distribution system behavior during normal situations requires a system model -- the composite of the circuit, equipment, and load models discussed above -- and an *electrical model* which represents the behavior and response of electric power flow.

Many of the approximations and analytical constraints forced on distribution planners are due to limitations of these system models. As mentioned previously, if the electric utility does not have phase-by-phase data then detailed three-phase analysis cannot be carried out, even if the computational technique can analyze such data. Likewise, load data may not be exact or specific, and the precise characteristics of some small equipment may not be known with high confidence, either. For these reasons *data accuracy* may limit the degree of *analytical accuracy* which can be applied in any meaningful way.

The electrical model is the representation of the electrical behavior of circuit and equipment elements, and can include numerous computational shortcuts and approximations of reality. For example, some three-phase load flow computation methods dispense with any consideration of mutual coupling (the influence of one phase on another), which greatly simplifies the calculations used and accelerates their completion, but which contributes noticeable inaccuracies in cases where imbalance in load or configuration is significant. As a more subtle example, most distribution load flow "engines" are based upon symmetrical component models of power flow -- but symmetrical component analysis itself makes certain assumptions about symmetry (balance) with respect to the neutral. While insignificant is some cases (particularly when applied to high voltage, delta-connected transmission), these assumptions of symmetry can develop into significant inaccuracies in imbalanced distribution situations.

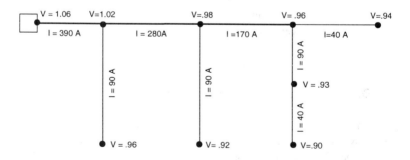

Figure 16.6 A load flow computes voltages at certain points in the system (nodes) and current flow through segments, based on data for loads and equipment, as shown above.

Distribution Feeder Analysis

As mentioned earlier, any one of many computational methods could be used to determine the values that result from application of an electrical model that represents three-phase unbalanced behavior but excludes mutual coupling and line charging. There are literally hundreds of load flow computational methods covered in the technical literature, but most of the differences described revolve around details of whether complex variables representing voltages and currents are represented in polar or rectangular coordinates, or whether real and complex power flow are "de-coupled" for computation, and how the convergence of the iterative, success-approximations method of solution is managed (i.e., Gauss or Newton method, etc.).

While such details of computational method are quite interesting technically, they are usually not relevant to practical planning applications. Planners should understand the limitations of the method and data being used, but assuming a distribution performance simulator's algorithm is well-written and tested, it should compute the voltage, current, and power flow accurately within the context of the electrical model being applied, be that a simple one-line DC model, or a very involved "five conductor" representation that includes all possible aspects of electric flow and interaction.

One-Line DC Model

By far the simplest representation possible with respect to both data and electric model is a "DC model" in which electrical behavior is not analyzed in a complex (phasor) manner, but represented simply as scalar quantities. The representation of AC electrical behavior with a DC model is only approximate, and for this reason generally applied only when computational resources are very limited or when the circuit models involved are only very approximate.

By dispensing with analysis of phasing or complex quantities, DC load flows gain tremendous computational speed advantages as well as immunity from a host of analytical complications inherent in comprehensive AC models (for example lack of convergence is never an issue). They are widely used where speed is more important than absolute accuracy, for example as the "inner loop" compute voltage and loading in optimization algorithms where thousands of load flow computations for various alternative designs may be necessary as part of the determination of the very best overall configuration.[7]

---

[7] Here, speed is of the essence -- a DC load flow may require only 1/1000th the time of a comprehensive unbalanced AC analysis. The approximate results are not important, for while the DC load flow may provide only very approximate results, it will still tend to rank various candidate configurations correctly with respect to losses costs and voltage drop.

One-Line AC Representation

Traditionally, this has been by far the most popular analysis method for distribution. It requires no phasing data in the circuit model, and no analysis of multiple phases, mutual coupling, etc., in the electrical model. Every element is represented with a one-line equivalent model. Three-phase elements are assumed to be perfectly balanced in both impedances and loadings, and a single-phase lateral branching from a three-phase trunk is represented as being served equally by each of the feeder's three phases, with its load distributed 33% on each of those phases -- a physical impossibility but nonetheless used in such analysis. While approximate, this type of analysis was traditionally chosen as the best compromise between available resources and required results.

Phase-by-Phase Computation Using a One-Line Electrical Model

Among the many benefits of three-phase analysis, by far the largest in terms of general accuracy is representation of unbalanced loading and configuration This is almost purely a circuit model (data and database) concern. In this approach, a one-line load flow algorithm is used to compute voltages and currents on each of the three phases individually, as if the other two did not exist, as shown in Figure 16.7. Three "phase load flows" are performed on the three-phase data model, by stripping out each phase's single-line "phase model" with its loads and computing voltages and currents for it with a single-line electrical model. Alternately, this may be done simultaneously with an algorithm whose off-diagonal elements (Z matrix) are zero. Either way, the simplification is in the electrical, not the circuit model.

While this approach does not acknowledge unbalanced electrical behavior particularly the impact of mutual coupling, it does compute the effects of uneven balancing of loads and configurations, both in aggregate and on a segment by segment basis. The gain in analytical detail provided is considerable, and the author considers this to be minimally sufficient in a majority of planning cases, particularly in urban and suburban situations, where roughly 90% to 95% of all unbalanced variation in voltage and current is due to imbalance of loadings.[8]

---

[8] In cases of long and lightly loaded feeders in rural areas, or where voltage drop constraints, not thermal capacity are the limiting factors in design, unbalanced mutual coupling can contribute significant voltage imbalance even if loads were balanced. Only in such situations is the computation of unbalanced mutual coupling critical.

Distribution Feeder Analysis

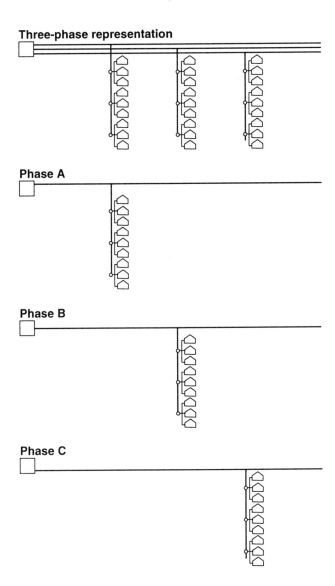

Figure 16.7 The unbalanced three-phase circuit representation (top) can be analyzed "a phase at a time" using a one-line equivalent electrical analysis applied to each of the three single-phase models developed from the top three-phase representation. This provides many of the benefits of full three-phase analysis while using only the simpler one-line analytical method.

Three-Phase Circuit Model with Three-Phase Electrical Analysis

More accurate analysis of the voltages and currents in a three-phase circuit model is done by using a computational method that simultaneously analyzes the three phases and their interactions (mutual coupling) in company with an accurate system model. There are a number of aspects relevant to the detail with which this can be done.

Symmetrical component representation

Assumptions of certain symmetries -- balance of a sort, are at the heart of symmetrical components analysis, and thus this very popular method of power systems analysis is approximate and not completely valid for circuits where real load is unbalanced, where reactive loading or injection in unbalanced, or where mutual impedances differ substantially among phases due to configuration. A popular approximation is to represent four-wire wye-connected circuits with only the three-phase components. This is usually done with symmetrical component models but in all cases is only approximate (although the difference is usually not significant).

Four-wire "Carson's equation" representation

An electrical behavior model based upon a full four-wire representation of wye-connected electrical behavior in a circuit will provide both a computation of the neutral current on a segment basis as well as the slight voltage imbalances that accompany such current. This type of model is sometimes referred to as a "Carson's equation" model in that it can apply directly the four-by-four impedance matrix derived from Carson's equations, which do not make assumptions about symmetry or balance (Carson, Kersting). This is generally considered the most accurate, practical distribution load flow method.[9]

Line charging

Line charging can be an important factor in determining voltages in underground cable distribution systems and in certain other cases -- for example it can lead to higher than one per unit voltages at the end of long, lightly loaded feeders in

---

[9] "Five-conductor models" that represent the effects and flows of ground as wel as all four wye-connected conductors have been developed, but ground flow is slight in most normal circumstances. Regardless, data on the exact impedance of (the hetrerogeneous soil) is so imprecise as to render such analysis imprecise. The majority of impact of ground -- its mutual coupling with the conductors, can be included in the four-by-four impedance matrixes and even its behavior as the return conductor in earth return systems can be approximated within close margin without resorting to such modeling.

Distribution Feeder Analysis

rural distribution systems. Line charging is typically represented in most AC load flows. However, several popular distribution analysis software packages do not render line charging well. Additionally, accurate values for line charging are sometimes inadvertently left out of cable and conductor databases. In either case, a significant, and occasionally, critical element of electrical performance has been neglected.

Node or segment load representation

In modern load flow analysis, loads are represented as associated with nodes. However, a number of older voltage drop computation procedures -- many still be in use at the time of this writing (1997) -- associate loads with segments, not nodes. Such "segment load" models represent the load as spread evenly along the feeder segment. They were popular (in fact, preferred) when computations were done mostly by hand or with limited computer resources as was the case until the late 1980s. In those cases, representation of an entire feeder with somewhere between five and twenty-five segments was typical. One segment in

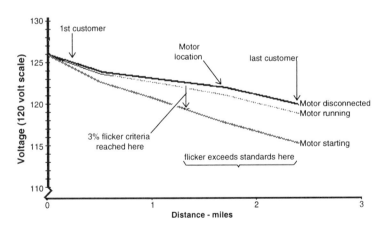

Figure 16.8 Motor-start study profiles for a feeder trunk and branch due to a 550 HP 3-Ph. induction motor (from Figure 4.3). Generally, a motor start analysis consists of three load flows, modeling the motor as off, as running at full speed with a maximum mechanical load, and as "locked-rotor" -- energized and at zero RPM, as at the very instant it is started. Profiles are compared to identify if and where voltage fluctuation during starting violates the utility's standards.

such a model might represent a mile of primary feeder, which had fifty service transformers along its length. Such a distributed load model was most appropriate under those circumstances. Modern distribution circuit databases use much greater node resolution, often to the point of representing each service transformer with an individual node. In such situations, representing load on a segment basis is not appropriate.

Motor Start Analysis

The temporary starting current from large motors can often be sufficient to cause undervoltages on the portion of the distribution system serving them, and "flicker" service quality problems for customers nearby (see Chapter 4). Motor-start studies to determine if a particular motor will cause such problems, and if so, how the problems can be mitigated, bridge the boundary between planning and engineering. Normally, "motor-start" or "flicker" studies consist of three load flow cases, so that engineer-planners can compare the voltage profiles along the feeder with the motor off, with it operating at full speed, and as it is starting, as shown in the three profiles in Figure 16.8. These studies consist of three "static" load flow cases, each set up with the same loads and conditions, stable for the representation of the motor, which is represented as zero load, full running load, and as a starting current.[10] Such load flow studies are fairly straightforward, however, validity rests on several important aspects of modeling:

> Set switched equipment such as regulators, switched capacitors, etc., so that they do not change status. Some load flows represent switched equipment with dynamic models that will "operate" the equipment, turning on capacitors and changing regulator taps, if needed to boost voltage, etc. Typically, control systems of regulators and capacitors are set to respond only to long-term changes (many seconds or minutes) in conditions, and thus they will not change status during a motor start up.

> Represent the loads' voltage sensitivity (see Section 16.4) as accurately as possible -- the difference in representing the rest of the loads on the feeder as constant power or constant impedance can make a substantial difference in the motor start case results.

---

[10] Representation of motors as they start is perhaps one reason to have a load flow that can model "constant current" loads (see section 16.4). Setting the motor to a constant current load equal to the motor's starting current is the easiest way to run the "locked rotor" case.

Distribution Feeder Analysis

The model of the motor itself is important -- its impedance, delta or Y connection, etc., particularly if it is starting in the face of any imbalance in voltage or other unusual situations (under some arrangements, three-phase motors can be run from a two-phase feed). In addition, large motors often have capacitor switching and other starting features to ease initial inrush current.

All the foregoing describes the most common type of motor start study, which is limited to an assessment of whether the motor start will cause overload, undervoltage, or flicker problems on any portion of the feeder. A more involved type of analysis is necessary in a few cases, when it must be established whether the motor will, in fact, start and run up to speed. When supply impedance is marginal, when power factor is already poor prior to starting, and when the motor must start against its full mechanical load, it may stall, or be unable to run up to full speed. Or it may take so long to accelerate up to full speed that it and other equipment overheats due to the high starting current. Transient motor start analysis can establish expected starting behavior to this level of detail: they step through a motor start simulation in short time slices (e.g., 1/180 second), modeling the motor with both electrical and mechanical (rotor inertia, and mechanical load, resistance and inertia, etc.) as they determine if, and how fast the motor will accelerate to full speed.

Load Flow Analysis Priorities

The foregoing discussion highlighted some of the differences in analytical method and data which distribution planners may face. If approximate data and method are all that is available, then they will have to suffice, but the planners must be aware of the limitations imposed on their planning as a result, and respect these limitations.

Every performance simulation method (i.e., every combination of a database and an algorithm) provides a certain level of dependable accuracy -- the computed voltages drops and currents are accurate within ± 10% or ± 5%, or ± 3%, or their values, etc. This range creates a "gray area" within which circuit performance is not exactly known, and this range -- whatever it is -- exists around every value computed in a distribution performance simulator, whether acknowledged or unacknowledged. Increasing levels of comprehensive detail in data representation and accuracy in electrical behavior computation, can reduce this approximation range. In fact, it is possible to reduce it to inconsequential levels by use of specific, timely and accurate data applied to highly accurate analytical techniques. However, in many cases the cost of providing more

Table 16.2 Recommended Factors to Include in Load Flows

Always Required

1. Phase-by-phase representation of circuit and loads (rather than one-line data) even if analyzed on a phase-by-phase basis.
2. AC rather than DC representation of electrical behavior.
3. Representation of loads as constant impedance, constant current, constant power, or any mixture of the three.
4. Representation of loads as associated with nodes, not segments.
5. Representation of line charging associated with both underground and overhead lines.
6. Accurate representation of the electrical behavior of equipment, including capacitors and regulators, etc.
7. Representation of mutual coupling among phases (e.g., use of an unbalanced electrical behavior model).
8. Four-wire model of electrical behavior, not based upon symmetrical component models.
9. Site-specific rather than generic data for line equipment, such as regulators, etc.
10. "Five conductor" rather than four-wire models of electrical behavior.
11. Representation of load voltage sensitivity as including constant current types.

Seldom Justifiable Based on Contribution to Accuracy

accuracy is not warranted by the improvement in results that are obtained. Planners must bear in mind that this issue should never be judged on the basis of level of accuracy, *per se,* but rather on the basis of how the accurate results contribute to the planner's goals of achieving satisfactory levels of service at the lowest possible cost. Table 16.2 lists the author's recommendation on the order of importance of various capabilities discussed in this section. The first six are generally necessary for accurate distribution planning; the last two are seldom, if ever, justifiable.

Distribution Feeder Analysis

Short-Circuit and Protection

A load flow analysis simulates the performance of a feeder under normal, or design conditions. A short circuit analysis simulates the performance under abnormal conditions that, while unlikely, must be anticipated. Its application is best supported by models and analysis of protective equipment including relays, breakers, fuses, and reclosers. Such tools, and their application, are important, but a part of engineering, not planning, and as all books must have boundaries, they are not discussed here. Several references give good discussions on short-circuit computation and protection (Burke, Grainger, and Stevenson).

Reliability Analysis

A completely different but increasingly important type of analysis involves assessment of the expected availability and power quality at points throughout the system. At the time of this writing (early 1997) reliability assessment on a routine basis is a relatively new feature of distribution planning, but with increased competition in the utility industry, and the growing importance of customer service quality that this can be expected to spawn, reliability analysis is very likely to become as common as load flow analysis.

Reliability analysis consists of two steps, independent of one another in both function and application -- some utilities perform only one of the other. To this, the author has added a third, and generally not recognized step required to combine the two types in a comprehensive study of reliability.

Historical reliability assessment, whose goal is to analyze system and historical operating data to assess the reliability "health" of the system, to identify problem areas, and to determine what caused any problems.

Predictive reliability assessment, whose goal is to predict expected future reliability levels on the system, both in general and at specific locations, by analyzing a specific candidate design for a part of the distribution system, and determining expected reliability. This is the type of analysis that is used to study how various "fixes" might solve a reliability problem.

Calibration is required to adjust a predictive model so that it correctly predicts past events on the system, if it is to be in any way reliable in predicting what will occur as changes are made to the system. Calibration is not easy -- it requires a good historical base of information, a good historical assessment model, and careful adjustment of a myriad of factors. But it is essential if the predictive assessment model is to be dependable.

The goal of most reliability analysis is to compute, for every node in the system, the expected frequency and duration of outages. In addition, a desirable, (but difficult) aspect is to compute the expected frequency and severity of voltage dips -- a dip in voltage to 70% of nominal may not represent an "outage" but to some customers and some electronic equipment it may have the same effect (Chapter 3).

Many different analytical approaches exist to estimate reliability levels on a distribution system. Most function somewhat like a load flow, in that they must work on a database of circuit and equipment models that provide connectivity, as well as details about equipment throughout the system (of course, reliability analysis needs details relevant to reliability, rather than electric flow analysis). In addition, some reliability analysis methods need the results of a load flow solution prior to their application, in order to track power flow. Those that compute expected frequency and severity of voltage dips also need access to a fault-current analysis of the system.

All reliability-analysis methods involve some application of probabilities, but there are several fundamentally different approaches. Those that seem to produce the best results analyze connectivity of equipment and power flow, and the operation/failure/repair of each unit of equipment explicitly. Such analysis will include the opening and closing of breakers and switches to isolate and then to pick up outaged portions of the system from alternate sources.

16.6 COINCIDENCE AND LOAD FLOW INTERACTION

Figure 16.9 shows typical peak-day load curves (winter) for large residential all-electric homes (about 2500 square feet) in one subdivision of a metropolitan area in the southwestern United States, illustrating the effects of load coincidence on load curve shape and peak load. As was discussed in Chapter 2 (Figure 16.9 shows data from Figures 2.4 and 2.8), the load curve shape and peak load per customer seen by the system varies depending on the number of customers in the group being observed.

A coincidence curve, showing "peak load as a function of group size," can be determined by measurement and analysis. Figure 16.10 shows the coincidence curve for the customers whose load curves are plotted in Figure 16.9.[11] Individual non-coincident peak load exceeds 22 kW each, whereas in a group of five, peak per customer is about 16 kW; and in groups of 25, 13 kW; while for a

---

[11] The load curves shown and the coincidence behavior shown are based on five-minute demand periods. Qualitatively identical behavior occurs at any demand period length -- 15 minutes, 30 minutes, 1 hour, or even daily.

Distribution Feeder Analysis

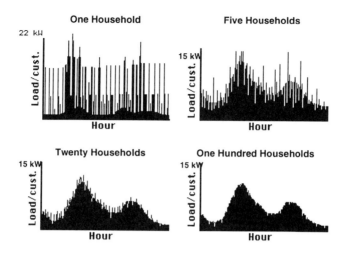

Figure 16.9 Daily load curves measured on a five minute basis for groups of two, five, twenty, and one hundred homes in a large suburban area. Note vertical scale is in "load per customer" for each group. *Peak load per customer decreases as the number of customers in the group becomes larger.*

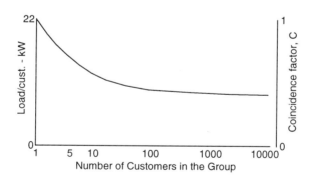

Figure 16.10 Coincidence curve for the residential load curves plotted in Figure 16.9. The peak load per customer drops as a function of the number of customers in a group.

group of 100 homes, the coincident peak load per customer is 9.9 kW. Coincidence can be given by the coincidence curve, which is the ratio of peak load per customer plotted against individual peak load

$$C(N) = \text{coincidence factor for group of N customers}$$

$$= \frac{\text{(observed peak for the group)}}{\Sigma \text{ (individual peaks)}} \qquad (16.6)$$

What peak load per customer should be used to model loads in a load flow study of peak conditions on the circuit feeding the customers whose loads are shown in Figure 16.9? To illustrate the limitations coincidence puts on load flow accuracy, consider a feeder which serves 500 customers of this type. The feeder's peak load would be about five hundred times 9.9 kW, or 4.95 MW.

Suppose that the planners wanted to study this feeder in as much detail as possible. As a result, they build a circuit model of both the primary and service levels, and represent every customer meter with an individual node. What loads should be input into the load flow for each customer node? A value of 22 kW is the most accurate estimate available of the actual load that would be metered at each customer point. However, using 22 kW/customer loads would produce load flow results that showed a feeder peak load of 11 MW, more than twice the actual feeder peak. The computed voltages would be much lower, and the computed currents much higher, than would ever occur during peak conditions.

Since the load flow effectively adds together all nodal loads, in order to produce the correct total feeder load, individual customer loads must be modeled at their coincident feeder-peak level -- in this case, essentially as 1/500th of the group total for 500 such customers, or about 9.9 kW each. It would be a mistake to represent loads as 22 kW/customer, or at any other level other than the feeder-peak coincident 9.9 kW/customer.

The most accurate estimate of peak load at a node is not *the load that should be used in a load flow.* Rather, the node's contribution to feeder peak should be used. The 22 kW value is the most accurate estimate of each individual customer-node's peak, but a value of 9.9 kW should be used as the load at that node. Similarly, load of 80 kW (five times 16 kW) is the best estimate of the peak load seen at a node with five customers behind it, but a load of 49.5 kW should be used in a load flow for any node representing five customers. This is the reason behind the recommendation given earlier (Section 16.4) that customer count data on a nodal should *not* be used to help adjust load data derived from billing data for coincidence. The load to use at a node is the best estimate of the customers' net contribution to feeder peak load, not the best estimate of the peak load seen at that node.

Distribution Feeder Analysis

All Load Flows Underestimate Maximum Voltage Drop and Current Flow for Equipment Near Customers

In the example cited above, the planners have little choice but to use feeder-coincident loads of 9.9 kW per customer for load-flow analysis. More generally, planners must use feeder- or substation-coincident peak loads throughout all distribution analysis -- anything greater results in a computed "overload" and an unrealistic case. With such loads in place, the feeder and substation peak loads, and the current flows and voltage drops for "large" parts of the feeder system (trunks, major branches, etc. -- any section serving more than about 50 customers) should be accurate. Assuming that other data is accurate and detailed, and the load flow engine produces a good solution, the computed voltages and currents will be quite close estimates of reality.

However, voltages and current flows computed will be underestimated on all "small" portions of the feeder -- those serving fewer than twenty five customers or so. Figure 16.11 shows a lateral which serves only five customers, modeled as five segments and their nodes. In the top drawing, in order to make the load

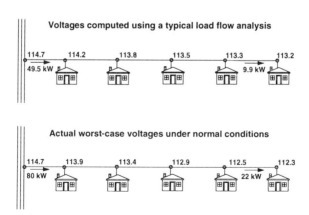

Figure 16.11 Top, voltages and currents computed based on coincident customer loads on a suburban feeder trunk and one of its a lateral branches. Bottom, actual worst voltages experienced due to customers and customer group peak loads.

flow "add up correctly," customer loads have been represented at their coincident contribution of 9.9 kW each, resulting in a flow of 49.5 kW on the lead segment of the lateral, and only 9.9 kW on the end segment. Primary-level voltages computed by the load flow are shown, and appear to meet the utility's requirements for at least 113 volts (120 volt scale) on the primary level.

The actual maximum flows and minimum voltages that can be expected to occur during peak conditions on this lateral are shown in the bottom diagram in Figure 16.11. Flow on the lateral's first line segment occasionally hits as much as 75 kW -- 50% more voltage drop along that segment than modeled by the load flow. All current flows and voltage drops computed on the lateral's downstream segments are low by between 60% to 120%. As a result, the voltages and loading calculated by the load flow do not represent the worst than can be expected. As shown, actual values fall below the utility's standards by a significant amount.

There is little that can be done with a normal load flow algorithm to correct this mismatch. If given non-coincident loads, it will overestimate total load and thus greatly overestimate voltage drop and loadings on the trunk and other portions of the feeder. Thus, its ability to then compute drops along all lateral sections correctly is rather a moot point.

"Coincidence Mismatch" is a Potential Problem

Coincidence mismatch is serious limitation in load flow applicability, of which many distribution planners are unaware. There is little that can easily be done about it (a work-around for some situations will be described later in this section). As a result, it creates three potential problems for planners who depend on a load flow evaluation to design their system

> Optimal equipment design and selection for portions of the system near the customer, particularly at the service level, is complicated because adjacent units of equipment may not see the same load curve shapes even though they are serving some of the same loads (see Chapter 13).

> Losses estimated based on load flow results are generally underestimates, often by embarrassingly large amounts.

> Voltage drop at some points of the system may fall sufficiently below predicted levels that service quality problems result. Worse, because the "needle peaks" of non-coincident load behavior are short, these problems may be intermittent, and voltage fluctuation caused by the load shifts may cause flicker problems.

The nature of the first two problems are discussed further in Chapter 13.

Distribution Feeder Analysis

In most utility systems, "coincidence mismatches" between computation and reality like those shown in Figure 16.11 create noticeable problems on the primary system only at the extreme ends of limited-capacity portions of the feeders, those that serve just a few customers. Unfortunately, such locations are where voltage drop is usually most extreme and service quality problems most expected.

The mismatch shown in Figure 16.11, of nearly a volt on a 120 volt scale, is about the worst that will be seen at the primary level due to coincidence. However, if voltages are computed by the load flow from the primary down through the service transformers, onto the service level and to the customer meter, coincidence-mismatch can becomes more serious. *Mismatches of up to two volts at the customer meter are not uncommon* between sound load flow cases (i.e., those based on good models, data, and algorithms) and reality.

Most distribution systems and their distribution voltage standards have a margin built into them to cover coincidence mismatch and a host of other "uncomputable" and "unforeseeable" problems. As a result, historically, this inability to estimate easily and quickly the maximum voltage drop at a customer location has been of little concern except in rare cases.

Problems usually occur when a new customer load is being added at the extreme end of a feeder, at a point where load flow analysis predicts the new load can be served, with the circuit barely staying within standards. Due to coincidence, the actual voltages stray out of standard ranges and customer complaints result once the new load is connected. Coincidence mismatch is most likely the cause anytime a customer on the end of a feeder complains of intermittent low voltage, and a good load flow case cannot duplicate such low voltages.[12]

As utilities continue the industry's long-term trend of gradually reducing capacity margin, and run systems with ever higher utilization levels, coincidence mismatch can be expected to be a more persistent problem.

Load Flow Set-Up to Model Non-Coincident "Worst Case" Voltages at One Node

A load flow case can be set up to estimate the worst case "non-coincident" voltage drop to any one branch end point. This requires adjusting the loads of nodes on the branch, as illustrated in Figure 16.12. Basically, the line of

---

[12] The needle peaks responsible for non-coincident peak loads usually last only 5 or 10 minutes, but they occur many times a day. As a result customers may experience only occasional, brief periods of low voltage during peak and near-peak conditions.

customers (or loads) from the end point back to the substation is adjusted it match the peak coincidence curve:

> The customer at the end of the feeder is set to the non-coincident load, C(1) × 22 kW.
>
> The next customer upstream (toward the substation) is set to a value of 22 kW × (2 × C(2) - C(1)), or 18 kW. This is the value that when added to the modeled load of the first customer, results in the proper coincident load *for the pair*.
>
> The next customer upstream (toward the substation) is set to a value of 22 kW × (3 × C(3)-2 × C(2)). This value, 14 kW, is that which when added to the modeled loads of the two, results in the proper coincident load *for the pair*.

Generally, the next customer upstream, N, is set to 22 kW × (N × C(N) - (N-1) × C(N-1)).

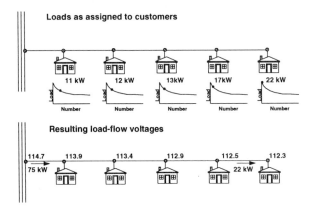

Figure 16.12 Method of setting up the load flow to model the worst case coincidence flows on the lateral consists of adjusting loads to represent the coincidence curve profile along the backpath from the lateral's end point. (Top) the values assigned to each customer are based upon, but are not exactly, the values given by the coincidence curve (see text). (Bottom) The voltages computed for the lateral with this case reflect the "worst case" (compare to Figure 16.11).

Distribution Feeder Analysis

This "non-coincident load allocation" is done for the stream of customers from the extreme end of the branch or lateral being studied back toward the substation, until a value of coincident load (9.9 kW in this example) is reached. The rest of the customers on the feeder are assigned their normal load flow case coincident load.

The result is a load flow case with a very slight increase in modeled load above coincident peak conditions (which is a valid representation of the worst-case situation at this customer location, by the way), arranged to replicate the profile of the coincidence curve to the point being analyzed. When the load flow is solved, it computes voltages and currents for what is legitimately a "worst case" condition with regard to this particular branch. The situation being modeled represents the loads at a time when the combination of loads on the branch is as high as normally expected.

It is possible to re-write any load flow program to apply this "work around" serially to all branches. However, even with its program logic cleverly structured, most algorithms will take much more computation time as a result. Load flow algorithms that use a "tree-walk" in a series of computation passes through each feeder are an exception -- this "trick" can be done with little increase in computation time. However, "tree walk" load flow algorithms are restricted in application to only radial circuits, and in general are not as accurate or applicable to many modern planning needs as full network load flows.

In addition, a load flow that applies this approach is more difficult to apply, because it requires additional input setup (the coincidence profile), and additional interpretation, and the accuracy of the results are dependent on good knowledge of customer types and their load characteristics. For this reason, the use of "non-coincident load flows" is not recommended for general planning application.

Recommended Procedures for Addressing Coincidence Mismatch

Service quality and small-equipment overload problems attributable to coincidence load mismatch can be expected to occur in distribution systems as loads grow, budgets, and design margins shrink, and utilization factors inevitably rise as a result. They will also become more common because of the trends in load flow usage (Figure 16.4). As node resolution gradually increases, the number of nodes representing only a handful of customers, and thus susceptible to the difficulty discussed here, can be expected to increase.

Planners have little choice but to use a load flow approach in the performance simulation of their systems. The "work-around" discussed above can be applied in cases where mismatch has been identified as the cause of service quality problems, but it is much too labor intensive for normal application. The recommended procedure is for planners to apply load flows based on coincident load, but to be aware of coincidence mismatch as a limitation of this study methodology, and to understand how and why it is likely to create problems. Work-arounds and special study should be applied only when necessary.

16.7 CONCLUSION

Distribution planners use a wide variety of analytical tools to simulate the performance of the future distribution system, in order to refine plans and establish that recommended additions to the system will do the job. As the industry continues to tighten operating margins, and to demand higher productivity levels from a continuously downsized workforce, the distribution planner's use and dependence upon analytical tools and automated design methods will increase. Among the key points made in this chapter are:

> *The goal is to perform the quickest, easiest, and lowest cost analysis that gets the job done,* not to apply a technically intricate and advanced method because it is more challenging, intellectually satisfying, and impressive to those who use it, and to their colleagues.[13]
>
> *A good circuit model is essential* -- it is the foundation of all planner detail and accuracy. Detailed, comprehensive representation of equipment is important, but many of the details involved in engineering-models are not essential to planning.
>
> *Load modeling is very important,* beginning with the allocation of load to nodes in a feeder, and determination of imbalance of load, and voltage sensitivity (proportion of constant power, constant impedance). This is important to planning and engineering results and worth devoting

---

[13] The author apologizes for stepping onto his soapbox to rant against his pet peeve, but all too often, utility planners and members of acedemia devote attention and resources to advanced and clever algorithms, at best making tiny improvements in potential accuracy of an algorithm, and occasionally doing nothing more than solving a problem in a newer and more intellectually challenging manner. Meanwhile, the basic models of circuitry and equipment, which if improved would yield great increases in planning effectiveness, go begging because mundane but standard field work, data/model maintenance, and engineering preparation are not done. Frankly, working on new algorithms and program development will always be more *fun,* but seldom as *effective,* as equivalent work put into refining models of the utility system, customers, loads, and equipment.

Distribution Feeder Analysis

significant effort to establishing standards and quality methods of model-building.

Reliability analysis is an increasingly critical part of planning and considerations to accommodate it should be part of any utility's vision of the future tools and procedures its planner's will apply.

Tracking loads and flows on a phase-by-phase basis is the single most important attribute of a distribution performance analysis method (Table 16.2).

Dynamic coincidence effects of load mean that load flow analysis underestimates 'worst" case voltage drop at all nodes with the equivalent of less than twenty households behind them.

REFERENCES

J. J. Burke, *Power Distribution Engineering: Fundamentals and Applications,* Marcel Dekker, New York, 1994.

J. R. Carson, "Wave Propagation in Overhead Wires with Ground Return," Bell System Technical Journal, New York, Vol. 5, 1926.

J. H. Grainger and R. L. Stevenson, *Power Systems Analysis,* McGraw-Hill, New York, 1993.

W. H. Kersting, "Distribution System Analysis," Chapter 6 in *IEEE Tutorial on Distribution Planning,* Institute of Electrical and Electronics Engineers, New York, 1992.

C. G. Veinott and J. E. Martin, *Fractional and Subfractional Horsepower Electric Motors,* McGraw Hill, New York, 1986.

H. L. Willis, *Spatial Electric Load Forecasting,* Marcel Dekker, New York, 1996.

H. L. Willis and G. B. Rackliffe, *Introduction to Integrated Resources T&D Planning,* ABB Guidebooks, Raleigh, NC, 1994.

17
Automated Planning Tools and Methods

17.1 INTRODUCTION

Decision support tools help utility planners select from among the available alternatives for the future distribution system layout, expansion scheduling, and operation. Unlike performance simulators (Chapter 16), which are used to predict the performance of a particular system design, decision support tools are used to help identify which one, out of all possible designs, should be selected. Decision support tools, implemented as computer programs, have two related but separate advantages. First, they are automated, performing some planning functions automatically, reducing effort required, and improving consistency and documentability. Second, they can apply optimization -- formal methods to find the best plan -- not just the a good plan.

Identification, Evaluation, and Selection

Planning consists of three functional steps: identifying alternatives; evaluating them against criteria and desired attributes; and selecting the best alternative (see Chapter 14). These steps are particularly difficult and "messy" with respect to distribution systems, which are highly interconnected and integrated: the various decisions about equipment selection, siting, sizing, and line routing throughout the distribution system are interrelated, and each must be made with consideration of how it will affect the others, and vice versa.

In addition, most distribution planning decisions involve numerous criteria and attributes (Chapter 4). As a result, planners routinely deal with problems in which there are millions of alternatives, each of which must be evaluated on the basis of a dozen or more criteria as well as initial and long-term costs.

While judgment and experience can help, evaluation of so many alternatives pertaining to so many characteristics, is a tedious and lengthy process, very time consuming and expensive, and prone to error or omission if done manually. Numerous computerized tools are available to help planners automate and streamline this process. They also standardize the planning process, and aid in documenting justification of the resulting plans.

This chapter begins with a tutorial discussion of optimization methods in section 17.2, which avoids lengthy examination of optimization algorithms or mathematics, focusing instead on issues related to selection of a method for planning, and its application. Section 17.3 looks at feeder and feeder system optimization -- tools to assist in determining the layout of distribution circuits. Section 17.4 focuses on strategic planning tools -- substation siting and distribution system design aids. Section 17.5 concludes with recommendations on the application of optimization-based tools.

17.2 FAST WAYS TO FIND GOOD ALTERNATIVES

Optimization methods are numerical or non-numerical procedures (algorithms) that essentially replicate the three functional steps involved in planning. Effectively, a good optimization method identifies all the alternatives or options for solving a problem, evaluates them on an equal basis, and selects the best one. The simplest optimization approach is the *exhaustive search* method -- a computer program cycles through all alternatives, evaluating each. When done it will have found the best design.

Exhaustive search is easy to program and apply, but even given the quite impressive computation speed of the many modern computers, it is often far too slow to be practical for some types of problems. For example, the problem of determining the lowest cost routes, conductor sizes, and connection of segments for three feeders serving 1,000 service transformers will have about 1 billion possible alternative layouts, of which perhaps 10 million are feasible (all nodes are electrically connected, the layout is radial, the alternative doesn't violate any rules about layout). Assuming that the exhaustive search program has been written to be "smart," it would check if an option violates any of the feasibility rules before doing a full evaluation performance and cost. If feasibility checking takes one microsecond, and full evaluation one millisecond, per alternative, then exhaustive search in this example would take 11,000 seconds to solve -- over three hours.

Automated Planning Tools and Methods

Practical optimization approaches use some type of mathematical and/or logical procedure to reduce the search process in some way. The best of them can yield spectacular results -- the fastest commercial feeder-circuit optimization method the author has seen can solve the 1,000 load point feeder problem given above in under three seconds on a personal computer (266 Mhz Intel processor).

There are many different types of fast search procedures (optimization algorithms). Most require using certain approximations (often an assumptions of linearity) in the representation of the distribution system interrelationships, or can be applied only when certain relationships are about the problem being solved. Thus, nearly every type of optimization method is limited in its application, and suitable to only certain types of problems.

Complicated Mathematics, but Simple *Concepts*

Most optimization algorithms involve complicated mathematics and logic in their programming, and the technical papers that describe them invariably are replete with lengthy equations, mathematical symbols, and intricate flow diagrams. The complexity required in their programming, and their intricate equations, disguise an often overlooked fact: optimization methods are always very simple in concept. Nearly every optimization method involves the use of what could be termed a "trick" -- a clever or imaginative step, or a resourceful way of ordering its search process, to greatly shorten the process of searching through millions or billions of possible alternatives to find the best. Table 17.1 lists several types of optimization methods. It is by no means exhaustive.

Tradeoffs in Optimization Application

Usually, the use of an optimization method for distribution planning (or any other purpose, for that matter) involves some amount of compromise among conflicting analysis goals. There is always a tradeoff between problem size (e.g., the number of feeders or substations that can be studied) and computation time. Most optimization methods have exponential run-time relationships -- double the problem size and computation time increases by a factor of four, or even eight or sixteen. In addition, methods that are fast and capable of large problems almost invariably purchase that speed by using approximations (usually linearization). Those that are precise are slow and limited in their application to one or two feeders at a time.

The important point is not speed, problem size, or accuracy, but whether the optimization method, developed appropriately and encapsulated in a good computer program, will prove to be a valuable tool. *Will it assist the planner in improving planning quality and reducing the time required to complete a plan?*

Table 17.1 Distinctions in Optimization Methods Applied to Distribution Systems

Adaptive Neural Nets (ANN) are an application of simultaneous equations or logic sets in a certain type of structured format, generally applied as a continuous method. ANNs can be set up with a type of feedback that causes them to change their coefficients in response to their results, theoretically producing better and better solutions as they are used. Mostly used in forecasting, they have some applications to decision-making. A trendy, "flavor of the month" algorithm in the academic circles at the time of this writing (1997).

Artificial intelligence (AI) consists of any of several different non-numerical approaches to structuring a set of rules based on human experience and manual approaches to a problem, so that they are applied in a legitimate way by computer --(heuristics done right, so to speak). Integer or contiguous. Linear or non-linear.

Branch and bound (BB) is an integer optimization approach in which the optimization starts out to look explicitly at every alternative -- for example, to evaluate possible configurations for the layout and sizing of a feeder circuit. As the BB method examines alternatives, it identifies characteristics that the optimal solution must have (e.g., the optimal solution with have a feeder trunk with conductor larger than 600 MCM conductor). As it identifies these bounds, it uses them to delete whole sets of alternatives from its considerations (i.e., don't look at any alternative with less than 795 MCM conductor as the trunk).

Continuous method: an optimization method that considers all possible values within a range as potential solutions, not just a few discrete values (see text).

Decomposition method: any of several methematical methods of breaking up a large problem into smaller, related ones, which can be optimized individually.

Dynamic programming (DP) methods "build" up a complete solution (e.g., a circuit configuration) a stage at a time, with a series approach in which part of the problem is optimized in isolation, then another part is evaluated, etc., until the entire problem has been studied and an "optimum found" The name "programming" doesn't refer to computer programming, but to scheduling (an alternative meaning of "programming"). DP applicability to a problem depends on whether it can be broken apart into a series of stages or steps which can be optimized serially, in a way that the stage N can be optimized after stages 1 through N-1 have been optimized, its optimization not affecting the conditions that led to the optimal solution in stages 1 through N-1. Integer or continuous, linear or non-linear, it does not handle constraints well.

Exhaustive search is a "brute force" optimization technique that simply examines every alternative to find the best one. Reliable, robust, but slow on all but problems of limited scope. Modern computers are fast enough that exhaustive search is a practical method for many smaller applications and "auxiliary optimization features." They can be developed as either integer or continuous format, addressing either linear or non-linear problems.

Expert system: an AI approach based upon the use of a set of expert rules.

Automated Planning Tools and Methods

Table 17.1 continued

Genetic algorithms (GA) are non-numerical, integer methods. They use a long binary string to represent possible alternatives, and modify the string values to try to produce a better solution. For example, a string of two hundred 0s and 1s could be used to represent the switching selections for two hundred switches in a feeder system (open = 0, closed = 1). The genetic algorithm randomly alters the 1s and 0s to try to find a good solution. Among good solutions it develops, it combines features, trying to come up with better features. GAs are "monte-carlo" optimization methods. they are easy to program, robust, flexible, and reasonably fast. They do no provide any assurance that the best solution will always be found.

Heuristic algorithms are methods based upon a procedure the programmer has created based on how the problem would be done manually (see text). They can be set up as integer or continuous, and linear or non-linear analysis.

Integer method: an optimization method that considers only certain discrete values as possible solutions (see text).

Linear programming (LP) is a method for solving a problem in which all relationships are linear. In such cases, mathematical theorems can prove that the optimum solution will have be among a small subset of all possibilities (vertices of the solution space) which can be found and checked quickly. LP methods can usually address constraints fairly well, but all relationships being considered must have linear coefficients. They work best with continuous problems, and not so well on integer problems.

Mixed integer method: an optimization method that can handle multi-faceted problems in which some aspects of the alternatives are integer, and other aspects are continuous.

Optimal power flows (OPF) are not an optimization algorithm *per se,* but an application using any one of several optimization algorithms to a load flow, which, while solving for voltages and flows throughout, can also adjust real and reactive output of generators, VAR and regulator switching to identify the operating configuration that will minimize some attribute of operations (usually losses).

Rule-based system: an expert system.

Simulated annealing method (SA): as a molten metal cools in a mold, it hardens in a way that minimizes internal stresses. A mathematical model of this cooling minimization process can be used to represent some varieties of non-linear, mixed-integer problems.

Trans-shipment method: an optimization approach that uses the fact that power can be "trans-shipped" moved from one node, through another node on its way to a third -- in the optimization. Applicable to distribution since this is, in fact, how power is distributed.

Unconstrained optimization method: an algorithm or procedure that cannot handle constraints imposed on the definition of an acceptable alternative.

Constraints

One important quality in determining if an optimization method is appropriate to a problem is the types of constraints -- limitations and requirements -- that it can address. Constraints are part of most planning situations -- voltage must be above 113 volts, load must not exceed capacity rating, configuration must be radial, etc. Optimization methods differ greatly in if, and how well they address constraints. Some methods can't respond to constraints. Others use approximations.

Integer versus continuous

"Integer" optimization methods address planning problems where the choices are restricted to a set of discrete values. For example, the substation transformers available to a utility may be 7.5, 10, 15, 20, and 24 MVA. An optimization method that can deal directly with these discrete sizes, that will pick from only among these sizes is called an *integer method*. This term does not mean that the sizes or values used must be integers or proportional to integers. It simply means the method addresses only a set of discrete sizes.

By contrast, some optimization methods are *continuous methods*. They work by considering any value within a certain range as permissible. They are appropriate for certain types of problems, for example when determining the optimal voltage target for CVR on a feeder, where the target voltage could be *any value* within the range 126 to 113 volts.[1]

Integer methods are best applied to planning problems where there are only discrete sizes involved, and continuous methods work best where the possible solutions cover all values over a range, but each can and has been applied to the other type of situation. For example, some substation planning programs apply a continuous optimization method to transformer sizing. It may compute a 8.8 MVA size as optimal, at which point some set of rules ("heuristics") is applied to pick from among the two nearest sizes -- 7.5 or 10 MVA. Similarly, an integer method could be applied selecting a target voltage by setting it to pick half volt increments from 125 to 113 volts.

---

[1] CVR (Conservation Voltage Reduction) is an energy conservation measure mandated by some state regulatory commissions. It requires that the utility operate every feeder at the lowest voltage that results in satisfactory service quality for all customers. This tends to reduce energy usage, and the optimal voltage for a feeder is usually determined by a complex process based on service quality, energy use, and feeder modifications. See Willis and Rackliffe, 1994, for a discussion of CVR applications and impact on energy use and cost.

Automated Planning Tools and Methods 735

An optimization method that can address a problem where simultaneous selection must be made from alternatives that both have integer and continuous features is called a *mixed integer* method.

Linear versus non-linear

Many of the fastest and most dependable optimization methods *are linear methods*: they apply only to situations or problems where the relationships being analyzed are linear (e.g., zero capacity costs zero; twice as much capacity costs twice as much, etc.). Distribution planning problems (as well as most problems where optimization methods are used) rarely involve linear relationships. Despite this, linear methods are often applied because they are fast, they can handle larger problems than non-linear methods, and they are easier to use. A planning problem is "linearized" in some fashion (Figure 17.1) usually creating a small approximation, which is judged acceptable in order to gain the speed and ease of use of a linear over a non-linear method. *Piecewise linear methods* represent non-linear relationships with a series of short straight-line segments.

Fixed costs

In many planning situations, the Y-intercept of a cost function is not zero, as for example in the conductor costs shown in Figure 17.1, A. Some optimization methods are capable of dealing with relationships that have such fixed costs -- buying any capacity has a certain minimum fixed cost, plus perhaps another cost. *Fixed cost optimization* techniques are often mixed integer algorithms.

Trans-shipment

Trans-shipment means moving a resource (e.g., power) to an intermediate location on its way to its final destination. For example, a food distributor might move loaves of bread from its factory to several local warehouses, before re-loading it on local delivery vans for transportation to its grocery stores. Moving large quantities in common for much of their journey permits a savings in economy of scale.

From the standpoint of optimization and cost reduction, the fact that there is an intermediate stop in the transportation of the product is almost irrelevant, what is important about "trans-shipment" is that for part of their journey, resources going to many destinations were moved as a *group*, producing a savings due to a larger economy of scale. Most distribution systems work in this manner -- the flow on any large segment consists of power heading to many different nodes farther downstream.

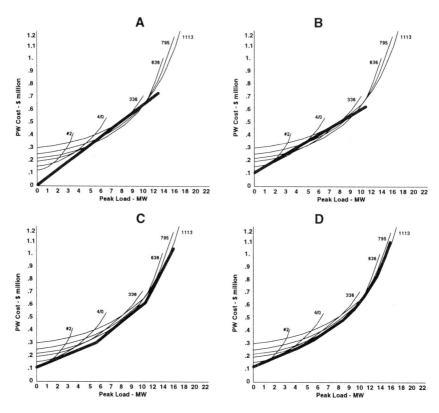

Figure 17.1 Linearization increases both analysis speed and the problem size that can be studied, but may affect accuracy. Feeder conductor selection economics can be linearized (A). Alternately, a fixed-cost linear method may be used (B) or a fixed-cost piece-wise linear method (C), either of which will take longer to solve. Slowest of all to solve, but most accurate, is a fixed-cost non-linear cost-capacity representation (D).

Some optimization algorithms cannot deal efficiently with situations where resources going to many final destinations can be added together for part of their journey and separated at other parts. They cannot simultaneously recognize the resources they are optimizing (loaves of bread, power) as single quantities assigned to individual destinations and as members of large groups. Methods that can make this distinction are often structured so that they *depend* on trans-shipment, algorithm structure (program flow) being built around the trans-shipment routing structure of circuits (or roads to grocery stores, etc.). These are

Automated Planning Tools and Methods

trans-shipment algorithms. Many feeder optimization programs use trans-shipment -- when developing the various layout permutation to consider during optimization, they work from a trans-shipment perspective.

How Well Does an Optimization Method Perform Each of the Three Steps?

Distribution planners do not need to master the intricate equations and program logic behind the optimization tools they use. However, they do need to understand the basic approach their optimization tools use (i.e., what "trick" it uses to obtain search speed -- if it is integer or continuous, linear or non-linear, etc.). More important, they need to understand its limitations with regard to application. From the applications standpoint, this is often best done by considering how well the optimization performs each of the three steps: search, evaluation, selection.

Step One: The Search Through *All* Alternatives

Given that there can be billions of alternatives involved, an optimization method will be of no use to planners if it requires that all alternatives to be evaluated be input by the user, listed, or in any way manually handled. A good optimization tool is capable of producing all possible permutations or variations *automatically*, from a simple description of the planning problem. Equally important, an optimization method is useful only if it generates and evaluates all possible alternatives -- if it skips some possibilities it is undependable: sometimes the optimal alternative would be among those it failed to consider and thus it would fail to find the best alternative.

Automatically generate alternatives. Suppose that planners have identified twelve possible sites for future substations, each of which could be built with one, two, three, or four transformers of 22 MVA each (the utility's sole standard size for substation transformers), or not built at all (a total of five possible "capacities" for each substation site. The planner needs to determine the optimal long-range substation plan: how many substations should be built eventually, which sites should be used, and what capacity should be built at each site used. With twelve sites and five possible capacities there are more than 244,000,000 permutations. A great majority of these will be unfeasible or unreasonable -- not enough capacity, too much capacity, no substation within required distances of new load, etc. However, depending on the exact planning situation, there will typically be from four to ten million feasible, reasonable alternatives among those nearly quarter billion permutations. At least one of those will be the least costly feasible alternative.

A good optimization tool would be capable of *automatically* generating and searching through all ten million of these feasible/reasonable alternatives, based only on a compact set of input data: descriptions of the twelve sites, the capacity levels available, costs involved for sites and capacity, any constraints (perhaps one of the sites is so constricted that no more than three transformers will fit, etc.), and other data used in the analysis (a "load map" as described in Chapter 15). There are three qualities sought in a useful optimization tool's search:

Minimal input: The user must input only a small amount of data. If the user had to input a list of all 244 million permutations, or anything approaching that, the method would be useless.

Completeness: The method must scan (at least implicitly, as will be discussed below) *all* feasible solutions, not just most of them.

Speed: This search must be done quickly.

Implicit searches. Most optimization methods would evaluate this example by using some sort of permutation loop that *could* generate all possible permutations (nearly a quarter billion) from the input description (twelve sites, five possible capacities, their costs, etc.). Note the use of the would "could" -- speed being of the essence, most of the search would be "implicit." The optimization would use one or more methods based on computation or logic to reject large groups of permutations from any consideration -- those that it "knows" cannot contain the optimal solution never have to be considered.

The method used to shorten the search process might be based on rules inferred during pre-processing ("no plan with less than two transformers at site 6 can possibly meet requirements") or by evaluation prior to beginning the search ("evaluation of fixed versus variable capacity costs shows that the optimal solution in this case will have either five or six substations built from among these twelve sites") or by application of some sort of mathematical construct that constrains the search. A really "smart" optimization method will refine and apply tests and rules like this so effectively, that it might have to look at only several thousand possibilities in the above example before it finally determined the optimal alternative.

But what is critical is that the search, before application of the rules and "tricks," *could have and would have defined its search space as including all permutations* (244 million in this example), and that the tricks applied to limit the search were entirely valid. Generally, it is here that crude and heuristic optimization methods fail -- the developers may believe the method covers all possibilities, but somehow it leaves some options unchecked. Unless there is no choice, *it is best to apply only formal optimization methods* -- those for which theoretical proofs of completeness and validity are available.

Automated Planning Tools and Methods

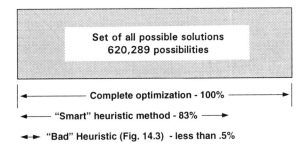

Figure 14.2 A feeder capacitor siting problem such as that discussed in Chapter 14 (Fig. 14.2), but for *three* possible banks instead of two, each of two possible sizes, has over 600,000 alternatives. A formal optimization method will study, implicitly, every option. A clever (but slow) heuristic method will average 83%, enough that it would provide a useful tool if optimization were unavailable. The heuristic algorithm described in Chapter 14 (Figure 14.3 and accompanying text) would nearly always fail to find the optimal solution, since it scan less than one percent of the space.

Formal or not, an automatic planning method's dependability -- whether it will produce good results from one case to another -- depends on what portion of the solution space it typically scans. Less effective methods will miss portions due to omissions, simplifications or approximations (Figure 17.2).

Step Two: Evaluating the Alternatives

Very often, an optimization procedure will combine the evaluation and selection functions. Regardless, evaluation is a key function of these methods. Optimization procedures have two aspects to their evaluation: application of *feasibility rules*, and the *performance index* -- the attribute or function of attributes which they are attempting to optimize. Feasibility rules or constraints might include requirements that all load be served, that the configuration be radial, and that major trunks be three-phase. The most common performance index is overall cost.

Traditional power delivery planning is *single attribute planning,* meaning that only one attribute (usually cost) is to be minimized. Evaluation consists of determining if each alternative meets all the feasibility rules (if it doesn't meet the rules, it's not acceptable no matter how low the cost), and if it is acceptable on all counts, determining the performance index value. The alternative with the lowest performance index (or in some cases, the highest) is selected as "optimal." Some types of optimization procedures can address multi-attribute problems, where the goal is to try to optimize two different indices when they cannot be put on a common basis -- such as cost and esthetic impact.

Mistakes or poor performance with optimization in the evaluation step generally occur because planners do not check to see if their methodology meets two related criteria for its application:

Does the evaluation consider all factors that are important to the goals? For example, if one requirement of a feeder layout is to assure that voltage at the end points is greater than 114 volts, then the optimization method must be able to check on voltage (some feeder layout optimization methods do not). Otherwise, there is no way to assure that goal is achieved, and certainly no way to separate options on the basis if they achieve that goal.

Does the planning method treat all options equitably? An evaluation method that considers all factors affecting some options but not all factors affecting others, provides a biased comparison of options. For example, a municipal utility doesn't pay taxes to itself, so taxes on new substations and property built inside the utility's municipal territory are zero. However, many municipal utilities serve areas outside their municipal boundaries, or occasionally find themselves having to locate facilities outside their boundaries, and in some cases do have to pay property taxes and permit and license fees in such cases. To be completely valid, the evaluation of all options must include the cost of any taxes that would have to be paid to other taxing authorities, even though this means simply saying "none" for those inside the municipal territory. Only in this way can the cost advantage of sites inside the territory be assessed properly against the advantages and costs of those outside.

Step Three: Selecting the Best Alternative

In many optimization methods, this function is combined with, or implicit in, the evaluation function, and it is difficult to identify where evaluation ends and selection begins. Regardless, planners should study their planning tools, to determine *how* the best alternative is identified, and *if* this is done in a valid manner with respect to their goals. Many planning mistakes can creep into the selection step because of approximations or mis-application.

Is everything that is evaluated and reported used in selecting the best alternative? The fact that all alternatives were evaluated on the basis of a particular attribute does not necessarily mean that that attribute was properly weighted in the selection phase. For example, a popular planning program used in the 1980s evaluated the cost of future losses for both feeders and substation

Automated Planning Tools and Methods

transformers, but then ignored those associated with substation transformers during the selection step (a compromise forced on its developers in order to gain the required speed and problem size they wanted). However, once an alternative was selected, the program reported transformer losses for the "optimal alternative" in its output, giving the impression they were considered in the selection. But as far as selection was concerned, transformer losses were "free" -- the program had an unrealistic bias toward building new substations in order to avoid having to build long feeders from existing ones.

Can the planning method validly distinguish among alternatives? Like any analytical method, computer programs for optimization can produce only so many "significant digits" of accuracy. Some optimization algorithms involve tremendous numerical computation, and in addition apply many approximations and short-cuts. As a result, in the worst case they may have only two significant digits of practical accuracy.

For example, an early EPRI distribution planning program for substation siting and sizing (EPRI RP-570) applied a branch and bound algorithm that the cost of load transfers between substations with only a very rough approximation that proved to have up to 10% error (EPRI EL-1198). A different approach -- service area optimization -- used a linear "synthetic feeder approach" and was accurate to within 3% (Crawford and Holt, Willis and Rackliffe). That accuracy advantage made service area optimization much better at *reliably* selecting the best site, one reason it is much preferred as a planning method today, but it is still far from perfect.

Planners often fail to consider this, and interpret minor differences in the evaluated values of alternatives as significant, when in fact they are probably meaningless. If the "best" alternative selected by a service area optimization approach is evaluated as 2% better than another alternative, then that alternative is not necessarily the best -- no doubt it's a good option, but there is no way without further study to prove it's the best. In many planning cases the margins in cost, environmental impact, or other attributes among alternatives can be very small. Error range in the evaluation and selection among cases must be small enough to assure that the determination of "which is best" is accurate.[2]

Is there any post-processing of the analysis after selection of the alternative, and does that affect the optimality of the result? Some optimization programs

---

[2] Accuracy in evaluating factors that *are the same* among alternatives is *not* necessary for selection. For example, an accurate estimate of right-of-way cost is not needed to decide between two alternative transmission designs that would use the same route and right-of-way width. Any error would impact both equally, and thus make no impact on the decision.

perform addition manipulation of their data after the actual optimization (selection) step is completed. In some cases this can slightly affect the result.

For example, many of the optimization algorithms that are considered best for feeder system configuration optimization (they are fast, reasonable accurate, and dependable) are unable to limit their analysis on only radial configurations. They typically produce feeder system configurations that have loops or even network portions to them.

For a variety of reasons beyond the scope of this discussion, modifying these algorithms to produce only radial configurations simply will not work -- the result are algorithms that are slow, inaccurate, and often just fail to produce any answer at all. Instead, program developers should go ahead and solve the optimization, then test it to see if it is radial. If not, they provide a "radialization" post-processor routine, which modifies the optimization algorithm's configuration to make it radial. Some of these radialization algorithms are very cleverly done (occasionally they represent more program code and computation that the optimization itself), and while they do a good job of radializing while not disturbing the configurations economics too much, there is no assurance that the configuration found is the optimal radial configuration. All that is known is that it is a good configuration, which is now radial.

No Optimization Method is Perfect:
Flaws Must Be Kept in Perspective

No optimization method considers all factors involved in distribution, and is completely devoid of approximations and shortcomings. What is important is to determine if a particular method provides a better tool than any other method available. In general, a well-studied planner can learn the approximations, linearizations, and limitations in the evaluation and selection steps of particular optimization methods, and largely account for them it its application, or at least be aware of their potential impact. But no amount of understanding and work can "fix" a method that does not span the set of all alternatives. Therefore, it is deficiencies in the first step -- the search, that are most serious -- methods that do not span *all alternatives* should be avoided.

Buyer Beware -- Many "Optimization Methods" Aren't

Any program that attempts to search for, evaluate, and identify the "best" alternative can be termed an "optimization algorithm," even if it does a wretchedly poor job in one or more of the three steps described above. As a result, many weak, incomplete, or faulty planning procedures have been portrayed by their developers as "optimization," when in fact there is no formal

Automated Planning Tools and Methods

(mathematical) assurance that they will find the best, or even a good, alternative. Many of these employ "heuristic algorithms" or computational procedures based on faulty assumptions.

A "heuristic algorithm" is a procedure that tries to reproduce the steps a human would take in manually addressing a problem. Generally, an engineer or planner familiar with the particular type of problem sits down and "reasons out" the approach she or he would use manually, and then writes a computer program to reproduce those steps. Heuristic algorithms often use simple rules of thumb formulae and tables as some part of their operation.

Heuristic algorithms are easy and quick to develop as computers program, and their developers will often be convinced that they are complete and valid. Generally, they give impressive results when demonstrated by the person who wrote them on the problem(s) that person had in mind. Although there are very rare exceptions, most heuristic programs prove unreliable in use, particularly when applied by people other than the developer, on other problems, or with even a slightly different performance index. The example capacitor optimization algorithm given earlier was a "heuristic algorithm."

Heuristic algorithms should not be confused with legitimate artificial intelligence (AI) techniques such as knowledge-based or rule based algorithms. The science of artificial intelligence -- sometimes called knowledge engineering -- has rigorous theorems, rules of verification, and formal procedures for gathering a planner's knowledge about "this is how I do it manually," testing those heuristic rules for completeness, and then putting them together into a logical construct within an AI program, so that the results *will* be dependable. Unfortunately, because of this similarity, many heuristic algorithms are often represented as "AI methods."

"Implicit" Siting and Routing Algorithms

A few distribution planning methods determine sites and routes in a plan *explicitly* -- they literally compute or develop them automatically, independent of input sites and routes. However, most siting and routing optimization methods determine optimal sites and routes *implicitly*, by assigning "zero capacity" to candidate sites and routes given as input by the user. For example, in addition to data about loads, costs, and planning goals, a substation siting optimization procedure might be given a list of twenty-five candidate locations for new substations, where its input data explains it can decide to install zero, one, two, three, or four "unit" substation transformers (see Chapters 10 for a discussion of transformer units) at each site. If it selects "zero" as the optimal capacity at a site, it is effectively rejecting that site as a substation site. This is *implicit siting*.

Similarly, feeder routing and layout programs assemble an optimal configuration by selecting segment capacities from candidate ROW and easement routes input by the user (or those inferred by the program, which might assume unless told otherwise that it can build a feeder along any street or road, etc.). The optimization assigns a decision of "build a capacity of zero" to any portion of routes it does not want.

No distribution optimization methods (other than few unsuccessful heuristic methods) compute sites or routes that are not originally input as a set of candidates by the user. Since strategic distribution optimization programs optimize only over candidates that are provided by the user -- some skill is required in selecting a good, all-inclusive set of sites.

Second, it is important that siting and routing programs be able to handle a "zero capacity for zero cost" option in all sites and routes. This restricts the choice of optimization algorithm slightly (some types of optimization method, particularly modified linear methods for fixed-cost feeder routing cannot model both zero capacity/zero cost options but must represent all possibilities as having some non-zero cost.)

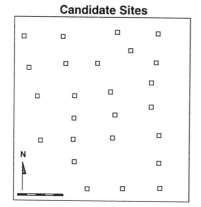 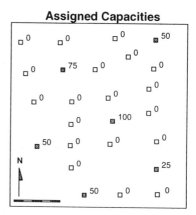

Figure 17.3 Most siting and routing programs determine optimal sites and routes implicitly, by assigning "zero capacity" to those sites not deemed desirable. At the left, a map of 25 candidate substation sites as input to an optimization that can assign 0, 25, 50, 75, or 100 MVA capacity to any site. At the right, the optimization has determined assigned non-zero capacity to five sites, assigning "0" to those it does not want.

Automated Planning Tools and Methods

17.3 AUTOMATED FEEDER PLANNING METHODS

To improve feeder system design, optimization must deal simultaneously with three aspects of distribution configuration -- layout, switching, and conductor size. Regardless of algorithm, these applications are best supported by database, editing, and display capabilities tailored specifically to the alternative database required by optimization, rather than the needs of the more typical load flow/short circuit application. Proper application of optimization methods to distribution system design can reduce the time required and improve the quality and cost of the resulting plan. Given the power industry's current trend toward labor cutbacks and increased cost-consciousness, the authors expect optimization methods to see increasing use in the next decade.

Three interrelated aspects of distribution configuration need to be simultaneously assessed in any attempt to minimize cost, and should be included in the span of any general optimization program for feeder planning. Chief among these is layout -- the manner in which the distribution pathways are split and re-split on their way from a few primary voltage sources (substation buses) to the utility's many customers. Most planners and engineers have a preference for a particular approach to feeder layout, but there is considerable design flexibility in most cases and there are no firm guidelines within the industry, as was discussed in Chapters 8 and 9.

Equally important is the selection of switching, beginning with the determination of the normally open points which together with layout determines how load is distributed among substations, feeders, and feeder branches during normal operation. Additional closed switches must be placed throughout the system to provide operating flexibility and to isolate small parts of the system during equipment outages.

Given that layout and switching are pre-determined and not subject to revision, selection of the most appropriate line type (size of conductor and number of phases) for each segment of a feeder system is straightforward (see Chapter 7). However, conductor size of key elements heavily influences what is the best layout and switching for any given situation, and in general, the least-cost configuration will not be found by serially determining layout, switching, and conductor sizing. They must be determined simultaneously, taking into account their influence on one another.

Thus, the generalized "feeder planning problem," illustrated in Figure 17.4, is to determine the best layout of a feeder system, from input data that describes the existing system and any areas that could be upgraded, available new ROW and easements, the loads, and the criteria and performance index. The planning method must determine what segments to build, of what line type, how to connect them (switching).

746 Chapter 17

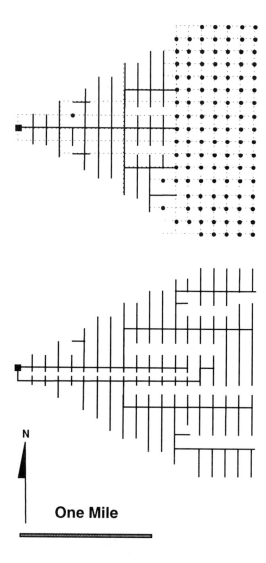

Figure 17.4 Automated feeder planning programs, utilizing optimization, work on the type of problem shown here. Top, an existing feeder (solid lines), the new loads that must be served (dots), the possible new routes (dotted lines) and segments that can be reinforced (solid in parallel with dotted lines). Feeder segments can be built in 3, 2, or 1 phase, in any of five conductor sizes. Radial configuration and switching can be changed. At the bottom, the plan found by an automated planning support program using constrained linear trans-shipment optimization.

Automated Planning Tools and Methods 747

This problem certainly can be solved manually, or by experienced distribution planners with tools no "smarter" than load flows -- tools that tell planners if a particular design will perform well, but cannot directly provide it if is inherently better than another possible design.

Most utilities use structured (formal written) procedures for all such manual planning studies. These procedures tend to be serial, starting with a standard recommended layout (e.g., something similar to a "large trunk" or "multi-branch" philosophy of layout, as presented in Chapters 8 and 9), calling for a certain set of design rules to be applied for the overall layout, then applying a set of segmentation guidelines to determine switching, followed by a segment-by-segment determination of the appropriate conductor size based on economic conductor sizing rules of the type analyzed in Chapter 7. Such procedures result in *workable* and *affordable* designs, but considerable published evidence suggests that costs can be reduced on average between five to ten percent by application of optimization to find the *best* configuration for each situation.

Combinatorial Design Challenges

The primary reason for such a large margin in performance of optimization over manual design is that the number of possible configurations to almost any real-world distribution design is so large that the probability of a smart, experienced engineer finding the very best configuration unaided is rather remote. Regardless, planners are rarely given the time it would require to do so unaided. The expansion planning of a single feeder, even one with much of the circuit already in place, as in Figure 17.4, can present over a million alternative layouts, most differing insignificantly from one another, but several hundred that are substantially different from one another, *and* that would be deemed sound plans. Decisions in selecting among them require balancing many factors, working around many constraints, and making comparisons among dozens, if not hundreds, of performance simulation cases. Judgment, experience, and structured design rules can assist planners in developing a good plan if working with manual procedures (those that do not use automated planning programs), but they are very unlikely to find the best plan.

The plan found by an automated procedure using constrained linear trans-shipment optimization (bottom in Figure 17.4) may not be the true optimum (see comments on the accuracy/round-off of optimization algorithms in section 17.2), but it meets all criteria, it was developed in two hours by one engineer, and it is 4% less costly than the manual plan developed by a team of two planners working one week. *Optimization methods save time, cut planning costs, and produce improved results.*

Characteristics Desirable in Feeder System Optimization Methods

Section 17.2 alluded to the very wide variety of innovative and clever algorithms, comprising more than 200 separate projects, that have been used to apply optimization to primary and service-level distribution layout. Regardless of algorithm, all face the same basic problem -- select the best combination of layout/switching/conductor sizing in problems of the type outlined in Figure 17.4. Desirable qualities in an automated feeder planning program are:

Problem size

Optimization has been widely demonstrated to provide greater benefits when it can be applied to a larger area of the system at one time. A program whose optimization engine can optimize ten feeders (e.g., 5,000 to 10,000 nodes) in one computation run, is much more valuable than one that can optimize only one feeder (e.g., 500-1000 nodes). Operating on a large-area rather than piecemeal basis, the optimization can shift load and re-configure more effectively, producing larger savings.

Trans-shipment

Trans-shipment (see Section 17.2) is such a predominant feature of flow on feeder systems that most practical automatic feeder and multi-feeder planning programs use a trans-shipment type of optimization.

Capacity constraints

Automated planning programs that cannot handle capacity constraints are of very limited value, to the extent that the author considers them valueless for practical application. Capacity constraints are what drive the need to reinforce, add circuits, and build much of the new additions that must be built. In modern distribution systems, where load is slowly growing and budgets are severely restricted, the ability of optimization to "work around" capacity constraints is essential. Generally, it permits planners to defer new additions through very creative load balancing (constrained trans-shipment methods are really good at this) and to find the minimum additions to allow working around capacity limits when load transfers alone will no longer do the trick.

Automated Planning Tools and Methods 749

Implicit routing

Generally, programs that use an optimization that works on an implicit routing basis (see Section 17.2) is ultimately easiest to use for feeder configuration planning, even though it means the user must provide all possible ROW and easements as input data, rather than have the program automatically generate routes, etc. Programs that do not use implicit routing tend to be difficult to control -- it is more difficult to limit them to not building feeders where they are not allowed, than it is to use an implicit routing method that requires identification of where they can be built.

Implicit routing is easiest to use if the program provides two useful utility features: a) an ability to produce a set of potential routes following all roads, b) a "connect" option where the user can specify a distance (e.g., 100 feet) and have the program assume that any two nodes less than this distance apart can be connected, even if no route was input between them.

Fixed cost

Optimization algorithms that can handle fixed costs take much longer to solve than those that cannot model fixed costs. They tend to have a much smaller program capacity, and to have more operational problems (non-convergence, round-off error). In general, non-fixed cost methods give quite satisfactory results for planning overhead and direct-buried distribution systems. But fixed cost methods are really necessary to apply fast, effective, and accurate optimization to underground systems that require duct banks.

Automated Planning Applications to Distribution

From the standpoint of computerized application, distribution planning problems fall into three categories, with differing analytical needs. These are new system expansion, augmentation of an existing system, and operational planning.

New expansion planning

Distribution expansion planning where there are no present facilities, but where load growth is expected, may seem to be the most challenging planning problem possible, but paradoxically it is the easiest to solve well using optimization. Such planning is often called "Greenfield" planning because the planner starts with nothing (a green field) and plans a completely new system in concert with the development of the region. The major challenge in these situations is to select the design of this new system from the overwhelming range of possibilities.

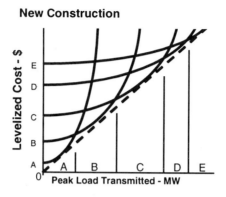
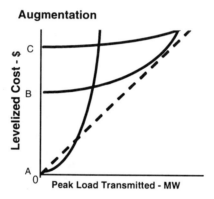
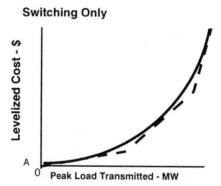

Figure 17.5 The three types of feeder planning situations described in the text each face different cost relationships among the feeder segments they analyze.

Automated Planning Tools and Methods

Fortunately, linearization of conductor economics permits use of a linear optimization method in these types of problems when the performance index is cost (as it usually is in distribution planning). Use of a linear method (with a zero Y-intercept) shortens computation times tremendously and permits very large problem size (75+ feeders can be analyzed in one computation), while introducing only a slight approximation error in the results for this type of problem.

The top drawing in Figure 17.5 illustrates such a linearization, a line from the origin is used to represent the "best" choice, among the five curve representing conductor economics for the distribution line type set (see Chapter 7 for details on line type sets). During optimization, the straight line is used to represent the cost versus capacity of any new segment considered for addition to the system -- the greater the power it will carry the greater it will cost. The optimization uses this linearized relationship in two ways:

1) During comparison of the virtues of one line's economics, or one configuration's economics, against an other's. The optimization builds, or solves for, a configuration it deems optimal using this cost/capacity relationship. When done, a flow value has been assigned to each segment in the system.

2) To determine the line type for each segment. The flow value assigned to each segment is used to determine line type for each segment. In Figure 17.5, top, this would be done based on which "bin" along the X axis the value assigned to the segment fell into -- A, B, C, D, or E.

Many distribution optimization algorithms use such linearization, representing each segment in the potential system with only two values -- a *linear* cost-versus kW slope (which may be different for each segment), and a capacity limit which constrains its maximum loading.

Applied in a well-designed program, the combination of linearization of conductor costs along with implicit routing provided very satisfactory performance (considering contemporary standards) when first developed in the 1970s. More than sixty utilities in the United States use this method routinely in the layout of major new distribution additions. Economic savings as large as 19% in comparison to good manual design practices have been reported in IEEE and EPRI publications, but the author believes that a three to eight percent margin of improvement is more typical.

Augmentation and reinforcement

Unfortunately, "Greenfield planning" represents only a minority of distribution planning needs. Much more common is a need to economically upgrade a distribution system already in place, perhaps in at established neighborhood where a slowly growing load means that some parts of the existing system will soon be overloaded. As is the case with most greenfield planning, the goal is to reduce overall cost to a minimum.

While such planning may appear to be much easier than greenfield planning because the range of options is limited and a feeder system is already in place, in fact "augmentation studies" are quite difficult to optimize, for two reasons. First, in most cases new routes and equipment sites and permitted upgrades of existing equipment are severely limited due to practical, operational, esthetic, environmental or community-relations reasons. The planners' challenge is to work around such limitations, balancing availability, and cost against impact on the system and its need. While in greenfield studies, the sheer combinatorial magnitude of the design problem defined the challenge, in augmentation studies the balancing of capacity constraints on the existing system, and these many constraints to any new additions or changes, is what provides the challenge. These limitations are almost always different in each case and location, and cannot be generalized. They must be represented well in the data and acknowledged accurately by the optimization algorithm.

Secondly, the options for upgrading existing lines cannot be linearized as well as in greenfield studies, as shown in the middle drawing of Figure 17.5. Despite this, linear trans-shipment programs have long been applied to augmentation studies, generally because they were in wide use for greenfield studies and nothing better was available. While a good planner using a linear trans-shipment program generally can reduce costs in such studies by two to five percent, interpretation of the results from linearized optimizations requires greater skill and experience than for greenfield planning. More important, the configuration computed is not necessarily optimal (a linearized computation will fail to recognize the greater than linear upgrade economic benefits that accrue in the middle of the X-axis in Figure 17.5, middle).

As a result, a number of "non-linear" algorithms have been developed specifically for augmentation studies, including piecewise linear and non-linear trans-shipment. However, in the author's opinion the algorithm is not the key factor in program usefulness for augmentation applications. The key for augmentation applications is to make the software usable by supporting quick entry, verification, and review of the database to reflect the unusual conditions and out-of-ordinary costs that are the real drivers in the augmentation planning. The real challenge in augmentation/reinforcement planning is in representing the existing system, and limitations on new construction accurately with little effort.

Whereas greenfield studies can be accomplished using generic editors and display (all "unbuilt" ROW are basically the same and have the same characteristics) that are based upon an expanded format as used in circuit

Automated Planning Tools and Methods

analysis (e.g., load flow) programs, the same does not apply to augmentation studies, where such programs become tedious to use and promote data errors. Here, a database-editor-display environment designed to work efficiently with entry, display, and verification of constraints, costs, and range of options on an exception and segment basis provides much greater ease of use. Routines to detect out-of-range entries, exceptions, or unusual and non-sensible data within the context of this type of problem are also very useful.

Operational planning

The only alternatives of system configuration in operational planning are changes that can be made in the radialization of the system. In some cases, this includes both switching and reconfiguration of open points such as double-dead ends, etc., in other cases, only changes in switch status are to be considered.

The performance index for some operational planning tasks will be minimization of cost (which are just losses costs, since there are no capital costs), but many operational planning situations address emergency requirements (such as during a storm when many line segments are down). The goal in these emergency situations is to maximize the amount of load that can be restored quickly, while minimizing excessive voltage drop, the amount of overloads, or number of switching operations.

In operational planning situations, sine there are no alternative line segment types, every line segment's cost curve is quadratic function (I^2R). Although application of linear optimization here can provide some benefit, a piecewise linear approximation (Figure 17.5, bottom) is really required to provide any reasonable accuracy. Non-linear trans-shipment methods (algorithms that represent I^2R losses without approximation) are more difficult to develop and to make user friendly, but give even better results.

Radialization

The vast majority of distribution feeder systems are operated in a radial manner, and therefore planners and operators of distribution systems usually desire optimization-driven design aids to compute the best *radial* configuration possible. Radialization is a constraint on the *type* of solution that is deemed feasible, and assuring that the optimization produces a radial configuration is an important part of the design of any practical distribution optimization program.

Forcing trans-shipment methods to reduce cost while meeting a radialization constraint can be a major challenge in algorithm design, particularly when non-linear costs are modeled. Radial configuration is difficult to enforce when the "cost-model" is non-linear, because the optimization algorithm senses (correctly)

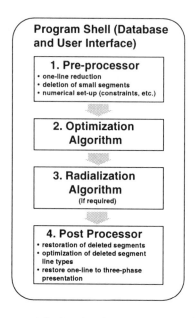

Figure 17.6 Many automated feeder planning programs consist of four stages shown here (see text for explanation).

that splitting a load between two feed paths will reduce losses -- given that either of two similar radial routes could serve a particular load, losses can be cut in half by leaving both connected to the load, splitting it between them. A number of methods exist to force radialization onto the computed network solution from an optimization (as in the null-point load flow and other applications), or to limit solutions considered by the optimization to only radial configurations. Some methods involve very lengthy analysis of configuration -- the radialization logic can exceed the length and complexity of the optimization routine itself.

Typical Feeder Optimization Algorithm Structure

Most automated feeder planning optimization algorithms work internally with one-line equivalent models of the distribution system. Usually, an automated planning program will have the structure shown in Figure 17.6, consisting of a shell providing database and user interface, and up to four functional stages. The first stage is a circuit pre-processor. This strips out all but the major trunks and

Automated Planning Tools and Methods

branches from the optimization, moving all load to main trunk nodes, much as was depicted for one-line reduction in Chapter 16 (Figure 16.3). Laterals, small branches, and any part of the circuit which the pre-processor determines cannot be subjected to any re-switching, can be deleted without loss of generality of the results, and optimized later (in stage 4). Deletion of laterals and small branches reduces the number of nodes of circuit representation that must be passed on to the optimization, which means a larger area of the system can be fit within any size-restriction the algorithm may have (all optimization algorithms have some upper limit on the size of the system they can handle), and that the optimization will solve much faster (since some algorithms have a cubic or worst run time to node number relationship, this alone is significant reason to have a pre-processor).

The second computation stage is the optimization itself, followed immediately by the third stage: radialization. The fourth and final stage is a post-processor, which adds the deleted laterals and small branches back into the completed circuit configuration, while simultaneously applying simple rules to determine their optimal conductor size. It also interprets the one-line representation (used internal to the optimization) back into a three-phase representation, if that is appropriate for the application.

Suitable Optimization Methods for Feeder System Planning

As mentioned in section 17.2, usually the selection of an optimization algorithm is a compromise between a number of conflicting requirements. Overall, despite some shortcomings, constrained linear trans-shipment has proven to be the most useful optimization approach for greenfield and augmentation planning, and non-linear trans-shipment best for switching studies.

Linear trans-shipment algorithms can handle very large (7,000+ node) problems, which means that with appropriate preprocessors to strip out laterals and small branches, layout/conductor size/switching optimization of an area of 15-20 substations can be done in one computation. The benefits of large problem size outweigh any limitations wrought by linearization, so on balance linear trans-shipment is regarded as the best proven optimization method for both greenfield and augmentation types of feeder system planning, (Willis et al., 1995 and 1996).

Genetic algorithms show great promise for greenfield and augmentation types of feeder system planning, but remain unproved in lengthy commercial application (Miranda). They are flexible, robust, and (when applied with linearized flow models) fast and dependable. The fact that GAs do not necessarily find the true optimum alternative is irrelevant in application; they produce results that are equivalent to linear trans-shipment (Willis et al., 1996).

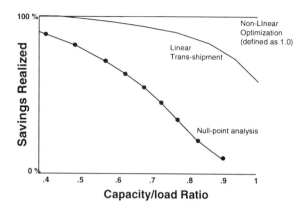

Figure 17.7 Capacity constraints make switching optimization a more challenging optimization. Ratio of realizable savings (i.e., as evaluated by detailed follow-up analysis) in losses due to switching changes found using null-point analysis (dotted line), linear trans-shipment, as compared to those found by non-linear trans-shipment optimization, as a function of the average equipment load/capacity ratio in the system. The more heavily loaded systems have more lines at or near maximum capacity -- switching is used mainly to balance load, not minimize losses.

Null-point load flows apply a network load flow to automated planning of radial feeder systems in a clever, heuristic manner. In the typical application, null-point analysis uses a network load flow program to "optimize" the switching of a radial feeder system in three steps:

1) Close all switches in the load flow feeder system model for the system being studied, making it a network.

2) Use the load flow algorithm to compute flows through the resulting network.

3) Due to its nature, this computed network flow will have "null points" -- where power flow is zero, throughout the system. These are interpreted as the 'optimal' locations for open switches.

The basic concept -- interpreting null points as the best places for open switches in a radial system -- has a certain intuitive appeal. While not a valid rule (a null point load flow makes *local* loss minimizations throughout, it does not accomplish a global least overall cost minimization), this approach is useful on systems that have few if any capacity constraints or voltage reach constraints.

Automated Planning Tools and Methods

Null point load flows give very poor results in the face of even a few capacity limitations in the feeder network, as shown in Figure 17.7. Most modern distribution systems are full of capacity limitations. The "solution of choice" for solving overloads caused by capacity limitations is load transfers, accomplished by re-switching. In such cases, "optimal" mostly means finding a way to work around such capacity limitations by switching, making load transfers without violating criteria, and service requirements. Losses are an important, but secondary issue -- the goal is not to find the switching configuration that lowers losses. It is to find the configuration that has the lowest losses from among those that limit flows to within capacity constraints. Null point load flows cannot do this, while most linear optimization methods (trans-shipment or otherwise) can.

The most effective tool is the best combination of problem size and accurate optimization. Figure 17.8 compares the savings found in "loss reduction analysis" on a typical system, using non-linear trans-shipment and linear-trans-shipment, and null point load flow, as a function of "program capacity." Although non-linear is the most accurate, linear trans-shipment's combination of problem size and fair accuracy makes it the most effective tool.

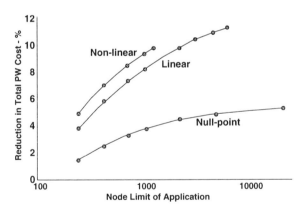

Figure 17.8 A linear, non-linear trans-shipment, and null-point algorithms were each limited to various program capacities (nodes) and used to plan a distribution system. With a small capacity, all had to be applied in a piecemeal fashion and each produced fewer results than when applied in larger format. Overall, best savings are found with the linear trans-shipment, which despite being a more approximate algorithm than non-linear trans-shipment can apply its optimization to much larger portions of the system.

17.4 SUBSTATION-LEVEL AND STRATEGIC PLANNING TOOLS

Chapters 11 and 12 discussed the performance and economics of the overall distribution system, particularly the interaction of the various levels of the power system. In particular, those Chapters stressed that the subtransmission, substation, and feeder levels must be compatibly designed if the system as a whole is to perform well and be economical. Studying the interaction of so many aspects of electrical and economic performance, among those three levels, and playing one level against another in a balanced tradeoff of cost, is the essence of long-range strategic distribution planning. The foremost goal of automated tools for long-range planning is to improve the planner's ability to perform this work: to evaluate alternative designs and decisions from the perspective of the combined, multi-level distribution system's performance and costs.

Chapter 11 and 12 also made clear that "substation siting" -- identification of the future substation sites, their capacities, and expansion schedules, is the function lying at the heart of strategic distribution system planning. Substation siting is strategic for three reasons:

- Substation locations and loads set the delivery requirements for the transmission system.

- Substations are expensive (both financially and politically). representing about one tenth to one fifth the total cost of power distribution system.

- Substation locations and capacities define the source locations and constraints for the distribution system. Their location relative to the load, neighboring substations and surrounding geography largely define the feeder system.

More than two hundred different automated procedures have been proposed for substation siting and sizing since the advent of computerized applications to power system planning in the late 1960s. Many of these are heuristic, and either inaccurate or applicable only in very limited circumstances. Others employ formal optimization procedures.

The Critical Focus of "Strategic" Substation Planning

While substations are the key strategic moves planners make in their chess game against the future, substations represent only a minority of overall cost. In particular, the potential impact that substation siting has on feeder system cost is generally greater than the substation cost itself (see Chapters 11 and 12). Both location and size of substations in relation to one another change the feeder

Automated Planning Tools and Methods 759

economics, often substantially. As a result, long-range planning, and "substation siting" programs must address feeder costs in order to provide maximum value to the planner.

Substation Capacity Optimization

Many automated substation siting and sizing tools focus on optimizing the assignment of capacity to substations. As represented by the tool the "substation planning problem" looks something like this:

- There are N sites available, of which some may already have substations built and capacity installed.
- Each site will have different costs and constraints associated with adding to its capacity -- new sites have to be cleared and prepared, etc.
- Each site will have a minimum load and a maximum load (the difference being what can be transferred to or from its neighbors)
- There is a global total load which must be served (i.e., the sum of all substation capacities must exceed a certain global total).
- There are various capacity margin and maximum size restrictions (design standards).

Typical of the many approaches to this particular problem was the "substation siting" program developed during EPRI project RP-570 (1979). Although crude with respect to today's database, graphics, and "program shells," the program applied a branch and bound algorithm to allocate capacity expansion among the N candidate sites in an overall minimum-cost solution that is no less effective in solving the problem than some procedures being published in academic journals at the time of this writing (1997). It outputs the expansion schedule of additions by site, and a list of site-to-site load transfers required to make the loads at each site fall within its capacity.

This particular program proved of little value as a planning tool, because it failed to address the feeder costs and constraints associated with the inter-substation load transfers. Load transfers between substations imply increased feeder capacities and losses, leading to cost increasess which always form a significant portion of the variable costs of any substation siting problem, and which occasionally outweigh the substation costs.

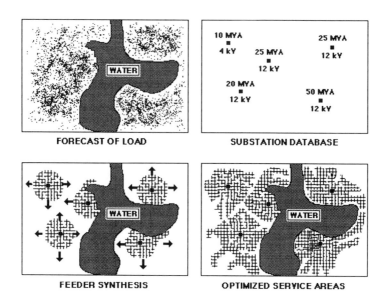

Figure 17.9 Service area optimization planning tools model substations as composed of both a substation site and a feeder network. They work by expanding the network out of each site until the collective set of networks covers all the load and meets all constraints, then adjusts substation locations and sizes to minimize cost.

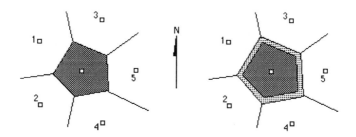

Figure 17.10 If a substation's capacity is increased, its service area must be expanded for that capacity to be utilized effectively, requiring a more extensive feeder system with heavier substation getaways and feeder trunks, a larger service area, and a greater overall expense. Service area optimization planning methods automatically include this cost element and similar impacts in their analysis when considering expansion options.

Automated Planning Tools and Methods 761

In addition, feeder system capacity and reach constraints, and geographic limitations, often interfere with the ability to transfer load beyond a certain amount, or dictate that load be transferred from an overloaded substation not to its least-loaded neighbor, but to another. Several attempts were made to fix this deficiency in the EPRI program. None worked well.

A number of other automated substation siting programs have been developed since 1980. Many work somewhat better than the EPRI program did, but the fact remains that programs that address substation siting as a problem whose primary variable cost is substation capacity cost miss the major variable cost at stake as siting and sizing decisions are made -- that of the feeder systems associated with the substations.

Service-Area Optimization

Service-area optimization (SAO) methods directly address the feeder-system impacts of substation siting, by representing substations as source locations, each with a feeder network. Although there are variations on the SAO approach, the general concept is shown in Figure 17.9. The optimization expands a set of substation service areas until they collectively cover all the load area. Well-designed SAO programs prove to be truly effective and valuable strategic planning tools.

Some SAO methods represent substations as "tiles" (see The "Game" of Substation Tile Planning, in Chapter 12), with a type of elastic pressure (economic cost differences) at the boundaries between substations pushing service area boundaries and substation locations into their lowest cost configuration. Other SAO methods represent a substation as a unit composed of a substation and a feeder network that connects the substation to surrounding load points. Some methods even model the subtransmission segment(s) required to move power from the high-voltage grid to the substation site.

Regardless, the common element of SAO representation is that a substation includes a feeder system leading from it to nearby load points, along with an operational constraint that this feeder system must reach every load point from one or another of the substations. Thus, if in the course of its optimization, the SAO represents a new substation as 50 MVA, it will simulate the construction of lines and losses for a 50 MVA feeder system built out of that substation to serve the load around it. If the substation is increased in size to 75 MVA, then its feeder system's cost and losses are similarly increased and considered in determining whether a net savings has been effected. In either case, the SAO applies optimization to balance the economics of using the 50 or 75 MVA capacity against the capacities, constraints, and economies of neighboring substations as shown in Figure 17.10.

Synthetic feeder system model

The SAO approach usually performs its analysis by explicitly routing feeders (i.e., it requires no input on potential feeder routes) on a small area grid basis. Often this is the same grid system used by the spatial forecast simulator (as described in Chapter 15). This grid approach approximates the feeder system with a *synthetic feeder system* representation, as shown in Figure 17.11. By using such a representation, the computer program can apply a dynamic or linear programming method, in a very rapid configuration (permitted because all links are of uniform length), allowing it to explore a multitude of options involving variations in the number, route, service area, primary voltage levels, and loading limits of feeders, as well as changes in substation capacity, within a reasonable amount of time. The existing system is represented in this manner, and models of the required future expansion are automatically produced by the program as substations are expanded and moved, and as load grows.

While not accurate enough for detailed feeder design, the grid-based synthetic feeder system representation has proven very accurate for long-range planning, where the major feeder planning goals are analysis of costs, and decisions on the voltage level, service area, and number of feeders (Crawford and Holt, Willis and Northcote-Green). Accuracy in estimating cost, and in identifying sensitivity to design constraints is generally within 3%.

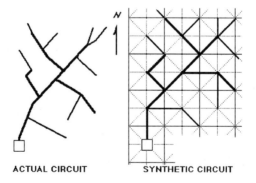

Figure 17.11 Service area T&D planning optimization approximates the feeder system with a "stick figure" network that allows routes only along vertical, horizontal, and diagonal lines through each small area. Although approximate, this method proves very accurate for long-range applications and allows very fast computation of large system optimization problems.

Automated Planning Tools and Methods

Explicit algorithm routing is useful in strategic planning

While implicit routing methods usually prove best for feeder planning purposes (see Section 17.3), explicit routing algorithms, that automatically generate feeder system route models with no user input, prove most useful for long-range planning. Of prime importance in the usefulness of this approach is the ability of the program to generate the synthetic feeder system from only a small description of the conductor sets and standards used by the electric utility. Thus supported by an explicit routing, the SAO approach does not require an extensive feeder database in order to perform its analysis. This means that it can be applied to distribution studies very economically, since such data does not have to be gathered. More important, this also means that its application to future years (for which feeder systems have yet to be built and for which, therefore, no database *could* exist) is just as valid.

Working in the same geographic data format as the spatial simulation of load growth often proves useful to a SAO method, too, because the program then has access to the GIS-type data used in land-use simulation analysis (see Chapter 15). Typically, a SAO will permit the user to define feeder-level constraints, restricting it from representing feeders as built across wide rivers, through cemeteries or wildlife preserves, etc., and so that it uses underground feeder costs in areas where that type of construction is necessary, etc.

SAO Method Application

While SAO programs differ, usually the resulting plan, displayed both graphically and as text, gives the optimal load and capacity, and the optimal service area for each substation and provides estimates on the performance, cost, and layout of the expected feeder system. Usually, this type of approach is applied with constraints that limit the solution it must find in order to produce a plan in which:

- All of the load is adequately served (voltage, reliability).

- No equipment is overloaded.

- Existing substations are utilized whenever feasible.

- Feeder system costs and its losses are minimized.

- Capacity exists to back up the loss of any one substation.

The optimization method is used to perform an analysis of expansion in which the system is configured to remain within the constraints listed above while achieving minimum cost. This determines:

- Which existing substations should be expanded and to what capacities?

- How many, where, and to what capacity should new substations be built?

- What is the best primary voltage for the feeder system at a new substation?

- Where operating problems due to extreme reach needs might plague the feeder system.

- An overall estimated substation-feeder expansion budget.

A distribution plan must conform to a long list of common-sense constraints and rules. The most useful SAO methods lay out the feeder/substation system while *automatically:*

- Constraining the plan to use only equipment within a certain set of design standards, such as limitations on the size of transformers.

- Not building feeders through cemeteries, across lakes, etc., such areas recognized from land use data provided by the small area forecast database (usually obtained from GIS systems, or as entered for the load forecast).

- Building only underground, or only overhead feeder systems, in areas designated by the user as solely one or the other.

- Building only a specified voltage in a specific region (e.g., 34.5 kV in one region, 25 kV in other regions).

- Meeting contingency criteria (e.g., outage any substation and still supply all loads through contingency feeder re-switching).

Automated Planning Tools and Methods 765

Finally, given a particular substation-level plan -- a complete description of the locations and capacities of all substations and their associated feeder networks -- a good SAO method can assess three important aspects of that plan when evaluated against any load forecast:

- Identify areas of the system where load cannot be served because the system does not have enough capacity or feeder reach.

- Areas of the system that can be served but only at a higher than average cost.

- Areas that cannot be served reliably, basing this on an estimate of reliability based on the synthetic feeder system.

An ability to evaluate these three aspects of any plan against any load forecast provides a very powerful tool in the first step of long-range planning. By comparing the present system to the future load pattern, the planning program identifies where the problems in the system are located and why they are problems (geographic constraints, cost, electric voltage drop, reliability).

Where is the capacity shortfall?

One useful application of a SAO program is to apply it to serve long-term (+10 year) loads using only the existing (+0 year) system, re-switching feeders and re-routing power as much as is practical, *but building no new facilities.* In most cases, and certainly if there is any substantial load growth, the optimization finds that this goal is impossible to meet within constraints it has been given. Being a cost minimization approach, the optimization *sheds the most-expensive-to-serve load first,* applying the existing system within operating limitations to pick up all the load it can serve, most economically. Plotting this "unserved load" as a map provides a very useful illustration to the planner, both in serving load and in reducing cost, as shown in Figure 17.12. By re-running the program with feeder reach distances set to infinity (constraints removed) the planner can identify what portion of the unserved load map is due to substation capacity limits. Similarly, by re-running it again with substation capacities set to infinity (constraints removed) the planner can identify what portion of the unserved load map is due to feeder reach limits alone. These studies help identify the type of expansion planning that will be needed to accommodate the load growth.

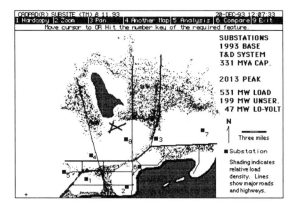

Figure 17.12 Unserved load map produced by a SAO method trying to serve year 2013 forecast loads with the 1993 system provides useful information at the start of the planning phase. Shading indicates locations and amount of load that cannot be served. This permits exploration of the system capability versus long-term needs.

Substation Durability and Utilization Planning Methods

The durability (expected reliability as a function of remaining life and loading) of a substation can be included in some substation siting and optimization approaches. As shown in Figure 17.13, the typical "bathtub" reliability curve for a substation's transformers and equipment can be used to estimate its expected remaining lifetime. The area under the curve is assumed to be 100%. The portion already "used up" is estimated based on age, past loading records, etc.

The PW value of the transformers serving various load levels in the future can be computed as shown in the lower left of Figure 17.13, which compares the utilization value if the transformers are limited to 23 MVA peak each, with their economic value at a loading limit of 28 MVA. The future PW value for either loading utilization plan is the area under its curve.

Higher loadings create an accelerated loss of life, and a higher likelihood that the transformers will fail in the short term. The expected PW cost of failures can be estimated (lower right in Figure 17.13). This can include only the cost of replacement, etc., or it may also include the cost of poor service for customers during the failure and repair (see the example feeder problems in Chapter 18, Section 18.4). Given enough time, every unit in service will fail, but lightly loaded equipment generally lasts longer, and hence has a lower PW failure cost.

Automated Planning Tools and Methods

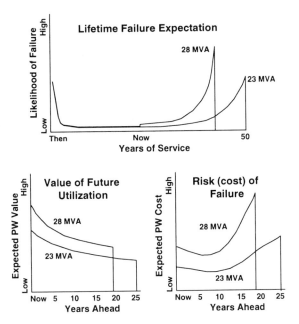

Figure 17.13 Expected remaining lifetime and failure likelihood of a 25 year old transformer is evaluated against loading plans calling for both a 23 MVA peak and 28 MVA peak (top). This leads to computation of both the PW value of its service at those loading levels (lower left) and the expected PW cost of failure (lower right) which can be used in determining optimal loading of each substation.

Both the value of utilization and the value of failure/customer service quality can be re-computed based on various assumed loading level plans and balanced against one another and other aspects of the substation utilization. Boosting the loading on a substation will increase its present value (the utility is making more use of it) but increase the future failure risk cost. These costs can be balanced against one another, and other costs, for any situation, to determine the optimal loading plan.

This approach tends to load new equipment to higher levels than older equipment, as shown in Figure 17.14. When applied in conjunction with assessment of the value of reliability to the customers, it can find the best compromise among loading levels, long-term use, and service reliability. Thus it is quite suitable for newer "budget constrained" planning paradigms that try to balance budget, loading standards, and customer reliability (see Chapter 18).

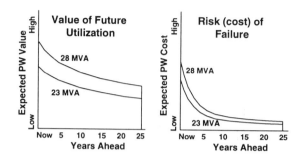

Figure 17.14 A new transformer evaluated in a manner similar to the older unit in Figure 17.13. This unit provides similar future utilization value but far different failure risk. As a result, optimization applications which include durability in service in their analysis tend to recommend loading newer equipment to higher levels than older equipment, particularly when the failure cost includes customer service valuation (see Section 18.4). They also balance load among substations in a way that "uses" lifetime optimally, which typically also means new transformers are loaded to higher levels than older units.

17.5 APPLICATION OF PLANNING TOOLS

Why Use Automated Planning Tools and Optimization?

Reducing cost is without doubt the primary justification for application of optimization -- savings of from five to twelve percent are realizable from the use of a well-designed automated, optimization-based programs for both short- and long-range planning. Claimed savings that accrue from optimization are always subject to a certain amount of interpretation -- would the original, non-optimal design have been built as planned or would an improved design have been found without optimization before construction was finished? Are the claimed savings realistic and have they been verified? The author believes that Figure 17.15, which is based on reported results from 25 users of a short-range feeder-system planning program utilizing a linear trans-shipment, realistically reflects what can be expected. (Similar results from long-range planning are unavailable). As can be seen, optimization generally gives larger savings in greenfield cases, probably because that type of problem is more amenable to today's optimization methods, but perhaps also because there is simply more to save in such situations.

Automated Planning Tools and Methods

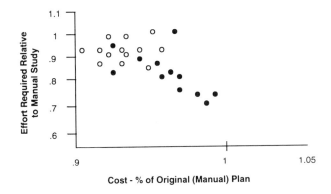

Figure 17.15 Percent reduction in overall levelized cost wrought by application of optimization versus the change in effort required to apply it, as reported by various users of a linear trans-shipment feeder system planning program. Hollow dots indicate results for new expansion studies, solid dots augmentation studies. Average results for new expansion are a 5.3% reduction in cost and an 8% reduction in effort required to complete a design, and an average 4.3% reduction in cost and 22% reduction in time for augmentation systems.

Reducing the time and effort required to develop, and document a distribution plan is also a consideration, particularly in light of staff downsizing and cost reduction pressures that all electric utilities face. Figure 17.15 indicated only 8 to 22% reduction in the time required for application -- most users reported that they used automated optimization to reduce system cost, not planning costs: While results equivalent to manual planning could be obtained in less than one-half the time required for manual work, when given automated tools it is judged best to invest time in using them to reduce cost to the minimum possible.

A learning tool. Use of optimization-based tools contributes to an engineer's understanding of the distribution system, the interplay of costs, performance, and tradeoffs within it, and the measures that can help reduce cost. It is often worthwhile to "play" with the input variables to learn how various aspects of a plan interact. For example, setting the cost of losses to zero in a planning program will lead to a plan that ignores their contributions to cost. Comparison of the resulting plan to the actual best plan with reveal characteristics and design practices in the plan that are due to losses.

Standardization and Credibility. Properly set up and applied, automated programs apply the same rules, costs, and evaluation methodology to all planning situations, standardizing the planning process and helping to assure that all planning requirements are meet. Optimization, if used correctly, ignores no viable alternative, which improves credibility of the resulting plan -- no bias was shown in the selection of the recommended plan, all options were considered on an equal basis.

Documentation. Automation (not optimization) aids considerably in documenting the need for additions and in demonstrating that diligent effort was devoted to examining all options and determining the best design.

Justification. Far beyond its ability to improve credibility and documentation, automated optimization can be used to perform justification studies, in defense of critical or controversial elements of a plan, as will be described later in this section.

Multi-Year versus Single-Year Application

Planners, and designers of automated planning software, usually must make a compromise between multiple years of optimization evaluation and the size of the system that can be evaluated. Most T&D plans cover a period of several years into the future, and the goal of the planning is to determine a schedule of additions over the next five to twenty years. It is possible to set up an optimization engine in a *multi-year framework,* in which it will automatically evaluate when as well as where to install capacity at a substation, expand a feeder segment, or make whatever change(s) are being considered. The optimization deals with a set of yearly plans, and analyzes changes from year to year as explicit objects in its optimization. By contrast, single-year applications work on a snapshot of the system and its loads for one particular year, optimizing the system plan to serve the load, and assuming in all evaluation of long-term economics that the loads and system remain fixed ever after.

If a particular optimization algorithm can optimize a system of 10,000 nodes in a single year analysis, it will be able to handle only a much smaller system -- perhaps only 1,000 nodes -- when performing a ten-year multi-year optimization. To date, most optimization applications have tended to be single-year programs, because experience has indicated that, within the limitations defined by current technology, having a large problem size, is more important than being able to optimize over time.

As a result, planners must usually apply optimization in a series of cases for different future years. Typically, such planning examines only selected years such as +1, +2, +3, +5, +10, +15, and +20 years ahead. The goal is to produce a smooth, economical, and effective schedule of T&D additions over the entire period, with everything added at the last moment possible, and everything that is added having long-term value. Approaches to performing a multi-year plan differ, there being two overall directions the planning in time can take.

Automated Planning Tools and Methods

Forward fill-in

The planners apply the T&D planning method to the first study year (+1) to determine the T&D additions required to serve the load growth in that year. From there, they study the second year's planning situation, using the optimization to make any additions required to serve that load forecast, for that year, and so forth, working forward to year +20, in each case determining what needed to be added in that time frame. In this way, they "fill-in" a schedule of additions on a year-by-year basis for the entire study period.

Backward pull-out

In this approach, planners start with the last year of the study period (e.g, 20 years ahead) and plan the T&D system backwards. They use optimization to produce an optimized plan for the final year of the plan, identifying what needs to be built sometime in the next 20 years in order to meet eventual needs. The planners then proceed backwards in time, analyzing the +15 year ahead timeframe, then the +10 year ahead time-frame, and so forth, to the present, determining what can be pulled out of the long-term plan in each time period. In this way they identify a schedule of additions over the 20-year period. This is illustrated in Figure 17.16.

Backward pull-out works best for long-range plans, particularly when facing great amounts of new customer growth. Forward fill-in works best for short-range planning and in cases where there are few additions or low new customer

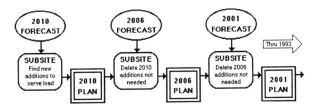

DISTRIBUTION SYSTEM EXPANSION PLANNING PROCESS

Figure 17.16 Most optimization for long-range planning are applied in single-year form, in reverse chronological order, starting with the horizon year and working backward.

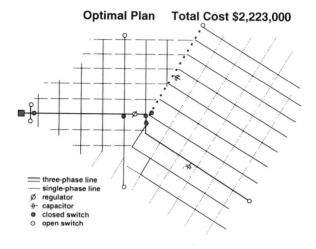

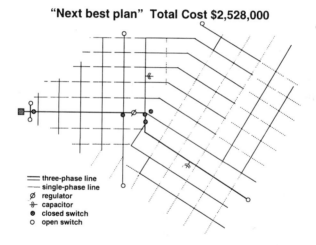

Figure 17.17 A justification study. The value of a controversial element in a completed least-cost feeder plan (top), in this case a trunk down a busy street (dotted line, top drawing) can be determined by deleting it from options available in the optimization's database and re-solving. The optimization re-structures the feeder and re-sizes a good number of segments, creating what is essentially the "next best plan" (bottom). Comparison of the costs for the two plans establishes that the line segment in questions saves a net $305,000. Adding that to its cost ($290,000), provides an estimate of its *value*, a total of $585,000.

Automated Planning Tools and Methods

growth rates. In either case, long-term economics (present value, levelized cost) must be used to evaluate costs from a decision-making perspective, even if year-to-year costs are reported in non-discounted, present dollars.

Justification Studies in Defense of the Plan

A valuable use optimization is for justification studies, which determine the value of a particular element of a distribution plan. Figure 17.17 shows a hypothetical example. A line in the new plan (top drawing in the figure) is both expensive and the subject of considerable community opposition due to its site. The value of this line segment can be determined by performing two additional optimization computations, in addition to those that led to the plan itself.

The savings this segment produces can be identified with one additional optimization run. The segment is deleted from the input database (nothing is allowed to be built in this location) and the optimization re-run. The optimization will find the "next best" option. Assuming the original plan was optimal, the cost of this new plan will be higher, its margin attributable to the savings that the line segment rendered. The value of the line segment is its cost plus this margin.

Producing a Plan that Has a Certain Element or Feature

Often planners may want to produce a design that includes a specific element, to force the optimization to use a certain substation site, or build a feeder system that includes a specific line segment. Optimization (can be forced to produce a plan using a specific element by setting the performance index penalty (usually cost) for that particular element to zero.

For example, if planners desire a plan that includes a 100 MVA substation at the 12th of 25 candidate sites shown in Figure 17.3, the input to the substation planning program being used can be given a cost of zero for a 100 MVA substation at that site. The optimization, whose goal is to reduce cost, will use this substation (since it costs nothing) as possible in lieu of any other options (all of which have a cost). The resulting plan both includes the 100 MVA substation and makes the maximum possible use of it.

Doing nothing should *always* be an option

One of the most serious mistakes made by many planners is to assume that the best alternative involves doing something. Occasionally, it is both less expensive and better to delay any commitment or changes in the T&D system -- to do nothing. Even when this is not the case, "nothing" should always be included as an option in all automated planning studies. By including "nothing," the planner explicitly shows why "doing nothing" is not the recommended option -- that it violates criteria or requirements, or that some other option delivers better overall economy.

Recommendations for Optimization Application

Use automated planning methods

Automated planning methods standardize and streamline the planning process. They improve planning credibility and the ability to document and justify plans, reduce the labor required, and improve the quality of the resulting plans.

Identify goals and needs prior to selecting a tool

Planners should explicitly study and document their goals and needs, and write down a prioritized list of their needs before picking a method, gathering data, or even setting a budget (Chapter 14). This list should be used to evaluate potential planning tools.

Focus on the planning, not the tools or the computers

Planners must remember that their job is to produce good plans, effectively and at low cost, and not necessarily to use advanced or exotic computer and mathematical methods. They should not let the quest for advanced technology, or often convoluted information systems requirements, drive the tool selection process.

Long-range planning tools should use the systems approach

Any long-range planning method must assess interaction of electrical and economic performance between the subtransmission, substation, and feeder levels of the system.

Long-range planning tools must evaluate the time value of money

The purpose of long-range planning is to study the long-term economics of new capital investments. Whether called present worth analysis, net present value analysis, levelized costing, or future versus present cost analysis, a method of comparing future versus present costs must be included in any long-range analysis if it is to be effective (Chapters 6 and 14).

Automated Planning Tools and Methods

Make certain that "do nothing" can be evaluated as an option

Some automated planning methods cannot evaluate costs and criteria for their "base" or "zero" cases. An automated planning tool is much more valuable if it can evaluate the absolute cost of a case, not just the cost of any additions or changes made to it.

Read and learn from past experience

The results of decades of work and experience by dozens of planners are available to anyone who is willing to do a little library research.

Don't re-invent existing methods

Many proven tools exist for every step in the T&D planning process, tools with the advantage of prior testing, documentation, and acceptance. Developing computer programs and new algorithms is fun, but it doesn't produce a plan and it is a waste of resources.

"New" algorithms may be no better than older ones

Any new optimization method will be the subject of intense publication in technical journals for several years after it is invented, simply because it *is* a new optimization approach, and researchers want to study it and report their findings, (that is how professors and researchers make their living). Many times the conclusions of such publications are "the new method works just about as well as older methods, but no better." Planners should ask *what* the new method does that existing programs and methods cannot: they should be a compelling reason to switch to a promising, new method instead of using a time-tested older one.

Be realistic about expected savings

A reduction in labor and time to complete plans of from 10% to 33%, along with a 5 to 12% savings in overall system costs (Figure 17.15) is realistic -- expectations of reductions beyond those values are not.

Expert to work hard

Optimization provides a tool, but it does not do the planner's job. Planners can expect to work harder, because automated planning tools reduce the amount of time spend on tedious, repetitious tasks of manual planning, allowing more time for productive work.

REFERENCES

D. M. Crawford and S. B. Holt, "A Mathematical Optimization Technique for Locating and Sizing Distribution Substations," *IEEE Transactions on Power Apparatus and Systems,* March 1975, p. 230.

J. K. Dillard, editor, *Transmission and Distribution Reference Book,* Westinghouse Electric Corporation, Pittsburgh, PA, 1964.

M. A. El-Kady, "Computer-Aided Planning of Distribution Substations and Primary Feeders, *IEEE Transactions on Power Apparatus and Systems,* June 1984, p. 1183.

M. V. Engel, et al, editors, *Tutorial on Distribution Planning,* IEEE Course Text EHO 361-6-PWR, IEEE, Hoes Lane, NJ, 1992.

V. Miranda et al, Genetic Algorithms in Optimal Multi-Stage Distribution Network Planning, *IEEE Transactions on Power Systems,* November 1994, p. 1927.

J. E. D. Northcote-Green *et al,* Research into Load Forecasting and Distribution Planning, Electric Power Research Institute, Palo Alto, CA, 1979

H. N. Tram and D. L. Wall, "Optimal Conductor Selection in Planning Radial Distribution Systems," *IEEE Transactions on Power Apparatus and Systems,* April 1987, p. 2310.

H. L. Willis and J. E. D. Northcote-Green, "Comparison of Several Distribution Planning Techniques," *IEEE Transactions on Power Appratus and Systems,* January 1985, p. 1278.

18
Traditional versus Competitive Industry Paradigms

18.1 CHANGING ENVIRONMENTS OFTEN REQUIRE NEW VALUE SYSTEMS

At the time of this writing (1997), governments worldwide are altering their view of the role electric utilities should play in their countries' energy infrastructures. Most are re-regulating their power industry into a framework of competition at several levels, so it provides a profit incentive for innovation and efficiency, as well as making certain the new structure fosters customer choice. Such a shift in the basic business and regulatory background requires a change in the utility's sense of identity and its basic value system: A change in paradigm. Usually, a paradigm shift will not require that planners adopt new tools or planning procedures, only that they use their existing methods in new ways.

A *paradigm* is the set of goals, assumed interrelationships ("truths") and values, along with the definition of the limits of their applicability, that forms a group or culture's framework of thinking. Often, a paradigm is conceptual and difficult for those involved to articulate precisely, being part of what "everybody knows." It is woven into the very fabric of the organization's sense of identity. Almost invariably, any pressures to change the prevailing paradigm are viewed as a threat to the group and its identity, rather than, as they should be, as a threat only to the way they have functioned or worked.

Organizations, particularly utilities facing the 21st century, must evolve as the conditions around them change and demands made of them grow. Often, they must change their paradigm. If they don't adapt, they will be replaced by organizations or groups who do, and who, as a result, out-perform the existing order to the point of extinguishing it.

This chapter compares several different power delivery paradigms for the distribution system. Section 18.2 discusses paradigms in general, and how and why they change, using as examples several historical cases from outside the power industry. The "traditional" utility T&D planning paradigm is then examined in Section 18.3, and a paradigm more appropriate for service-quality conscious but budget-constrained utilities is introduced in Section 18.4. The purely profit-based paradigm is explored in Section 18.5.

18.2 PARADIGM SHIFTS

A shift in paradigm is often driven by a change in external conditions. For example, most military organizations go through a paradigm shift every time they switch from peace to war or back again. During war, an army or navy has one purpose: Win the war. Procedure and priorities quickly adjust from peacetime practice to those focused on lean wartime efficiency, innovation in tactics, and appropriate strategic use of technology. All else is secondary. After the war, the transition back to peacetime conditions and purpose can be complex, confusing, and chaotic, as priorities must be re-set and basic values shifted.[1]

Sudden paradigm shifts are also driven by new technology. An often-cited example is the impact of electronic watches on the Swiss watch-making industry. Prior to the development of electronic watches, the Swiss dominated the watch market worldwide. They had a firm paradigm: A watch was a mechanical device, assembled from many moving parts, in a low fixed-cost/high variable labor business structure, with testing, standards, and quality judged by Swiss-mandated metrics appropriate for mechanical devices. Even non-Swiss manufacturers followed this paradigm.

The Swiss failure to see electronic watches as a threat to their market dominance was at least partly due to their paradigm, which did not recognize that the market would consider a wrist-mounted device built from an integrated chip, a digital display, and a battery as a suitable substitute for a mechanical watch. As a result, many businesses failed, and the Swiss lost market share while competitors from Japan and other countries took command of their market.

---

[1] See for example, *The Last Great Victory,* by Stanley Weintraub, Dutton, New York, 1995, for details of the difficulties in reconverting to peacetime after WW II.

Traditional versus Competitive Industry Paradigms 779

But the new paradigm had its own inadequacies, which led to its partial downfall within two decades. Manufacturer's of electronic watches mapped their industry's paradigm onto watch-making. They used their concept of high fixed-cost and very low variable-cost ("The first chip costs eighty million dollars, every subsequent one only a penny") as their business model. There are vast differences in complexity between the simple clock-circuit/quartz crystal needed for a watch and the integrated chip set of a computer, but executives at these companies never re-examined their own assumptions.

Their designers and stylists also thought in terms of the efficient, but antiseptic industrial designs they used for computers and electronic equipment. After the novelty wore off and the new watches wore out, the public realized that the new generation of watches was boring and impersonal -- there were only a few dozen types, each offered in slight varieties of metal color and style, like computers and televisions. Roughly 20 years after electronic watches took over the market, new European watchmakers like Swatch and Fossil regained dominance through their own paradigm shift: Watches are *apparel, not apparatus*. They found ways to reduce the fixed cost of limited production runs and treated watches as fashion. They began to offer colors and nonsensical designs, and to deliberately create short-lived fashion fads. They prevailed.

One Paradigm's Values Are Another's Nonsense

Major paradigm shifts are often controversial and tumultuous, because the assumptions and values inherent in one successful paradigm can appear utterly nonsensical from the perspective of another. Returning to the military analogy mentioned above, in a time of peace, a military organization's purpose is to preserve the peace and maintain itself in readiness. While during war, it is supposed to *change* the status quo (i.e., win the war and bring peace), its mission during peacetime is to *maintain* the status quo. This tends to make it conservative. Drill and "spit and polish" become not just the norm, but the very sense of identity for the organization.

Thus, during the Victorian era, the British navy, which had not fought a major war since the Napoleonic period, was populated by admirals and captains to whom the thought of actually testing the gigantic guns on their battleships was unthinkable. The heat of even one practice round would blister the flawless enamel finish on the gun barrels, cover their sparkling ships with soot, which would take a week to clean away, and divert resources from polishing the brass and keeping the decks scoured to a smooth, even luster.

Armchair strategists of the 20th and 21st century can smugly poke fun at this value system, but the prevailing naval paradigm in the last half of the 19th century held that a ship's and crew's appearance was *the* measure of its captain's potential for further promotion. Furthermore, it was correct for the time. How else was the British Admiralty to judge the quality of a captain in time of peace? His ability to pay attention to the myriad details of his ship's appearance, to juggle limited materials and labor, to select officers, organize his crew, and motivate his men to clean the decks, paint the metalwork, sand and varnish the wood, and polish the brass fittings, was as good a measure as existed of his ability to command. It was *assumed* that outstanding achievement in this activity would translate to an outstanding ability to wage war.

Moreover, the paradigm was *strategically* correct. During this lengthy period of peace, battleships were instruments of diplomacy, whose mission was to visit foreign ports as visible proof of the British empire's strength and intimidatingly long reach. A dingy gray warship covered with signs of recent battle practice might have impressed a few foreign admirals; but glittering white paint, shiny brass, and a crew turned out in spotless uniforms left more lasting impressions on those attending the diplomatic receptions routinely held on the ships' quarterdecks. Many a British captain never fired his guns in battle, but found them indispensible as the supports for the canvas awnings rigged to shade the canapés tables and orchestra podiums during afternoon diplomatic receptions in the tropics.

The ambassadors, presidents, and ministers entertained upon those ships never questioned the prevailing paradigm either. The last time the British navy had fought a war, it had savaged Napoleon's French and Spanish navies with ruthless efficiency. It was *assumed* that it could do it again, and the perfection of paint and drill was taken as evidence that it would do it almost perfectly.

Paradigms defined by and appropriate for their times should never be judged from the perspective of another era. The Victorian captains who abhorred the idea of gunnery practice *were* incapable of fighting and winning a major war, but they maintained "Pax Britannica" for a century. They and their paradigm did the job. The naval warriors who won WWI 30 years later by following a completely different paradigm, could not have performed that peacetime function well. Each generation of naval officers had the right paradigm for their time.

Large Organizations Can Change Their Paradigms

Having the the correct paradigm leads to success, but failure to recognize and shift paradigm as conditions or technology change can be the death of an organization or culture. It certainly hurt the Swiss watch industry and a similar failure to perceive aircraft as competition for passenger trains, doomed railroads throughout North America to be "cargo only" enterprises ever since the 1960s.

It has become popular myth that a paradigm can never recognize its own shortcomings, that change will inevitably overcome it from outside and that the organization it infuses will die as a result, particularly if that organization is large, conservative, and has been in the lead for a long time. Historical fact does not support this belief: While there are many examples where complacency and stupidity did lead to failure, there are many examples of successful paradigm shifts by large organizations.

One such example is the Victorian British navy cited earlier. In 1900, it was indisputably the largest organization and culture of its time, and arguably the most conservative. But for all of its self-centered satisfaction and preoccupation with peacetime routine, as the 20th century dawned it began to pro-actively alter itself. It developed metrics to measure captains on their ship's rate and accuracy of gunfire. It set up a promotion system based on annual target practice competitions among its warships. Ultimately, it designed a new class of battleships (the dreadnought battleship) to revolutionize naval warfare.

These changes occurred because the British navy realized that both its technology and its environment were about to change, and decided that it had to adapt before any opposing navies did. The resulting paradigm shift affected every aspect of its organization, culture, and material. It was not without internal tumult and controversy. Many officers and seamen lost their jobs. But many others found opportunity and rapid advancement as this large, very conservative organization recognized its need to change, identified what changes were necessary, and managed that change in what in retrospect can be seen as a marvelously well-managed alteration of personality. When war came in 1914, the British navy was more capable and ready than its enemies. It won easily.[2,3]

Critical Self-Examination Is the key to Managing Paradigm Transitions

The British navy survived, and in fact prevailed, because it had a deep commitment to quality (whether of "spit and polish" or gunfire) and a nearly paranoid sense of survival that led to critical self-examination. Utilities most likely to survive the transition to de-regulation, and to prevail in an environment of competition, are those who similarly re-examine their values and assumptions, who pro-actively change their culture, and who challenge themselves to take what are, from within their old paradigm, bold and perhaps even brash risks.

---

[2] Its losses amounted to less than 1%, and it captured its enemy's *entire* navy.
[3] This organization had much more difficulty shifting back to peacetime status. Parkinson's Law ("Work expands to fill the time allotted to it.") was a serious "rule" coined shortly after WWI by management consultant C. Northcote Parkinson, hired by the British Admiralty to investigate why, with their navy reduced to a third of its wartime complement, it still required more officers and paperwork than ever.

Paradigm Shifts

A new paradigm does not have to abandon the precepts of previous paradigms: It may involve re-interpretation of what they mean, or how they are interrelated, or it might *add* additional requirements or considerations. What a change in paradigm does indicate, however, is a shift in an organization's thinking about why it exists, and what is important to it. Changing the paradigm will not necessarily change the planning process or the tools used to analyze alternatives, but it will change the way those results are used and the decisions they lead to.

Example problems

Table 18.1 shows several alternatives for a feeder being planned in a growing part of a utility service territory. Something must be built to extend service to new customers, in an area where substantial load is expected very soon and long-term growth prospects are high. This feeder can be built in any of several sizes and designs, as shown, using conductors and layouts of the type discussed in Chapters 7 through 9. The details of exactly what each alternative is, and the differences in costs are unimportant to this discussion. What matters is that there *are* alternatives, that they differ in capacity and ability to handle future growth, and that those with more capacity and more ability to handle growth cost more.

Table 18.1 Example New Feeder Problem Alternatives Evaluated on the Basis of Traditional Least-Cost Present Worth Value System at 11% Discount Rate - $ × 1000

Feeder Option		Meets Criteria?	Year 1 Capital	PW Cost	Comments
1	4/0 large trunk	No	$2,250	$7,550	Must be rebuilt for load growth in year 2.
2	4/0-4/0 branch	Almost	$2,350	$7,150	Must be rebuilt for load growth in year 2.
3	336-4/0 branch	Barely	$2,775	$6,360	High losses. Branch rebuilds yrs. 7 & 12.
4	336 large trunk	Barely	$2,850	$6,100	Must be rebuilt for load growth in year 8.
5	636 large trunk	Suffices	$3,125	$5,100	Reinforcement needed in year 14.
6	795 large trunk	Yes	$3,500	$4,900	Reinforcement needed in year 17.
7	636-336 branch	Yes	$3,550	$4,800	Partial rebuild in yr. 17.
8	795-336 branch	Yes	$3,675	$4,750	Partial rebuild, yr. 24. Lowest PW cost.
9	795-636 branch	Yes	$3,800	$4,800	Lower losses don't justify larger cond.
10	1113 large trunk	Yes	$3,900	$5,100	Lower losses don't justify larger cond.
11	1113-633 branch	Yes	$4,250	$4,950	Lower losses don't justify larger cond.

Traditional versus Competitive Industry Paradigms

Table 18.2 Refurbishment Alternatives for Service Quality Refurbishment of an Aging Feeder in a Stable Developed Area at 11% Discount Rate - $ × 1000- $ × 1000

Feeder Option	Capital cost	O&M PW	Failure risk	Total PW	Comments
Base Do nothing		$900	$5,145	$6,045	Tolerate failures and poor service.
1 Fix only serious	$650	$700	$3,900	$5,250	Repair only the most critical items.
2 Rebuild plan A	$1650	$400	$2,000	$4,050	Rebuild trunk, repl. switches & regul.
3 Rebuild plan B	$2250	$350	$1,200	$3,800	"A" + pole insp./repl. Lowest PW.
4 Rebuild plan C	$3175	$290	$771	$5,161	"B" plus serv. trans. insp./replcmt.

Table 18.2 shows the alternatives for refurbishment of a 45-year-old feeder in an area of the system where equipment aging and slow load growth (.3% per capita increase annually) have lead to below-average service quality. The table shows the costs of various rebuild options, the expected PW O&M savings and failure risk (PW cost) due to future failures, calculated by analysis of age, equipment condition, load curves, and failure-likelihood based on equipment statistics for this and other utilities (see for example, the IEEE "Gold" Book).

18.3 LEAST-COST PRESENT WORTH PLANNING (TRADITIONAL UTILITY PLANNING)

Traditionally, electric utilities have applied a planning paradigm that includes several fundamental assumptions:

1. All new customers will be connected to the system

2. All new designs will meet standards

3. Long-term (e.g., present worth) costs will be minimized.

4. Capital funding for any project that meets the above requirements will be forthcoming.

This can be labeled an engineering-driven paradigm, because engineering requirements (1 and 2 above) drive the decision-making process (3) and define the financial requirements (4). This is illustrated in Figure 18.1.

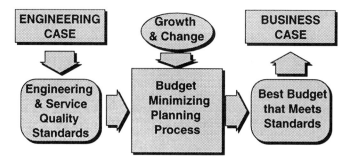

Figure 18.1 Traditionally, utility budgets were largely engineering driven. Rigorous (and inflexible) standards and long-term present worth cost-minimization defined budget needs, which drive the utility's subsequent business planning. Not surprisingly, most engineers and planners are particularly fond of the process, and the distribution planning paradigm which grew up around it.

The alternatives in Tables 18.1 and 18.2 would be evaluated for total present worth cost over a long period (usually 30 years) or alternatively on a lifetime basis, adding together present worth costs for both initial and continuing operating costs over the period. If load projections and planning criteria showed that the feeder had to be upgraded during the period, the present worth cost of the future capital cost for that upgrade would typically be included.

In each case, the alternative with the lowest PW cost would be selected. In the example shown in Table 18.1, this process would select the 795/336 branched feeder design for the new feeder. Among alternatives that completely meet the utility's standards, this option's combination of losses, and no need for reinforcement in the near term, gives the lowest PW cost. The refurbishment project in Table 18.2 would be evaluated similarly on a present worth basis and "Plan B" selected because it has the lowest overall PW cost. The utility would need a total of $5,935,000 in capital for the two projects.

18.4 BUDGET-CONSTRAINED SERVICE VALUE PLANNING

Sensible and commendable as the goals and assumptions of traditional distribution planning may be, many elements of the power industry have or will be de-regulated by the 21st century, and very different paradigms may need to be applied to some aspects of power system planning.

Traditional versus Competitive Industry Paradigms

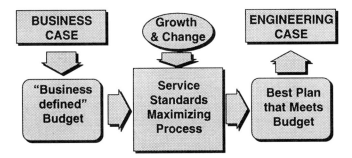

Figure 18.2 Many utilities have shifted to a business-driven structure, in which business needs define the available budget. The traditional planning paradigm has difficulty adjusting to this external change in cases where the reduction from traditional budget levels is significant. Although this paradigm treats some traditionally rigid standards as "elastic," it obtains both better reliability and better economy from the limited budget.

Like many other paradigms, the traditional distribution planning paradigm has its assumptions and "rules" inextricably intertwined. The four basic tenets listed in Section 18.3 are so dependent on one another, and so fundamental to the engineering-driven planning process and the application of planning tools within that structure, that a change in any one shakes the foundation of planning methods based on the other three.

A Business- Rather than Engineering-Driven Planning Process

Even though a utility may still operate within a regulatory framework, it may face limits on the amount of money it can borrow or other financial pressures that preclude as much capital spending as it would like. The fact is that many utilities are forced to shift to the business-driven structure shown in Figure 18.2, in order to plan to meet the pressures of a competitive marketplace and industry. A survival and expansion plan for the utility, based on business necessities and goals, defines the available T&D budget. Good or bad, that's all there is, and unfortunately, experience through the first three quarters of the 1990s has proven that in general it is 30% to 40% less than traditional levels.

This is completely incompatible with the fourth precept listed for the traditional paradigm in Section 18.3, which assumed that sufficient capital funding would be available for all projects that justify themselves on the basis of present worth economics.

Trying to handle budget constraints within the traditional paradigm

One possible adjustment within the traditional paradigm to accommodate budget constraints is to change the time-value of money used in the PW analysis. The discount rate can be raised to 20%, 30%, or even higher. Each increase reduces the perceived value of future savings, and biases the decision-making process in favor of alternatives with lower initial costs. The capital budget required by the planning process is reduced. However, there are three problems with this approach:

1. It does not provide a mechanism to force the budget below any specific limit. By trial and error, the utility can find a discount rate that "works out," but it is not simple and does not lead to a simple rule that can be "aimed" at a specific target.

2. No ability to make really difficult cuts. As discussed in Chapter 6, PW analysis never says "no" to essential projects. In the face of severe budget cuts, the traditional paradigm still decrees that all new construction must be done, and work must be done to standards.

    What if the utility cannot afford this minimum? Raising discount rate changes only the relative importance of the future savings against which present costs are evaluated: The planning method still assumes that the goal is to balance long and short-term costs. Truly deep budget cuts need to be addressed by looking at how necessary the various projects are today, not at how today's costs weigh against tomorrow's savings, in any measure. (This will be discussed at length in a few paragraphs.)

3. Using a high discount rate is inappropriate. High discount rates means future savings are unimportant, which is not true: The utility's current financial situation precludes it from *affording* long-term savings, but this is not the same as saying it does not want them. (The importance of this distinction will become clear in several more pages).

Suppose the utility that wanted to do both projects listed in Tables 18.1 and 18.2 were forced to reduce its budget by 40%. This would reduce the $5,925,000 total for the two projects in Tables 18.1 and 18.2 to an allowable $3,550,000). A discount rate 38% is necessary to force the selection process for these two projects to $3,550,000. This changes the selection for the new feeder to option 7, 636-336 branched design, with a cost of $3.55 million (just meeting the budget) and selects "do nothing" for the refurbishment project.

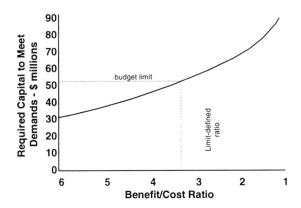

Figure 18.3 Cumulative capital requirements versus benefit/cost ratio for all "worthwhile" projects. The utility needs $88 million. A reduction of 40% (to $53 million) means it can afford projects only up to 3.4:1, and only those above that limit should be approved. Benefit/cost ratio provides a way to limit spending to within a severe budget constraint, but it does not spend that budget most effectively (see text).

Using Benefit/Cost Ratio

An often applied method to decide among project alternatives, approving some and rejecting others, is to rank them on the basis of benefit/cost ratio (i.e., savings over costs) or to evaluate them on the basis of pay-back period (which is a comparable analysis that produces very similar results). With all proposed projects ranked in this manner, the utility can then determine how much capital it would need to finance all worthwhile projects, or only those above any arbitrary benefit/cost limit.

Figure 18.3 shows the results of such analysis. This particular utility needs $88 million to finance all worthwhile projects.[4] For example, it could achieve a budget reduction of 40%, limiting capital spending to only $53 million, by approving only projects with a benefit-cost ratio of 3.4:1 or better.

---

[4] Any project that passes the "lowest PW cost" test within the traditional paradigm is considered "worthwhile." Usually, all worthwhile projects have a benefit/cost ratio better than 1:1, although not all projects with a ratio better than 1:1 are worthwhile (see Tables 18.3 and 18.4).

Benefit/cost ratio leads to "non-optimal" spending

Benefit-cost analysis is a valid way to compare alternatives within any one project or program, as for example in evaluation of options for a DSM program (Willis and Rackliffe), but using it to compare *options across project boundaries* can "waste" capital budget. Table 18.3 provides an example based on two hypothetical substation expansion projects. (Rather than use the two example feeder problems, the author prefers the comparison of these two very similar cases (both involve the expansion of an existing substation) in Table 18.3, which proves rather conclusively that the problem in evaluation is not due to problems in evaluating "old versus new" of "construction versus refurbishment" but due to a basic structural weakness in how the method evaluates projects.)

Both projects encompass nine alternatives priced in $100,000 increments. Each is ranked against a "do nothing" base ($3,000,000 in one case, $4,000,000 in the other). In the traditional planning paradigm, alternative 5 would be selected for Project A, and alternative 7 for Project B. This would spend a total of $1,200,000.

If the utility's budget must be limited to $53 million, and it decides to approve only projects with a benefit/cost ratio of 3.4:1 or better, it would select option 5 for Project A, and option 3 for Project B. In total it spends $800,000 ($500,000 for Project A, option 5, and $300,000 for Project B, option 3). It receives a total of $2,720,000 in benefits. Total benefit/cost ratio is exactly 3.4:1.

This spending can be improved. Suppose the utility switches from option 5 to option 4 for Project A, freeing $100,000 in capital, which it spends by selecting option 4 over 3 for Project B. This means it gives up $115,000 in benefit on Project A (the difference between the benefit listed for option 5 versus that for option 4), but it gains $275,000 of benefit by moving from option 3 to 4 in Project B. The net improvement is $160,000 in overall benefit without any increase in net spending.

The utility can repeat this "trick" one more time, moving to option 3 from 4 in Project A, and spending the capital freed by that decision by selecting option 5 over option 4 for Project B. Again, this makes no change in the $800,000 total it will spend on the two projects, but it increases net benefit by $50,000. Benefits now total $2,930,000. The utility is getting 8% more for its money than when it selected options based on benefit to cost ratio. As a result, overall benefit/cost ratio for its $800,000 capital expenditure has risen, to 3.66:1.

> *Using benefit/cost ratio as an approval rule does not lead to optimal use of a limited budget*

Traditional versus Competitive Industry Paradigms

Table 18.3 Two Hypothetical Projects Evaluated on a Minimum PW Cost, and Benefit/Cost Ratio Basis (thousands of dollars)

Project A Option	Capital Cost	Total PW	Net Benefit	B/C Ratio
Base		3000	0	
1	100	2400	600	7.0
2	200	2050	1150	5.8
3	300	1900	1400	4.7
4	400	1800	1600	4.0
5	500	1785	1715	3.4
6	600	1800	1800	3.0
7	700	1805	1895	2.7
8	800	1845	1955	2.4
9	900	1945	1955	2.2
Project B Option	Capital Cost	Total PW	Net Benefit	B/C Ratio
Base		4000	0	
1	100	3725	375	3.8
2	200	3510	690	3.5
3	300	3295	1005	3.4
4	400	3120	1280	3.2
5	500	2970	1530	3.1
6	600	2820	1780	3.0
7	700	2795	1905	2.7
8	800	2885	1915	2.4
9	900	2945	1955	2.2

One way to see the problem created by using the benefit/cost ratio as a decision-making yardstick among projects is to realize that the initial option in Project A had such a high benefit/cost ratio (7:1) that any option that built upon it could contribute very little benefit and the cumulative benefit/cost ratio could still be fairly high.[5] In the case of project A, at the point where three additional "serial" decisions (options 2 through 5) have been added, so the benefit/cost ratio is only 3.4:1 (option 5), the last $100,000 brings a net savings of only $15,000, only 15 cents savings for each dollar spent. Worthwhile, but not large.

---

[5] In fact, one could add a marginal option that cost $100,000 but contributed nothing at all to option 1, and this combination would still have a benefit/cost ratio of 3.5:1, so it would be approved under this budget-restricted guideline, for a waste of $100,000!

By contrast, Project B's economics are more linear. The first in B's case has a much less impressive benefit/cost ratio of 3.8:1. But each successive option drops only a little in the savings per dollar it contributes, so that by the time a ratio of 3.4:1 is reached (option 3), the $100,000 increment spent brings an additional savings of $150,000 *fully ten times as much* per dollars as money spent on Project A's option 5 at this same benefit-cost ratio.

Such differences in the distribution of benefit among alternatives in projects are common. The example given here is hypothetical, but typical of what is commonly encountered when using benefit/cost ratio as a project approval rule.

Equitable service quality spending

Adhering to the traditional utility paradigm in the presence of budget restrictions usually leads to concerns about non-equitable treatment of old and new customers, regardless of whether it is applied in the standard least-cost PW manner, with high discount rates, or on a benefit/cost basis. In some utilities, this is because the traditional paradigm simply decrees: "We have to connect all new customers." This decision to "do something" being fundamental and immutable. The planning method decrees that the "sensible thing" is to select the lowest PW cost among all options for the new construction. As a result, all new construction projects are done just about as they would be in an unconstrained budget situation, and all budget reductions must come from what would normally be spent on refurbishment of older parts of the system.

In other cases, utilities use a "revenue requirements" method in their present worth analysis: New customer projects have a projected revenue (from the sales to future customers) which is weighed against expenses, while refurbishment projects do not. New extension projects get approved because they "look good" from a financial perspective. Meanwhile, capital spending to bolster reliability in aging parts of the system does not look nearly so good, because there are no new future revenues against which the project can be justified.

Either way, the result is the same. Customers in older areas of the system can suffer poor service due to aging equipment which is not replaced, while the limited capital budget is devoted entirely to new construction to provide new customers with robust equipment built to high, long-term standards.

Budget-Constrained Planning Method

A planning method capable of addressing both of the foregoing problems, the non-optimality of the spending and the old versus new customer issue, can be applied by combining marginal benefit/cost analysis, "zero-base" planning, and adding valuation of service quality to the economic evaluation in the planning.

Traditional versus Competitive Industry Paradigms

Marginal analysis of a series of decisions

Table 18.4 shows the Project A and B options from Table 18.3 evaluated on a *marginal* benefit/cost ratio basis. In each project, the additional present worth savings delivered by moving up from one option to the next is evaluated on a marginal basis (i.e., as a function of its cost over the preceding lowest cost alternative). The choices that maximize PW savings, as discussed earlier (option 3 in Project A, option 5 in Project B), are obtained when the marginal ratio for both is set at 1.5:1. Evaluation and use of marginal benefit cost ratio leads to improved spending.

Figure 18.4 shows the budget-requirements when all projects the utility is considering are re-ranked on the basis of their marginal present worth/marginal capital ratio. Again, slightly more than $88 million is required to meet all projects with a ratio above zero. As indicated, when limited to $50 million, the utility must approve only projects with a MPW/C ratio of 1.5:1 or better. But, unlike the case where it was using benefit/cost ratio as the guiding rule, it will now always "buy" an equivalent (and biggest overall) PW savings -- the most it can afford.

> Selection of project alternatives on the basis of the MPW/C ratio leads to decisions that maximize the benefit that can be obtained from a limited budget.

Table 18.4 The Two Hypothetical Projects from Table 18.3 Evaluated on a Marginal Benefit/Cost Ratio Basis (thousands of dollars)

Project A Project Option	Δ Cost	Δ PW	Marginal B/C Ratio	Project B Project Option	Δ Cost	Δ PW	Marginal B/C Ratio
Base				Base			
1	$$100	$600	6.0	1	$100	$275	2.8
2	$$100	$350	3.5	2	$100	$315	3.1
3	$$100	$150	1.5	3	$100	$225	2.2
4	$$100	$100	1.0	4	$100	$175	1.8
5	$100	$15	0.2	5	$100	$150	1.5
6	$100	-$15	-0.2	6	$100	$50	1.5
7	$100	-$5	-0.1	7	$100	$30	.3
8	$100	-$40	-0.4	8	$100	-$90	-.9
9	$100	-$100	-1.0	9	$100	-$60	-.6

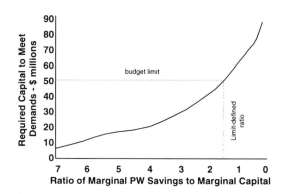

Figure 18.4 Cumulative capital requirements for the utility in Figure 18.1 versus marginal PW/capital ratio. The utility's $50 million budget allows it to spend up to a 1.5:1 ratio. This ratio provides a much better guiding rule for project approval.

This marginal benefit/cost ratio method can be applied to the two feeder examples given earlier in the chapter. The decision to build the 795/336 MCM option (selected as the "best" for the new feeder in Table 18.1) is viewed as a *series* of decisions, beginning with a decision to "do something" by building the least costly capital alternative (4/0 large trunk), and continuing with a series of decisions to "upgrade" this decision to the next level (4/0-4/0 branched, costing $100,000 more), and to upgrade that, and so forth to the final 795/336 alternative.

Table 18.5 ranks the effectiveness of the money committed in each of these "marginal decisions" for the example new feeder. The decision to "do something" spends $2,250,000 in capital with a PW cost of $7,575,000. This is the "base" decision, because under the traditional paradigm, something must be done. The decision to "upgrade" that to 4/0-4/0 spends another $100,000, but saves $400,000 in PW (the additional $100,000 not only pays for itself in terms of PW but lowers long-term costs enough to reduce PW by $400,000). The ratio of *marginal PW savings to marginal capital expenditure*, which will be called the MPW/C ratio here, is 4:1 for this upgrade decision.

Similarly, the next decision in the series spends a further $300,000 to move from 4/0-4/0 to a 336-4/0 branched design, but reduces PW cost by $925,000, for a ratio of 3.1:1. The rest of the series of decisions leading up to the "best" alternative within the traditional paradigm (i.e., that with the lowest PW) have

Traditional versus Competitive Industry Paradigms

yields of 2.0, 1.55, 1.77, .75, and .5 respectively. In this example, the process stops at this point, because further steps would lose money.

Similarly, the alternatives for the refurbishment project can be evaluated as shown in Table 18.5. In both the new and refurbishment feeder examples, this method would lead to the same overall option selection as the traditional paradigm in the unconstrained case, because the marginal analysis determines that every decision in the series, up to a marginal ratio of zero, "buys" more future PW than it costs. For any project, the decision ends at the same point as traditional PW analysis would determine, as can be discerned from the data in Tables 14.18 and 14.19.

Table 18.5 Example Feeder Problem Alternatives Evaluated on the Basis of Marginal Capital Cost versus Marginal Present Worth - $ × 1000

Feeder Option	Meets Criteria?	Year 1 Capital	Δ Capital	PW Cost	Δ PW	MPW/C Ratio
4/0 large trunk	No	$2,250	-	$7,575	-	-
4/0-4/0 branched	Almost	$2,350	$100	$7,175	$400	4.0
336/4/0 branched	Barely	$2,650	$300	$6,250	$925	3.1
336 large trunk	Barely	$2,850	$200	$5,850	$400	2.0
636 large trunk	Yes	$3,125	$275	$5,425	$425	1.5
795 large trunk	Yes	$3,450	$325	$4,850	$575	1.8 (1.6)
636/336 branched	Yes	$3,650	$200	$4,700	$150	.75
795/336 branched	Yes	$3,750	$100	$4,650	$50	.5
795-636 branched	Yes	$3,950	$200	$4,800	-$50	
1113 large trunk	Yes	$3,950	$0	$4,900	-$100	
1113/633 branched	Yes	$4,250	$300	$4,800	$100	

Table 18.6 Refurbishment Alternatives for an Existing Feeder in a Developed Area, Evaluated by BCP Including Value of Service Credits - $ × 1000

Feeder Option	Δ capital	O&M PW	failure risk (est.)	Total PW	Δ Capital	Δ PW	MPW/C ratio
Do nothing		$900	$5,145	$6,045			
Fix most serious	$650	$700	$3,900	$5,250	$650	$795	1.22
Rebuild plan A	$1,650	$400	$2,000	$4,050	$1,000	$1,100	1.25
Rebuild plan B	$2,250	$350	$1,200	$3,800	$600	$250	.42
Rebuild plan C	$3,175	$290	$771	$5,161	$1,850	-$1,361	-

This MPW/C ratio method is the basis of the budget-constrained marginal decision method, hereafter referred to as budget-constrained planning (BCP). It avoids the difficulty cited earlier. *It will force the selection process below any budget limit, in a way that maximizes PW savings.* For the example new feeder, the basic BCP method would select the 795 large-trunk marginal design option. That option has a 1.8 ratio above the preceding alternative, but since the preceding alternative has a lower ratio than it, (this happens frequently in the real world), its marginal ratio against the option two lines above it is evaluated, yielding a 1.66 marginal ratio over that option, good enough for selection.

Unfortunately, while marginal benefit/cost evaluation provides a way to accommodate severe budget restrictions, it does not, of itself, solve the customer service concerns cited earlier. Evaluation of the refurbishment project yields a decision to "do nothing," as can be seen in Table 18.6. None of the options shown passes the budget-constrained 1.5:1 ratio test.

Adding customer reliability and service value to accommodate refurbishment

A "fairer" way to allocate a limited budget, when reliability and equitable service are both an issue, is to include some value for customer service reliability and quality in *all* evaluations. While this change would seem to favor new construction even more than refurbishment projects (since not doing new construction would result in the ultimate "reliability problem," no service at all), when done within a *marginal* decision framework it re-allocates *marginal* spending away from new construction and toward refurbishment, in an organized, quantitative and balanced manner.

This change begins with two major paradigm shifts:

1. *Zero-base planning.* No project, even a new customer extension, is absolutely necessary. All planning is done on a "zero base" basis, by beginning with "do nothing" as the base assumption.

2. *Service-quality rather than standards-driven planning.* If the cost penalty due to poor reliability is high enough to justify "doing something," then money will be spent. Otherwise nothing is approved no matter how far below standards the resulting design will be (excepting safety standards, which must be met).

Tables 18.7 and 18.8 show evaluation of the alternatives for the two feeder problems discussed earlier in this section evaluated in this manner. Both now have a base "do nothing" option. Both include a PW cost of outages and poor service calculated in a "value-based planning" manner. For the sake of brevity, details behind these figures are not given here. Each alternative for the new

Traditional versus Competitive Industry Paradigms

Table 18.7 Example Feeder Problem Alternatives Evaluated on the Basis of Marginal Capital Cost versus Marginal Present Worth Including Value of Service - $ × 1000

Feeder Option	Δ cost	Hours/yr. interrupt.	kWhr non-stan.	Serv. Cost	Total PW	Δ PW	MPW/C Ratio
Do nothing		8,760	0	$841,000	$841,000		
4/0 large trunk	$2,250	8	25	$1,440	$9,015	$832,000	370.
4/0-4/0 branched	$100	4	10	$600	$7,775	$1,240	12.4
336/4/0 branched	$300	2	4	$252	$6,502	$1,273	4.24
336 large trunk	$200	2	2	$145	$5,995	$507	2.54
636 large trunk	$275	1	1	$80	$5,505	$490	1.78
795 large trunk	$325	1	1	$79	$4,929	$576	1.77
636/336 branched	$200	1	1	$78	$4,778	$151	.76
795-336 branched	$100	1	1	$72	$4,722	$56	.56
795-636 branched	$200	1	1	$72	$4,872	-$150	-
1113 large trunk	$0	1	1	$72	$4,972	-$100	-
1113/633 brnchd.	$300	1	1	$72	$4,872	$100	-

extension was evaluated for expected problems. Each option in the refurbishment project was similarly evaluated against the current outage rate (3.9 hours/year) and poor service rate (about 12 hours year).

For the extension project, the base, or "do nothing," option means 8,760 hours of service interruption per year, giving that alternative a tremendously high "cost." As a result, option 1 looks tremendously good from a marginal ratio standpoint, because even though it does not meet traditional utility standards, it reduces expected lack of service from 8,760 hours to less than 8.[6] Every other alternative is similarly evaluated with a value for its estimated improvement in service quality value.[7] Similarly, the various options for the refurbishment project are evaluated based on their costs and service quality value.

---

[6] For this reason, this type of planning must be done on a marginal decision basis, using something like the MPW/C ratio. Using B/C ratio or payback period in the evaluation would overvalue options that subsequently build on this first option, with selection results similar to those for the example discussed in Table 18.3.

[7] Service quality values used here were $5/kWhr for interruptions and $1/kWhr for non-standard service hours. Such values are difficult to determine precisely, but rather easy to estimate sufficiently for planning purposes. Any set of reasonable values will enable this method to be applied and planners can experiment with various values until a set that seems to provide stable and acceptable results is obtained.

Table 18.8 Refurbishment Alternatives for an Existing Feeder in a Developed Area, Evaluated by BCP Including Value of Service - $ × 1000

Feeder Option	Δ cost	Hours/yr. interrpt.	kWhr non-stan.	Serv. Cost	Total PW	Δ PW	MPW/C Ratio
Do nothing		6	20	$2,286	$8,331		-
Fix only serious	$650	4	12	$1,398	$6,648	$1,683	2.59
Rebuild plan A	$1000	1	3	$130	$4,130	$2,518	2.52
Rebuild plan B	$600	1	2	$89	$3,889	$241	1.2
Rebuild plan C	$1850	1	1	$78	$5,239	$1,350	-.7

Effect of including service quality value

Using a value-of-service based budget-constrained planning approach has several effects. First, it boosts the PW value of *all* "do something" options. Each is now given a "credit" for additional savings due to its contribution to improved service quality, increasing its perceived savings. Some options are raised in value more than others, but *all* increase in the savings they deliver and hence their overall value, their B/C ration, and their marginal PW/cost ration.[8]

This means that no matter how the utility subsequently looks at its budget needs, in the unconstrained situation, this decision-making rule will mean a larger bigger capital budget. Consider an option that was just at the point of approval under the old method: It had benefits and costs that exactly matched. There was no reason to build it, but neither was there a reason not to do it. In such a case the utility would say "no." But now, this project will have been given some credit, at least a small credit, and anything will push it over the limit so its B/C ratio, marginal ratio, or present worth says "built it." [9]

However, in the presence of budget constraints, both benefit/cost ratio or MPW/C-ratio analysis methods compensate for this swelling in recommended budget by raising their cutoff ratio. Figure 18.5 shows the result for the utility example cited in Figures 18. The total budget required to afford "all worthwhile projects" increases to $113 million. However, the MPW/C ratio limit to stay within the $50 million budget limit rises to 2.5:1.

---

[8] In this case there is no change in the order of ranking, but this is not always the case.

[9] However, in this case, the recommendations of the traditional paradigm would not change, even with service value added. Note that in Tables 14.20 and 14.21, the lowest present worth is still as before -- 795-336 branched and "Plan B." Within the traditional paradigm, this would be interpreted as meaning "we're doing things well without using service value analysis," which is a valid interpretation, *given no budget constraints*.

Traditional versus Competitive Industry Paradigms

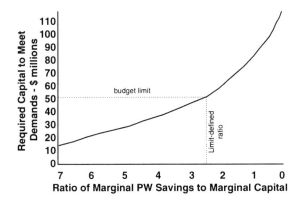

Figure 18.5 Cumulative capital requirements for the utility in Figure 18.3 versus MPW/C ratio. The $50 million budget allows it to spend up to a 2.5:1 ratio. This ratio provides a good guiding rule for budget-constrained project approval.

Refurbishment alternatives receive a "marginal" advantage

The net effect of these changes on the ultimate decisions made about what to build are complicated. First, although this paradigm does not assume that new service extension will be done, it's impossible to argue with the value of service contribution that *doing something* delivers in all new construction cases (nearly 8760 hours improvement). As a result, from any practical perspective, this planning method will always decide to build something to serve new customers.

But there are substantial differences in what this process will decide to build. Taking into account the value of service, this paradigm will no longer build the new service extension project to a very high level of service quality, just because it must meet standards or minimize PW. Instead, it adjusts both new customer and refurbishment decisions toward equivalent levels of reliability: Money is spent on making *some* refurbishment additions to raise poor service levels, and the capital is obtained by cutting what is built in new areas so that service levels there may be less than if built to traditional "lowest PW" levels.

Tables 18.7 and 18.8 indicate the results. Restricted by budget constraints to projects with a 2.5:1 marginal ratio, the new feeder project is built with only a 336 large trunk design, for a capital reduction of $900,000 over the unconstrained (lowest PW cost) situation. For the refurbishment project, its option 2 (plan A) is selected, a reduction from the "plan B" recommendation that flowed out of unconstrained PW cost minimization, for a capital reduction of $600,000. Note that with this planning method both projects get some of the limited funds, though neither gets as much as in the unconstrained budget (minimize present worth).

From the perspective of the traditional paradigm, the decision made for the new feeder project in this situation makes no sense: It is built with a capacity and style so that it is stressed at peak conditions from the very day it is built. However, from this new paradigm's standpoint, the service quality provided by the new feeder is no worse (in fact it is slightly better) than the service quality delivered to customers in older areas of the system.

Interestingly, the ultimate result of this planning method is not a drop in overall reliability compared to continuing with the traditional paradigm. The traditional paradigm can result in areas of very good (new, robust) and very poor (old, stressed) service. The new paradigm results in old and new areas being more balanced in capital spending and service quality results. Due to the non-linearities in reliability-versus-cost relationships, very often the utility's service quality indices, such as SAIFI, actually improve slightly with this value system.[10]

This planning method, and paradigm, are recommended in situations where a distribution utility has capital budget limitations that must be addressed and there is a concern that inequities in service quality may become a real or perceived problem. It is the *combination* of the "zero-base" basis for all projects, the inclusion of service quality value as the only "standard" that must be met, *and* marginal decision evaluation that avoids the difficulty cited earlier (the capital budget is spent optimally) while also addressing the potential service controversies (all customers, old and new, are treated equivalently).

18.5 PROFIT-BASED PLANNING PARADIGMS

It is unlikely that T&D systems will ever operate in an environment where the basic decision-making process is made within a purely profit-based value system. However, in a de-regulated utility industry, some ancillary and optional services and equipment may be treated as profit-making enterprises. For completeness, it is worth examining how the profit-motive paradigm would view one of the example feeder projects presented earlier in this section.

A purely profit-motivated paradigm focuses on trying to maximize the return on the investment. Electric utilities try to do this, too, but while working within a regulatory framework, and subject to a number of constraints and the precepts of the traditional paradigm. While there are variations on this theme, in general the

---

[10] They improve with respect to any other budget-constrained selection, but of course they do not match the levels that good use of an unconstrained budget would provide.

Traditional versus Competitive Industry Paradigms

profit paradigm has several specific differences from traditional utility planning paradigms. These and other major factors affecting planning may not be identified explicitly, but the key differences that change how planning decisions are made are:

1. Serving only the most profitable portion of the demand is in the business's best interests.
2. Standards are defined by whatever quality and service characteristics yield the highest profits.
3. Profit is the goal. Specifically, the goal is usually to maximize the return/investment ratio or the return to cost *ratio*.

A basis for many profit-driven corporate value systems is a belief that if a good return on investment cannot be found among the opportunities involved in a project, other potential avenues of investment are available. Thus, the third goal given above is not to maximize the profit, but *to maximize the profit as a percentage of investment:* Earning $10 million profit on a $50 million investment is much better than earning $17 million profit on a $100 million investment.

Example from outside the electric utility industry

A purely profit-motivated perspective often determines that the "smart" decision is to underbuild slightly. As a simple example, consider the options an airline faces in deciding what size commercial jet to buy and put into service on a shuttle route between two cities like Chicago and Tulsa. Analysis of demand, competition, and other factors might establish that while demand varied by season and day of the week, between 100 to 150 people would normally want to travel this route on any particular flight.

Table 18.9 shows the airliners available, which range from 70- to 300-seat capacity, along with their initial costs (a 300-seat airliner is more expensive than a 150-seat airliner, but not twice as expensive), annual operating costs (much more linear as a function of size), and total annual cost (a function of capital costs plus operations costs).

Any airliner with less than 100 seats will always travel fully-loaded. Thus, below the 100-seat limit, revenues will be proportional to number of seats. But any plane with more than 100 seats will fly only partially loaded at times. The 300-seat airliner will never fly with more than 50% of a load, or generate more than half its potential revenue.

Table 18.9 Profit Analysis of an Airliner for Shuttle Service -$ × 1000

Airliner Seats	Initial Cost	Operating Cost	Annual Cost	Annual Revenues	Profit $	Percent Profit
70	$7,000	$2,100	$2,975	$3,340	$365	12%
85	$7,277	$2,550	$3,460	$4,056	$597	17%
100	$7,815	$3,000	$3,977	$4,772	$795	20%
125	$8,776	$3,750	$4,847	$5,560	$843	17%
150	$10,221	$4,500	$5,778	$5,726	$10	0%
225	$12,910	$6,750	$8,364	$5,726	-$2,576	-31%
300	$17,271	$9,000	$11,159	$5,726	-$5,371	-48%

The 125-seat airliner generates the most profit, $843,000 per year. However, the 100-seat airliner generates a higher *percentage* profit, 20% as opposed to 17%. Most businesses would buy the smaller jet, the assumption being that the money saved by not buying a larger one could be invested in another jet on another route that would similarly earn 20%, not 17%.

Obviously, many more factors are part of any actual airline purchase decision, but the point is that an option far below peak capacity will look very good to an unregulated company looking for the maximum profit margin. The fact that customers may have to be turned away at peak periods would be of tertiary importance -- the airliner would be making a good return on investment.

Table 18.10 Profit Analysis of a Distribution Feeder - $ × 1000

Feeder Seats	Initial Cost	Oper. Cost	Annual Cost	Annual Revenues	Peak/ Cap.	Profit $	Percent Profit
4/0 large trunk	$2,250	$400	$775	$807	116%	$32	4%
4/0-4/0 branched	$2,350	$355	$747	$832	110%	$85	10%
336-4/0 branched	$2,775	$300	$763	$975	102%	$212	22%
336 large trunk	$2,850	$310	$785	$999	97%	$214	21%
636 large trunk	$3,125	$325	$846	$1,070	82%	$224	21%
795 large trunk	$3,500	$335	$918	$1,150	57%	$232	20%
636-336 branched	$3,550	$347	$939	$1,150	67%	$211	18%
795-336 branched	$3,675	$355	$968	$1,150	55%	$183	16%
795-636 branched	$3,800	$365	$998	$1,150	50%	$152	13%
1113 large trunk	$3,900	$370	$1,020	$1,150	45%	$130	11%
1113-633 brnchd.	$4,250	$375	$1,083	$1,150	40%	$67	6%

Traditional versus Competitive Industry Paradigms

Profit-analysis of a feeder example

The result of the profit paradigm is directly at odds with the traditional paradigm used in utility planning. The concept of *deliberately* planning for less than peak capacity will seem a strange concept to planners who have worked in an environment where falling short of capacity is never considered an option and is deemed a sure sign that the planners failed in their mission. Table 18.10 shows the example new-feeder options from earlier in this section (Table 18.1) evaluated on a purely profit-motivated basis in a manner similar to the airliner example discussed above. All of the data required to develop the numbers shown is not given here, the point being to demonstrate the concept, not the details of computing business costs, etc. Ratio of peak demand to minimum capacity (of whatever element or factor limits capacity) is also shown.

Revenues are projected from an 8760-hour analysis of sales of feeder delivery services under open distribution access (retail wheeling. See Willis, 1996, Chapter 12), and account for the fact that feeders with insufficient capacity cannot deliver all the load during peak periods. They also include: a) revenues from marginal sales when extra capacity is available, and b) penalties for non-delivery or poor service under performance-based pricing. Thus, the "minimal" options have lower revenues, but capacity above a certain point does not increase revenue. The greatest amount of profit, $232,000 per year, is for the 795 MCM large-trunk design. However, the highest profit margin, 22%, is from a much lower-cost option, the 336-4/0 layout, a feeder that is overloaded at peak.

In a manner very similar to the airliner's case, and for the same reasons, under a profit-paradigm there is little incentive to build a large feeder. Particularly from the standpoint of maximizing percent profits, the best policy is to underbuild slightly.

> *From a purely profit-based perspective, a feeder that is slightly overloaded at peak is usually a very profitable feeder, even when the cost penalties of performance-based pricing are considered.*

Of course, many more considerations often enter into business decisions of the type discussed here, but the point is that a profit-paradigm provides significantly different incentives and leads to quite different results than either the traditional or budget-constrained utility paradigms discussed earlier.

The conclusion stated above is one reason why it is unlikely that T&D systems and their operation as a delivery network will ever be fully de-regulated. What is in the best interests of a profit-motivated business would not necessarily be in the best interests of the public, when sufficient electric capacity is viewed as a near-essential item for proper functioning of society and the common good. Regulation provides both an assurance that such needs will be addressed, and a mechanism for the utility to recover the costs required to meet these goals.

18.6 CONCLUSION

The first century of electric power provided a stable and relatively unchanging background against which the electric power industry could grow into a mature infrastructure, one of the basic cornerstones of the first-world economy. Unfortunately, the stability afforded by a constantly growing demand for power and a protective regulatory climate lulled many utilities into a sense of complacency and indifference to change.

It is unlikely that the future will provide an environment anything nearly as tranquil or secure: The power industry *is* a necessary fixture for our society's continued prosperity and cultural growth, but the electric utilities are not. One of the implicit "rule changes" of de-regulation is a change in the rules: If a utility company can't keep up, *let* them fail -- a more capable successor will replace it.

Those electric utilities who not only survive, but prevail, will be flexible, focus on almost draconian levels of cost reduction, and rely on good planning, not just at the distribution level, but throughout the organization. One key to their success will be their management of their paradigm. They will have to:

- *Know* their organization's purpose and values, not just its goals.

- *Understand* and constantly re-examine their most cherished values and assumptions.

- *Study* the future, to identify if and when tomorrow's needs will depart from today's capabilities and values.

- *Be willing to change.*

REFERENCES

S. Weintraub, *The Last Great Victory,* Dutton, New York, 1995.

H. L. Willis, *Spatial Electric Load Forecasting,* Marcel Dekker, New York, NY, 1996.

H. L. Willis and G. B. Rackliffe, *Introduction to Integrated Resource T&D Planning,* ABB Guidebook series, ABB Power T&D Company, Raleigh, NC, 1994.

Index

AC (alternating current), 3, 60
  model of power flow, 710
  as "first" harmonic, 91
Airlines (example of profit-
  motivated company), 799
Algorithms, 688
Allocation of load (see Load)
American type of system layout, 238
  compared to European, 239
  secondary systems, and costs, 539
Appliances, definition, 52
  climatological impact on AC, 227
  and duty cycle load impact, 59
  as harmonic problem causes, 105
  impact of harmonics on, 90
  recommended harmonic limits for, 90
  voltage sensitivity of, 131
  also see End-uses of electricity

Backwards trace (method of
  laying out feeders), 351
Balancing loadings on feeders
  between phases, 245
  of loading among feeders, 342
Bang-bang control (appliances), 59
Benefit/cost analysis, 213, 787
  non-optimality of in some cases, 788

Budget-constrained planning, 784
  examples, 791, 794

C-message curve (in harmonics), 95
Cable aging and reliability, 78
CAIDI and CAIFI (reliability), 166
Capacitors (shunt), 22
  action of on circuit, 370
  application to reduce VAR flow, 369
  decreasing marginal effectiveness, 370
  optimal location for, 371
  multiple banks on one circuit, 373
  guidelines for, 377
  representation in circuit models, 699
  switching of, 384
  strategic planning of, 388
Carson's equations, 241, 712
CBEMA curves (power quality), 83
Circuit Analysis
  and coincidence interaction, 718
    method to minimize error from, 723
  detail used in, 692
    diagram, 694
    industry trend toward more detail, 696
  methods of analysis, 707-720
  models, 690
  recommendations for (table), 716

803

804 Index

Circuit breakers, 12, 21, 31
  low-side in substation, 419
  and substation configuration, 400
Coincidence, definition, 65
  and appliance duty cycles, 59
  between classes of customer, 70
  creating economy of scale, 562
  curves (diagram), 65
  example load curve behavior, 60-64
  factor for loads, C, 65
  factor for harmonics (HC), 99
  formula for, 65
  interaction with design tables, 560
  and load flow inaccuracies, 718
  and service level interactions, 545
Collector point (load allocation), 694
Combinatorics, 747
Commonest pitfall of planning, 548
Conductors
  limitations and impact on design, 470
  number of types used in system, 273
  table of typical types, 231
  also see Lines and Line types
Conductor sets, definition, 259
  and economic load reach, 262
  guidelines for, 271
  need to review periodically, 268
  and voltage level, 260
Contingency margin and planning, 150
  policies for feeders, 351
  and feeder switching, 449
  ratings for transformers, 280
Conservation Voltage Reduction, 143
Corporate forecasts used for T&D, 674
Corporate mission statements, 567
Cost and cost analysis
  for a complete T&D system, 437-444
  continuing, 187
  effectiveness, 213
  embedded, 190
  of feeder contingency schemes, 328
  fixed, 187
  of typical equipment, 24-29
  incremental, 190
  of interruptions, 110
    by customer class (diagram), 121

[Cost and cost analysis, interruptions]
    as a function of duration, 114
    included in planning, 794
    and time of use (diagram), 119
    of surges and harmonics, 121
    typical customer values (table), 116
    values used in example, 795
  of harmonics, 110
  of losses, 31-34
    as a function of operating cond., 222
  present worth analysis of, 242,
  for maintenance and operation, 28
  marginal, 190
  reduction methods
    lowering f_v (feeder system), 364
    of a substation (substation only), 424
    of a substation (complete), 491
    of subtransmission, 453
  spatial distributions of, 34, 227
  sunk, 190
  variability of
    by equipment type, 220
    by time, 221
    by level of system, 222
    by location, 224
Credibility of plans and methods, 679
Criteria and Standards, definitions, 130
  design criteria, 151
  economic, 185
  example criteria, 152
  equipment, standard types, 151
  loading standards and ratings, 148
  and rhinoceroses, 152
  safety standards, 145
  voltage, 132
    imbalance, 145
    typical standards, 135
    application of, 145
Customer billing records, use of, 57, 705
Customer classes, 57, 652
  as land-use classes, 652
  table of examples, 652
Customer level planning, 607
Customer value of service
  assessing negative elements of, 109
  as a cost in selection of options, 794

Index

[Customer value of service]
 of interruptions, 113
 of power quality, 109
 of prior warning of interruption, 116
 see also Value-based planning
CVR, Conservation voltage reduction
 form of DSM, 143

DC power as "zeroeth" harmonic, 91
DC model (of power flow), 709
Decision support tools, 729
 application of, 768
 for feeder planning, 745
 optimization recommendations, 774
 steps involved in, 737
 for substation planning, 758
 also see Optimization
Delta-connected equipment, 234
 compared to wye-connected, 235
Demand (for electricity), 49
Demand periods, definition, 56
 sampling of load curves, 71
 illustrations, 55, 71
Design criteria, 151
 table of, 152
 also see Criteria and Standards
Design flexibility (feeders), 309-315
Direct burial of distribution cable, 297
Discrete sampling (load curves), 69
 illustration of errors caused by, 71
 and manual or SCADA records, 70
Distribution planning
 see Planning
Distributed resources, definition, 2
Distribution system
 delta or wye connected, history, 234
 design interrelationships, 445-476
 design style (Amer. vs. Euro.), 238, 313
 earth return (one wire) design, 239
 eight important characteristics, 47
 elements of economy, 5
 examples analyzed, 433-444, 477-486
 levels of, 7
  common characteristics of levels, 9
  interruptions as function of level, 10

[Distribution system]
 mission of, 2, 4
 optimum layout of, 466
 in rural areas, 477
 system example, in detail, 432-444
 vs. transmission, definition, 17
Diversity
 of customer types, 652
 in harmonics generation, 98
 also see Coincidence
Doing nothing (as an option in
  planning). 575, 773
Dreadnought battleship (example of
  technology-driven paradigm), 781
Dual voltage feeders, 313
 American types, 315
 Diagrams of, 314
 European loop type, 313
Duct banks (UG construction), 232, 297
Durability vs. loading (of substation
  transformers), 283, 767
Duration of interruptions
 see Reliability
Duty cycles of appliances, definition, 59
 of AC and HP, 61
 illustration, 59, 62
 interaction with coincidence, 68

Earth return (one-wire distribution), 239

Economics (engineering evaluation)
 of basic line types, 241-251
 using benefit/cost analysis, 215
 of conductor sets, 266
 effectiveness tests, 216
 guidelines, 194
 integrated resource tests, 216
 net present value
  see Present worth analysis
 and paradigm shift changes, 782-796
 payback period, 215
 and revenue requirements, 191
 and time value of money, 193
  factors influencing, 198-202
  application example, 203
 also see Present worth analysis

Electrical behavior modeling, 707
Electrical foam (analogy for VARS), 23
End-uses of electricity, 50
End use analysis and modeling, 51, 658
Explicit routing (in optimization), 763
Extent of an outage, 158
  also see Reliability

Fault current (factor in design), 147
Feeder analysis, 685-727
  see also Feeder planning
Feeders and feeder systems, 14, 343
  costs, 26, 497, 439-462
  cost reduction, 362
  dendrilic configuration of, 14, 289
  dual voltage feeders, 313
  frequency response of (harmonics), 96
  layout
    see Feeder layout
  line types, 230
  mission of, 286
  modern perspective on layout, 345
  must cover dist. between subs., 288
  phases: one, two, or three, 234
  service areas, 287
  switching (line) for contingencies, 329
  underground, 297
Feeder getaway, 15, 442
Feeder layout, 306
  classical perspective on layout, 343
    diagram, 344
    flaws in, 343
    see also modern perspective
  and contingency support, 319
  flexibility, 309
  formula to estimate cost and savings
    from changes in design, 362
  interaction with substation siting, 509
  interlaced feeder layout, 38
  large trunk vs. multi-branch, 309
  and reliability (extent), 159
  types of layout design
    delta vs. wye, 234
    loop, 294
    network, 295
    radial, 293

Feeder planning, 347, 603
  automated methods for, 745
  backwards trace method, 351
  key features of (table), 604, 605
  goals of, 348
  optimization method selection and
    application, 748, 755
  reducing cost -- methods of, 362
  risk minimization trick, 360
  splitting feeders, 359
  tools for, 687
  process of, 349
Filters
  feeders, as harmonics filter, 96
  for harmonics reduction, 107
  as spatial frequency domain planning
    methodology, 677
Flicker
  see Voltage flicker
Frequency of interruption
  see Reliability
Frequency scanning
  see Harmonics, analysis methods

Genetic algorithm, 733, 755
Grids (street) and feeder routing, 291
Grid like layout (UG feeders), 300

Harmonics, definition, 90
  analysis methods, 102
    frequency scanning, 102
    linearized analysis, 103
    non-linear, 103
  attenuation of, 98
  causes of, 92
  coincidence-like behavior
    phase diversity, 98
    temporal diversity, 105
  costs of, 121
  daily curves of distortion, 97
  and delta-wye transformers, 102
  fixes for, 104
  frequency response, 96
  standard limits for, 100
  problems caused by, 90
  waveshapes, voltage and current, 94

Index

Heuristic algorithm, definition, 733
  as a flawed type of optimization, 743
  example application to capacitors, 572

Imbalance of load, 145
  effect on line cost, 245
  formula, 145
  standards for, 146
Impedance, 240
  and spacing, 241
  models of, 696
Implied present worth factor, 208
Interlaced feeder system, 38
Interruptions, definition, 76
  customer costs of, 114-119
    diagram, 114
    typical values (table), 116,118
  fixes for, 78
  frequency and duration of, 113
  types (table), 163

Judgment in the planning process, 575
Justification (of plans)
  and documentation, 678
  using optimization "backwards," 773

K-factor (harmonics measure), 95
K-factor (feeder voltage evaluation), 253

Land-use forecasting methods
  see Forecasting, simulation
Large trunk feeder layout, 40, 306
Laterals, 14
"Laws of T&D", 4
Layout of feeders
  See Feeder layout
Lead time for planning, 57
  defines short-range period, 582
  in planning, 571
Lebesque metric (distance measure), 291
  classical errors due to ignoring, 343
  "cheat" wherever possible, 358
Levels of the system, 7
  also see Distribution system
Levelized value
  see Present worth analysis

Line drop compensators, 414
Line types, 230
  changeout decisions, 246
  economics of, 241
  linear cost approximation, 249
  ratings of, 255
  also see Conductors
Linearization
  of conductor economics, 249, 751
  in feeder optimization methods, 735
  and trans-shipment methods in
    optimization, 755
Loads
  frequency dependent, 92
  non-linear causing harmonics, 92
  types of voltage sensitivity, 702
Load allocation (load flow), 704
Load balancing, 342
  also see Switching and Phasing
Load curves, 54-69
  illustration, 54, 57
  and sampling rates, 71
Load density
  methods for forecasting, 667
  impact on system layout, 467
Load factor, definition and formula, 56
Load flow methods, 707-720
  and criteria, 145
  and coincidence interaction, 718
Load forecast for T&D, 615
  comparison of methods, 671
  criteria important in, 672-679
  and feeder planning, 354
  first step in T&D planning, 595, 610
  geometric methods, 649
  multivariate trending methods, 649
  simulation methods, 650
  and substation planning, 529
  trending methods, 634
  vacant area inference method, 644
Load growth, 615
  effect on conductor economics, 244
  effect on substation timing, 529
  interaction with load reach limits, 533
  "S" curve dynamics, 623
  two causes of, 618

Load models, 701
  constant power, impedance, or, 702
  constant power loads
Load reach, definition, 251
  compared to K-factor, 255
  economic, 252
  in feeder layout, 348
  guidelines for, 263
  and primary voltage level, 262
  and splitting feeders, 359
  and substation construction timing, 532
  and substation spacing, 459
  thermal, 251
  and voltage drop, 255
Load-related losses, 31, 282
  also see Losses
Loading limits of equipment, 148
  also see Conductors, Line types,
    and Transformers
Loop circuit configurations, 36, 313
Loss of life (transformers), 766
Losses
  cost of, 33
  part of detailed examples, 442, 482
  viewed as power to run T&D, 30
  of transformers, 282
Low footprint transformers, 277
Low loss transformers, 276

MAIFI (reliability index), 167
Maintenance
  see Operation and Maintenance
Marginal cost analysis, 109, 791, 794
Mobile substations, 426
Models (as distinct from computer
  algorithms), 688
Momentary interruptions, definition, 79
  caused by voltage sags, 111
  customer costs of, 115
  fixes for, 79
  illustration, 112
  also see Reliability
Motor starting analysis, 141, 715
Multi-branch feeder layout, 40, 306
Multi-scenario planning, 587
Multi-year planning, 595, 617, 771

Networks, 11, 295
No-load losses, 31, 282
  also see Losses
Non-harmonic, non-60 HZ power, 91-93
Null point load flow, 756

One-line models, 691, 696, 709
Operations and Maintenance
  and capacitor placement, 378
  costs, 28
  in examples (tables), 425-440
  criteria and standards, 145
  planning of, 753
OPF (Optimal Power Flow), 733
Optimal feeder layout, 345
Optimization, 730
  choosing an appropriate method, 755
  comparison of methods, 737
  forcing optimization to produce a
    plan with certain elements in it, 773
  tradeoffs in application, 731
  types of method, 734-736
  recommendations for application, 744
Organization change, 780
Outages, 157
  also see Interruptions and Reliability

Paradigms, definition, 777
  also see Planning
Parkinson's Law (and organizational
  resistance to change), 781
Performance simulators (analysis), 687
Perpendicular bisector rule, 493
Phasing, 234
  American vs. European styles, 237
  in analytical tools, 690
  evaluation of economics, 248
  single-phase components, 236
Phase-by-phase analysis (load flow), 711
Planning
  attribute, single or multiple, 186
  backward fill-in (multi-year), 771
  coordination among levels, 608
  for customer level, 606
  for feeder level, 604
  forward fill-in (multi-year), 771

Index

[Planning]
  goals, 569
  long-range planning, 583
  periods, 594, 673
  process, 566
    diagrams of, 594, 611
    and paradigms, 777-783
    short range planning, 581
    steps in planning, 566
    for substation level, 600
    tools and methods, 768
    for transmission level, 598
    uncertainty in planning, 587
Polynomial curve fit (forecasting), 636
Power factor, 22, 369
  also see VARS
Power quality, definition, 76
  end-use modeling of, 122
  also see Interruptions, Harmonics, and Voltage control
Present worth analysis, 195
  for approval of projects, 197
  for conductors, 244
  discount rate, 196
  factor (PWF), 195
  factors in determining, 199-201
  example application, 203
  formula, 195
  levelized analysis, 211
  lifetime analysis, 212
  also see Economics
Profit-motivated planning, 799
Protection Engineering, 31, 147
  also see Circuit Breakers

Quick method of estimating feeder system costs, 362

Radial type distribution systems, 35, 293
Radialization (in optimization), 753
Rate impact measure (RIM) test, 218
Reducing cost
  feeder estimation formula and its application, 362
  feeder cost reduction, 350-354, 359, 363, 388, 455, 483, 486-487

[Redducing costs]
  substation cost reduction tips, 424, 428, 496, 508
  and substation siting, 496, 520-521
  and site cost maps, 512-514
  and substation spacing, sizing, and system layout, 463-465
  and the system's approach, 341, 486, 491
  transformer units and lower cost, 427
Reduce voltage and split path character of T&D system design, 7
Reinforcement planning
  in budget-constrained planning, 791
  conductor sizing trick to avoid, 360
  and contingency planning, 318-324
  higher cost than new construction, 246
  economic analysis of, 271
  examples, 355
  low cost in rural planning, 306
  splitting feeders to avoid, 359
  switching may avoid, 342
Reliability, definition, 155
  analysis of, 169-171, 171
  and budget-constrained planning, 794
  and contingency margin, 174
  cost of, 451, 462, 463
  also see Reliability planning
  criteria for, 171
  example system, tables of, 442, 443
  system reliability
  extent of an outage, 158
  first level of design for, 78
  frequency and duration, 158
  historical assessment, 178
  indices and use of, 164
    comparison of utility results, 175
    differences in reporting, 176
  interaction with layout, 447-451, 462
  by level of the system, 10, 442
  map of reliability needs, 125
  temporal curves of reliability needs, 124
  tiered service reliability areas, 173
  and value based planning, 109, 180
  also see customer value of service

Reliability planning
  balancing treatment of old and new parts of the system, 794
  durability (transformer "use of life" planning), 766-767
  tips and reliability/cost concepts, 346, 351, 399, 417, 446-447, 462, 466, 473, 483, 499, 542
  tools for reliability planning, 717-719, 765, 766
  also see Value-based planning
Retail sale of energy, 1
Revenue requirements method, 192
Rhinoceros, and design criteria, 152
Rhino Ridge substation (example), 422
RIM (rate impact measure test), 216
Ripple control system (automation), 93
Robustness (of planning tools), 677
Rural feeders, 296
  characteristics of, 302
  planning for, 305

Safety standards, 145
SAIDI (reliability index), 166
SAIFI (reliability index), 165
Sampling by integration, 69
Secondary level
  see Service level
Service area optimization T&D, 761
  system planning method
Service areas of equipment, 41
Service level, definition, 17, 537
  American vs. European practice, 540
  coincidence interaction with, 545
  complicated level due, 563
    to interactions
  costs of, 27
  design tables as guides, 554
  diagrams of, 538
  layout styles and variations, 544
  "long drops" type layout, 544
  "many transformers" type layout, 544
  planning and layout, 552
  "wimp" voltage, and impact of, 540
Service quality
  see Reliability

Simulation
  see Load Forecasting, simulation
Single-phase circuit portions, 145
  constitute majority of American type system circuit miles, 435, 481
  and laterals, 236
Site cost maps (substations), 512-514
Siting of substations
  see Substation planning
Spatial resolution (and forecasting), 673
Spikes
  see Voltage spikes
Splitting feeders, 359
Standards
  see Criteria
Standard equipment types, 150
Substations, definition, 12
  central location in service area, 287
  combined T&D substations, 394
  costs, 25, 491, 496
  elements in, 25
  mobile substations, 426
  parts of, 392
  planning considerations, 426
    also see Substation planning purpose, 391
  reliability of, 400, 404, 411
  site, and considerations for, 420, 496
    also see Substation planning
  size and its interaction with
    cost and economy, 446
    load density, 472
    substation between substations, 445
    reliability, 447
  spacing between substations, 445
    rule of thumb, 460
  method of optimization,
    also see Service Area Optimization
Substation planning, 489-531, 600
  automatic methods for, 758
  formula, 523
  tables and figures describing, 520, 603
  interaction with service-area boundaries, 522
  as a "Tile planning" game, 515
  reasons to build a new substation, 526
  also see Substation siting and sizing

Index

Substation sites, 420, 492, 496
Substation siting and sizing, 495
  cost maps based on, 512
  planning methods for, 758
  service areas, 490
  sensitivities (table), 496
  guidelines/ rules of thumb, 493, 505
Sub-transmission, definition, 12, 395, 452
  cost interaction with substation, 453
  features of, 404
  radial types, 397
  reliability of, 404
  types of configuration, 398, 401
Surges
  see Voltage surges
Switching (of capacitors), 385
Switching (of feeder systems), 21
  fix for interruptions, 78
  locations of switches, 378
  rule of thumb (contingency zones), 328
  switch to separate paths, 333
  time to switch, 178
  zones (contingency planning), 325
  also see radialization
Synthetic feeder system (model), 762
  also see Service area optimization
Systems approach, 43, 341, 491
  diagram of cost interactions, 45
  example, in detail, 433-476
  feeder-at-a-time myopia, 350
  importance of fitting together, 431
  in strategic T&D planning, 758
  also see Service area optimization

T&D system
  see Distribution system
THD (total harmonic distortion), 95
Three-phases, 234
  delta or wye connection, 234
  imbalance in, 145
  circuit model algorithms, 711
  also see single-phase circuits
Three winding transformers, 276
Time value of money
  see Present worth analysis
Total resource cost (TRC) test, 216

Transformers
  capacity used in load allocation, 704
  costs and economics, 451, 552
  economics and selection of, 276
  features of, 412
  feeders as contingency backup, 338
  losses in, 20
  also see Costs
  number at substation, 408-412
  and cost and reliability, 449-451
  "many transformers" way of
    laying out service level, 543
  loading standards and equipment load
    ratings, 149, 279
  reliability of, 412, 449
  three-rating nomenclature, 278
  types of transformers, 149
  unit (packaged) transformers, 426
Transformer units (planning), 427
  table and diagram, 428
  planning with, 429
Transmission, definition, 11
  planning of, 598, 611
  also see sub-transmission
TRC (total resource cost) test, 216
Tree-trimming and reliability, 78
Trending
  see Load Forecasting
TVSS (transient voltage surge
  suppressers), 88
Two-thirds rule
  see Capacitors

Unbalanced voltage and current
  see Imbalance
Underground systems, 232, 297
  direct burial of cables, 297
  economics of, 249
  and grid-like layout, 300
  and hierarchical European design, 313
  feeder systems, 297
Upgrading a feeder often costs more
  than new construction, 29, 246
Utilization voltage, 17, 130, 538
  distance it can move power, 540
  various standards for, 132

Value-based planning (VBP), 110, 180
  also see Reliability planning
Variation in delivery cost by level, time
  location, and other factors, 220-227
VARS, 22, 367
  effect on feeder economics 368
  electrical foam analogy 23
  plots of distribution along a feeder, 371
Voltage
  computation methods
    see Load flow methods
  control and power quality, 81
  definitions of various terms, 131
  drop, 130, 139, 230
    example plots, 137, 141, 144
    and line types, 251
  flicker (motor start) 139, 142, 714
  fluctuation of service voltage, 133
  hierarchical levels of, 6
  for primary distribution, 260
  profile (along feeder), 134, 414
  regulation of, 22, 413
  sags, 83
  spikes, 87
  spread of voltages on a feeder, 139
  standards and criteria, 130
  surges, 87
Voltage flicker
  see Motor starting and Voltage
Voltage regulators, 389
Voltage sags
  see Voltage, sags
Voltage spikes
  see Voltage, spikes
Voltage surges
  see Voltage, surges

Whole T&D system, 7, 433
  detailed example (tables), 437-438
  planning method for, 598, 611
  power prices throughout, 223
  also see Systems approach
Worst-case lateral models in load
  flow analysis, 699
Wye-connected systems, 234
  compared to delta-connected, 235

X/R ratio, 254
  and design, 350
  feeder layout limitations due to, 470
  and power factor interaction, 369

Y-axis intercept analysis of economics
  for minimization of feeder costs, 364

Zero-base planning, 790, 794